AF601754

Birkhäuser

Trends in Mathematics

Trends in Mathematics is a series devoted to the publication of volumes arising from conferences and lecture series focusing on a particular topic from any area of mathematics. Its aim is to make current developments available to the community as rapidly as possible without compromise to quality and to archive these for reference.

Proposals for volumes can be submitted using the Online Book Project Submission Form at our website www.birkhauser-science.com.

Material submitted for publication must be screened and prepared as follows:

All contributions should undergo a reviewing process similar to that carried out by journals and be checked for correct use of language which, as a rule, is English. Articles without proofs, or which do not contain any significantly new results, should be rejected. High quality survey papers, however, are welcome.

We expect the organizers to deliver manuscripts in a form that is essentially ready for direct reproduction. Any version of TeX is acceptable, but the entire collection of files must be in one particular dialect of TeX and unified according to simple instructions available from Birkhäuser.

Furthermore, in order to guarantee the timely appearance of the proceedings it is essential that the final version of the entire material be submitted no later than one year after the conference.

More information about this series at http://www.springer.com/series/4961

Marcello D'Abbicco • Marcelo Rempel Ebert •
Vladimir Georgiev • Tohru Ozawa
Editors

New Tools for Nonlinear PDEs and Application

Editors
Marcello D'Abbicco
Mathematics
University of Bari
Bari, Italy

Marcelo Rempel Ebert
Department of Computing and Mathematics
University of São Paulo
Ribeirão Preto, São Paulo, Brazil

Vladimir Georgiev
Department of Mathematics
University of Pisa
Pisa, Italy

Tohru Ozawa
Department of Applied Physics
Waseda University
Tokyo, Japan

ISSN 2297-0215 ISSN 2297-024X (electronic)
Trends in Mathematics
ISBN 978-3-030-10936-3 ISBN 978-3-030-10937-0 (eBook)
https://doi.org/10.1007/978-3-030-10937-0

This book is published under the imprint Birkhäuser, www.birkhauser-science.com by the registered company Springer Nature Switzerland AG.
The registered company address is: Gewerbestrasse 11, 6330 Cham, Switzerland

Preface

The theory of evolution partial differential equations (PDEs) has made considerable strides in the last several years. This rapid development was driven on by the connections between this theory and other fields of the mathematics, e.g., the harmonic analysis, and by its strong ties to problems from mathematical physics. This volume includes 13 papers highlighting recent results in mathematics, and focusing on nonlinear PDEs and their applications. Readers will find, e.g., contributions on the qualitative properties of solutions of linear and nonlinear evolution models, as well as results concerning well-posedness, asymptotic profiles of solutions, blow-up behavior, and the influence of low regular coefficients.

We employed a strict blind review process, in the course of which each contribution was evaluated by two anonymous referees. The papers provide a broad range of ideas and include detailed proofs of their results.

Most of the contributors attended the sessions "Recent progress in evolution equations" and "Nonlinear PDEs" during the 11th ISAAC congress, which was held in Växjö, Sweden, in 2017. Some speakers were invited to deliver their talks during a joint day of these two sessions. Though the event is what initially provided the idea of creating a special volume of selected research papers, the present volume is not merely a collection of proceedings, but a stand-alone project gathering original contributions from active researchers on the latest trends in nonlinear evolution PDEs.

The International Society for Analysis, its Applications and Computation (ISAAC) has organized the biennial ISAAC congress at venues around the globe since 1997. The 2017 congress continued the successful series of meetings: in Delaware, USA (1997), Fukuoka, Japan (1999), Berlin, Germany (2001), Toronto, Canada (2003), Catania, Italy (2005), Ankara, Turkey (2007), London, UK (2009), Moscow, Russia (2011), Krakow, Poland (2013), Macau, China (2015). ISAAC is home to nearly 300 members from all regions of the world, as well as eight special interest groups focusing on different areas of analysis and computation.

Bari, Italy — Marcello D'Abbicco
Ribeirão Preto, São Paulo, Brazil — Marcelo Rempel Ebert
Pisa, Italy — Vladimir Georgiev
Tokyo, Japan — Tohru Ozawa
November 2018

Contents

On Effective PDEs of Quantum Physics

Ilias Chenn and I. M. Sigal

Abstract The Hartree-Fock equation is a key effective equation of quantum physics. We review the standard derivation of this equation and its properties and present some recent results on its natural extensions – the density functional, Bogolubov-de Gennes and Hartree-Fock-Bogolubov equations. This paper is based on a talk given at ISAAC2017.

1 Introduction

The Hartree-Fock equation (HFE) is a (if not the) key effective equation of quantum physics. It plays a role similar to that of the Boltzmann equation in classical physics. It gives a fairly accurate and yet sufficiently simple description of large (and not so large) systems of quantum particles. The trade-off here is the high dimension for nonlinearity: while the $n-$particle Schrödinger equation

$$i\hbar\frac{\partial\Psi}{\partial t} = H_n\Psi \tag{1}$$

is a linear equation in $3n+1$ variables, the Hartree-Fock one is a nonlinear one in $3+1$ variables. Here $\hbar$ is the Planck constant divided by 2π and H_n is the Schrödinger operator or (quantum) Hamiltonian of the $n-$particle system, it is given in (14) below.

The HFE involves an orthonormal system of n functions, $\{\psi_i\}$, on $\mathbb{R}^3$, or the projection operator $\gamma := \sum_i |\psi_i\rangle\langle\psi_i|$ acting on $L^2(\mathbb{R}^3)$, and can be written in the latter case as

$$i\hbar\frac{\partial\gamma}{\partial t} = [h_\gamma, \gamma] \tag{2}$$

I. Chenn · I. M. Sigal (✉)
Department of Mathematics, University of Toronto, Toronto, ON, Canada
e-mail: il.chen@mail.utoronto.ca; im.sigal@utoronto.ca

M. D'Abbicco et al. (eds.), *New Tools for Nonlinear PDEs and Application*, Trends in Mathematics, https://doi.org/10.1007/978-3-030-10937-0_1

where $h_\gamma := h + v * \rho_\gamma + ex(\gamma)$, with h a one-particle Schrödinger operator (say $h := -\frac{\hbar^2}{2m}\Delta + V(x)$, where $V(x)$ is an external potential), $\rho_\gamma(x,t) := \gamma(x,x,t) = \sum_i |\psi_i(x)|^2$ and $ex(\gamma)$ ("exchange term") is the operator with the integral kernel

$$ex(\gamma)(x,y) := -v(x-y)\gamma(x,y) \tag{3}$$

$$= -\sum_i \psi_i(x)v(x-y)\bar{\psi}_i(y). \tag{4}$$

(Here and in what follows, $A(x,y)$ stands for the integral kernel of an operator A.) Furthermore, to deal with quantum statistics (where the number of particles is not fixed but is a quantum observable), (2) is extended to arbitrary non-negative, trace class operator γ on $L^2(\mathbb{R}^3)$ satisfying $\gamma \leq \mathbf{1}$ (expressing the Pauli exclusion principle). This describes fermions. For bosons, one drops the exchange term $ex(\gamma)$ and the condition $\gamma \leq \mathbf{1}$.

Replacing $ex(\gamma)$ given above by a local function $\mathrm{xc}(\rho_\gamma)$ of the function $\rho_\gamma(x,t) := \gamma(x,x,t)$ leads to the Kohn-Sham equation underlying the density functional theory (DFT) which is exceptionally effective in the computations in Quantum Chemistry and in particular, of the electronic structure of matter.

It was discovered by Bardeen, Cooper and Schrieffer for fermions and by Bogolubov, for bosons, that for quantum fluids (superconductors and superfluids, respectively)

- the HFE falls short
- there are natural generalizations of the HFE describing these phenomena.

It turns out that this generalization is mathematically very natural and was overlooked in the mathematics literature, though the framework for it existed.

To explain how this generalization arises, we go back to the HFE and present its alternative derivation. We just indicate main steps; for details, see [3] and for background, [9, 36].

In abstract formulation, which applies also to statistical mechanics and quantum field theory, the states are defined as positive linear ('expectation') functionals on a C^* algebra, $\mathscr{A}$, elements of which are called observables, and the evolution of states is given by the von Neumann-Landau equation

$$i\hbar\partial_t\omega_t(A) = \omega_t([A,H])\,,\ \forall A \in \mathscr{A}, \tag{5}$$

where H is a quantum Hamiltonian which is affiliated with $\mathscr{A}$.

Technically, one takes for $\mathscr{A}$, an algebra of bounded operators (namely the Weyl algebra, $\mathfrak{W}$) on the fermionic/bosonic Fock space, which for spinless particles is written as

$$\mathscr{F} := \sum_0^\infty \circledast_1^n L^2(\mathbb{R}^d),\ d = 1,2,3, \tag{6}$$

where ⊕ stands either for the wedge product, $\wedge$, or symmetric product, Ⓢ. For a many-body system, the quantum Hamiltonian H on Fock space, $\mathscr{F}$ is given by $H := \oplus_0^\infty H_n$, with the n−particle Schrödinger operators H_n defined in (14) below. If one introduces annihilation and creation operators, $\psi(x)$ and $\psi^*(x)$ on $\mathscr{F}$, which map the n-particle sector in (6) into $(n-1)$- and $(n+1)$-sectors, respectively, then H is written in terms of these operators as

$$H = \int dx\ \psi^*(x)h\psi(x) + \frac{1}{2}\int dxdy\ v(x-y)\psi^*(x)\psi^*(y)\psi(x)\psi(y), \qquad (7)$$

with h a one-particle Schrödinger operator acting on the variable x and v a pair potential of the particle interaction (see (14) below).

We can think about the algebra of observables as generalized by (unbounded) operators $\psi(x)$ and $\psi^*(x)$. The Hartree-Fock approximation is obtained by restricting the evolution to the states, φ, determined by the expectation

$$\gamma(x, y) := \varphi[\psi^*(y)\,\psi(x)], \qquad (8)$$

provided $\varphi[\psi(x)] = 0$, in the following way. Let $\psi^\#(x)$ stands for either $\psi(x)$ or $\psi^*(x)$. We require that $\varphi[\psi^\#(x_1)\ \ldots\ \psi^\#(x_k)]$ to be zero if the number of ψ^*'s and ψ are not equal and is expressed in terms of sums of products of $\varphi[\psi^*(x_i)\,\psi(x_j)]$ according to the Wick theorem (see [9]), exactly as for the Gaussian processes in probability; such states are called the *quasifree states*.[1]

However, the property of being quasifree is not preserved by the dynamics (5) and the main question here is how to project the true quantum evolution onto the class of quasifree states. Following [3], we do this by restricting the evolution,

$$i\hbar\partial_t\varphi_t(A) = \varphi_t([A, H]) \qquad (9)$$

to observables A, which are at most quadratic in the creation and annihilation operators. Then we arrive at a closed, self-consistent dynamics for φ_t. When expressed in terms of the operator γ with the integral kernel $\gamma(x, y)$, it gives exactly the Hartree-Fock equation, (2).

The point here is that states determined by the expectations (8) are not the most general quasifree states. The most general quasifree states φ determine and are determined by expectations of all possible pairs of $\hat{\psi}^\sharp(x) := \psi^\sharp(x) - \varphi(\psi(x))$:

$$\begin{cases} \gamma(x, y) := \varphi[\hat{\psi}^*(y)\,\hat{\psi}(x)], \\ \alpha(x, y) := \varphi[\hat{\psi}(x)\,\hat{\psi}(y)]. \end{cases} \qquad (10)$$

[1] For application of the quasifree states in the classical kinetic theory see [46].

Mathematically, these are exactly the states discovered by Bardeen, Cooper and Schrieffer for fermions and by Bogolubov, for bosons, and for which the former received and the latter should have received the Nobel prize.

Now, let γ and α denote the operators with the integral kernels $\gamma(x, y)$ and $\alpha(x, y)$. After peeling off the spin components, definition (10) implies that

$$0 \leq \gamma = \gamma^* \quad (\leq \mathbf{1}) \quad \text{ and } \alpha^* = \bar{\alpha}, \tag{11}$$

where $\bar{\sigma} = C\sigma C$ with C being the complex conjugation and the condition $\gamma \leq \mathbf{1}$ applies only to fermions (as was mentioned above, it is an expression of the Pauli exclusion principle).

The operator γ can be considered as a one-particle density operator (matrix) of the system, so that $\rho_\gamma(x) := \gamma(x, x)$ is the particle density. The operator α gives the particle pair coherence ($\alpha(x, y)$ is a two-particle wave function). (For *confined* systems, γ and α are trace class and Hilbert-Schmidt operators, respectively, with $\text{Tr}\gamma = \int \gamma(x, x)dx < \infty$ giving the particle number, while for *thermodynamic* systems, they are only locally so.)

Following [3], we define self-consistent approximation as the restriction of the many-body dynamics to quasifree states. More precisely, we map the solution ω_t of (5), with an initial state ω_0, into the family φ_t of quasifree states satisfying

$$i\hbar\partial_t\varphi_t(A) = \varphi_t([A, H]) \tag{12}$$

for all observables A, which are at most quadratic in the creation and annihilation operators. As the initial condition, φ_0, for (12) we take the 'quasifree projection' of ω_0. We call this map the *nonlinear quasifree approximation* of equation (5).

We expect φ_t to be a good approximation of ω_t, if ω_0 is close to the manifold of quasifree states.

The BdG equations give an equivalent formulation of the Bardeen-Cooper-Schrieffer (BCS) theory of superconductivity.

Evaluating (12) for monomials $A \in \{\psi(x), \psi^*(x)\psi(y), \psi(x)\psi(y)\}$, yields a system of coupled nonlinear PDE's for (ϕ, γ, α) where $\phi(x) := \varphi(\psi(x))$ and γ and α are defined in (10). For the standard many-body hamiltonian, (7), these give the (time-dependent) *Hartree-Fock-Bogolubov (HFB) or Bogolubov-de Gennes (BdG) equations*, depending on whether we deal with bosons or fermions (see (99), (100) and (101) or (108), (109) and (110) below). In the latter case, one takes $\phi(x, t) := \varphi_t(\psi(x)) = 0$. As was mentioned above, the HFB equaitons describes Bose-Einstein condensation and superfluidity while the BdG equations describes superconductivity, the remarkable quantum phenomena.

HFB and BdG equations provide a more faithful description of quantum systems going beyond the Gross-Pitaevski (i.e. the nonlinear Schrödinger) and Ginzburg-Landau equations, which can be derived from them in certain regimes. While the latter equations accumulated quite a substantial literature (see e.g. [16, 19, 54, 55] and [53] for recent books and a review), the research on the former ones is just beginning.

* * * * **

There are many fundamental problems about the HFB and the BdG equations which are completely open. Generally, there are three types of questions one would like to ask about an evolution equation:

- Derivation;
- Well-posedness;
- Special solutions (say, stationary solutions or traveling waves) and their stability.

Some rigorous results on the derivation of the Hartree-Fock-Bogolubov (HFB) equations can be found in [34, 40, 48] (see also [6, 7, 30–32, 52] for earlier results and references). The well-posedness (or existence) for the time-dependent HFB equations for confined systems (see above) was proven in [4]. The well-posedness theory for the time-dependent Bogolubov-de Gennes (BdG) equations is developed in [5]. For thermodynamics systems (see above), it is open. Some important stationary solutions of the BdG and HFB equations were found in [22, 37] and [3, 49, 50], respectively.

In this contribution, we recall the standard derivation and properties of the HF (and H) equations and discuss recent work on the Kohn-Sham (KS), HF, BdG and HFB equations [3, 22, 23]. To fix ideas, we concentrate mostly on the BdG equations.

There is a considerable physics literature on the subject. As for rigorous works, the three fundamental contributions to the subject, [2, 33, 37], deal with foundational issues (relation to quasifree states and quadratic hamiltonians on the Fock space and the general variational problem), with the critical temperature and the superconducting solutions and with the derivation of the Ginzburg-Landau equations respectively. For more references, and discussion see some recent papers [3, 5, 22, 23] and reviews [38, 39]. The object of these and other works on the subject is the time-independent theory. The results we discuss are complementary to this work.

2 Hartree and Gross-Pitaevski Equations

2.1 Origin and Properties

In what follows we use the *units in which the (normalized) Planck constant* $\hbar$ *and the speed of light* c *are both equal to* 1 *and the typical particle mass is set to* $1/2$. With this agreement, the evolution of quantum n-particle system is given by the Schrödinger equation

$$i\frac{\partial \Psi}{\partial t} = H_n \Psi. \tag{13}$$

Here H_n is the Schrödinger operator or Hamiltonian of the physical system. For the system of n identical particles (say, electrons or atoms) of mass $1/2$, interacting with each other and moving in an external potential V the Hamiltonian is

$$H_n := \sum_{i=1}^{n} h_{x_i} + \frac{1}{2}\sum_{i\neq j} v(x_i - x_j), \tag{14}$$

where $h_x = -\Delta_x + V(x)$ and v is the interaction potential. For spinless fermions/bosons, it acts on the state space, which in the spinless case can be written as

$$\text{\textcircled{\#}}_1^n L^2(\mathbb{R}^d),\ d = 1, 2, 3.$$

The Schrödinger equation is an equation, (13), in $dn+1$ variables, $x_1, \ldots, x_n$ and t. Even for a few particles it is prohibitively difficult to solve. Hence it is important to have manageable approximations.

One such an approximation, which has a nice unifying theme and connects to a large areas of physics and mathematics, is the self-consistent (or mean-field) one. In it one approximates solutions of n-particle Schrödinger equations by products of n one-particle functions (i.e. functions of $d+1$ variables) appropriately symmetrized. This results in a single nonlinear equation in $d+1$ variables, or several coupled such equations. The trade-off here is the number of dimensions for the nonlinearity. This method is especially effective when the number of particles, n, is sufficiently large.

We give a heuristic derivation of the self-consistent approximation for the Schrödinger equation above. (See [36] for details and references to rigorous results.) First, we observe

Proposition 1 *The Schrödinger equation is the Euler-Lagrange equation for stationary points of the action functional*

$$S(\Psi) := \int \big\{ - Im\langle \Psi, \partial_t \Psi\rangle - \langle \Psi, H_n \Psi\rangle\big\} dt, \tag{15}$$

Now, for bosons, we consider the the action functional (15) on the space (not linear!)

$$\{\Psi := \otimes_1^n \psi | \psi \in H^1(\mathbb{R}^3)\}, \tag{16}$$

where $(\otimes_1^n \psi)$ is the function of $3n+1$ variables defined by $(\otimes_1^n \psi)(x_1, \ldots, x_n, t) := \psi(x_1, t) \ldots \psi(x_n, t)$. For fermions, we take

$$\{\Psi := \wedge_1^n \psi_j : \psi_i \in H^1(\mathbb{R}^3)\ \forall i = 1, \ldots, n\} \tag{17}$$

Here $(\wedge_1^n \psi_j)(x_1, \ldots, x_n, t) := \det[\psi_i(x_j, t)]$ is the determinant of the $n \times n$ matrix $[\psi_i(x_j, t)]$, called the *Slater determinant.*

We begin with bosons. We have the following elementary result:

Proposition 2 *Let* $\|\psi\|^2 = n - 1 \approx n$ *and* $S_H(\psi) := \frac{n-1}{n} S(\otimes_1^n \psi)$ *('H' stands for the Hartree). Then we have*

$$S_H(\psi) = \int\int \{ - Im\langle \psi, \partial_t \psi \rangle - |\nabla \psi|^2 - V|\psi|^2 - \frac{1}{2}|\psi|^2 v * |\psi|^2 \} dx dt. \tag{18}$$

We see that the quadratic terms on the r.h.s. of (18) are of the order $O(n)$, while the quartic ones, are $O(vn^2)$ The regime in which these terms are of the same order, $O(n^2)$, i.e. for which, $v = O(1/n)$ is called the *mean-field* regime.

The Euler-Lagrange equation for stationary points of the action functional (18) considered on the first set of functions is

$$i\frac{\partial \psi}{\partial t} = (h + v * |\psi|^2)\psi, \tag{19}$$

with the normalization $\|\psi\|^2 = n - 1 \approx n$. This nonlinear evolution equation is called the *Hartree equation* (HE).

If the inter-particle interaction, v, is significant only at very short distances (one says that v is very short range, which technically can be quantified by assuming that the "particle scattering length" a is small), one replaces $v(x) \to 4\pi a\delta(x)$ and Equation (19) becomes

$$i\frac{\partial \psi}{\partial t} = h\psi + \kappa|\psi|^2\psi, \tag{20}$$

where $\kappa := 4\pi a$ (with the normalization $\|\psi\|^2 = n$). This equation is called the *Gross-Pitaevski equation* (GPE) or the *nonlinear Schrödinger equation*. It is derived using the Gross-Pitaevski approximation to the original quantum problem for a system of n bosons. The Gross-Pitaevski equation is widely used in the theory of superfluidity, and in the theory of Bose-Einstein condensation (see [36, 41] and references therein).

Proofs of the local and global existence for (19) and (20) can be found in [19, 21, 55].

2.1.1 Properties of the Hartree and Gross-Pitaevski Equations

We say that the map T on a space of solution is a *symmetry* of an equation iff the fact that ψ is a solution of the equation implies that $T\psi$ is also a solution. It is straightforward to prove the following

Proposition 3 *The Hartree and Gross-Pitaevski equations have the following symmetries*

1. *the time-translations,* $\psi(x,t) \to \psi(x,t+s),\ s \in \mathbb{R}$,
2. *the gauge transformations,*

$$\psi(x,t) \to e^{i\alpha}\psi(x,t),\ \alpha \in \mathbb{R},$$

3. *for* $V = 0$*, the spatial translations,* $\psi(x,t) \to \psi(x+y,t),\ y \in \mathbb{R}^3$,
4. *for* $V = 0$*, the Galilean transformations,* $v \in \mathbb{R}^3$,

$$\psi(x,t) \to e^{i(\frac{1}{2}v\cdot x - \frac{v^2 t}{4})}\psi(x - vt, t),$$

5. *for* V *spherically symmetric, the spatial rotations,* $\psi(x,t) \to \psi(Rx,t),\ R \in O(3)$,

As the result of the time-translational and the gauge symmetries, the energy and the number of particles functionals

$$E(\psi) := \int \left\{ |\nabla\psi|^2 + V|\psi|^2 + G(|\psi|^2) \right\} dx, \tag{21}$$

where $G(|\psi|^2) := \frac{1}{2}|\psi|^2 v * |\psi|^2$ for HE and $G(|\psi|^2) := \frac{1}{2}\kappa|\psi|^4$ for GPE, and

$$N(\psi) := \int |\psi|^2 dx,$$

are independent of time, t. Moreover, for $V = 0$, the field momentum,

$$P(\psi) := \int \bar{\psi}(x,t)(-i\nabla_x)\psi(x,t)dx,$$

and, for V spherically symmetric, the field angular momentum,

$$L(\psi) := \int \bar{\psi}(x,t)(x \wedge (-i\nabla_x))\psi(x,t)dx,$$

are conserved. These conservation laws impose constraints on the dynamics leading to qualitative understanding of possible scenarios and are used in the proofs of the global existence, existence and stability of stationary solutions and traveling waves; for definitions and a review see [36].

We also note that HE and GPE are Hamiltonian systems (see Section 19.1 of [36]).

2.2 *Particles Coupled to the Electromagnetic Field*

We start with the action

$$S(\psi) = \int\int \{-\mathrm{Im}\langle\psi, \partial_t \psi\rangle \tag{22}$$

$$- |\nabla\psi|^2 - V|\psi|^2 - G(|\psi|^2)\}dxdt, \tag{23}$$

where $G(|\psi|^2)$ is given after (21), and use the principle of minimal coupling in which one replaces the usual derivatives ∂_t and ∇ by covariant ones, $\partial_{t\phi} = \partial_t + ie\phi$ and $\nabla_a = \nabla - iea$, where ϕ and a are the electric and magnetic potentials and e is the electric charge of ψ, and adds the action,

$$S_{\mathrm{EM}}(a,\phi) := \int\int \left\{|\partial_t a + \nabla\phi|^2 - |\operatorname{curl} a|^2\right\} dxdt,$$

of the the electro-magnetic field (for the latter, see e.g. [36], Sections 19.1.1 and 19.6). Then, assuming the external potential $V = 0$, the total action becomes

$$S(\psi, a, \phi) := \int\int \{ - \mathrm{Im}\langle\psi, \partial_{t\phi}\psi\rangle - |\nabla_a\psi|^2 - G(|\psi|^2)\}dxdt$$
$$+ S_{\mathrm{EM}}(a,\phi). \tag{24}$$

for a triple $(\psi, a, \phi) : \mathbb{R}^d \to \mathbb{C} \times \mathbb{R}^d \times \mathbb{R}$, of complex and real functions and a vector field. The Euler-Lagrange equations for this action are given by

$$i\frac{\partial\psi}{\partial t} = h_{a\phi}\psi + g(|\psi|^2)\psi, \tag{25a}$$

$$-\partial_t(\partial_t a + \nabla\phi) = \operatorname{curl}^* \operatorname{curl} a - \mathrm{Im}(\bar{\psi}\nabla_a\psi), \tag{25b}$$

$$-\operatorname{div}(\partial_t a + \nabla\phi) = e|\psi|^2, \tag{25c}$$

where $h_{a\phi} := -\Delta_a + e\phi + V$, with $\Delta_a = \nabla_a^2$, the covariant Laplacian, $g(s) = G'(s)$ and the vector quantity $J(x) := \mathrm{Im}(\bar{\psi}\nabla_a\psi)$ is the electric current, while $|\psi|^2$ is the charge density (remember we omit the charge of the particle), so that the second and third equations are Ampère's and Gauss law part of the Maxwell equations.

Moreover, curl* is the L^2–adjoint of curl, so that for $d = 3$, we have $\operatorname{curl}^* = \operatorname{curl}$ and for $d = 2$, $\operatorname{curl} a := \partial_1 a_2 - \partial_2 a_1$ is a scalar, and for a scalar function, $f(x)$, $\operatorname{curl}^* f = (\partial_2 f, -\partial_1 f)$ is a vector.

It is straightforward to prove that (25) are the Euler-Lagrange equations for action (24). Now, in addition to translation and rotation invariance (if $V = 0$), equations (25) are invariant under the *local gauge transformations*: for any sufficiently regular function $\chi : \mathbb{R}^2 \times \mathbb{R} \to \mathbb{R}$,

$$T_\chi^{\text{gauge}} : (\psi(x,t), a(x,t), \phi(x,t)) \mapsto (e^{i\chi(x,t)}\psi(x,t), a(x,t) + \nabla_x \chi(x,t), \phi(x,t) - \partial_t \chi(x,t)). \tag{26}$$

Using this gauge invariance, we can choose χ so that a and/or ϕ satisfy certain additional conditions. This is called gauge fixing. For instance, we can choose χ so that $\operatorname{div} a = 0$ (the Coulomb gauge), or ϕ satisfies $\phi = 0$ (the temporal gauge). Both conditions break gauge invariance. The gauge fixing which preserves the gauge invariance is the Lorentz (or radiation) gauge

$$\operatorname{div} a + \partial_t \phi = 0.$$

Note that in the Coulomb gauge, $\operatorname{div} a = 0$, Eq. (27b) becomes the familiar Poisson equation, $-\Delta\phi = e|\psi|^2$.

Neglecting in (25) the magnetic field produced by changing charge distribution (and the electric field), we arrive at the Schrödinger-Poisson system

$$i\frac{\partial \psi}{\partial t} = h_\phi \psi + g(|\psi|^2)\psi, \tag{27a}$$

$$-\Delta\phi = e|\psi|^2, \tag{27b}$$

where $h_\phi := -\Delta + e\phi + V$

One can derive (25) from the many-body Schrödinger equation coupled to the quantized electromagnetic field.

3 The (Generalized) Hartree-Fock Equations

3.1 *Formulation and Properties*

The Euler-Lagrange equation for stationary points of the action functional (15) considered on the Hartree-Fock states, (17), is a system of nonlinear, coupled evolution equations

$$i\frac{\partial \psi_j}{\partial t} = (h + v * \sum_i |\psi_i|^2)\psi_j - \sum_i (v * \psi_i \bar{\psi}_j)\psi_i, \tag{28}$$

where, recall, $h := -\Delta + V$, for the unknowns $\psi_1, \ldots, \psi_n$. This system plays the same role for fermions as the Hartree equation does for bosons. Equation (28) is called the *Hartree-Fock equations* (HFE).

Properties of HFE The Hartree-Fock equations are

1. invariant under the time-translations and gauge transformations, and, for $V = 0$, the spatial translations, $\psi_j(x) \to \psi_j(x+y)$, $y \in \mathbb{R}$, and the Galilean transformations, $v \in \mathbb{R}^3$, and, for V spherically symmetric, the rotations.
2. invariant under time and space independent unitary transformations of $\{\psi_1, \dots, \psi_n\}$.
3. a Hamiltonian system (see Sections 24.6 and 24.7 of [36]).

Again, similarly to HE, as the result of the time-translational and the gauge symmetries, the energy and the number of particles functionals

$$E(\psi) := \int \{\sum_i (|\nabla\psi_i|^2 + V|\psi_i|^2) + \frac{1}{2}(\sum_i |\psi_i|^2) v * (\sum_i |\psi_i|^2)$$

$$- \frac{1}{2}\int v(x-y)|\sum_i \psi_i(x)\psi_i(y)|^2 dy\} dx, \tag{29}$$

$$N(\psi) := \sum_i \int_{\mathbb{R}^3} |\psi_i|^2 dx \tag{30}$$

are conserved, similarly, for linear and angular momenta. Moreover, HFE conserve the inner products, $\langle\psi_i, \psi_j\rangle$, $\forall i, j$. For a rigorous theory, see [8, 20, 42, 44, 45, 47].

The item (2) above shows that the natural unknown for HFE is the subspace spanned by $\{\psi_i\}$, or the corresponding projection $\gamma := \sum_i |\psi_i\rangle\langle\psi_i|$. HFE can be rewritten as an equation for γ:

$$i\frac{\partial\gamma}{\partial t} = [h_\gamma, \gamma] \tag{31}$$

where $h_\gamma := h + v * \rho_\gamma + ex(\gamma)$, with $\rho_\gamma(x) := \gamma(x, x) = \sum_i |\psi_i(x)|^2$ and $ex(\gamma)$ is the operator with the integral kernel

$$ex(\gamma)(x, y) := -v(x-y)\gamma(x, y) = -\sum_i \bar{\psi}_i(x) v(x-y)\psi_i(y). \tag{32}$$

Recall that $A(x, y)$ stands for the integral kernel of an operator A.

This can be extended to arbitrary non-negative density operators γ satisfying (for fermions) $\gamma \leq 1$, and leads to a new class of nonlinear differential equations. (The properties $0 \leq \gamma$ and $\gamma \leq 1$ as well as all eigenvalues of γ as conserved under the evolution.)

Finally, note that the energy and the number of particles in the new formulation is given by

$$E(\gamma) := \mathrm{Tr}((h + \frac{1}{2}v * \rho_\gamma)\gamma) + Ex(\gamma), \tag{33}$$

$$N(\gamma) := \mathrm{Tr}\gamma = \int \rho_\gamma, \tag{34}$$

where, recall, $h := -\Delta + V$, $\rho_\gamma(x) := \gamma(x,x)$ and $Ex(\gamma) := -\frac{1}{2}\mathrm{Tr}(\gamma v\sharp\gamma)$, where $v\sharp\gamma$ is the operator with the integral kernel $v(x-y)\gamma(x,y)$. Note that

$$\mathrm{Tr}((v * \rho_\gamma)\gamma) = \int \rho_\gamma v * \rho_\gamma dx = \int\int \rho_\gamma(x)v(x-y)\rho_\gamma(y)dxdy,$$

$$\mathrm{Tr}(\gamma v\sharp\gamma) = \int\int v(x-y)|\gamma(x,y)|^2 dxdy.$$

It is straightforward to show that that equations (28), (29) and (30) can be rewritten as (31), (32), (33) and (34), respectively.

Note that the HE can be also formulated with γ being a rank one projection times n and extended to operators γ with no constraint on the size. In this case, the exchange terms $ex(\gamma)$ and $Ex(\gamma)$ should be omitted from the definition of h_γ and the energy.

γ is called the (one-particle) density operator and $\gamma(x,x)$ (or $\gamma(x,x,t)$) is interpreted as the one-particle density, so that $\mathrm{Tr}\gamma = \int \gamma(x,x)dx$ is the total number of particles. It should satisfy

$$0 \le \gamma = \gamma^* \ (\le 1) \tag{35}$$

where the second inequality is required only for fermions. The HF flow preserves these properties.

3.1.1 Exchange Energy Term

We extend Eq. (31) by allowing different exchange terms in the definition of h_γ, rather than just (32). Specifically, we let the exchange energy term, $ex(\gamma)$, to take the following forms:

- $ex(\gamma) := 0$ for the Hartree (or reduced Hartree-Fock, if $\gamma \le 1$) model,
- $ex(\gamma) := -v\sharp\gamma$ for the Hartree-Fock case and
- $ex(\gamma)$ is a local function, $ex(\gamma) = xc(\rho_\gamma)$, of the function $\rho_\gamma(x) := \gamma(x,x)$, say, coming from $Ex(\rho) = -c\int \rho^{4/3}$, in the density functional theory (DFT).

We call (31) with a general exchange energy term, $ex(\gamma)$, the *generalized Hartree-Fock equation* (gHFE).

3.2 Static gHF Equations

Clearly, γ, is a static solution to (31) iff γ solves the equation

$$[h_\gamma, \gamma] = 0. \tag{36}$$

For any reasonable function f and $\mu \in \mathbb{R}$, solutions of the equation

$$\gamma = f(\beta(h_\gamma - \mu)), \tag{37}$$

solves (36). Under certain conditions, the converse is also true. (The reason for introducing the parameters $\beta = 1/T$, $\mu > 0$ (the inverse temperature and chemical potential) will become clear later.)

Under certain conditions on f satisfied by our choice below, the chemical potential μ is determined by the condition that $\mathrm{Tr}\gamma = n$.

The physical function f is selected by either a thermodynamic limit (Gibbs states) or by a contact with a reservoir (or imposing the maximum entropy principle). For fermions, it is given by the Fermi-Dirac distribution

$$f(\lambda) = (e^\lambda + 1)^{-1}, \tag{38}$$

and for bosons, by the Bose-Einstein one

$$f(\lambda) = (e^\lambda - 1)^{-1}. \tag{39}$$

(One can also consider the Boltzmann distribution $f(\lambda) = e^{-2\lambda}$.) Inverting the function f and letting $f^{-1} =: s'$, we rewrite the stationary gHFE as

$$h_{\gamma,\mu} - \beta^{-1} s'(\gamma) = 0, \tag{40}$$

Here, recall, $h_{\gamma,\mu} := h_\gamma - \mu = -\Delta + V + ex(\gamma) - \mu$ and $0 < \beta \leq \infty$ (inverse temperature) and $\mu \geq 0$ (chemical potential). It follows from the equations $s' = f^{-1}$ and (38) that, up to a constant, the function s is given by

$$s(\lambda) = -(\lambda \ln \lambda + (1 - \lambda) \ln(1 - \lambda)), \tag{41}$$

for fermions, and by

$$s(\lambda) = -(\lambda \ln \lambda - (1 + \lambda) \ln(1 + \lambda)), \tag{42}$$

for bosons, so that for fermions, we have

$$s'(\lambda) = -\ln \frac{\lambda}{1 - \lambda}. \tag{43}$$

3.3 Coupling to the Electromagnetic Field

We couple the gHFE to the electromagnetic field. We assume that the particles are carry the unit charge density $e = -1$, so that the charge of density is $-\rho_\gamma$.

As before, we use the principle of minimal coupling assuming the inter-particle potentials and external potentials are of the electromagnetic nature. This gives the system of self-consistent equations for γ and the vector and scalar potentials a and ϕ:

$$i\partial_t\gamma = [h_{\phi,a,\gamma}, \gamma], \tag{44}$$

$$-\operatorname{div}(\partial_t a + \nabla\phi) = 4\pi(\kappa - \rho_\gamma), \tag{45}$$

$$-\partial_t(\partial_t a + \nabla\phi) = \operatorname{curl}^* \operatorname{curl} a - j(\gamma, a), \tag{46}$$

where $\kappa(x)$ is an external (positive) charge distribution, $j(\gamma, a)$ is the current given by $j(\gamma, a)(x) := -4\pi[-i\nabla_a, \gamma]_+(x, x)$, with $[A, B]_+ := AB + BA$,

$$h_{\phi,a,\gamma} = -\Delta_a - \phi + ex(\gamma). \tag{47}$$

Since $e = -1$, we have that $\nabla_a = \nabla + ia$ and $\Delta_a = \nabla_a^2$. We call (44), (45), (46) and (47) the *gHFem equations*.

We will discuss symmetries of this system in a more general context later on. Here we only note briefly that, in addition to the rigid motion symmetries, it has the gauge symmetry which did not make its appearance so far and which plays a central role in quantum physics.

As above, the energy and the number of particles are conserved and are given by

$$E(\gamma, a, \phi) := \operatorname{Tr}(h_a\gamma) + Ex(\gamma) + E_{\text{em}}(a, \phi), \tag{48}$$

$$N(\gamma) := \operatorname{Tr}\gamma = \int \rho_\gamma, \tag{49}$$

where $h_a := -\Delta_a$ and $E_{\text{em}}(a, \phi)$ is the energy of the the electro-magnetic field, given by

$$E_{\text{em}}(a, \phi) := \frac{1}{8\pi}\int \left\{|\partial_t a + \nabla\phi|^2 + |\operatorname{curl} a|^2\right\} dx. \tag{50}$$

The conservation of N is obvious. To prove the conservation of E, we use the definition $j := -4\pi d_a \operatorname{Tr}\big((-\Delta_a)\gamma\big)$ and the relation $dEx = ex$, to compute

$$\partial_t(\operatorname{Tr}(h_a\gamma) + Ex(\gamma)) = \operatorname{Tr}(h_{a,\gamma}\dot\gamma) - \frac{1}{4\pi}\int j\dot a \tag{51}$$

where $h_{a,\gamma} = -\Delta_a + ex(\gamma)$. By (44) and $h_{a,\gamma} = h_{\phi,a,\gamma} + \phi$, we have $\mathrm{Tr}(h_{a,\gamma}\dot{\gamma}) = \mathrm{Tr}(\phi\dot{\gamma}) = \int \phi\dot{\rho}_\gamma$, this gives

$$\partial_t(\mathrm{Tr}(h_a\gamma) + Ex(\gamma)) = \int \phi\dot{\rho}_\gamma - \frac{1}{4\pi}\int j\dot{a}. \tag{52}$$

Next, using that $E = -\dot{a} - \nabla\phi$, we compute

$$\partial_t E_{\mathrm{em}}(a, \phi) = \frac{1}{4\pi}\int \big[- (\dot{a} + \nabla\phi)\cdot \dot{E} + \mathrm{curl}^* \,\mathrm{curl}\, a \cdot \dot{a}\big]dx. \tag{53}$$

Combining the last two relations and and integrating by parts gives

$$\begin{aligned}\partial_t E(\gamma, a, \phi) =& \frac{1}{4\pi}\int \big(\phi(4\pi\dot{\rho}_\gamma + \mathrm{div}\,\dot{E}) \\ &- (\dot{E} + j - \mathrm{curl}^*\,\mathrm{curl}\, a)\dot{a}\big).\end{aligned} \tag{54}$$

Now, using (45) and (46) ($\mathrm{div}\, E = 4\pi(\kappa - \rho_\gamma)$, and $\dot{E} = \mathrm{curl}^*\,\mathrm{curl}\, a - j(\gamma, a)$) yields $\partial_t E(\gamma, a, \phi) = 0$. □

Above, we assumed the external magnetic field is zero.

To describe crystals we take κ to be either periodic (crystals) or uniform (jellium).

If κ and ρ_γ are $\mathscr{L}$-periodic, then integrating (45) over a fundamental cell, Ω, of the lattice $\mathscr{L}$, we arrive at the solvability condition (the charge conservation law)

$$\int_\Omega \rho_\gamma = \int_\Omega \kappa. \tag{55}$$

3.4 *Static gHFem Equations*

It is easy to see that (γ, a, ϕ) is a static solution to (44), (45) and (46) if and generically only if (γ, a, ϕ) solves the equations

$$\gamma = f(\beta(h_{\phi,a,\gamma} - \mu)), \tag{56}$$

$$\Delta\phi = 4\pi(\kappa - \rho_\gamma), \tag{57}$$

$$\mathrm{curl}^*\,\mathrm{curl}\, a = j(\gamma, a), \tag{58}$$

where, recall, $h_{\phi,a,\gamma} := -\Delta_a - \phi + ex(\gamma)$ and f is a sufficiently regular function f. Physically relevant f are given by either (38) or (39), depending on whether the particles in question are fermions or bosons. (Remember that the unit charge of γ is $e = -1$.)

To this we add the solvability condition (55), which determines the chemical potential μ.

3.4.1 Free Energy

The static gHF equations (56), (57) and (58) arise as the Euler-Lagrange equations for the free energy functional

$$F_\beta(\gamma, a) := E(\gamma, a) - \beta^{-1}S(\gamma) - \mu N(\gamma), \tag{59}$$

where $S(\gamma) = -\mathrm{Tr}(\gamma \ln \gamma + (1-\gamma)\ln(1-\gamma))$ is the entropy, $N(\gamma) := \mathrm{Tr}\gamma$ is the number of particles and $E(\gamma, a)$ is the static part of energy (48), with ϕ expressed in terms of ρ_γ by solving the Poisson equation (57) for ϕ,

$$E(\gamma, a) = \mathrm{Tr}\big((-\Delta_a)\gamma\big) + \frac{1}{2}\int (\kappa - \rho_\gamma)4\pi(-\Delta)^{-1}(\kappa - \rho_\gamma)dx$$
$$+ \frac{1}{8\pi}\int dx|\operatorname{curl} a(x)|^2 + Ex(\gamma). \tag{60}$$

This, not quite trivial, fact is proven in [22]. (For a formal statement in a more general situation see Theorem 4 below.)

We demonstrate informally that (56), (57) and (58) are the Euler-Lagrange equations for (59). By the definitions of $E(\gamma, a)$, $Ex(\gamma)$ and $S(\gamma)$, we have

$$d_\gamma E(\gamma, a)\xi = \mathrm{Tr}(h_\gamma \xi) \tag{61}$$

and

$$d_\gamma S(\gamma) = \mathrm{Tr}(s(\gamma)\xi), \tag{62}$$

which implies (56) with ϕ given by (57). Next, using the definition $j_a := -4\pi d_a \mathrm{Tr}\big((-\Delta_a)\gamma\big)$, we find

$$d_a E(\gamma, a)\alpha = \frac{1}{4\pi}\int \big(j_a - \operatorname{curl}^* \operatorname{curl} a\big)\alpha, \tag{63}$$

which yields (58).

3.4.2 Electrostatics

We describe the important case of electrostatics here, i.e. the time-independent case with $a = 0$. In this case, Eqs. (56), (57) and (58) become

$$\gamma = f(\beta(h_{\phi,\gamma} - \mu)), \tag{64}$$

$$\Delta\phi = 4\pi(\kappa - \rho_\gamma), \tag{65}$$

where $h_{\phi,\gamma} := -\Delta - \phi + ex(\gamma)$, which after solving Eq. (57) for ϕ, gives

$$\gamma = f(\beta(h_\gamma - \mu)), \tag{66}$$

where $h_\gamma := -\Delta - \phi_{\rho_\gamma} + ex(\gamma)$, with $\phi_\rho = \Delta^{-1}4\pi(\kappa - \rho)$. To this we add the solvability condition (55), which determines the chemical potential μ. Moreover, we associate with the charge density, $\kappa - \rho$, the potential

$$\phi_\rho = 4\pi(-\Delta)^{-1}(\kappa - \rho), \tag{67}$$

satisfying the Poisson equation (65).

The energy and free energy for (66) are given by

$$E(\gamma) := \mathrm{Tr}((-\Delta)\gamma) \tag{68}$$

$$+\frac{1}{2}\int(\kappa - \rho_\gamma)(x)4\pi(-\Delta)^{-1}(\kappa - \rho_\gamma)(x)dx + Ex(\gamma), \tag{69}$$

$$F_\beta(\gamma) := E(\gamma) - \beta^{-1}S(\gamma) - \mu N(\gamma). \tag{70}$$

4 Density Functional Theory

The starting point of the (time-dependent) density functional theory (DFT) are the equations (44), (45) and (46) but with the exchange term $ex(\gamma)$ is taken to be of the form $\mathrm{xc}(\rho_\gamma)$, where $\mathrm{xc}(\lambda)$ is a local function combining contributions of the exchange and correlation energy. For the former one usually take the expression $-c\rho^{4/3}$, going back to Dirac, and the latter is found empirically. This simple but profound modification opens an incredible computational potential of the theory.

We concentrate on the simplest case of electrostatics. In this case Eq. (66) becomes

$$\gamma = f(\beta(h_{\rho_\gamma} - \mu)), \tag{71}$$

where f is given by (38) and, with $\phi_\rho = (-\Delta)^{-1}4\pi(\kappa - \rho) =: v * (\kappa - \rho)$,

$$h_\rho := -\Delta - \phi_\rho + \mathrm{xc}(\rho). \tag{72}$$

Equation (71) is an extension of the key equation of the DFT – the Kohn-Shan equation – to positive temperature $T = 1/\beta > 0$. The energy and free energy for (71) are given by

$$E(\gamma) := \mathrm{Tr}((-\Delta)\gamma) + \frac{1}{2}\int(\rho_\gamma - \kappa)v * (\rho_\gamma - \kappa) + \mathrm{Xc}(\rho_\gamma), \tag{73}$$

$$F_\beta(\gamma) := E(\gamma) - \beta^{-1}S(\gamma) - \mu N(\gamma). \tag{74}$$

Let den be the map from operators, A, into functions $\rho_A(x) = \mathrm{den}[A](x) := A(x,x)$ with $A(x,y)$ being generalized kernel of A ('den' stands for 'density'). Taking the diagonal of (71), we arrive at the following equation for ρ

$$\rho = \mathrm{den}[f(\beta(h_\rho - \mu))]. \tag{75}$$

Equation (75) gives an equivalent formulation of the Kohn-Sham equation (71). For κ (and ρ) $\mathscr{L}$-periodic, we add to equation (75) the charge conservation law (cf. (55)), which determines the chemical potential μ,

$$\int_\Omega \rho = \int_\Omega \kappa, \tag{76}$$

where Ω is a fundamental cell of the lattice $\mathscr{L}$.

Conversely, starting from (75) and (76), we define the potential $\phi = (-\Delta)^{-1}4\pi(\kappa - \rho)$ produced by the charge distribution $\kappa - \rho$. Then ϕ satisfies

$$-\Delta\phi = 4\pi(\kappa - \rho). \tag{77}$$

Note that because of the minimal coupling, there is *no (pure) DFT theory* when the system in question is coupled to the magnetic field.

4.1 Crystals

Here one deals with the electrostatics, (64), or, in the DFT context, (71) (or (75)). for an ideal crystal, one assumes that $\kappa = \kappa_{\mathrm{per}}$ is periodic w.r. to some lattice $\mathscr{L}$, representing an $\mathscr{L}$ periodic charge distribution of crystal ions. An example of such an κ_{per} is

$$\kappa_{\mathrm{per}}(x) = \sum_{l\in\mathscr{L}} \kappa_a(x - l)\,. \tag{78}$$

where κ_a denotes an ionic ('atomic') potential.

The simplest special case of periodic κ is κ constant. Such a system is called the *jellium*. For $\kappa = \kappa_{\mathrm{jel}}$ constant, (75) has the solution $(\rho_{\mathrm{jel}} = \kappa_{\mathrm{jel}}, \mu_{\mathrm{jel}})$. Indeed, (76) reduces to $\rho_{\mathrm{jel}} = \kappa_{\mathrm{jel}}$ and (75) to one equation for μ, which has a unique solution for μ near μ_{jel} [23].

The existence (without uniqueness) of a certain periodic, trace class solution to equation (71) (or (75)) with certain class of density terms xc is obtained in [1] via variation techniques. (See [17, 18] for earlier results for the Hartree and Hartree-Fock equations. We present a somewhat different proof of the latter result in Sect. 7.) The next result proven in [23], establishes, under more restrictive conditions, uniqueness and quantitative bounds needed for the next result.

Let Ω be a fundamental cell of the lattice $\mathcal{L}$ and $|\Omega|$ denote its area. Denote by $H^s_{\text{per}}(\mathbb{R}^d)$ the locally Sobolev space of $\mathcal{L}$-periodic functions with the inner product given by that of $H^s(\Omega)$. We have:

Theorem 1 (Ideal crystal) *Let $T > 0$, $d = 2$ or 3, $\beta = 1/T$ be sufficiently large and $|\Omega|$ be sufficiently small. We assume that*

1. *κ_{per} is the $\mathcal{L}$−periodic background charge distribution s.t.*
   (a) *$\kappa_{\text{per}} \in H^s_{\text{per}}$ for $s \geq 2$ and $||\kappa_{\text{per}}||_{H^s}$ is sufficiently small;*
   (b) *$\kappa_{\text{jel}} = \frac{1}{|\Omega|}\int_\Omega \kappa_{\text{per}}$ and $\kappa'_{\text{per}} = \kappa_{\text{per}} - \kappa_{\text{jel}}$ satisfy*
   $|xc(\kappa_{\text{jel}})| < \frac{\kappa^2_{\text{jel}}}{w^2_{d-1}}$ and $\kappa'_{\text{per}} \in H^s_{\text{per}}$ for $s \geq 2$, where w_d is the volume of the d-sphere;
2. *$xc \in W^{s,\infty}$ for $s \geq 2$ and $\|xc\|_{W^{s,\infty}}$ is sufficiently small.*

Then the Kohn-Sham equation (75) *has a unique solution $(\rho_{\text{per}}, \mu_{\text{per}}) \in H^s_{\text{per}}(\mathbb{R}^d) \times \mathbb{R}_+$ satisfying*

$$\|\rho_{\text{per}} - \kappa_{\text{per}}\|_{H^s} \lesssim \|\kappa'_{\text{per}}\|_{H^s_{\text{per}}}, \tag{79}$$

$$|\mu_{\text{per}} - \mu_{\text{jel}}| \lesssim \|\kappa'_{\text{per}}\|_{H^s_{\text{per}}}. \tag{80}$$

where $(\rho_{\text{jel}} = \kappa_{\text{jel}}, \mu_{\text{jel}})$ is a solution to (75) *with $\kappa = \kappa_{\text{jel}}$.*

Proof (Idea of proof of Theorem 1*)* We write (75) as a fixed point problem

$$\rho = \Phi(\rho, \mu), \;\; \Phi(\rho, \mu) := \text{den}[f(\beta(h_\rho - \mu))]. \tag{81}$$

To this we add the charge conservation law (76) with Ω a fundamental cell of $\mathcal{L}$.

To handle the constraint (76), we let P denote the projection onto constants, $Pf := \frac{1}{|\Omega|}\int_\Omega f$, and let $\bar{P} = 1 - P$ and split (81) into two equations

$$\rho' = P\Phi(\rho' + \rho'', \mu), \tag{82}$$

$$\rho'' = \bar{P}\Phi(\rho' + \rho'', \mu). \tag{83}$$

where $\rho' := P\rho = \frac{1}{|\Omega|}\int_\Omega \rho$ and $\rho'' := \bar{P}\rho = \rho - \rho'$. By the constraint (76), we have $\rho' = \frac{1}{|\Omega|}\int_\Omega \kappa$. Hence (82) and (83) are equations for μ and ρ''. We first solve (83) for ρ'' by a fixed point theorem and then (82) for μ, by an implicit function argument.

A *central* open problem here is to determine whether the (locally) free energy minimizing solution *breaks spontaneously symmetry* or not. The spontaneous symmetry breaking means that ρ_γ has lower (coarser) symmetry than κ ('spontaneous symmetry breaking').

4.2 Macroscopic Perturbations

A key problem in solid state physics is derivation of an effective, macroscopic equations for crystals from microscopic ones. In the full generality this problem is far from our reach. However one can reasonably hope to derive such equations starting from the DFT microscopic theory.

We consider macroscopic perturbations (say, local deformations) of ideal crystals and the dielectric response to them. At the first step, one would like to prove existence of solutions under local deformation of crystals. The appropriate spaces for our analysis are the homogenous Sobolev spaces:

$$\dot{H}^s(\mathbb{R}^3) = \left\{ f : \|f\|^2_{\dot{H}^s} := \int |p|^{2s} |\hat{f}|^2(p) < \infty \right\}. \tag{84}$$

We note that $\dot{H}^s$ and $\dot{H}^{-s}$ are dual spaces under the usual $L^2(\mathbb{R}^3)$ pairing $\langle \cdot, \cdot \rangle$ and that $\dot{H}^s$, unlike H^s, contains only s-order derivative in its norm.

We state some of the assumptions used below. To begin with we assume $d = 3$.

[A1] (regularity of κ)

$\kappa = \kappa_{\text{per}} + \kappa'$, where

κ_{per} is $\mathscr{L}$-periodic and satisfies

$\kappa_{\text{per}} \in H^2_{\text{per}}(\mathbb{R})^3$

and $\kappa' \in (H^2 \cap \dot{H}^{-2})(\mathbb{R}^3)$,

[A2] (regularity of xc**)**

$\text{xc} \in C^4(\mathbb{R}_+)$ together with its derivatives

is bounded near the origin as

$|\text{xc}(\lambda)| < \epsilon\lambda$ for ϵ small.

Since κ' is not periodic, constraint (76) does not apply here. Let $(\rho_{\text{per}}, \mu_{\text{per}})$ be the periodic solution to the Kohn-Sham equation (75), with the $\mathscr{L}-$periodic background charge density κ_{per} given in Theorem 1. The next result shows that the periodic solutions of Theorem 1 are stable under local perturbations.

Theorem 2 (Stability under local perturbations) *Let $d = 3$ and the constraints of Theorem 1 be obeyed and assume [A1] and [A2]. In addition, let $\|\kappa\|_{H^2 \cap \dot{H}^{-2}} \ll 1$*

and $\|\kappa'\|_{H^2} \ll 1$. *Then the Kohn-Sham equation* (75)*, with* $\kappa = \kappa_{\rm per} + \kappa'$ *and* $\mu = \mu_{\rm per}$*, has a unique solution* ρ *satisfying*

$$\rho = \rho_{\rm per} + \rho' \;\; \text{with} \;\; \rho' \in (H^2 \cap H^{-2})(\mathbb{R}^3) \;\text{and} \tag{85}$$

$$\|\rho'\|_{H^2 \cap H^{-2}} \lesssim \|\kappa'\|_{H^2 \cap H^{-2}}. \tag{86}$$

Theorem 2 is proven in [23]. Similar results for $T = 1/\beta = 0$ were proven in [1, 11, 12, 15, 17] (see [14, 39] for very nice reviews).

Dielectric response We consider Eq. (75) in the macroscopic variables at $1 << \beta < \infty$. Let $\mathscr{L}_\delta := \delta\mathscr{L}$ be a microscopic crystalline lattice (on the microscopic scale 1) with a fundamental domain Ω_δ centered at the origin. Let $\kappa^\delta_{\rm per}$ be $\mathscr{L}_\delta-$periodic microscopic charge distribution of the form

$$\kappa^\delta_{\rm per}(x) = \delta^{-d}\kappa_{\rm per}(\delta^{-1}x) \tag{87}$$

where $\kappa_{\rm per}$ is a $\mathscr{L}-$periodic function on $\mathbb{R}^d$. Note that under this scaling, the L^1-norm is preserved.

We consider a macroscopically perturbed background charge distribution (written in the macroscopic coordinate x)

$$\kappa_\delta(x) = \kappa^\delta_{\rm per}(x) + \kappa'(x), \tag{88}$$

where $\kappa'(x) \in L^2(\mathbb{R}^d)$ is a small local perturbation living on the macroscopic scale (1), producing macroscopically deformed crystal. To study the macroscopic behavior, we rescale the Kohn-Sham equations (75) to obtain

$$\rho_\delta = \text{den}[f_{FD}(\beta(h_{\phi_\delta} - \mu)], \tag{89}$$

where $h_{\phi_\delta} = -\delta^2\Delta - \delta\phi_\delta(x)$ with the potential ϕ_δ given by

$$\phi_\delta := (-\Delta)^{-1}4\pi(\kappa_\delta - \rho_\delta)\,. \tag{90}$$

Given $\kappa^\delta_{\rm per}$, Theorem 8 implies that (89) has a $\mathscr{L}_\delta$-periodic solution $\rho^\delta_{\rm per} = \delta^{-3}\rho_{\rm per}(\delta^{-1}x)$, with associated potential $\phi^\delta_{\rm per} = \delta^{-1}\phi_{\rm per}(\delta^{-1}x)$. We list additional assumptions needed for the next and key result.

Let $h_{\rm per} = -\Delta - \phi_{\rm per}$. Let $\xi(\mathbb{R}^3)$ denote the size of the spectral gap of $h_{\rm per}$ at μ on $L^2(\mathbb{R}^3)$ and $\xi(\Omega)$ denote the size of the spectral gap of $h_{\rm per}$ at μ on $L^2(\Omega)$ with periodic boundary condition.

[A3] (spectral gap condition)

$$\xi := \xi(\mathbb{R}^3) - \frac{5}{6}\xi(\Omega) > 0.$$

[A4] (scaling condition)

$$\delta \ll 1 \text{ and } \beta \geq C\xi^{-1} \ln(1/\delta) \text{ for } C \text{ large.}$$

We now present the main result of [23] on the derivation of the effective Poisson equation:

Theorem 3 *Suppose that $d = 3$ and fix a solution ρ_{per} as above. Let assumptions [A1]–[A4] hold. Then the rescaled Kohn-Sham equation* (89)*, with background charge distribution defined in* (88) *and $\mu = \mu_{\text{per}}$, has a unique solution ρ_δ in $L^2_{\text{per}} + \dot{H}^{-1} + \dot{H}^{-2}$ with associated potential ϕ_δ of the form*

$$\phi_\delta = \phi^\delta_{\text{per}} + \phi_0 + \phi_{\text{rem},1} + \phi_{\text{rem},2}, \tag{91}$$

where ϕ^δ_{per} is the potential associated to the periodic solution ρ^δ_{per}, $\phi_{\text{rem},i}$, $i = 1, 2$, obey the estimates

$$\|\phi_{\text{rem},1}\|_{\dot{H}^1(\mathbb{R}^3)} \lesssim \delta^{1/2} \text{ and } \|\phi_{\text{rem},2}\|_{L^2(\mathbb{R}^3)} \lesssim \delta \tag{92}$$

and ϕ_0 satisfies the equation

$$-\operatorname{div} \epsilon_0 \nabla \phi_0 = \kappa' \tag{93}$$

with a real positive 3×3 matrix, ϵ_0, given in (94), (95), (96) *and* (97) *below.*

A similar result for $T = 1/\beta = 0$ was proven in [13, 14] (see also [27–29]).

Remark 1

1. We note that in general $\xi(\mathbb{R}^3) \leq \xi(\Omega)$. One sees this by passing to Bloch-Floquet decomposition of h_{per} and noting that $\xi(\mathbb{R}^3)$ is the inf of all spectral gaps of the fiber decomposed operators on $L^2(\Omega)$.
2. The number $\frac{5}{6}$ comes from Hardy's inequality. In dimension $d = 3$, Hardy's inequality is $\|f\|_{L^6(\mathbb{R}^3)} \lesssim \|\nabla f\|_{L^2(\mathbb{R}^3)}$. We note that if $p = 6$, then its conjugate is $q = \frac{6}{5}$.
3. The constant C appearing in [A4] can be taken to be any number $C > 100$.
4. The 3×3 matrix ϵ_0 in (93) is of the form

$$\epsilon_0 = \mathbf{1}_{3\times 3} + \epsilon_0', \tag{94}$$

$$\epsilon_0' = \frac{1}{|\Omega|} \operatorname{Tr}_{L^2(\Omega)} \oint r^2_{\text{per}}(z)(-i\nabla) r_{\text{per}}(z)(-i\nabla) r_{\text{per}}(z) \tag{95}$$

$$- \frac{1}{|\Omega|} \|\rho_1\|^2_{\dot{H}^{-1}(\Omega;\mathbb{C}^3)}, \tag{96}$$

where $r_{\text{per}}(z) = (z - h_{\text{per}})^{-1}$, $h_{\text{per}} = -\Delta - \phi_{\text{per}} + \text{xc}(\rho_{\text{per}})$, and

$$\rho_1 = 2\chi_{\mathbb{R}^3 \setminus \Omega^*}(-i\nabla)\,\text{den} \oint r_{\text{per}}^2(z)(-i\nabla) r_{\text{per}}(z)\,. \tag{97}$$

Here χ_Q denotes a characteristic function of the set Q and Ω^* stands for a fundamental cell of the reciprocal lattice.

5 Hartree-Fock-Bogoliubov Equations

For appropriate spaces, it is shown in [3] that, for the Hamiltonian H given in Eq. (7), φ_t satisfies (12) if and only if the triple (ϕ, γ, α) of 1st- and 2nd-order truncated expectations of φ_t, defined by (cf. (10))

$$\begin{cases} \phi(x,t) := \varphi_t(\psi(x)), \\ \gamma(x,y,t) := \varphi_t[\psi^*(y)\,\psi(x)] - \varphi_t[\psi^*(y)]\,\varphi[\psi(x)], \\ \alpha(x,y,t) := \varphi_t[\psi(x)\,\psi(y)] - \varphi[\psi(x)]\,\varphi_t[\psi(y)]\,, \end{cases} \tag{98}$$

satisfies the time-dependent Hartree-Fock-Bogoliubov equations

$$i\partial_t\phi = h(\gamma)\phi + |\phi|^2\phi + k(\alpha)\bar{\phi}\,, \tag{99}$$

$$i\partial_t\gamma = [h(\gamma^\phi), \gamma]_- + [k(\alpha^\phi), \alpha]_-\,, \tag{100}$$

$$\begin{aligned} i\partial_t\alpha =& [h(\gamma^\phi), \alpha]_+ + [k(\alpha^\phi), \gamma]_+ \\ & + k(\alpha^\phi), \end{aligned} \tag{101}$$

where the subindex t is not displayed, $[A_1, A_2]_\pm = A_1 A_2^{T/*} \pm A_2 A_1^{T/*}$, $\gamma^\phi := \gamma + |\phi\rangle\langle\phi|$ and $\alpha^\phi := \alpha + |\phi\rangle\langle\bar{\phi}|$, and

$$h(\gamma) = h + v * d(\gamma) + v \sharp \gamma\,, \tag{102}$$

$$k(\alpha) = v \sharp \alpha\,, \quad d(\alpha)(x) := \alpha(x,x). \tag{103}$$

In these equations, $v \sharp \alpha$ is the operator with the integral kernel $v \sharp \alpha\,(x;y) := v(x-y)\alpha(x;y)$.

Here, ϕ describes the Bose-Einstein condensed atoms, γ, thermal atomic cloud and σ, the superfluid component of the atomic gas.

For the pair potential $v(x-y) = g\delta(x-y)$, the HFB equations in a somewhat different form have first appeared in the physics literature; see [26, 35, 51] and, for further discussion, [3, 4].

Note that if we drop the third terms in (99) and (100), then we arrive at, essentially, the Gross-Pitaevski and Hartree equations, respectively. If we drop the last term on the r.h.s. of (101), then equations (99), (100) and (101) have solutions of the form $(\phi, 0, 0)$ and $(0, \gamma, 0)$, where ϕ and γ solve the Gross-Pitaevski and Hartree equations, $i\partial_t\phi_t = h\phi_t + |\phi_t|^2\phi_t$ and $i\partial_t\gamma_t = [h(\gamma_t), \gamma_t]_-$, respectively. The last term on the r.h.s. of (101) prevents the 100% condensation.

Equations (99), (100) and (101), with the last term on the r.h.s. of (101) dropped, form the no quantum depletion model. Equations (99) and (100), with $\alpha = 0$, are called the two-gas model.

Given appropriate spaces, here are some key properties of (99), (100) and (101) at a glance [3, 4]:

(A) *Conservation of the total particle number*: If φ_t solves Eq. (12) then the number of particles,

$$\mathcal{N}(\phi_t, \gamma_t, \sigma_t) := \varphi_t(N), \tag{104}$$

where N is the particle-number operator, is conserved.

(B) *Existence and conservation of the energy*: If φ_t solves (12) then the energy

$$\mathcal{E}(\mu(\varphi_t)) := \varphi_t(H) \tag{105}$$

is conserved. Moreover, $\mathcal{E}$ is given explicitly by the expression

$$\mathcal{E}(\phi, \gamma, \alpha) = \mathrm{Tr}[h(\gamma^\phi) + b[|\phi\rangle\langle\phi|]\gamma + \frac{1}{2}b[\gamma]\gamma] + \frac{1}{2}\int v(x-y)|\alpha^\phi(x, y)|^2 dxdy. \tag{106}$$

(C) *Positivity preservation property*: If $\Gamma = \begin{pmatrix} \gamma & \alpha \\ \bar{\alpha} & 1+\bar{\gamma} \end{pmatrix} \geq 0$ at $t = 0$, then this holds for all times.

(D) *Global well-posedness of the HFB equations*: If the pair potential v is in the Sobolev space $W^{p,1}$, with $p > d$, and satisfies $v(x) = v(-x)$ and the initial condition $(\phi_0, \gamma_0, \alpha_0)$ is in a certain mixed functional – operator space and satisfies $\begin{pmatrix} \gamma_0 & \alpha_0 \\ \bar{\alpha}_0 & 1+\bar{\gamma}_0 \end{pmatrix} \geq 0$, then the HBF equations (99), (100) and (101) have a unique global solution in the same space.

6 Bogoliubov-de Gennes Equations

6.1 Formulation

We assume for simplicity that the external potential is zero, $V = 0$. Since the Bogoliubov-de Gennes (BdG) equations describe the phenomenon of superconductivity, they are naturally coupled to the electromagnetic field. We describe the latter by the vector and scalar potentials a and ϕ.

It is convenient to organize the operators γ and α (see (10)) into the self-adjoint matrix-operator

$$\eta := \begin{pmatrix} \gamma & \alpha \\ \alpha^* & \mathbf{1} - \bar{\gamma} \end{pmatrix}. \tag{107}$$

Assuming γ carries electric charge in units of -1 (i.e. the charge density is $-\rho_\gamma$, the time-dependent BdG equations can be written as (see e.g. [22, 24, 25])

$$i\partial_t \eta = [\Lambda(\eta, a), \eta], \tag{108}$$

with $\Lambda(\eta, a) = \begin{pmatrix} h_{\gamma a} & v\sharp\alpha \\ \overline{v\sharp\bar{\alpha}} & -\overline{h_{\gamma a}} \end{pmatrix}$, where $v(x)$ is a pair potential, the operator $v \sharp \alpha$ is defined through the integral kernels as $v \sharp \alpha\,(x; y) := v(x - y)\alpha(x; y)$, and

$$h_{\gamma a} := h_a + v * \rho_\gamma - v \sharp \gamma\,,\ \rho_\gamma(x) := \gamma(x; x). \tag{109}$$

Above $h_a = -\Delta_a$ and the terms $v * \rho_\gamma$ and $-v \sharp \gamma$ describe the self-interaction and exchange energies. Equation (108) is coupled to the Ampère's law part of the Maxwell equations

$$-\partial_t(\partial_t a + \nabla\phi) = \mathrm{curl}^* \,\mathrm{curl}\, a - j(\gamma, a), \tag{110}$$

where ϕ is the scalar potential and $j(\gamma, a)$ is the superconducting current, given by

$$j(\gamma, a)(x) := [-i\nabla_a, \gamma]_+(x, x).$$

Here, recall, $[A, B]_+ := AB + BA$.

Finally, recall that γ and α satisfy (11). In fact, one has the stronger property

$$0 \leq \eta = \eta^* \leq 1. \tag{111}$$

Remarks

(1) In general, h_a might contain also an external potential $V(x)$, due to the impurities.
(2) For $\alpha = 0$, Eq. (108) becomes the time-dependent Hartree-Fock equation (44) for γ. Thus the HFE is the special diagonal case of the BdG equations.
(3) We may assume that the physical space is either $\mathbb{R}^d$ or a finite box in $\mathbb{R}^d$ and γ and α are gauge periodic operators trace-class and Hilbert-Schmidt operators w.r. to trace per volume. For a detailed discussion of spaces see [22].
(4) One should be able to derive (108) and (110) from hamiltonian (7) coupled to the quantized electro-magnetic filed.

Connection with the BCS theory Equation (108) can be reformulated as an equation on the Fock space involving an effective quadratic hamiltonian (see [3] for the bosonic version). These are the effective BCS equations and the effective BCS hamiltonian (see [24, 25, 38]).

6.2 Symmetries

The equations (108), (109) and (110) are invariant under the *gauge* transformations and, if the external potential V is zero, also under *translations* and *rotations*, defined as

$$T_\chi^{\text{gauge}} : (\gamma, \alpha, a, \phi) \mapsto (e^{i\chi}\gamma e^{-i\chi}, e^{i\chi}\alpha e^{i\chi}, a + \nabla\chi, \phi - \partial_t\chi), \tag{112}$$

for any sufficiently regular function $\chi : \mathbb{R}^d \to \mathbb{R}$, and

$$T_h^{\text{trans}} : (\gamma, \alpha, a, \phi) \mapsto (U_h\gamma U_h^{-1}, U_h\alpha U_h^{-1}, U_h a, U_h\phi), \tag{113}$$

for any $h \in \mathbb{R}^d$,

$$T_\rho^{\text{rot}} : (\gamma, \alpha, a, \phi) \mapsto (U_\rho\gamma U_\rho^{-1}, U_\rho\alpha U_\rho^{-1}, \rho U_\rho a, U_\rho\phi), \tag{114}$$

for any $\rho \in O(d)$. Here U_h and U_ρ are the standard translation and rotation transforms $U_h : f(x) \mapsto f(x+h)$ and $U_\rho : f(x) \mapsto f(\rho^{-1}x)$. In terms of η, say the gauge transformation, T_χ^{gauge}, could be written as

$$\eta \to \hat{T}_\chi^{\text{gauge}}\eta(\hat{T}_\chi^{\text{gauge}})^{-1}, \quad \text{where } \hat{T}_\chi^{\text{gauge}} = \begin{pmatrix} e^{i\chi} & 0 \\ 0 & e^{-i\chi} \end{pmatrix}. \tag{115}$$

Notice the difference in action of this transformation on the diagonal and off-diagonal elements of η.

The invariance under the gauge transformations can be proven by using the relation

$$\hat{T}_\chi^{\text{gauge}} g'(\eta)(\hat{T}_\chi^{\text{gauge}})^{-1} = g'(\hat{T}_\chi^{\text{gauge}}\eta(\hat{T}_\chi^{\text{gauge}})^{-1}),$$

proven by expanding $g'(\eta)$ (or $g^{\#}(\beta H_{\eta a})$), and the gauge covariance of $\Lambda(\eta, a)$:

$$(\hat{T}_\chi^{\text{gauge}})^{-1}(\Lambda(\hat{T}_\chi^{\text{gauge}}\eta, a)) = \Lambda(\eta, a). \tag{116}$$

The gauge symmetry is not a physical one, but rather an invariance of the solution space (or the covariance of the equations) under ‘reparametrizations’. Therefore the natural objects are gauge-equivalent classes of solutions. This leads to the notion of gauge or *magnetic translations* (mt, below) and gauge or magnetic rotations. The former are given by the transformations

$$T_{bs} : (\eta, a) \to (T^{\text{gauge}}_{\chi_s})^{-1} T^{\text{trans}}_s (\eta, a), \tag{117}$$

for any $s \in \mathbb{R}^d$, where $\chi_s(x) := x \cdot a_b(s)$, where $a_b(x)$ is the vector potential with the constant magnetic field, $\text{curl}\, a_b = b$. The invariance under these transformations will be called the *magnetic translation* (mt) symmetry. The latter is given by the transformations

$$T_{b\rho} : (\eta, a) \to (T^{\text{gauge}}_{\chi_\rho})^{-1} T^{\text{rot}}_\rho (\eta, a), \tag{118}$$

for $\rho \in O(d)$. We remark that in general T_{bs} and $T_{b\rho}$ are only projective representations of $\mathscr{L}$ and $O(d)$, respectively.

Finally, the equations (108), (109) and (110) are invariant under the transformations (see [2])

$$\eta \to \mathbf{1} - \eta \quad \text{and} \quad \eta \to -\overline{J^* \eta J} \quad \text{(the particle-hole symmetry)}.$$

Here $J := \begin{pmatrix} 0 & \mathbf{1} \\ -\mathbf{1} & 0 \end{pmatrix}$. The second relation follows from (the particle-hole symmetry)

$$J^* \Lambda J = -\bar{\Lambda}. \tag{119}$$

The form (107) of the matrix operator η is characterized by the relation

$$J^* \eta J = \mathbf{1} - \bar{\eta}. \tag{120}$$

By the above, the evolution preserves this relation, i.e. if an initial condition has this property, then so does the solution.

6.3 Conservation Laws

The Bogolubov-de Gennes equations (108), (109) and (110) form a hamiltonian system with the conserved energy functional

$$E(\eta, a) = \text{Tr}_\Omega\big(h_a \gamma\big) + \frac{1}{2}\text{Tr}_\Omega\big((v * \rho_\gamma)\gamma\big) - \frac{1}{2}\text{Tr}_\Omega\big((v \sharp \gamma)\gamma\big) \tag{121}$$

$$+ \frac{1}{2}\text{Tr}_\Omega\big(\alpha^*(v \sharp \alpha)\big) + \frac{1}{2}\int_\Omega dx |\,\text{curl}\, a(x)|^2. \tag{122}$$

where Ω is either $\mathbb{R}^d$ or a fundamental cell of a macroscopic lattice in $\mathbb{R}^d$ (see Sect. 6.6).

The energy $E(\eta, a)$ can be derived from the total quantum hamiltonian, H_{tot}, of the many body system coupled to the quantum electromagnetic field, through quasifree reduction as $E(\eta, a) := \varphi(H_{\text{tot}})$, where φ is a quasifree state in question (see (10) and [3] or [38]). Its conservation law is related to the conservation of the total energy $\varphi(H_{\text{tot}})$. (The combinatorial coefficients of each term result from restriction to $SU(2)$ invariant states and peeling of spin variables (cf. [38]).)

Conservation of (121)–(122) can be also proven directly similarly to the proof of the conservation law of (48).

6.4 Stationary Bogoliubov-de Gennes Equations

We consider stationary, rather than static, solutions to (108) of the form

$$\eta_t := \hat{T}_\chi^{\text{gauge}} \eta, \tag{123}$$

with η and $\dot{\chi} \equiv \mu$ independent of t and a independent of t and $\phi = 0$. We have

Proposition 4 *Equation* (123)*, with* η *and* $\dot{\chi} \equiv \mu$ *independent of* t*, is a solution to* (108) *iff* η *solves the equation*

$$[\Lambda_{\eta a}, \eta] = 0, \tag{124}$$

where $\Lambda_{\eta a} \equiv \Lambda_{\eta a \mu} := \Lambda(\eta, a) - \mu S$*, with* $S := \begin{pmatrix} 1 & 0 \\ 0 & -1 \end{pmatrix}$*, and is given explicitly*

$$\Lambda_{\eta a} := \begin{pmatrix} h_{\gamma a} - \mu & v \sharp \alpha \\ v \sharp \alpha^* & -\bar{h}_{\gamma a} + \mu \end{pmatrix}, \tag{125}$$

with $h_{\gamma a} := -\Delta_a + v * \rho_\gamma - v \sharp \gamma$ *and, recall,* $v \sharp \alpha$ *is an operator with the integral kernel* $v(x - y)\alpha(x, y)$.

Proof Plugging (123) into (108) and using that for χ independent of x,

$$\partial_t \eta_t = i \dot{\chi} \hat{T}_\chi^{\text{gauge}} [S, \eta]$$

and (116), we obtain

$$-\dot{\chi}[S, \eta] = [\Lambda(\eta, a), \eta]. \tag{126}$$

Since $\dot{\chi} \equiv \mu$, the latter equation can be rewritten as (124).

For any reasonable function f, solutions of the equation

$$\eta = f(\beta \Lambda_{\eta a}), \tag{127}$$

solve (124) and therefore give stationary solutions of (108). Under certain conditions, the converse is also true.

The physical function f is selected by either a thermodynamic limit (Gibbs states) or by a contact with a reservoir, or imposing the maximum entropy principle. It is given by the Fermi-Dirac distribution (38), i.e.

$$f(h) = (1 + e^h)^{-1}. \tag{128}$$

Inverting the function f, one can rewrite (127) as $\beta \Lambda_{\eta a} = f^{-1}(\eta)$. Let $f^{-1} =: s'$. Then the static Bogoliubov-de Gennes equations can be written as

$$\Lambda_{\eta a} - \beta^{-1} s'(\eta) = 0, \tag{129}$$

$$\mathrm{curl}^* \, \mathrm{curl}\, a - j(\gamma, a) = 0. \tag{130}$$

Here $0 < \beta \leq \infty$ (inverse temperature) and $s(\eta) := -(\eta \ln \eta + (1 - \eta) \ln(1 - \eta))$ (see (41)).

Remarks

(1) One can express these equations in terms of eigenfunctions of the operator $\Lambda_{\eta a}$, which is the form appearing in physics literature (see [2, 3]).
(2) If we drop the direct $v * \rho_\gamma$ and exchange self-interaction $-v\sharp\gamma$, then the operator $h_{\gamma a \mu}$ and therefore $\Lambda_{\eta a}$ are independent of γ and consequently Eq. (127) defines γ in terms of α and a:

$$\eta_{\beta a} = f(\beta \Lambda_{\alpha a}), \quad \text{where } \Lambda_{\alpha a} := \Lambda_{\eta a}\big|_{\gamma=0}. \tag{131}$$

(3) For (127) to give η of the form (107), the function $f(h)$ should satisfy the conditions

$$f(\bar{h}) = \overline{f(h)} \text{ and } f(-h) = \mathbf{1} - f(h). \tag{132}$$

The function $f(h)$ given in (128) satisfies these conditions. From now on, *we assume $f(h)$ has explicit form* (128).

6.5 Free Energy

The stationary Bogoliubov-de Gennes equations (129) and (130) arise as the Euler-Lagrange equations for the free energy functional

$$F_\beta(\eta, a) := E(\eta, a) - \beta^{-1} S_1(\eta) - \mu N(\eta), \tag{133}$$

where $S(\eta) = \mathrm{Tr} s(\eta)$ is the entropy, $N(\eta) := \mathrm{Tr}\gamma$ is the number of particles, and $E(\eta, a)$ is the energy functional given in (121)–(122) with η and a time-independent.

It is shown in [22] that on carefully chosen spaces

(a) The free energy functional F_β is well defined;
(b) F_β is continuously (Gâteaux) differentiable;
(c) If $0 < \eta < 1$ and (η, a) is even in the sense of [22], Eq. (1.17), then critical points of F_β satisfy the BdG stationary equations (129) and (130);
(d) Minimizers of F_β are its critical points.

Now, we define the partial gradients $\partial_\eta F_\beta(\eta, a)$ and $\partial_a F_\beta(\eta, a)$ by $d_\eta F_\beta(\eta, a)\eta' = \mathrm{Tr}(\eta' \partial_\eta F_\beta(\eta, a))$ and $d_a F_\beta(\eta, a)a' = \int a' \cdot \partial_a F_\beta(\eta, a)$, respectively. (Though the expression for $F_\beta(\eta, a)$ is often formal, $\partial_\eta F_\beta(\eta, a)$ and $\partial_a F_\beta(\eta, a)$ could be well-defined on appropriate spaces.)

Theorem 4 *Minimizers of the free energy $F_\beta(\eta, a)$ are critical points of $F_\beta(\eta, a)$, i.e. they satisfy the Euler-Lagrange equations*

$$\partial_\eta F_\beta(\eta, a) = 0 \text{ and } \partial_a F_\beta(\eta, a) = 0, \tag{134}$$

for some β and μ (the latter are determined by fixing $S(\eta)$ and $\mathrm{Tr}(\gamma)$). The Gâteaux derivatives, $\partial_\eta F_\beta(\eta, a)$ and $\partial_a F_\beta(\eta, a)$, are given by

$$\partial_\eta F_\beta(\eta, a) = \Lambda_{\eta a} - \beta^{-1} g'(\eta), \tag{135}$$

and

$$\partial_a F_\beta(\eta, a) := \mathrm{curl}^* \,\mathrm{curl}\, a - j(\gamma, a), \tag{136}$$

where, recall, $j(\gamma, a)(x) := [-i\nabla_a, \gamma]_+(x, x)$, with $[A, B]_+ := AB + BA$. Consequently, the equations (134) *can be rewritten as* (129) *and* (130).

For the translation invariant case, the corresponding result is proven in [37]. In general case, but with $a = 0$ (which is immaterial here), the fact that BdG is the Euler-Lagrange equation of BCS was used in [33], but seems with no proof provided.

By (134), (135) and (136), we can write the equations (129) and (130) as

$$F'_\beta(\eta, a) = 0, \tag{137}$$

where $F'_\beta(\eta, a) = (\partial_\eta F_\beta(\eta, a), \partial_a F_\beta(\eta, a))$.

Remarks

(1) Due to the symmetry (120), $S(\eta) = \mathrm{Tr} s(\eta) = -Tr \eta \ln \eta$, with $s(\lambda)$ given in (41).
(2) $F_\beta(\eta, a)$ is a Helmholtz free energy. This energy depends on the temperature and the average magnetic field, $b = \frac{1}{|Q|} \int_Q \mathrm{curl}\, a$ (for a sample occupying a finite domain Q), in the sample, as thermodynamic parameters. Alternatively,

one can consider the free energy depending on the temperature and an applied magnetic field, h. For a sample occupying a finite domain Q, this leads (through the Legendre transform) to the Gibbs free energy

$$G_{\beta Q}(\eta, a) := F_{\beta Q}(\eta, a) - \Phi_Q h,$$

where $\Phi_Q = b|Q| = \int_Q \operatorname{curl} a$ is the total magnetic flux through the sample. The parameters b or h do not enter the equations (129) and (130) explicitly.

6.6 Ground/Gibbs States

We are looking for stationary states which minimize the free energy per unit volume. More precisely, with some license, we say that (η_*, a_*) is a ground/Gibbs state (depending on whether $\beta = \infty$ or $\beta < 0$), if there is a macroscopic lattice $\mathscr{L}^{\text{macro}}$, s.t. (η_*, a_*) satisfies

- $T_s^{\text{trans}}(\eta, a) = \hat{T}_{\chi_s}^{\text{gauge}}(\eta, a)$, $\forall\, s \in \mathscr{L}^{\text{macro}}$ and for some function $\chi_\cdot : \mathscr{L}^{\text{macro}} \times \mathbb{R}^d \to \mathbb{R}$,

for some lattice $\mathscr{L}^{\text{macro}} \subset \mathbb{R}^d$ with macroscopic fundamental cell Ω^{macro}, and (η_*, a_*) minimizes $F_{\beta \Omega^{\text{macro}}}(\eta, a)$ among states having the above property. This is equivalent to considering the equations on a large twisted torus.

In what follows, we will deal with $\beta < \infty$, i.e. with the Gibbs states only.

In general, equations (129) and (130) have the following stationary solutions which are candidates for the Gibbs state:

1. Normal state: $(\eta_*, 0)$, with $\alpha_* = 0$.
2. Superconducting state: $(\eta_*, 0)$, with $\alpha_* \neq 0$.
3. Mixed state: (η_*, a_*), with $\alpha_* \neq 0$ and $a_* \neq 0$.

One expects that the Gibbs state has the maximal possible symmetry. If the external fields are zero, then the equations are magnetically translationally invariant. Thus, one expects that the Gibbs state is magnetically translational invariant.

We have the following general result

Proposition 5 ([22]) *If η is mt-invariant, then $\alpha = 0$ (i.e. the state (η, a) is normal).*

In the opposite direction we have

Conjecture 5 *For $\beta < \infty$ sufficiently small, a Gibbs, normal state is mt-invariant and therefore unique.*

A stronger form of this conjecture is

Conjecture 6 *A Gibbs, normal state is mt-invariant and therefore unique.*

6.7 Symmetry Breaking

Theorem 7 ([22]) *Let $d = 2$. Suppose that $b > 0$ and assume that the interaction potential $v \leq -C|x|^{-\kappa}$, $\kappa < 2$. Then $\exists 0 < \beta_c''(b) \leq \beta_c'(b) < \infty$ s.t.*

- *If $\beta < \beta_c''(b)$, then any Gibbs state is normal;*
- *If $\beta > \beta_c'(b)$, then the ground/Gibbs state is a mixed state.*

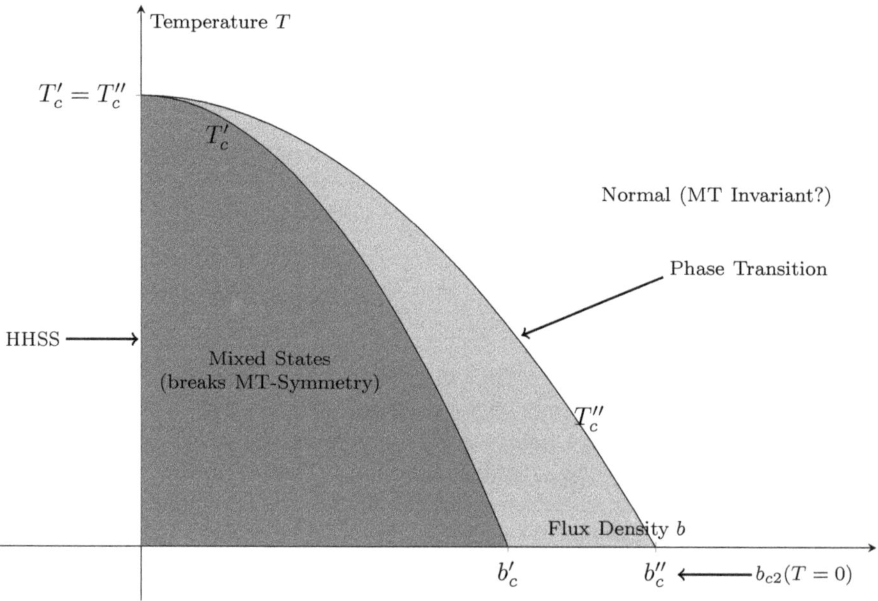

In view of Proposition 5 and Conjecture 5 above, this result suggests that under the stated conditions and as the temperature is lowered, the symmetry of the Gibbs state is broken spontaneously.

The corresponding result for $b = 0$ was proved in [37]. In this case, there are no mixed states and the 'mixed state' in the statement should be replaced by the 'superconducting state'. Consequently, there are no symmetry breaking in this case.

6.8 Stability

To formulate the next result, we need some definitions. Recall that $F_\beta'(\eta, a)$ is the *gradient* of $F_\beta(\eta, a)$ in the metric

$$\langle(\eta', a'), (\xi', c')\rangle := \mathrm{Tr}((\eta')^* \xi') + \int a' \cdot c'.$$

Consequently, the Gâteaux derivative $dF'_\beta(\eta, a)$ is the Hessian of $F_\beta(\eta, a)$ at (η, a) and therefore is formally symmetric. It can be shown that it is self-adjoint.

Let $u = (\eta, a)$. We say that a solution u_* to (137) is (linearly or energetically) *stable* iff the linearization $dF'_\beta(u_*)$ of the map $F'_T(u)$ (i.e. the *hessian*, $F''_T(u_*)$, of the functional $F_T(u)$) at u_* is non-negative, i.e.

$$dF'_\beta(u_*) \geq 0,$$

and *unstable* otherwise.

Note that the stability implies the energy minimization property locally in space (i.e. on a sufficiently large twisted torus).

We also consider a weaker notion of stability – the stability w.r. to generation of the superconducting α-component, which we call the *α-stability*.

Proposition 6 ([22]) *Let $b > 0$. The mt invariant (normal) state is α-stable for $\beta < \beta''_c(b)$ and, if $v(r) < -|r|^{-\kappa}$ with $\kappa < 2$, unstable for $\beta > \beta'_c(b)$.*

6.8.1 Normal States

For $b = 0$ we can choose $a = 0$ and the magnetic translation invariance becomes the usual translation invariance. In this case, if we drop the direct and exchange self-interactions from $h_{\gamma a \mu}$, then, as was mentioned above, the normal state is given by (131), with $a = 0$. If the direct and exchange self-interactions are present, then the existence of the normal states is established in [10].

These are normal translationally invariant states. For $b \neq 0$, the simplest normal states are the magnetically translation (mt-) invariant ones. The existence of the mt-invariant normal states for $b \neq 0$ is proven in [22]. They are of the form ($\eta = \eta_{\beta,b}$, $a = a_b$), where $a_b(x)$ is the magnetic potential with the constant magnetic field b ($\operatorname{curl} a_b = b$) and (cf. (131))

$$\eta_{\beta b} := \begin{pmatrix} \gamma_{\beta b} & 0 \\ 0 & \mathbf{1} - \bar{\gamma}_{\beta b} \end{pmatrix}, \tag{138}$$

with $\gamma_{\beta b}$ a solution to the equation

$$\gamma = s^\sharp(\beta h_{\gamma, a_b}),$$

with $s^\sharp := (s')^{-1}$. (For $s(x) = -(x \ln x + (1 - x) \ln(1 - x))$, we have $s^\sharp(h) = (e^h + 1)^{-1}$ and therefore $\gamma_{\beta b}$ solves the equation $\gamma = \left(e^{\beta h_{\gamma, a_b}} + 1\right)^{-1}$.)

6.8.2 Superconducting States

The existence of superconducting, translationally invariant solutions is proven in [37] (see this paper and [38] for the references to earlier results and [22], for a somewhat different approach).

6.8.3 Mixed States

For the mixed states, in the cylinder geometry, which means effectively $d = 2$, there is the following specific possibility:

- Vortex lattices: For a mesoscopic lattice $\mathscr{L}$ (i.e. much finer that $\mathscr{L}^{\mathrm{macro}}$), the state (η, a) satisfies $T_s^{\mathrm{trans}}(\eta, a) = \hat{T}_{\chi_s}^{\mathrm{gauge}}(\eta, a)$, for every $s \in \mathscr{L}^{\mathrm{meso}}$ and for some maps $\chi_s : \mathscr{L} \times \mathbb{R}^2 \to \mathbb{R}$.

  The map $\chi_s : \mathscr{L} \times \mathbb{R}^2 \to \mathbb{R}$ satisfies the co-cycle conditions,

$$\chi_{s+t}(x) - \chi_s(x+t) - \chi_t(x) \in 2\pi\mathbb{Z}, \ \forall s, t \in \mathscr{L}, \tag{139}$$

and are called the *summands of automorphy* (see [53] for a relevant discussion). (The map $e^{i\chi} : \mathscr{L} \times \mathbb{R}^2 \to U(1)$, where $\chi(x, s) \equiv \chi_s(x)$ is called the *factor of automorphy*.)

Excitations of the ground state are given by magnetic vortices, which are defined by the condition

- $T_\rho^{\mathrm{rot}}(\eta, a) = \hat{T}_{g_\rho}^{\mathrm{gauge}}(\eta, a)$ for every $\rho \in O(2)$ and some functions $g_\rho : O(2) \times \mathbb{R}^2 \to \mathbb{R}$.

  The existence of vortex lattices is proven in [22]). One might be able to prove the existence of vortices by making lattices coarser (or $b \to 0$) in the vortex lattice solutions.

6.8.4 Magnetic Flux Quantization

Denote by $\Omega_{\mathscr{L}}$ a fundamental cell of $\mathscr{L}$. One has the following results

(a) Magnetic vortices: $\frac{1}{2\pi}\int_{\mathbb{R}^2} \operatorname{curl} a = \deg g \in \mathbb{Z}$;
(b) Vortex lattices: $\frac{1}{2\pi}\int_{\Omega_{\mathscr{L}}} \operatorname{curl} a = c_1(\chi) \in \mathbb{Z}$.

Here $\deg g$ is the degree (winding number) of the map $e^{ig} : O(2) \to U(1)$ (which is map of a circle into itself, here we assume that $g(\rho) \equiv g_\rho$ is independent of x) and $c_1(\chi)$ is the first Chern number associated to the summand of automorphy $\chi : \mathscr{L} \times \mathbb{R}^2 \to \mathbb{R}$ (see [53]).

7 Existence of Periodic Solutions by the Variational Technique

Let $d = 2$ or 3. An operator A on $L^2(\mathbb{R}^d)$ is said to be ($\mathscr{L}-$) periodic iff $U_s A U_s^* = A, \forall s \in \mathscr{L}$, where U_s is the translation operator by $s \in \mathbb{R}^d$. In what follows for any periodic operator A, the trace is understood as the trace per volume

$$\mathrm{Tr} A := \mathrm{Tr}_{L^2(\mathbb{R}^3)} \chi_\Omega A \chi_\Omega \tag{140}$$

where χ_Ω is the indicator function on a fundamental domain Ω of $\mathscr{L}$. Let $L^2_{\mathrm{per}}(\mathbb{R}^d)$ denote the local L^2 space of $\mathscr{L}$-periodic functions with the inner product of $L^2(\Omega)$. We define the spaces

$$I^{s,p} = \{\gamma \in \mathscr{B}(L^2_{\mathrm{per}}(\mathbb{R}^d)) : \|\gamma\|_{s,p} := \|M^s \gamma M^s\|_p < \infty\}, \tag{141}$$

where $M = \sqrt{-\Delta}$ and $\|\cdot\|_p$ is the usual Schatten tracial p-norm. Set

$$\begin{aligned} I_0^{s,p} =& I^{s,p} \cap \{\mathrm{Tr}\,\gamma = Z\} \cap \{0 \le \gamma = \gamma^* \le 1\} \\ & \cap \{\|(-\Delta)^{-1/2}(\rho_\gamma - \kappa)\|_{L^2(\Omega)} < \infty\} \end{aligned} \tag{142}$$

In this section we use the variational approach and the fact that (71) (or (75)) is the Euler-Lagrange equations for free energy (74) to prove the following (see [22])

Theorem 8 *Let $\beta < \infty$. Let $\kappa = \kappa_{\mathrm{per}}$ is $\mathscr{L}-$periodic (an ideal crystal) and Xc assume is smooth bounded below, and C^1 on with Xc$'$ bounded. Then there exists $\mu \in \mathbb{R}$ such that the KS equation* (71) *on $I_0^{1,1}$ have an $\mathscr{L}-$periodic, energy minimizing solution γ satisfying $\int_\Omega \gamma(x,x) = \int_\Omega \kappa$.*

Since we minimize the free energy for $\mathrm{Tr}\gamma$ constant, we drop the term $-\mu\mathrm{Tr}\gamma$ from (74) to arrive at the free energy functional to be minimized

$$\begin{aligned} \mathscr{F}_\beta(\gamma) =& \mathrm{Tr}((-\Delta)\gamma) + \frac{1}{2}\langle(\rho_\gamma - \kappa), (-\Delta)^{-1}(\rho_\gamma - \kappa)\rangle_{L^2(\Omega)} \\ & + \int_\Omega \mathrm{Xc}(\rho_\gamma) - \beta^{-1} S(\gamma). \end{aligned} \tag{143}$$

Moreover, recall $\rho_\gamma(x) = \gamma(x,x)$ and $\mathrm{Xc}'(s) = \mathrm{xc}(s)$ and

$$S(\gamma) = \mathrm{Tr}\, s(\gamma), \quad s(x) = -(x \ln(x) + (1-x)\ln(1-x)). \tag{144}$$

We set $\mathscr{F}_\beta(\gamma) = \infty$ if any of the terms is not defined.

Theorem 9 (Main Result) *Under the conditions of Theorem 8, $\mathscr{F}_\beta(\gamma)$ has a minimizer on the set $I_0^{1,1}$. Moreover, this minimizer satisfies KS equation* (71).

We prove this theorem in a series of steps. We will use standard minimization techniques to prove that $\mathcal{F}_\beta(\gamma)$ is coercive and weakly lower semi-continuous, and $I_0^{1,1}$ weakly closed.

Part 1: coercivity

Lemma 1 *Assume that* $\operatorname{Tr}\gamma = Z$. *We have the lower bound*

$$\mathcal{F}_\beta(\gamma) \geq \frac{1}{2}\operatorname{Tr}((-\Delta\gamma) + \frac{1}{2}\langle(\rho-\kappa), (-\Delta)^{-1}(\rho-\kappa)\rangle - C\,. \tag{145}$$

for some constant C.

Proof Recall that $f_{FD}(\lambda) = \left(e^\lambda + 1\right)^{-1}$. First observe that $\frac{1}{2}\operatorname{Tr}(-\Delta\gamma) - \beta^{-1}S(\gamma)$ with $\operatorname{Tr}\gamma = Z$ has minimizer

$$\gamma = f(\beta(-\frac{1}{2}\Delta - \mu)) \tag{146}$$

for a suitable Lagrangian multiplier, μ, from $\operatorname{Tr}\gamma = Z$. Evaluating $\frac{1}{2}\operatorname{Tr}(-\Delta\gamma) - \beta^{-1}S(\gamma)$ at this minimizer gives some constant, say, C_1.

Recalling definition (143) and using that Ex is bounded below, say by C_2, gives (145).

Part 2: Convergence We follow the ideas of [18]. By Part 1, we note that each term on the r.h.s. of (145) is either positive or constant. Thus, $\mathcal{F}_\beta$ is bounded below. Let γ_n be a minimizing sequence of $\mathcal{F}_\beta(\gamma)$. Then we see that $\|\gamma_n\|_{I^{1,1}} = \operatorname{Tr}(-\Delta)\gamma_n$ and $\|\nabla^{-1}\rho_{\gamma_n}\|_{L^2(\Omega)}$ are uniformly bounded. We look for a limit of the sequence (γ_n). The non-abelian Hölder inequality show that

$$\|\gamma_n\|_{I^{0,2}} \leq \|\gamma_n\|_\infty \|\gamma_n\|_{I^{0,1}} \leq Z < \infty \tag{147}$$

is bounded. Hence, upto a subsequence, the kernels $\gamma_n(x, y)$ are in $L^2_{per}(\mathbb{R}\times\mathbb{R})$ (the space of $\mathcal{L}$-periodic under the action $(x, y) \to (x+s, y+s)$, $s \in \mathcal{L}$), locally L^2 functions on $(\mathbb{R}^2 \times \mathbb{R}^2)$ and converges weakly to some $\gamma_0'(x, y) \in L^2_{per}(\mathbb{R}^2 \times \mathbb{R}^2)$. We extend $\gamma_0'(x, y)$ to all of $\mathbb{R}^2 \times \mathbb{R}^2$ by periodicity. Let γ_0' denote the operator whose kernel is $\gamma_0'(x, y)$. Clearly, $\gamma_n \rightharpoonup \gamma_0'$ weakly in $I^{0,2}$.

Now, we show that $\gamma_0' \in I_0^{1,1}$. That is, $\gamma_0' \in I^{1,1}$ and $\operatorname{Tr}(\gamma_0') = Z$, $(\gamma_0')^* = \gamma_0'$, and $0 \leq \gamma_0' \leq 1$. Using the Bloch-Floquet decomposition, we see that

$$\int_{\Omega^*} d\hat{\xi}\, \operatorname{Tr}_{L^2(\Omega)}[(1-\Delta_\xi)^{1/2}(\gamma_n)_\xi(1-\Delta_\xi)^{1/2}] \tag{148}$$

$$= \int_{\Omega^*} d\hat{\xi}\, \operatorname{Tr}_{L^2(\Omega)}[(1-\Delta_\xi)\gamma_n] \tag{149}$$

$$= \operatorname{Tr}(1-\Delta)\gamma_n < \infty\,. \tag{150}$$

where the second line follows by expanding the traces on $L^2(\Omega)$ in an orthonormal basis of eigenfunctions of $-\Delta_\xi$ and the fact $0 \leq (\gamma_n)_\xi$. This shows that $(1 - \Delta_\xi)^{1/2}(\gamma_n)_\xi(1 - \Delta_\xi)^{1/2}$ is trace class (hence HS) for almost every $\xi \in \Omega^*$. It follows that the full operator $(1 - \Delta)^{1/2}\gamma_n(1 - \Delta)^{1/2}$ is HS in trace per volume norm and whose trace is equal to (150). Hence a weak limit exists and necessarily is $(1 - \Delta)^{1/2}\gamma_0'(1 - \Delta)^{1/2}$. We see that

$$Z = \lim_{n\to\infty} \mathrm{Tr}(\gamma_n) \tag{151}$$

$$= \lim_{n\to\infty} \mathrm{Tr}((1 - \Delta)^{1/2}\gamma_n(1 - \Delta)^{1/2}(1 - \Delta)^{-1}) = \mathrm{Tr}(\gamma_0') \tag{152}$$

since $1 - \Delta$ is HS (in trace-per-volume norm) for $d = 2, 3$. The fact $\gamma_0' \in I_0^{1,1}$ is proved by using a compactness argument pointwise in the fiber decomposition through a Bloch-Floquet argument similar to one used in (148), (149) and (150). Note that the fact $\gamma_0' = \gamma_0'^*$ and the bound $0 \leq \gamma_0' \leq 1$ is preserved by weak HS (per volume) convergence.

Finally, we show that $\sqrt{\rho_n} \in H^1(\Omega)$ and converges to some $\rho_0'' \in H^1(\Omega)$ weakly. Let $\varphi_\lambda(\xi, x)$ denote the eigenvectors of γ_ξ with eigenvalue λ in its Bloch-Floquet-Zack decomposition. Since the map $f \mapsto \int_\Omega |\nabla\sqrt{f}|^2$ is convex, we see that

$$\int_\Omega |\nabla\sqrt{\rho(x)}|^2 dx = \int_\Omega \left|\nabla\left(\int_{\Omega^*} d\xi \sum \lambda_\xi |\varphi_\xi(x)|^2\right)^{1/2}\right|^2 dx \tag{153}$$

$$\lesssim \int_\Omega \int_{\Omega^*} dx d\hat{\xi} \sum \lambda_\xi |\nabla|\varphi_{\lambda_\xi}(\xi, x)||^2 \tag{154}$$

$$\lesssim \int_\Omega \int_{\Omega^*} dx d\hat{\xi} \sum \lambda_\xi |\nabla\varphi_{\lambda_\xi}(\xi, x)|^2 \tag{155}$$

$$= \mathrm{Tr}(-\Delta\gamma) \tag{156}$$

This shows that $\sqrt{\rho_n}$ are bounded in $H^1(\Omega)$ and thus converges weakly, in H^1 to $\sqrt{\rho_0''} \in H^1(\Omega)$. Compactness of $H^1(\Omega)$ in $L^2(\Omega)$ shows that $\sqrt{\rho_n}$ converges to $\sqrt{\rho_0''}$ in L^2, hence $\rho_n \to \rho_0''$ in $L^1(\Omega)$. It follows that for any smooth bounded periodic function f

$$\langle\rho_0'', f\rangle = \lim_{n\to\infty} \langle\rho_n, f\rangle = \lim_{n\to\infty} \mathrm{Tr}\gamma_n f \tag{157}$$

$$= \lim_{n\to\infty} \mathrm{Tr}(\gamma_0' f) \tag{158}$$

$$= \langle\rho_0', f\rangle \tag{159}$$

Thus, we denote the common limit as $\gamma_0 := \rho_0' = \rho_0''$ and $\rho_0 := \text{den}[\gamma_0]$. We summarize the types of convergences here:

$$(1-\Delta)^{1/2}\gamma_n(1-\Delta)^{1/2} \rightharpoonup (1-\Delta)^{1/2}\gamma_0(1-\Delta)^{1/2} \text{ weakly in } I^{0,2} \tag{160}$$

$$\sqrt{\rho_n} \rightharpoonup \sqrt{\rho_0} \text{ in } H^1(\Omega) \tag{161}$$

$$\rho_n - \kappa \to \rho_0 - \kappa \text{ in } H^{-1}(\Omega) \tag{162}$$

for some $\gamma_0 \in I_0^{1,1}$ and $\rho_0 := \text{den}[\gamma_0]$. The last line follows by compact embedding theorem on Ω.

Part 3: Weak lower semi-continuity

Lemma 2 *The functional $\mathscr{F}_\beta$ is weakly lower semi-continuous with respect to convergence* (160), (161) *and* (162).

Proof We study the functional $\mathscr{F}_\beta(\gamma)$ term by term. For the first term on the r.h.s. of (143), it satisfies $\text{Tr}(h\gamma) = \|\gamma\|_{I^{1,1}}$ and is linear, it is $\|\cdot\|_{I^{1,1}}$-weakly lower semi-continuous. The Coulumb term $\langle(\kappa-\rho_\gamma), (-\Delta)^{-1}(\kappa-\rho_\gamma)\rangle$ is quadratic and easily seen to be $\dot{H}^{-1}(\Omega)$-weakly lower semi-continuous. The exchange-correlation term is weakly lower semi-continuous by (161) (which implies that $Xc(\rho_n) \to Xc(\rho_0)$ a.e.) and Fatou's lemma.

Thus, we study the term $-\beta^{-1}S(\gamma)$. We use an idea from [43] which allows to reduce the problem to a finite-dimensional one. To the latter end, we recall that $S(\gamma) = \text{Tr}(s(\gamma))$ for $s(x) = -x\ln x$. In Bloch-Floquet decomposition, this term is

$$-S(\gamma_n) = -\int_{\Omega^*} d\hat{\xi}\; S((\gamma_n)_\xi) = -\int_{\Omega^*} d\hat{\xi}\; \text{Tr}(s((\gamma_n)_\xi)) \tag{163}$$

where $s(x) = \frac{1}{2}(-x\ln(x) - (1-x)\ln(1-x))$. We define the relative entrop of A and B to be

$$S(A|B) := \text{Tr}(s(A|B)), \quad s(A|B) := A(\ln(A) - \ln(B))\,. \tag{164}$$

Then we see that

$$S(A) = S(B) - S(A|B) - \text{Tr}[(A-B)\ln(B)]\,. \tag{165}$$

Using this formula, writing $A = (\gamma_n)_\xi$ and $B = (\gamma_*)_\xi = \left(\frac{C}{1+e^{\beta\sqrt{-\Delta}}}\right)_\xi$ where C is chosen so that $\text{Tr}(g_*) = Z$,

$$-S(\gamma_n) = \int_{\Omega^*} d\hat{\xi}\; (-1)S((\gamma_*)_\xi) + \text{Tr}((\gamma_n)_\xi - (\gamma_*)_\xi \ln(\gamma_*)_\xi) \tag{166}$$

$$+ S((\gamma_n)_\xi|(\gamma_*)_\xi) \tag{167}$$

We note that $\ln((\gamma_*)_\xi) \lesssim 1 + \sqrt{-\beta \Delta_\xi}$ and $|S(\gamma_*)| < \infty$. By (160) and linearity (hence convexity), (166) converges in the limit to $-S(\gamma_*) + \mathrm{Tr}((\gamma_0 - \gamma_*)\ln(\gamma_*)$. So it suffices that we control the last term (167). We improve convergence for a.e. ξ. By considering $\sqrt{(\gamma_n)_\xi}$ and dropping to a subsequence, (148) shows that $(1 - \Delta_\xi)^{1/2}\sqrt{(\gamma_n)_\xi}$ converges weakly in HS norm for almost every $\xi \in \Omega^*$. Similarly,

$$\int_{\Omega^*} d\hat{\xi}\, \mathrm{Tr}_{L^2(\Omega)}[\sqrt{(\gamma_n)_\xi}(1 - \Delta_\xi)\sqrt{(\gamma_n)_\xi}] \tag{168}$$

$$= \int_{\Omega^*} d\hat{\xi}\, \mathrm{Tr}_{L^2(\Omega)}[(1 - \Delta_\xi)\gamma_n] \tag{169}$$

$$= \mathrm{Tr}(1 - \Delta)\gamma_n < \infty \tag{170}$$

by expanding the trace using an orthonormal basis of $(\gamma_n)_\xi$. Thus, weak convergence is also obtained for $\sqrt{(\gamma_n)_\xi}(1 - \Delta_\xi)^{1/2}$. Regarding $\sqrt{(\gamma_n)_\xi}$ as an kernel in $L^2(\Omega \times \Omega)$, and since Ω is compact, we may assume that $(\gamma_n)_\xi \to (\gamma_0)_\xi$ in HS norm for almost every $\xi \in \Omega^*$. Now, by [43], we can write

$$\begin{aligned} S((\gamma_n)_\xi|(\gamma_*)_\xi) &+ \mathrm{Tr}((\gamma_*)_\xi - (\gamma_n)_\xi) \\ &= \sup_{\lambda \in (0,1)} \mathrm{Tr}(s_\lambda((\gamma_n)_\xi|(\gamma_*)_\xi)) \end{aligned} \tag{171}$$

where $s_\lambda(x)(A|B) = \lambda^{-1}(s(\lambda A + (1-x)B) - \lambda s(A) - (1-\lambda)s(B))$. Moreover, $s_\lambda(A|B) \geq 0$ for any A, B since the entropy function s is concave. Hence, we may write

$$\begin{aligned} S((\gamma_n)_\xi|(\gamma_*)_\xi) &+ \mathrm{Tr}((\gamma_*)_\xi - (\gamma_n)_\xi) \\ &= \sup_{\lambda \in (0,1)} \sup_P \mathrm{Tr}(P s_\lambda((\gamma_n)_\xi|(\gamma_*)_\xi)) \end{aligned} \tag{172}$$

where the $\sup_P$ is taken over all finite rank projections P. It follows that for any λ sufficiently small and any finite rank projection P,

$$S((\gamma_n)_\xi|(\gamma_*)_\xi) + \mathrm{Tr}((\gamma_*)_\xi - (\gamma_n)_\xi) \geq \mathrm{Tr}(P s_\lambda((\gamma_n)_\xi|(\gamma_*)_\xi)) \tag{173}$$

Taking $n \to \infty$, since P is finite rank and $(\gamma_n)_\xi \to (\gamma_0)_\xi$ in HS norm (hence operator norm) for almost every $\xi \in \Omega^*$,

$$\begin{aligned} \liminf_{n\to\infty} S((\gamma_n)_\xi|(\gamma_*)_\xi) &+ \mathrm{Tr}((\gamma_*)_\xi - (\gamma_0)_\xi) \\ &\geq \mathrm{Tr}(P s_\lambda((\gamma_0)_\xi|(\gamma_*)_\xi)) \end{aligned} \tag{174}$$

Now taking $\limsup_{\lambda \to 0^+}$ and $\sup_P$, we see that

$$\liminf_{n\to\infty} S((\gamma_n)_\xi | (\gamma_*)_\xi) + \mathrm{Tr}((\gamma_*)_\xi - (\gamma_0)_\xi) \geq \mathrm{Tr}(s((\gamma_0)_\xi | (\gamma_*)_\xi)) \,. \tag{175}$$

The proof is complete by Fatou's Lemma applied to the integral $\int_{\Omega^*} d\hat{\xi}$ and the fact

$$\int d\hat{\xi}\,\mathrm{Tr}((\gamma_*)_\xi - (\gamma_0)_\xi) = \mathrm{Tr}(\gamma_*) - \mathrm{Tr}(\gamma_0) = 0\,. \tag{176}$$

Proof of Theorem 9: Existence of Minimizer. With the results above, the proof is standard. Let $(\gamma_n) \in I_0^{1,1}$ be a minimizing sequence for $\mathscr{F}$. Lemma 1 shows that $\mathscr{F}_\beta$ is coercive. Hence $\|\gamma_n\|_{(1)}$ is bounded uniformly in n. By Sobolev-type embedding theorems, (γ_n) converges strongly in $I_0^{s,1}$ for any $s < 1$. Moreover, together with the Banach-Alaoglu theorem, the latter implies that (γ_n) converges weakly in $I_0^{1,1}$. Hence, denoting the limit by γ_0, we see that, by Lemma 2, $\mathscr{F}_\beta$ is lower semi-continuous:

$$\liminf_{n\to\infty} \mathscr{F}_\beta(\gamma_n) \geq \mathscr{F}_\beta(\gamma_0). \tag{177}$$

Hence, γ_0 is indeed a minimizer. To show that minimizer satisfies the gHF equation, we start with some lemmas.

Lemma 3 *Let $\gamma \in I_0^{1,1}$ be such that $s(\gamma) := -(\gamma \ln \gamma + (1-\gamma)\ln(1-\gamma))$ is trace class and γ' satisfy*

$$\mathrm{Tr}\gamma' = 0 \text{ and } (\gamma')^2 \lesssim (\gamma(1-\gamma))^2\,. \tag{178}$$

Then, $\mathscr{F}_\beta(\gamma)$ is Gâteaux differentiable at γ with respect to variations γ' and

$$d_\gamma \mathscr{F}_\beta(\gamma) g' = d_\gamma F_\beta(\gamma)\gamma' = \mathrm{Tr}[(h_\phi - \beta^{-1}s'(\gamma))\gamma']\,. \tag{179}$$

Proof We consider first the variation in $I_0^{1,1}$ of the form $\gamma + \epsilon\gamma'$ for $\epsilon > 0$ small. Note that if γ' satisfies (178), then for ϵ small enough, $\gamma + \epsilon\gamma' \in I_0^{1,1}$. Let $d_\gamma F_\beta(\gamma, a)\gamma' := \partial_\epsilon F_\beta(\gamma + \epsilon\gamma', a)\mid_{\epsilon=0}$, if the r.h.s. exists. From (133) and (121) and the assumption that Xc' is bounded, we see that

$$d_\gamma F_\beta(\gamma, a)\gamma' = \mathrm{Tr}(h_\phi\gamma') - \beta^{-1} dS(\gamma)\gamma', \tag{180}$$

where $-\Delta\phi = 4\pi(\kappa - \rho)$ provided $dS(\gamma)\gamma' := \partial_\epsilon S(\gamma + \epsilon\gamma')\mid_{\epsilon=0}$ exists. Differentiability of S is proved in the next lemma.

Lemma 4 *Let $\gamma \in I_0^{1,1}$ be such that $s(\gamma) := -(\gamma \ln \gamma + (1-\gamma)\ln(1-\gamma))$ is trace class and γ' satisfy the second condition in* (178). *Then S is Gâteaux differentiable* *and its derivative is given by*

$$dS(\gamma)\gamma' = \mathrm{Tr}(s'(\gamma)\gamma'). \tag{181}$$

Proof For simplicity, we will only consider the case $s(\lambda) = -\lambda \ln(\lambda)$ as the full case is similar. Denote $\gamma'' := \gamma + \epsilon\gamma'$. We write

$$\begin{aligned} S(\gamma'') - S(\gamma) &= -\mathrm{Tr}(\gamma(\ln\gamma'' - \ln\gamma) \\ &\quad + \epsilon\gamma'(\ln\gamma'' - \ln\gamma) + \epsilon\gamma'\ln\gamma) \qquad (182) \\ &=: A + B - \epsilon\mathrm{Tr}(\gamma'\ln\gamma). \qquad (183) \end{aligned}$$

Using the formula $\ln a - \ln b = \int_0^\infty [(b+t)^{-1} - (a+t)^{-1}]dt$ and the second resolvent equation, we compute

$$\begin{aligned} A :=& -\mathrm{Tr}(\gamma(\gamma'' - \ln\gamma)) = -\mathrm{Tr}\int_0^\infty \{\gamma[(\gamma+t)^{-1} - (\gamma''+t)^{-1}]\}dt \\ =& -\mathrm{Tr}\int_0^\infty \{\gamma(\gamma+t)^{-1}\epsilon\gamma'(\gamma''+t)^{-1}\}dt \\ =& -\mathrm{Tr}\left(\int_0^\infty \{\gamma(\gamma+t)^{-1}\epsilon\gamma'(\gamma+t)^{-1}\}dt\right. \\ &\left. - \int_0^\infty \{\gamma(\gamma+t)^{-1}\epsilon\gamma'(\gamma+t)^{-1}\epsilon\gamma'(\gamma''+t)^{-1}\}dt\right). \end{aligned} \tag{184}$$

Similarly, we have

$$\begin{aligned} B :=& -\mathrm{Tr}(\epsilon\gamma'(\ln\gamma'' - \ln\gamma)) \qquad (185) \\ =& -\mathrm{Tr}\int_0^\infty \{\epsilon\gamma'[(\gamma+t)^{-1} - (\gamma''+t)^{-1}]\}dt \\ =& -\mathrm{Tr}\int_0^\infty \{\epsilon\gamma'(\gamma+t)^{-1}\epsilon\gamma'(\gamma''+t)^{-1}\}dt. \qquad (186) \end{aligned}$$

Combining the last two relations with (183), we find

$$S(\gamma + \epsilon\gamma') - S(\gamma) = \epsilon S_1 + \epsilon^2 R_2 \tag{187}$$

$$S_1 := -\mathrm{Tr}\gamma'\ln\gamma - \mathrm{Tr}\int_0^\infty \{\gamma(\gamma+t)^{-1}\gamma'(\gamma+t)^{-1}\}dt \tag{188}$$

$$\begin{aligned} R_2 :=& -\mathrm{Tr}\int_0^\infty \{\gamma(\gamma+t)^{-1}\gamma'(\gamma+t)^{-1}\gamma'(\gamma''+t)^{-1} \\ & - \gamma'(\gamma+t)^{-1}\gamma'(\gamma''+t)^{-1}\}dt \end{aligned} \tag{189}$$

The estimates below show that the integrals on the r.h.s. converge. We can compute the integral

$$\mathrm{Tr}\int_0^\infty \{\gamma(\gamma+t)^{-1}\gamma'(\gamma+t)^{-1}\}dt \tag{190}$$

$$=\mathrm{Tr}\int_0^\infty \{\gamma(\gamma+t)^{-2}\gamma'\}dt = \mathrm{Tr}\gamma' \tag{191}$$

in the expression for S_1. Moreover, using $\gamma(\gamma+t)^{-1}-1=-t(\gamma+t)^{-1}$, we can rewrite the expression for R_2. Together, we obtain

$$S_1 := \mathrm{Tr}\{\gamma'\ln\gamma+\gamma'\}, \tag{192}$$

$$R_2 := \mathrm{Tr}\int_0^\infty \{t(\gamma+t)^{-1}\gamma'(\gamma+t)^{-1}\gamma'(\gamma''+t)^{-1}\}dt. \tag{193}$$

Using $(\gamma')^2 \lesssim (\gamma(1-\gamma))^2$ and γ is trace class, we see that (192) is well defined and finte. To demonstrate the convergence in (193), we estimate the integrand on the r.h.s. of (193). we can formally write

$$(\gamma+t)^{-1}\gamma'(\gamma+t)^{-1}\gamma'(\gamma''+t)^{-1} \tag{194}$$

$$=(\gamma+t)^{-1}\gamma'(\gamma+t)^{-1}\gamma'(\gamma+t)^{-1}\sum_{n\geq 0}\epsilon^n[-\gamma'(\gamma+t)^{-1}]^n. \tag{195}$$

Since γ' and γ are bounded. We see that

$$t\|(\gamma+t)^{-1}\gamma'(\gamma+t)^{-1}\gamma'(\gamma+t)^{-1}\epsilon^n[-\gamma'(\gamma+t)^{-1}]^n\|_{I^{0,1}} \tag{196}$$

$$\leq \epsilon^n t\|(\gamma+t)^{-1}\|_\infty\|\gamma'(\gamma+t)^{-1}\gamma'(\gamma+t)^{-1}\|_{I^{0,1}}\|\gamma'(\gamma+t)^{-1}\|_\infty^n \tag{197}$$

$$\leq \epsilon^n\|\gamma'(\gamma+t)^{-1}\gamma'(\gamma+t)^{-1}\|_{I^{0,1}}\|\gamma'(\gamma+t)^{-1}\|_\infty^n \tag{198}$$

Thus, if $\epsilon < \frac{1}{2}\|\gamma'(\gamma+t)^{-1}\|_\infty^{-1}$ for all $t\in[0,\infty)$ and $\int_0^\infty \|\gamma'(\gamma+t)^{-1}\gamma'(\gamma+t)^{-1}\|_{I^{0,1}}dt < \infty$, then we have convergence. By the condition in (178) on γ', we have

$$\|\gamma'(\gamma+t)^{-1}\|_\infty \leq \|\gamma(\mathbf{1}-\gamma)(\gamma+t)^{-1}\|_\infty \leq \|\eta(\eta+t)^{-1}\|_\infty$$

where $\eta\gamma(\mathbf{1}-\gamma)$. Since $0\leq\gamma\leq 1$, so does η. Hence

$$\|\gamma'(\gamma+t)^{-1}\|_\infty \leq 1$$

Next,

$$\begin{aligned}&\|\gamma'(\gamma+t)^{-1}\gamma'(\gamma+t)^{-1}\|_{I^{0,1}}\\&\leq\|\gamma'(\gamma+t)^{-1}\|_{I^2}\|\gamma'(\gamma+t)^{-1}\|_{I^{0,2}}\end{aligned}\tag{199}$$

Now, we show that (199) is $L^2(dt)$. By the condition in (178) on γ', we have

$$\|\gamma'(\gamma+t)^{-1}\|_{I^2}\leq\|\gamma(\mathbf{1}-\gamma)(\gamma+t)^{-1}\|_{I^2}\leq\|\eta(\eta+t)^{-1}\|_{I^2},$$

where $\eta:=\gamma(\mathbf{1}-\gamma)$. Thus,

$$\|\gamma'(\gamma+t)^{-1}\|_{I^2}\lesssim\mathrm{Tr}(\eta^2(t+\eta)^{-2})=\int_{\Omega^*}d\hat{\eta}\,\mathrm{Tr}(\eta_\xi^2(t+\eta_\xi)^{-2})$$

Let $\mu_{\xi,n}$ be the eigenvalues of the operator $\eta_\xi:=\gamma_\xi(\mathbf{1}-\gamma_\xi)$. Then we have

$$\|\eta_\xi(\eta_\xi+t)^{-1}\|_{I^2}^2=\sum_n\mu_{\xi,n}^2(\mu_{\xi,n}+t)^{-2},\tag{200}$$

and therefore

$$\begin{aligned}\int_0^\infty\|\eta_\xi(\eta_\xi+t)^{-1}\|_{I^2}^2dt&=\int_0^\infty\sum_n\mu_{\xi,n}^2(\mu_{\xi,n}+t)^{-2}dt\\&=\sum_n\mu_{\xi,n}=\mathrm{Tr}\eta_\xi.\end{aligned}\tag{201}$$

Since $\gamma(\mathbf{1}-\gamma)$ is a trace class operator, this proves the claim and, with it, the convergence of the integral in (193).

To sum up, we proved the expansion (187) with S_1 given by (192), which is the same as (181), and R_2 bounded. In particular, this implies that S is C^1 and its derivative is given by (181).

And finally, we have the following:

Lemma 5 *Suppose that γ is a minimizer of $\mathscr{F}_\beta$ on $I_0^{1,1}$, then $0<\gamma<1$.*

Proof We prove that γ cannot have eigenvalues 0 and 1 simultaneously. The case where only 0 or only 1 is an eigenvalue is treated similarly. If not, decomposing into Bloch-Floquet decomposition γ_ξ, we see that γ_ξ has a kernel for a subset, $S_0\subset\Omega^*$, and eigenspace of 1 on $S_1\subset\Omega^*$, both of positive measure. For $\lambda=0,1$, let $P_{\lambda,\xi}$ denote the projection onto the λ-eigenvector for each $\xi\in S$ in a way such that $P_{\lambda,\xi}$ is measurable in ξ. Let

$$P=\int_{\Omega^*}d\hat{\xi}\,f(\xi)(P_{0,\xi}-P_{1,\xi}).\tag{202}$$

where $f(\xi) \geq 0$ is chosen so that $\mathrm{Tr} P = 0$. Since $0 \leq \gamma \leq 1$, it is not hard to sees that P satisfies (178). Following the proof of Lemmas 3 and 4, we compute

$$\mathscr{F}_\beta(\gamma + \epsilon P) - \mathscr{F}_\beta(\gamma) \tag{203}$$

$$= \beta^{-1} \int_{\Omega^*} d\hat{\xi} (\epsilon f(\xi)) \ln(\epsilon f(\xi)) P_{0,\xi} \tag{204}$$

$$+ (1 - \epsilon f(\xi)) \ln(1 - \epsilon f(\xi)) P_{1,\xi}) + O(\epsilon) \tag{205}$$

By choosing $f(x) = \frac{|S_1|}{|S_0|}\chi_{S_0} + \frac{|S_0|}{|S_1|}\chi_{S_1}$, for example, we note that the first term is of order $O(\epsilon \ln \epsilon) \gg O(\epsilon)$ and negative. This contradicts minimality of γ.

Proof (Proof of Theorem 9: Solution to KS equation (71)*)* By the minimizer existence part of Theorem 9, let $\gamma_0 \in I_0^{1,1}$ denote the minimizer of the free energy $\mathscr{F}_\beta$. For notational convenience let $A := d_\gamma \mathscr{F}(\gamma_0)$. We show that A is multiple of the identity. Let

$$v_0 := \gamma_0(1 - \gamma_0) \int_{\Omega^*} d\hat{\xi}\, 1, \tag{206}$$

and let

$$v := \gamma_0(1 - \gamma_0) \int_{\Omega^*} d\hat{\xi} u_\xi \tag{207}$$

where $u_\xi \in L^2_\xi(\Omega)$ is an arbitrary elements of the fiber space in the Bloch-Floquet decomposition and $\|u_\xi\|_2$ is uniformly bounded upto a null set in Ω^* and v is orthogonal to v_0. By Lemma 5, we see that $0 < \gamma_0 < 1$. This shows that $\gamma(1-\gamma)$ is a (possibly unbounded) bijection. Hence the linear space spanned by all such v's is dense in $L^2(R^3)$. Let

$$\gamma' = P_v - \frac{\|v\|_2^2}{\|v_0\|_2^2} P_{v_0}\,. \tag{208}$$

where P_x is the orthogonal projection onto x. Then we note that γ' satisfies the condition (178). Hence, by minimality of γ_0, Lemma 3 shows that

$$\mathrm{Tr}(A\gamma') \geq 0\,. \tag{209}$$

We note that if γ' satisfies condition (178), so does $-\gamma'$. It follows that

$$\mathrm{Tr}(A\gamma') = 0\,. \tag{210}$$

It follows that

$$0 = \int_{\Omega^*} d\hat{\xi}\, \mathrm{Tr}(A_\xi g'_\xi) \tag{211}$$

$$= \int_{\Omega^*} d\hat{\xi}\, \mathrm{Tr}(A_\xi (P_v)_\xi) - \frac{\|v\|_2^2}{\|v_0\|_2^2} \mathrm{Tr}(A_\xi (P_0)_\xi) \tag{212}$$

$$= \langle v, Av\rangle - \frac{\|v\|_2^2}{\|v_0\|_2^2} \langle v_0, Av_0\rangle \,. \tag{213}$$

Let $\hat{x} = x/\|x\|$, then we see that

$$\langle \hat{v}, A\hat{v}\rangle = \langle \hat{v}_0, A\hat{v}_0\rangle \tag{214}$$

for all v orthogonal to v_0 of the form (207). Since the space of v_0 and all such v's are dense, we conclude that A is a multiple of the identity, which we denote by μ. This shows that

$$0 = A - \mu = d_\gamma \mathscr{F}(\gamma_0) - \mu \mathbf{1} = h_{A,\mu,\phi} - \beta^{-1} s'(\gamma_0) \,. \tag{215}$$

The case for $d_a \mathscr{F}_\beta(\gamma_0) = 0$ is much easier. Its proof is standard and can be found, for example, in [22].

Finally, to see that $\mu \in \mathbb{R}$, we simply note that $\mu \mathbf{1} = h_\phi - \beta^{-1} s'(\gamma_0)$ is symmetric.

Acknowledgements The second author is grateful to Volker Bach, Sébastien Breteaux, Thomas Chen and Jürg Fröhlich for enjoyable collaboration, and both authors thank Dmitri Chouchkov, Rupert Frank, Christian Hainzl, Jianfeng Lu, Yuri Ovchinnikov, and especially Antoine Levitt, for stimulating discussions. The authors are grateful to the anonymous referees for useful remarks and suggestions.

The research on this paper is supported in part by NSERC Grant No. NA7901. The first author is also in part supported by NSERC CGS D graduate scholarship.

References

1. A. Anantharaman, E. Cancès, Existence of minimizers for Kohn-Sham models in quantum chemistry. Ann. Inst. Henri Poincarè (C) **26**, 2425–2455 (2009)
2. V. Bach, E. Lieb, J.P. Solovej, Generalized Hartree-Fock theory and the Hubbard model. J. Stat. Phys. **76**, 3–89 (1994)
3. V. Bach, S. Breteaux, T. Chen, J. Fröhlich, I.M. Sigal, The time-dependent Hartree-Fock-Bogoliubov equations for Bosons, arXiv 2016 (https://arxiv.org/abs/1602.05171v1) and (revision) arXiv 2018 (https://arxiv.org/abs/1602.05171v2)
4. V. Bach, S. Breteaux, T. Chen, J. Fröhlich, I.M. Sigal, On the local existence of the time-dependent Hartree-Fock-Bogoliubov equations for Bosons. arXiv 2018 (https://arxiv.org/abs/1805.04689)

5. N. Benedikter, J. Sok, J.P. Solovej, The Dirac-Frenkel principle for reduced density matrices, and the Bogoliubov-de-Gennes equations. arXiv:1706.03082
6. N. Benedikter, M. Porta, B. Schlein, Hartree-Fock dynamics for weakly interacting fermions, in *Mathematical Results in Quantum Mechanics* (World Scientific Publishing, Hackensack, 2015), pp. 177–189
7. N. Benedikter, M. Porta, B. Schlein, Mean-field evolution of fermionic systems. Commun. Math. Phys. **331** 1087–1131 (2014)
8. A. Bove, G. Da Prato, G. Fano, On the Hartree-Fock time-dependent problem. Commun. Math. Phys. **49**, 25–33 (1976)
9. O. Bratteli, D.W. Robinson, *Operator Algebras and Quantum Statistical Mechanics*, vol. 2 (Springer, Berlin, 1997)
10. G. Bräunlich, C. Hainzl, R. Seiringer, Translation-invariant quasifree states for fermionic systems and the BCS approximation. Rev. Math. Phys. **26**(7), 1450012 (2014)
11. E. Cancès, A. Deleurence, M. Lewin, A new approach to the modelling of local defects in crystals: the reduced Hartree-Fock case. Commun. Math. Phys. **281**(1), 129–177 (2008)
12. E. Cancès, A. Deleurence, M. Lewin, Non-perturbative embedding of local defects in crystalline materials. J. Phys. Condens. Mat. **20**, 294–213 (2008)
13. E. Cancès, M. Lewin, The dielectric permittivity of crystals in the reduced Hartree-Fock approximation. Arch. Ration. Mech. Anal. **197**, 139–177 (2010)
14. E. Cancès, M. Lewin, G. Stolz, The microscopic origin of the macroscopic dielectric permittivity of crystals: a mathematical viewpoint. arXiv 2010 (https://arxiv.org/abs/1010.3494)
15. E. Cancès, N. Mourad, A mathematical perspective on density functional perturbation theory. Nonlinearity **27**(9), 1999–2033 (2014)
16. R. Carles, *Semi-classical Analysis for Nonlinear Schrödinger Equations* (World Scientific, Singapore, 2008)
17. I. Catto, C. Le Bris, P.-L. Lions, On some periodic Hartree type models. Ann. Inst. H. Poincaré Anal. Non Lineaire **19**(2), 143–190 (2002)
18. I. Catto, C. Le Bris, P.-L. Lions, On the thermodynamic limit for Hartree-Fock type models. Ann. Inst. H. Poincaré Anal. Non Lineaire **18**(6), 687–760 (2001)
19. T. Cazenave, *Semilinear Schrödinger Equations* (AMS, Providence, 2003)
20. J.M. Chadam, The time-dependent Hartree-Fock equations with Coulomb two-body interaction. Commun. Math. Phys. **46**(2), 99–104 (1976)
21. J. Chadam, R. Glassey, Global existence of solutions to the Cauchy problem for time dependent Hartree equations. J. Math. Phys. **16**, 1122 (1975)
22. I. Chenn, I.M. Sigal, Stationary states of the Bogolubov-de Gennes equations. arXiv:1701.06080 (2019)
23. I. Chenn, I.M. Sigal, On density functional theory (2019, In preparation)
24. M. Cyrot, Ginzburg-Landau theory for superconductors. Rep. Prog. Phys. **36**(2), 103–158 (1973)
25. P.G. de Gennes, *Superconductivity of Metals and Alloys*, vol. 86 (WA Benjamin, New York, 1966)
26. R.J. Dodd, M. Edwards, C.W. Clark, K. Burnett, Collective excitations of Bose-Einstein-condensed gases at finite temperatures. Phys. Rev. A **57**, R32–R35 (1998). https://doi.org/10.1103/PhysRevA.57.R32
27. W. E, J. Lu, Electronic structure of smoothly deformed crystals: Cauchy-Born rule for the nonlinear tight-binding model. Commun. Pure Appl. Math. **63**(11), 1432–1468 (2010)
28. W. E, J. Lu, The electronic structure of smoothly deformed crystals: Wannier functions and the Cauchy-Born rule. Arch. Ration. Mech. Anal. **199**(2), 407–433 (2011)
29. W. E, J. Lu, *The Kohn-Sham Equation for Deformed Crystals* (Memoirs of the American Mathematical Society, 2013), 97 pp.; Softcover MSC: Primary 74; Secondary 35
30. E. Elgart, L. Erdös, B. Schlein, H.-T. Yau, Nonlinear Hartree equation as the mean field limit of weakly coupled fermions. J. Math. Pures Appl. **83**(9), 1241–1273 (2004)

31. E. Elgart, L. Erdös, B. Schlein, H.-T. Yau, Derivation of the cubic non-linear Schrödinger equation from quantum dynamics of many-body systems. Inv. Math. **167**, 515–614 (2006)
32. E. Elgart, L. Erdös, B. Schlein, H.-T. Yau, Derivation of the Gross-Pitaevskii equation for the dynamics of Bose-Einstein condensate. Ann. Math. (2) **172**(1), 291–370 (2010)
33. R. Frank, C. Hainzl, R. Seiringer, J.-P. Solovej, Microscopic derivation of Ginzburg-Landau theory. J. Am. Math. Soc. **25**, 667–713 (2012) and arXiv 2011 (https://arxiv.org/abs/1102.4001)
34. M. Grillakis, M. Machedon, Pair excitations and the mean field approximation of interacting Bosons, II. arXiv 2015 (http://arxiv.org/abs/1509.05911)
35. A. Griffin, Conserving and gapless approximations for an inhomogeneous Bose gas at finite temperatures. Phys. Rev. B **53**, 9341–9347 (1996). https://doi.org/10.1103/PhysRevB.53.9341
36. S.J. Gustafson, I.M. Sigal, *Mathematical Concepts of Quantum Mechanics*. Universitext, 2nd edn. (Springer, Berlin, 2011)
37. C. Hainzl, E. Hamza, R. Seiringer, J.P. Solovej, The BCS functional for general pair interactions. Commun. Math. Phys. **281**, 349–367 (2008)
38. C. Hainzl, R. Seiringer, The Bardeen-Cooper-Schrieffer functional of superconductivity and its mathematical properties. J. Math. Phys. **57**, 021101 (2016)
39. C. Le Bris, P.-L. Pierre-Louis Lions, From atoms to crystals: a mathematical journey. Bull. Am. Math. Soc. **42**, 291–363 (2005)
40. M. Lewin, P.T. Nam, B. Schlein, Fluctuations around Hartree states in the mean-field regime. Am. J. Math. **137**(6), 1613–1650 (2015)
41. E.H. Lieb, R. Seiringer, J.P. Solovej, J. Yngvason, *The Mathematics of the Bose Gas and its Condensation*. Oberwolfach Seminars Series, vol. 34 (Birkhaeuser Verlag, Basel, 2005)
42. E.H. Lieb, B. Simon, The Hartree-Fock theory for Coulomb systems. Commun. Math. Phys. **53**(3), 185–194 (1977)
43. G. Lindblad, Expectations and entropy inequalities for finite quantum systems. Commun. Math. Phys. **39**, 111–119 (1974)
44. P.L. Lions, Solutions of Hartree-Fock equations for Coulomb systems. Commun. Math. Phys. **109**, 33–97 (1987)
45. P.L. Lions, Hartree-Fock and related equations. Nonlinear partial differential equations and their applications. Collège de France Seminar, vol. IX. Pitman Research Notes in Mathematics Series, vol. 181 (Pitman Advanced Publishing, Boston, 1988), pp. 304–333
46. J. Lukkarinen, H. Spohn, Not to normal order – notes on the kinetic limit for weakly interacting quantum fluids. J. Stat. Phys. **134**, 1133–1172 (2009)
47. P.A. Markowich, G. Rein, G. Wolansky, Existence and nonlinear stability of stationary states of the Schrödinger – Poisson system. J. Stat. Phys. **106**(5), 1221–1239 (2002)
48. P.T. Nam, M. Napiórkowski, Bogoliubov correction to the mean-field dynamics of interacting bosons. [arXiv:1509.04631]
49. M. Napiórkowski, R. Reuvers, J.P. Solovej, The Bogoliubov free energy functional I. Existence of minimizers and phase diagram. Arch. Ration. Mech. Anal. 1–54 (2018)
50. M. Napiórkowski, R. Reuvers, J.P. Solovej, The Bogoliubov free energy functional II. The dilute limit. Commun. Math. Phys. **360**(1), 347–403 (2018)
51. A.S. Parkins, D.F. Walls, The physics of trapped dilute-gas Bose-Einstein condensates. Phys. Rep. **303**(1), 1–80 (1998)
52. M. Porta, S. Rademacher, C. Saffirio, B. Schlein, Mean field evolution of fermions with Coulomb interaction. J. Stat. Phys. **166**, 1345–1364 (2017)
53. I.M. Sigal, Magnetic vortices, Abrikosov lattices and automorphic functions, in *Mathematical and Computational Modelling* (With Applications in Natural and Social Sciences, Engineering, and the Arts) (Wiley, 2014)
54. C. Sulem, J.-P. Sulem, *The Nonlinear Schrödinger Equation* (Springer, New York, 1999)
55. T. Tao, *Nonlinear Dispersive Equations: Local and Global Analysis*. CBMS Regional Conference Series in Mathematics, vol. 106 (AMS, Providence, 2006)

Critical Exponents for Differential Inequalities with Riemann-Liouville and Caputo Fractional Derivatives

Marcello D'Abbicco

Abstract We find the critical exponents for global in time solutions to differential inequalities with power nonlinearities, supplemented by an initial data condition. The operator for which the differential inequality is studied contains a Caputo or Riemann-Liouville time derivative of fractional order and a sum of homogeneous spatial partial differential operators. In the special case of a fractional diffusive equation, the obtained critical exponents are sharp. In particular, global existence of small data solutions to the fractional diffusive equation with Caputo and Riemann-Liouville time derivative of order in $(0, 1)$ and in $(1, 2)$, holds for supercritical powers. The existence result for the superdiffusive case ($\alpha \in (1, 2)$), which interpolates a semilinear heat equation and a semilinear wave equation, was recently obtained in the general setting by the author and his collaborators. We use a simple representation of Mittag-Leffler functions to show that global existence of small data solutions for supercritical powers also holds for to the subdiffusive equation with Caputo and Riemann-Liouville time derivative ($\alpha \in (0, 1)$).

1 Introduction

We consider the fractional differential inequalities

$$ {}^{\mathrm{C}}\mathrm{D}_{0+}^{\alpha} u + A(x, \partial_x) u \geq |u|^p, \qquad t \geq 0, \ x \in \mathbb{R}^n, \tag{1} $$

and

$$ {}^{\mathrm{RL}}\mathrm{D}_{0+}^{\alpha} u + A(x, \partial_x) u \geq |u|^p, \qquad t \geq 0, \ x \in \mathbb{R}^n, \tag{2} $$

M. D'Abbicco (✉)
Department of Mathematics, University of Bari, Bari, Italy
e-mail: marcello.dabbicco@uniba.it

M. D'Abbicco et al. (eds.), *New Tools for Nonlinear PDEs and Application*,
Trends in Mathematics, https://doi.org/10.1007/978-3-030-10937-0_2

where $\alpha \in \mathbb{R}_+ \setminus \mathbb{N}$, $p > 1$, and ${}^{\mathrm{C}}\mathrm{D}^{\alpha}_{0+}$ and ${}^{\mathrm{RL}}\mathrm{D}^{\alpha}_{0+}$ respectively denote the Caputo and the Riemann-Liouville (forward) fractional derivatives of order α, with starting time 0.

We assume that

$$A(x, \partial_x) = \sum_{1 \le |\beta| \le m} a_\beta(x)\, \partial_x^\beta, \tag{3}$$

is a differential operator of order $m \ge 1$, with a_β smooth in $\mathbb{R}^n \setminus \{0\}$ and homogeneous of degree $r_\beta < |\beta|$, that is,

$$\forall \beta, \quad \exists r_\beta < |\beta| : \qquad \forall x \neq 0, \quad a_\beta(x) = |x|^{r_\beta} a_\beta(x/|x|). \tag{4}$$

The differential operator $A(x, \partial_x)$ is homogeneous of degree h if $|\beta| - r_\beta = h$, for any β. In the general case, we will denote by h the lowest degree of the homogeneous terms $a_\beta \partial_x^\beta$ (see later, definition (13)).

Moreover, we assume that

$$\forall \beta : \qquad \partial_x^\beta a_\beta(x) = 0. \tag{5}$$

Thanks to condition (5), the adjoint operator $A^*(x, \partial_x)$ contains no zero order terms (see later, Definition 1).

The study of differential inequalities for evolution equations in the space-time is inspired by the study of differential inequalities in the space $\mathbb{R}^n$ (see, for instance, [2]). In particular, the method of the test function used to prove differential inequalities in space can be adapted to operators in the space time $[0, \infty) \times \mathbb{R}^n$ (see, in particular, [6], for general variable coefficients operators). The technique employed to study differential inequalities is often sharp even when applied to the corresponding equality, replacing the inequality $Lu \ge |u|^p$ by the equation $Lu = |u|^p$. That is, for several models (for instance, heat equations, damped wave equations, and related systems) the counterpart of a nonexistence result for a differential inequality (or a system of differential inequalities) in some range for power nonlinearities, is given by the existence result for differential equations (or system of differential equations), out of the previous range for power nonlinearities.

Here and in the following, we set $\kappa = \lceil \alpha \rceil = -\lfloor -\alpha \rfloor$, the smallest integer which is greater or equal than α. For any $\beta > 0$, the (forward) Riemann-Liouville fractional integral of order β is given by

$$J^\beta_{0+} f(t) = \frac{1}{\Gamma(\beta)} \int_0^t (t-s)^{\beta-1}\, f(s)\, ds,$$

for any $t > 0$. For any $\alpha \in \mathbb{R}_+ \setminus \mathbb{N}$, the (forward) Caputo and Riemann-Liouville fractional derivatives of order $\alpha > 0$, are given by the expressions

$$ {}^{\mathrm{C}}\mathrm{D}^{\alpha}_{0+} f(t) = J^{\kappa-\alpha}_{0+}\big(\partial_t^{\kappa} f\big)(t), \tag{6} $$

$$ {}^{\mathrm{RL}}\mathrm{D}^{\alpha}_{0+} f(t) = \partial_t^{\kappa}\big(J^{\kappa-\alpha}_{0+} f\big)(t), \tag{7} $$

for any $t > 0$. We remark that, due to $\alpha \notin \mathbb{N}$, the definition of Caputo fractional derivative given in (6) is equivalent to the more general one (see Theorem 2.1 in [11]):

$$ {}^{\mathrm{C}}\mathrm{D}^{\alpha}_{0+} f(t) = {}^{\mathrm{RL}}\mathrm{D}^{\alpha}_{0+} \tilde{f}(t), \qquad \tilde{f}(t) = f(t) - \sum_{j=0}^{\kappa-1} \frac{f^{(j)}(0)}{j!}\, t^j. $$

At $t = 0$, the previous definitions are intended as the limit:

$$ {}^{\mathrm{C}}\mathrm{D}^{\alpha}_{0+} f(0) = \lim_{t\to 0} {}^{\mathrm{C}}\mathrm{D}^{\alpha}_{0+} f(t), \tag{8} $$

$$ {}^{\mathrm{RL}}\mathrm{D}^{\alpha}_{0+} f(0) = \lim_{t\to 0} {}^{\mathrm{RL}}\mathrm{D}^{\alpha}_{0+} f(t), \tag{9} $$

$$ J^{\beta}_{0+} f(0) = \lim_{t\to 0} J^{\beta}_{0+} f(t). \tag{10} $$

We supplement the inequalities with an initial condition, respectively,

$$ \partial_t^{\kappa-1} u(0,x) = u_{\kappa-1}(x), \tag{11} $$

for (1), and

$$ {}^{\mathrm{RL}}\mathrm{D}^{\alpha-1}_{0+} u(0,x) = u_{\alpha-1}(x), \tag{12} $$

for (2) (initial condition (12) is intended in the sense of (9)). In initial condition (12) we formally set

$$ {}^{\mathrm{RL}}\mathrm{D}^{\alpha-1}_{0+} u(0,x) = J^{1-\alpha}_{0+} u(0,x), $$

when $\alpha \in (0,1)$ (in the sense of (10)).

We derive a necessary condition on the exponent p in (1) and, respectively, (2), which has to be satisfied to have global in time solutions, provided that suitable sign assumptions are verified by the initial data defined in (11) and, respectively, (12). For the ease of reading, we postpone the definition of global (weak) solution to Sect. 2.

Theorem 1 *Let $\alpha \in \mathbb{R}_+ \setminus \mathbb{N}$. Assume that $A(x, \partial_x)$ verifies* (4) *and* (5)*, and fix $h > 0$ as*

$$h = \min_{a_\beta \neq 0} (|\beta| - r_\beta). \tag{13}$$

Assume that there exists a non-trivial global weak solution u to (2) *(in the sense of Definition* 4*), with*

$$u_{\alpha-1} \geq 0, \quad \text{or } u_{\alpha-1} \in L^1 \text{ with} \quad \int_{\mathbb{R}^n} u_{\alpha-1}(x)\, dx > 0, \tag{14}$$

where $u_{\alpha-1}$ is the initial condition in (12)*. Then*

$$n > h\left(1 - \frac{1}{\alpha}\right),$$

and $p > \tilde{p}(n, \alpha)$, where

$$\tilde{p}(n, \alpha) = 1 + \frac{h}{n - h(1 - 1/\alpha)}. \tag{15}$$

Assume that there exists a global weak solution u to (2) *(in the sense of Definition* 4*), and that there exist $\varepsilon > 0$, $R > 0$, such that*

$$u_{\alpha-1}(x) \geq \varepsilon\, |x|^{-\theta}, \qquad \forall |x| \geq R, \tag{16}$$

for some $\theta \in (-\infty, n)$, where $u_{\alpha-1}$ is the initial condition in (12)*. Then*

$$\theta > h\left(1 - \frac{1}{\alpha}\right),$$

and $p > \tilde{p}(\theta, \alpha)$, where

$$\tilde{p}(\theta, \alpha) = 1 + \frac{h}{\theta - h(1 - 1/\alpha)}. \tag{17}$$

We remark that the critical exponent in (15) for problem (2) was the same under both the assumption $u_{\alpha-1} \geq 0$ and $\int_{\mathbb{R}^n} u_{\alpha-1}\, dx > 0$. Namely, the result remain valid even if $u_{\alpha-1} = 0$ (and the solution is non trivial, which implies $\alpha > 1$). On the other hand, for problem (1), the stronger sign assumption on the initial data $u_{\kappa-1}$ brings the benefit of a larger critical exponent.

Theorem 2 *Let $\alpha \in \mathbb{R}_+ \setminus \mathbb{N}$ and set $\kappa = \lceil \alpha \rceil$. Assume that $A(x, \partial_x)$ verifies* (4) *and* (5)*, and fix $h > 0$ as in* (13)*.*

Assume that there exists a non-trivial global weak solution u to (1) *(in the sense of Definition* 3*), with*

$$u_{\kappa-1} \geq 0, \tag{18}$$

where $u_{\kappa-1}$ *is the initial condition in* (11)*. Then*

$$n > h\left(1 - \frac{1}{\alpha}\right),$$

and $p > \tilde{p}(n, \alpha)$*, where* $\tilde{p}(n, \alpha)$ *is given by* (15)*.*

Assume that there exists a global weak solution u to (1) *(in the sense of Definition* 3*), with*

$$u_{\kappa-1} \in L^1, \qquad \int_{\mathbb{R}^n} u_{\kappa-1}(x)\, dx > 0, \tag{19}$$

where $u_{\kappa-1}$ *is the initial condition in* (11)*. Then*

$$n > \frac{h(\kappa - 1)}{\alpha},$$

and $p \geq \bar{p}(n, \alpha)$*, where*

$$\bar{p}(n, \alpha) = 1 + \frac{h}{n - h(\kappa - 1)/\alpha}. \tag{20}$$

Assume that there exists a global weak solution u to (1) *(in the sense of Definition* 3*), and there exist* $\varepsilon > 0$*,* $R > 0$*, such that*

$$u_{\kappa-1}(x) \geq \varepsilon\, |x|^{-\theta}, \qquad \forall |x| \geq R, \tag{21}$$

for some $\theta \in (-\infty, n)$*, then*

$$\theta > \frac{h(\kappa - 1)}{\alpha},$$

and $p > \bar{p}(\theta, \alpha)$*, where*

$$\bar{p}(\theta, \alpha) = 1 + \frac{h}{\theta - h(\lceil\alpha\rceil - 1)/\alpha}. \tag{22}$$

Remark 1 It is clear that

$$\bar{p}(\theta, \alpha) > \tilde{p}(\theta, \alpha),$$

for any $\alpha \in \mathbb{R}_+ \setminus \mathbb{N}$, and for any $\theta \leq n$.

The critical exponents in Theorems 1 and 2 have a special interest, in view of the fact that for $\alpha \in (0, 1)$ and for $\alpha \in (1, 2)$, we can provide examples of Cauchy-type problems with power nonlinearity $|u|^p$, for which global small data solutions exist in the supercritical range of p.

Indeed, in the limit case $\alpha = 1$, both the critical exponents in (15) and (20) tend to $1 + h/n$. For integer, even, values of h, this latter is Fujita exponent for the diffusive equation (see, in particular, [10])

$$u_t + (-\Delta)^{h/2} u = |u|^p, \quad t \geq 0.$$

In [7, 8], it has been shown that global solutions to the Cauchy-type problem

$$\begin{cases} {}^{\mathrm{C}}\mathrm{D}_{0+}^{\alpha} u + (-\Delta)^{h/2} u = |u|^p, & t \geq 0, \\ u(0, x) = u_0(x), \\ u_t(0, x) = u_1(x), \end{cases} \tag{23}$$

for $\alpha \in (1, 2)$, exist, for any $p \geq \bar{p}(n, \alpha)$, if initial data are assumed to be small in L^1. Moreover, if the second data u_1 vanishes, then global small data solutions exist for any $p > \tilde{p}(n, \alpha)$.

Similarly, global solutions to the Cauchy-type problem

$$\begin{cases} {}^{\mathrm{RL}}\mathrm{D}_{0+}^{\alpha} u + (-\Delta)^{h/2} u = |u|^p, & t \geq 0, \\ J_{0+}^{2-\alpha} u(0, x) = 0, \\ {}^{\mathrm{RL}}\mathrm{D}_{0+}^{\alpha-1} u(0, x) = u_{\alpha-1}(x), \end{cases} \tag{24}$$

for $\alpha \in (1, 2)$, exist, for any $p > \tilde{p}(n, \alpha)$. Here the initial condition is intended in the sense of (9).

Moreover, global solutions to (23) exist assuming small data in L^m, with $m \in (1, 2]$, if $p > \bar{p}(n/m, \alpha)$, and if $p > \tilde{p}(n/m, \alpha)$ when u_1 vanishes, and global solutions to (24) exist assuming small data in L^m, with $m \in (1, 2]$, if $p > \tilde{p}(n/m, \alpha)$.

The previous results show that the exponents $\tilde{p}(n, \alpha)$ and $\bar{p}(n, \alpha)$, in Theorems 1 and 2 are sharp for $\alpha \in (1, 2)$. Indeed, Theorems 1 and 2 are valid, in particular, if the equality is verified in (2) and (1), and for the constant coefficients, homogeneous, operator $A = (-\Delta)^{h/2}$, with h even integer.

The next statements will also prove the sharpness of the exponent $\bar{p}(n, \alpha)$ in Theorem 2 and of the exponent $\tilde{p}(n, \alpha)$ in Theorem 1 for $\alpha \in (0, 1)$. For the sake of brevity, we fix $h = 2$ and we only consider L^1 smallness of the initial data.

Theorem 3 *Let $\alpha \in (0, 1)$, $n \geq 1$, and*

$$p \geq \bar{p}(n) \doteq 1 + \frac{2}{n} = \bar{p}(n, \alpha).$$

Moreover, let $p < 1 + 2/(n-2)$*, if* $n \geq 3$*. Then there exists* $\varepsilon > 0$ *such that for any* $u_0 \in \mathscr{A} = L^1 \cap L^p$*, with*

$$\|u_0\|_{\mathscr{A}} = \|u_0\|_{L^1} + \|u_0\|_{L^p} \leq \varepsilon,$$

there exists a unique solution $u \in \mathscr{C}([0, \infty), L^p)$ *to the Cauchy-type problem*

$$\begin{cases} {}^{\mathrm{C}}\mathrm{D}^{\alpha}_{0+}u - \Delta u = |u|^p, & t \geq 0, \\ u(0, x) = u_0(x). \end{cases} \tag{25}$$

Moreover, the solution verifies the following long time decay estimate

$$\|u(t, \cdot)\|_{L^p} \leq C\,(1+t)^{-\frac{n\alpha}{2}\left(1-\frac{1}{p}\right)}\,\|u_0\|_{\mathscr{A}}, \tag{26}$$

for any $t \geq 0$*, and for some* $C > 0$*, independent of* u_0*.*

Remark 2 For $\alpha \in (0, 1)$, the critical exponent $\bar{p}(n) = 1 + 2/n$ in Theorem 3 is the same critical exponent for the heat equation. However, in the critical case $p = 1 + 2/n$, we have the existence of global small data solutions for the subdiffusive equation, whereas finite time blow-up holds for the heat equation.

Before stating the corresponding result with the Riemann-Liouville fractional derivative, we need to introduce the following space. For any $\gamma \in (0, 1)$, we define $\mathscr{C}_\gamma(I, X)$, where $I = [0, T]$ or $I = [0, T)$, and X is a functional space, as the space of functions $f(t, x)$ such that $t^\gamma f(t, \cdot) \in \mathscr{C}(I, X)$.

Theorem 4 *Let* $\alpha \in (0, 1)$*,* $n \geq 1$*, and*

$$p > \tilde{p}(n, \alpha) = 1 + \frac{2}{n + 2(1/\alpha - 1)}.$$

Moreover, let $p < 1 + 2/(n-2)$*, if* $n \geq 3$*, and assume the following restriction on* p*:*

$$\left(\frac{1}{\alpha} - 1\right)(p-1) + \frac{n}{2}\left(1 - \frac{1}{p}\right) \leq 1. \tag{27}$$

Then there exists $\varepsilon > 0$ *such that for any* $u_{\alpha-1} \in \mathscr{A} = L^1 \cap L^\infty$*, with*

$$\|u_{\alpha-1}\|_{\mathscr{A}} = \|u_{\alpha-1}\|_{L^1} + \|u_{\alpha-1}\|_{L^\infty} \leq \varepsilon,$$

there exists a unique solution $u \in \mathscr{C}_{1-\alpha}([0, \infty), L^p)$ *to the Cauchy-type problem*

$$\begin{cases} {}^{\mathrm{RL}}\mathrm{D}^{\alpha}_{0+}u - \Delta u = |u|^p, & t \geq 0, \\ J^{1-\alpha}_{0+}u(0, x) = u_{\alpha-1}(x). \end{cases} \tag{28}$$

Moreover, the solution verifies the following long-time decay estimates

$$\|u(t,\cdot)\|_{L^p} \leq C\, t^{\alpha-1}\,(1+t)^{-\frac{n\alpha}{2}\left(1-\frac{1}{p}\right)}\, \|u_{\alpha-1}\|_{\mathscr{A}}, \tag{29}$$

for any $t > 0$, *and for some* $C > 0$, *independent of* $u_{\alpha-1}$. *In* (28), *the initial condition is intended in the sense of* (10).

Remark 3 Condition (27) will be used to control the singular behavior of the power nonlinearity at $t = 0$. We remark that the range of powers $p > \tilde{p}(n,\alpha)$ which satisfy (27) is not empty. Indeed, $\tilde{p}(n,\alpha)$ is the solution of the equation

$$\left(\frac{1}{\alpha}-1\right)(\tilde{p}-1)+\frac{n}{2}(\tilde{p}-1)=1.$$

Condition (27) may also be written as a second order inequality in p:

$$\left(\frac{1}{\alpha}-1\right)(p-1)p+\frac{n}{2}(p-1)\leq p.$$

The equations in problems (23), (24), (25) and (28), are generally called *fractional diffusive equations*. To distinguish among the cases $\alpha \in (1,2)$ or $\alpha \in (0,1)$, we may say that the equations in (23), (24) are *superdiffusive equations*, and that the equations in (28) and in (25) are *subdiffusive equations*, to mean that the fractional order of these equations is above, or below, the order 1 of the classical diffusive equation (heat equation).

Global existence of small data solutions to the fractional subdiffusive equation in integral form

$$u(t,x) = u(0,x) + \int_0^t |u(\tau,x)|^{p-1}u(\tau,x)\,d\tau + J^{\alpha}_{0+}\Delta u(t,x), \tag{30}$$

with $\alpha \in (0,1)$ and $p > 1 + 2/(n\alpha)$, have been recently studied in [1]. The main difference between this model and ours is that a classical integral of order 1 is applied to the power nonlinearity in (30), whereas only a fractional integral of order α is applied to the power nonlinearity in the integral formulation of (25). Indeed, applying J^{α}_{0+} to both sides of (25), we get the integral formulation of (25):

$$u(t,x) - u(0,x) = \int_0^t u_t(\tau,x)\,d\tau = J^1_{0+}u(t,x) = J^{\alpha}_{0+}(\Delta u + |u|^p).$$

The problem to find critical exponents for the differential inequalities (1) and (2), or for the Cauchy-type problem associated, has some analogy with the problem to find critical exponents for partial differential equations with nonlinear memory of power nonlinearities. In particular, in [3], the authors proved that the critical exponent for the Cauchy problem

$$\begin{cases} u_t - \Delta u = J^\gamma_{0+}|u(t,x)|^p, & t>0,\ x\in\mathbb{R}^n, \\ u(0,x) = u_0(x), \end{cases}$$

is given by

$$\max\left\{\hat{p}(n,\alpha)\,,\ \frac{1}{1-\alpha}\right\}, \qquad \hat{p}(n,\alpha) = 1 + \frac{2(1+\alpha)}{n-2\alpha}. \tag{31}$$

Namely, small data global solutions exist for $p > \max\{\hat{p}(n,\alpha), 1/(1-\alpha)\}$, and any solution blows up in finite time if $1 < p \leq \max\{\hat{p}(n,\alpha), 1/(1-\alpha)\}$, provided that $u_0 \geq 0$ is non-trivial. The same critical exponent remains valid for damped waves with nonlinear memory (see [4])

$$\begin{cases} u_{tt} + u_t - \Delta u = J^\gamma_{0+}|u(t,x)|^p, & t>0,\ x\in\mathbb{R}^n, \\ u(0,x) = u_0(x), \\ u_t(0,x) = u_1(x), \end{cases}$$

and a modified version of this critical exponent comes into play for waves with fractional damping $(-\Delta)^\delta u_t$ and nonlinear memory [5]. The result in [3] has also a special interest, since it provides an example of Cauchy problem, for which the critical exponents is not the one predicted by scaling arguments. For problems (1) and (2), the critical exponent is the one predicted by scaling arguments, in general.

Remark 4 The critical exponent $\tilde{p}(n,\alpha)$ obtained in Theorem 1 for (2) with initial condition (12), under the assumption (14), is given by scaling arguments, when $A(x,\partial_x)$ is homogeneous (or quasi-homogeneous, as in [9]). The same happens for the critical exponent $\bar{p}(n,\alpha)$ obtained in Theorem 2 for (1) with initial condition (11), under the assumption (19).

Indeed, if $A(x,\partial_x)$ is homogeneous of degree h, given a solution u to the equation in (2) or in (1), the function $\lambda^{\frac{h}{p-1}}\,u(\lambda^{\frac{h}{\alpha}}t,\lambda x)$ is a solution to (2) or to (1) for any $\lambda\in(0,+\infty)$. We notice that

$$\begin{aligned} {}^{\mathrm{RL}}\mathrm{D}^{\alpha-1}_{0+}\big(u(\lambda^{\frac{h}{\alpha}}t,\lambda x)\big)\big|_{t=0} &= \lambda^{\frac{h(\alpha-1)}{\alpha}}\,u_{\alpha-1}(\lambda x), \\ \partial_t^{\kappa-1}\big(u(\lambda^{\frac{h}{\alpha}}t,\lambda x)\big)\big|_{t=0} &= \lambda^{\frac{h(\kappa-1)}{\alpha}}\,u_{\kappa-1}(\lambda x), \end{aligned}$$

and

$$\|\lambda^{\frac{h}{p-1}}\varphi(\lambda\cdot)\|_{L^q} = \lambda^{\frac{h}{p-1}-\frac{n}{q}}\,\|\varphi\|_{L^q},$$

with $\varphi = u_{\alpha-1}, u_{\kappa-1}$. Therefore, the scaling exponent for (2) with initial condition (12), that is, the solution to

$$\frac{h(\alpha-1)}{\alpha} + \frac{h}{p-1} - \frac{n}{q} = 0,$$

is

$$q_{sc} = \frac{n(p-1)}{h} \frac{\alpha}{(\alpha-1)(p-1)+\alpha}.$$

Indeed, the critical exponent $\tilde{p}(n, \alpha)$ is the solution to $q_{sc} = 1$.

On the other hand, the scaling exponent for (1) with initial condition (11), that is, the solution to

$$\frac{h(\kappa-1)}{\alpha} + \frac{h}{p-1} - \frac{n}{q} = 0,$$

is

$$q_{sc} = \frac{n(p-1)}{h} \frac{\alpha}{(\kappa-1)(p-1)+\alpha}.$$

Indeed, the critical exponent $\bar{p}(n, \alpha)$ is the solution to $q_{sc} = 1$.

If one replaces assumption (19) by (18), then Theorem 2 only gives the critical exponent $\tilde{p}(n, \alpha)$. This latter is, indeed, the critical exponent obtained for $\alpha \in (1, 2)$ and $A = (-\Delta)^{\frac{h}{2}}$, when $u_1 = 0$ and $u_0(x) \in L^1$ (see (23)). However, this critical exponent is not given by scaling arguments.

1.1 Notation

In the following, we write $f_1 \lesssim f_2$ when there exists $C > 0$ such that $f_1 \leq C f_2$. We write $f_1 \approx f_2$ when $f_1 \lesssim f_2$ and $f_2 \lesssim f_1$.

2 Global Weak Solutions

To deal with weak solutions, we shall investigate how integration by parts work with respect to both the space and time variable.

Definition 1 For a given operator $A(x, \partial_x)$, as in (3), its adjoint operator is obtained by

$$\begin{aligned}A^*(x, \partial_x)f(x) &= \sum_{1\leq|\beta|\leq m} (-1)^{|\beta|}\partial_x^\beta\big(a_\beta(x)\, f(x)\big)\\ &= \sum_{1\leq|\beta|\leq m} (-1)^{|\beta|}\sum_{\gamma\leq\beta}\binom{\beta}{\gamma}\big(\partial_x^{\beta-\gamma}a_\beta(x)\big)\,\partial_x^\gamma f(x)\\ &= \sum_{|\gamma|\leq m} b_\gamma(x)\,\partial_x^\gamma f(x),\end{aligned}$$

where

$$b_\gamma = \sum_{\substack{\beta\geq\gamma\\ |\beta|\leq m}}\binom{\beta}{\gamma}\partial_x^{\beta-\gamma}a_\beta(x).$$

In particular, if condition (5) holds for $A(x, \partial_x)$, then $b_0 = 0$ in $A^*(x, \partial_x)$.

We provide some examples of operators, for which conditions (4) and (5) are valid.

Example 1 If

$$A = A(\partial_x) = \sum_{1\leq|\beta|\leq m} a_\beta\partial_x^\beta,$$

is an operator with constant coefficients, then condition (5) trivially holds, h in (13) is given by

$$h = \min\{|\beta| : a_\beta \neq 0\},$$

and $A^*(\partial_x) = A(-\partial_x)$.

Example 2 If

$$A(x, \partial_x) = |x|^2\Delta^2,$$

then condition (5) trivially holds, $h = 2$ in (13), and

$$\begin{aligned}A^*(x, \partial_x) &= \Delta(|x|^2\Delta + 4x\cdot\nabla + 2n)\\ &= (|x|^2\Delta + 4x\cdot\nabla + 2n)\Delta + (4x\cdot\nabla + 2n)\Delta + 8\Delta\\ &= A + 8x\cdot\nabla\Delta + 4(n+2)\Delta.\end{aligned}$$

Example 3 If

$$A(x, \partial_x) = x_1^2 \partial_{x_1}^4 - \partial_{x_2}^2,$$

then condition (5) trivially holds, $h = 2$ in (13), and

$$A^*(x, \partial_x) = A + 8x_1 \partial_{x_1}^3 + 12 \partial_{x_1}^2.$$

Example 4 Let us consider

$$A = \sqrt{|x_2|} \partial_{x_1} + \sqrt{|x_1|}\, x_2 \partial_{x_2}^2 + x_1^{-2} \partial_{x_1} \partial_{x_2}.$$

Then condition (5) holds, $h = 1/2$ in (13), and

$$A^* = A + 2(\sqrt{|x_1|} - x_1^{-3}) \partial_{x_2}.$$

Remark 5 We notice that, for any $\lambda > 0$, it holds

$$a_\beta(x)\, \partial_x^\beta \big(f(\lambda x)\big) = \lambda^{|\beta| - r_\beta}\, (a_\beta\, \partial_x^\beta f)(\lambda x),$$

for any sufficiently smooth f. Namely, setting

$$A_\beta = a_\beta(x) \partial_x^\beta$$

we have

$$A_\beta(f(\lambda x)) = \lambda^{|\beta| - r_\beta} (A_\beta f)(\lambda x).$$

By homogeneity,

$$A_\beta^*(f(\lambda x)) = \lambda^{|\beta| - r_\beta} (A_\beta^* f)(\lambda x).$$

The proof of this property in the more general setting of quasi-homogeneous operators $L(x, y, \partial_x, \partial_y)$, and the application of the test function method to Liouville problems with these operators, can be found in [9]. A generalization of the definition of quasi-homogeneous operators and the application of the test function method to Liouville and Cauchy problems for these operators, is given in [6]. The statements in this paper may be conveniently improved using the definition of quasi-homogeneous operator, but our interest is more focused to study the influence of the fractional derivatives in time on a differential inequality in the space time $[0, \infty) \times \mathbb{R}^n$.

As a consequence of (13) and Definition 1, we derive

$$A_\beta(f(\lambda x)) = O(\lambda^h), \qquad A_\beta^*(f(\lambda x)) = O(\lambda^h),$$

as $\lambda \to 0$.

In order to give our definition of weak solution to problem (1) with initial condition (11) and to problem (2) with initial condition (12), we need the backward in time analogous of the fractional integration, and Caputo and Riemann-Liouville derivatives.

Definition 2 For any $\beta > 0$ and $T \in (-\infty, +\infty]$, we define the (backward) Riemann-Liouville fractional integral of order β,

$$J^{\beta}_{T-} f(t) = \frac{1}{\Gamma(\beta)} \int_t^T (s-t)^{\beta-1} f(s)\, ds,$$

for any $t \in (-\infty, T)$. Then, we define

$$^{\mathrm{C}}\mathrm{D}^{\alpha}_{T-} f(t) = J^{\kappa-\alpha}_{T-}\big((-\partial_t)^{\kappa} f\big)(t), \tag{32}$$

$$^{\mathrm{RL}}\mathrm{D}^{\alpha}_{T-} f(t) = (-\partial_t)^{\kappa}\big(J^{\kappa-\alpha}_{T-} f\big)(t), \tag{33}$$

for any $t \in (-\infty, T)$, the (backward) Caputo and Riemann-Liouville fractional derivatives of order $\alpha > 0$.

Remark 6 We considered the possibility to take $T = +\infty$ in Definition 2 (see (2.2.2) in [11]), but trough this paper we will always consider compactly supported functions. We remark that if $\operatorname{supp} f \subset (-\infty, T_1)$, for some $T_1 < T$, then obviously

$$J^{1-\gamma}_{T-} f(t) = J^{1-\gamma}_{T_1-} f(t),$$

in particular, the fractional integrals and derivatives of functions compactly supported in $[0, b)$ are zero for any $t \in [b, T)$. For this reason, we choose $T = +\infty$ to avoid to fix, time by time, a sufficiently large T such that f is compactly supported in $[0, T)$.

Definition 3 We say that $u \in L^p_{\mathrm{loc}}(\mathbb{R}_+ \times \mathbb{R}^n)$ is a global weak solution to (1) if for any test function $\varphi \in \mathscr{C}_c^{\max\{\kappa,m\}}(\mathbb{R}_+ \times \mathbb{R}^n)$, it holds

$$\begin{aligned}
&\int_0^\infty \int_{\mathbb{R}^n} u(t,x)\,(^{\mathrm{RL}}\mathrm{D}^{\alpha}_{\infty-} + A^*(x,\partial_x))\varphi(t,x)\,dx\,dt \\
&- \int_{\mathbb{R}^n} (\partial_t^{\kappa-1} u(0,x))\,(J^{\kappa-\alpha}_{\infty-}\varphi)(0,x)\,dx \\
&- \sum_{j=0}^{\kappa-2} \int_{\mathbb{R}^n} (\partial_t^{j} u(0,x))\,(^{\mathrm{RL}}\mathrm{D}^{\alpha-1-j}_{\infty-}\varphi)(0,x)\,dx \\
&\geq \int_0^\infty \int_{\mathbb{R}^n} |u(t,x)|^p\,\varphi(t,x)\,dx\,dt.
\end{aligned}$$

Definition 4 We say that $u \in L^p_{\mathrm{loc}}(\mathbb{R}_+ \times \mathbb{R}^n)$ is a global weak solution to (2) if for any test function $\varphi \in \mathscr{C}_c^{\max\{\kappa,m\}}(\mathbb{R}_+ \times \mathbb{R}^n)$ it holds

$$\begin{aligned}
&\int_0^\infty \int_{\mathbb{R}^n} u(t,x)\,({}^{\mathrm{C}}\mathrm{D}^\alpha_{\infty-} + A^*(x,\partial_x))\varphi(t,x)\,dx\,dt \\
&\quad - \sum_{j=0}^{\kappa-2} \int_{\mathbb{R}^n} ({}^{\mathrm{RL}}\mathrm{D}^{\alpha-1-j}_{0+} u)(0,x)\,(-\partial_t)^j \varphi(0,x)\,dx \\
&\quad - \int_{\mathbb{R}^n} (J^{\kappa-\alpha}_{0+} u)(0,x)\,(-\partial_t)^{\kappa-1}\varphi(0,x)\,dx \\
&\quad \geq \int_0^\infty \int_{\mathbb{R}^n} |u(t,x)|^p\,\varphi(t,x)\,dx\,dt.
\end{aligned}$$

In order to motivate the definition of global weak solution to (1) and (2) given in Definitions 3 and 4, we will employ the following fractional integration by parts result.

Lemma 1 (Lemma 2.7 in [11]) *Let $b > 0$, $f \in L^{p_1}([0,b])$, $g \in L^{p_2}([0,b])$, and either $p_1, p_2 \geq 1$ with $1/p_1 + 1/p_2 < 1+\beta$, or $p_1, p_2 > 1$ and $1/p_1 + 1/p_2 = 1+\beta$. Then we have the following:*

$$\int_0^b (J^\beta_{0+} f)(t)\,g(t)\,dt = \int_0^b f(t)\,J^\beta_{b-} g(t)\,dt. \tag{34}$$

We are now ready to show that classical solutions to (1) are weak solutions, according to Definition 3.

Proposition 1 *Let $u \in \mathscr{C}^{\max\{\kappa,m\}}(\mathbb{R}_+ \times \mathbb{R}^n)$ be a classical solution to* (1). *Then u is a global (weak) solution to* (1), *according to Definition* 3.

Proof Let $\varphi \in \mathscr{C}_c^{\max\{\kappa,m\}}(\mathbb{R}_+ \times \mathbb{R}^n)$. We fix $b > 0$ such that φ is supported in $[0,b) \times \mathbb{R}^n$.

After multiplying Eq. (1) by φ and integrating over $\mathbb{R}_+ \times \mathbb{R}^n$, we get

$$\begin{aligned}
&\int_0^\infty \int_{\mathbb{R}^n} ({}^{\mathrm{C}}\mathrm{D}^\alpha_{0+} u(t,x))\varphi(t,x)\,dx\,dt \\
&\quad + \int_0^\infty \int_{\mathbb{R}^n} \big(A(x,\partial_x)u(t,x)\big)\,\varphi(t,x)\,dx\,dt \\
&\quad \geq \int_0^\infty \int_{\mathbb{R}^n} |u(t,x)|^p \varphi(t,x)\,dx\,dt.
\end{aligned}$$

By Definition 1, we may write

$$\int_0^\infty \int_{\mathbb{R}^n} \big(A(x,\partial_x)u(t,x)\big)\,\varphi(t,x)\,dx\,dt \\ = \int_0^\infty \int_{\mathbb{R}^n} u(t,x)\,\big(A^*(x,\partial_x)\varphi(t,x)\big)\,dx\,dt.$$

On the other hand, recalling the definition of Caputo fractional derivative (6), we have

$$\int_0^\infty \int_{\mathbb{R}^n} ({}^{\mathrm{C}}\mathrm{D}^\alpha_{0+}\,u(t,x))\varphi(t,x)\,dx\,dt \\ = \int_0^\infty \int_{\mathbb{R}^n} (J^{\kappa-\alpha}_{0+}\,(\partial_t^\kappa u(t,x)))\varphi(t,x)\,dx\,dt.$$

Being φ supported in $[0,b)\times\mathbb{R}^n$, due to the fact that $\partial_t^\kappa u$ and φ are continuous, we may apply first fractional integration by parts (34) and then classical integration by parts, to obtain:

$$\begin{aligned}
&\int_0^\infty \int_{\mathbb{R}^n} (J^{\kappa-\alpha}_{0+}\,(\partial_t^\kappa u(t,x)))\varphi(t,x)\,dx\,dt \\
&= \int_0^b \int_{\mathbb{R}^n} (J^{\kappa-\alpha}_{0+}\,(\partial_t^\kappa u(t,x)))\varphi(t,x)\,dx\,dt \\
&= \int_0^b \int_{\mathbb{R}^n} (\partial_t^\kappa u(t,x))(J^{\kappa-\alpha}_{b-}\,\varphi(t,x))\,dx\,dt \\
&= \int_0^\infty \int_{\mathbb{R}^n} (\partial_t^\kappa u(t,x))(J^{\kappa-\alpha}_{\infty-}\,\varphi(t,x))\,dx\,dt \\
&= \int_0^\infty \int_{\mathbb{R}^n} u(t,x)((-\partial_t)^\kappa(J^{\kappa-\alpha}_{\infty-}\,\varphi(t,x)))\,dx\,dt \\
&\quad - \sum_{j=0}^{\kappa-1} \int_{\mathbb{R}^n} (\partial_t^j u(0,x))((-\partial_t)^{\kappa-1-j}(J^{\kappa-\alpha}_{\infty-}\,\varphi(0,x)))\,dx.
\end{aligned}$$

The proof follows, noticing that

$$(-\partial_t)^{\kappa-1-j}(J^{\kappa-\alpha}_{\infty-}\,f) = (-\partial_t)^{\kappa-1-j}(J^{(\kappa-1-j)-(\alpha-1-j)}_{\infty-}\,f) = {}^{\mathrm{RL}}\mathrm{D}^{\alpha-1-j}_{\infty-}\,f,$$

for any $j=0,\ldots,\kappa-2$, according to (33), due to $\lceil\alpha-1-j\rceil = \lceil\alpha\rceil-1-j = \kappa-1-j$.

Similarly, we may show that classical solutions to (2) are weak solutions, according to Definition 4.

Proposition 2 *Let $u \in \mathscr{C}^{\max\{\kappa,m\}}(\mathbb{R}_+ \times \mathbb{R}^n)$ be a classical solution to* (2)*. Then u is a global (weak) solution to* (2)*, according to Definition* 4.

Proof Let $\varphi \in \mathscr{C}_c^{\max\{\kappa,m\}}(\mathbb{R}_+ \times \mathbb{R}^n)$. We fix $b > 0$ such that φ is supported in $[0, b) \times \mathbb{R}^n$.

After multiplying Eq. (2) by φ, integrating over $\mathbb{R}_+ \times \mathbb{R}^n$ and employing Definition 1, we get

$$\begin{aligned}\int_0^\infty \int_{\mathbb{R}^n} ({}^{\mathrm{RL}}\mathrm{D}^\alpha_{0+}\, u(t,x))\varphi(t,x)\, dx\, dt \\ + \int_0^\infty \int_{\mathbb{R}^n} u(t,x)\, \big(A^*(x,\partial_x)\varphi(t,x)\big)\, dx\, dt \\ \geq \int_0^\infty \int_{\mathbb{R}^n} |u(t,x)|^p \varphi(t,x)\, dx\, dt.\end{aligned}$$

Recalling the definition of Riemann-Liouville fractional derivative (7) and using classical integration by parts, we obtain

$$\begin{aligned}&\int_0^\infty \int_{\mathbb{R}^n} ({}^{\mathrm{RL}}\mathrm{D}^\alpha_{0+}\, u(t,x))\varphi(t,x)\, dx\, dt \\ &= \int_0^\infty \int_{\mathbb{R}^n} (\partial_t^\kappa (J^{\kappa-\alpha}_{0+}\, u(t,x)))\varphi(t,x)\, dx\, dt \\ &= \int_0^\infty \int_{\mathbb{R}^n} (J^{\kappa-\alpha}_{0+}\, u(t,x))((-\partial_t)^\kappa \varphi(t,x))\, dx\, dt \\ &\quad - \sum_{j=0}^{\kappa-1} \int_{\mathbb{R}^n} (\partial_t^j (J^{\kappa-\alpha}_{0+}\, u))(0,x)((-\partial_t)^{\kappa-1-j}\varphi(0,x))\, dx\, dt.\end{aligned}$$

Being φ supported in $[0, b) \times \mathbb{R}^n$, due to the fact that u and $\partial_t^\kappa \varphi$ are continuous, we may apply fractional integration by parts (34), obtaining

$$\begin{aligned}&\int_0^\infty \int_{\mathbb{R}^n} (J^{\kappa-\alpha}_{0+}\, u(t,x))((-\partial_t)^\kappa \varphi(t,x))\, dx\, dt \\ &= \int_0^b \int_{\mathbb{R}^n} (J^{\kappa-\alpha}_{0+}\, u(t,x))((-\partial_t)^\kappa \varphi(t,x))\, dx\, dt \\ &= \int_0^b \int_{\mathbb{R}^n} u(t,x)(J^{\kappa-\alpha}_{b-}\, ((-\partial_t)^\kappa \varphi(t,x)))\, dx\, dt \\ &= \int_0^\infty \int_{\mathbb{R}^n} u(t,x)(J^{\kappa-\alpha}_{\infty-}\, ((-\partial_t)^\kappa \varphi(t,x)))\, dx\, dt.\end{aligned}$$

The proof follows by noticing that

$$\partial_t^j (J_{0+}^{\kappa-\alpha} u(0,x)) = {}^{\mathrm{RL}}\mathrm{D}^{\alpha+j-\kappa} u(0,x),$$

according to (7), and that

$$J_{\infty-}^{\kappa-\alpha} ((-\partial_t)^\kappa \varphi(t,x)) = {}^{\mathrm{C}}\mathrm{D}_{\infty-}^{\alpha} \varphi(t,x),$$

according to (32).

3 A Suitable Test Function

According to Definitions 3 and 4, we need, in general, κ initial conditions to supplement Eqs. (1) and (2). Namely, to get a Cauchy-type problem, we have to assign the initial values

$$u(0,x), \qquad \partial_t^{\kappa-1} u(0,x),$$

if we consider Eq. (1), and the initial values

$$J_{0+}^{\kappa-\alpha} u(0,x), \qquad {}^{\mathrm{RL}}\mathrm{D}^{\alpha+1-\kappa} u(0,x), \qquad \ldots, \qquad {}^{\mathrm{RL}}\mathrm{D}^{\alpha-1} u(0,x),$$

if we consider Eq. (2).

However, in order to derive a nonexistence result which is independent of the first $\kappa-1$ initial conditions, we may choose a suitable test function $\varphi(t,x)$. For problem (2), the task is trivial, as in the case of Cauchy problems for operators with integer derivatives in time.

Remark 7 Let $u \in L^p_{\mathrm{loc}}(\mathbb{R}_+ \times \mathbb{R}^n)$ be a global weak solution to (2), supplemented by (12). Let $\varphi \in \mathscr{C}_c^{\max\{\kappa,m\}}(\mathbb{R}_+ \times \mathbb{R}^n)$ be a test function, with $\varphi(t,x)$ independent of t in a neighborhood of the line $\{t=0\}$. Then it holds

$$\begin{aligned}\int_0^\infty \int_{\mathbb{R}^n} & u(t,x)\, ({}^{\mathrm{C}}\mathrm{D}_{\infty-}^{\alpha} + A^*(x,\partial_x))\varphi(t,x)\, dx\, dt \\ & - \int_{\mathbb{R}^n} u_{\alpha-1}(x)\, \varphi(0,x)\, dx \\ & \geq \int_0^\infty \int_{\mathbb{R}^n} |u(t,x)|^p\, \varphi(t,x)\, dx\, dt,\end{aligned}$$

as a consequence of $\partial_t^j \varphi(0,x) = 0$, for any $j = 1, \ldots, \kappa-1$. In the previous inequality, the only initial condition appearing is the one assigned in (12).

In order to obtain the analogous of Remark 7, we shall find a test function $\varphi(t,x)$, such that

$$ {}^{\mathrm{RL}}\mathrm{D}_{\infty-}^{\alpha-1-j}\varphi(0,x)=0, \qquad j=0,\ldots,\kappa-2. $$

Moreover, we want a nonnegative test function $\varphi(t,x)$.

Lemma 2 *Let $\ell\geq 1$ and $g\in\mathscr{C}^\ell([0,\infty))$, positive, non-increasing, with* $\operatorname{supp} g=[0,1]$, *be such that*

$$ g(t)=c_0(1-t)^{\ell+1}, \quad \text{in a left neighborhood of } t=1, \tag{35} $$

for some $c_0>0$. Let $\gamma\in(0,1)$. Then

$$ f(t)={}^{\mathrm{RL}}\mathrm{D}_{\infty-}^{\gamma}g(t)={}^{\mathrm{C}}\mathrm{D}_{\infty-}^{\gamma}g(t) $$

verifies $f\in\mathscr{C}^{\ell-1}([0,\infty))$, is supported in $[0,1]$, is nonnegative, and

$$ J_{\infty-}^{\gamma}f(t)=g(t). \tag{36} $$

Proof Due to the fact that $\operatorname{supp} g=[0,1]$, it holds

$$ {}^{\mathrm{RL}}\mathrm{D}_{\infty-}^{\gamma}g(t)=\begin{cases} {}^{\mathrm{RL}}\mathrm{D}_{1-}^{\gamma}g(t) & \text{if } t\leq 1,\\ 0 & \text{if } t>1,\end{cases} $$

$$ {}^{\mathrm{C}}\mathrm{D}_{\infty-}^{\gamma}g(t)=\begin{cases} {}^{\mathrm{C}}\mathrm{D}_{1-}^{\gamma}g(t) & \text{if } t\leq 1,\\ 0 & \text{if } t>1.\end{cases} $$

Due to the fact that $g(1)=0$, we get

$$ {}^{\mathrm{RL}}\mathrm{D}_{1-}^{\gamma}g(t)={}^{\mathrm{C}}\mathrm{D}_{1-}^{\gamma}g(t), \qquad t\in[0,1). $$

As a consequence of (35), we obtain (see (2.1.19) in [11])

$$ {}^{\mathrm{RL}}\mathrm{D}_{1-}^{\gamma}g(t)=\frac{\Gamma(\ell+2)}{\Gamma(\ell+2-\gamma)}\,c_0\,(1-t)^{\ell+1-\gamma}, $$

in a left neighborhood of $t=1$. In particular, ${}^{\mathrm{C}}\mathrm{D}_{1-}^{\gamma}g(1)={}^{\mathrm{RL}}\mathrm{D}_{1-}^{\gamma}g(1)=0$ (as usual, the values in $t=1$ are intended as the limits as $t\to 1$, see also (8), (9)). It follows that f is well-defined, it belongs to $f\in\mathscr{C}^{\ell-1}([0,\infty))$ and it is supported in $[0,1]$. We notice that

$$ f(t)={}^{\mathrm{C}}\mathrm{D}_{1-}^{\gamma}g(t)=-J_{1-}^{1-\gamma}g'(t), $$

and so it is nonnegative, being $g'(t)$ nonpositive. Equality (36) follows by (see also (2.4.43) in [11])

$$J^{\gamma}_{1-} f(t) = -J^{\gamma}_{1-} J^{1-\gamma}_{1-} g'(t) = -\int_t^1 g'(s)\,ds = g(t) - g(1) = g(t).$$

Remark 8 Let g be as in Lemma 2, with $\ell - 1 \geq \max\{\kappa, m\}$. Moreover, assume that g is constant in a right neighborhood of $t = 0$, and set f as in Lemma 2, with $\gamma = \kappa - \alpha$. Let $\psi \in \mathscr{C}_c^{\max\{\kappa,m\}}(\mathbb{R}^n)$, nonnegative. Then $\varphi(t,x) = f(t)\,\psi(x)$ is a nonnegative test function for which Definition 3 applies.

Let $u \in L^p_{\mathrm{loc}}(\mathbb{R}_+ \times \mathbb{R}^n)$ be a global weak solution to (1), supplemented by (11). Then it holds

$$\begin{aligned}
&\int_0^\infty \int_{\mathbb{R}^n} u(t,x)\,({}^{\mathrm{RL}}\mathrm{D}^{\alpha}_{\infty-} + A^*(x,\partial_x))\varphi(t,x)\,dx\,dt \\
&\quad - \int_{\mathbb{R}^n} u_{\kappa-1}(x)\,(J^{\kappa-\alpha}_{\infty-}\varphi)(0,x)\,dx \\
&\quad \geq \int_0^\infty \int_{\mathbb{R}^n} |u(t,x)|^p\,\varphi(t,x)\,dx\,dt,
\end{aligned}$$

as a consequence of

$${}^{\mathrm{RL}}\mathrm{D}^{\alpha+1-\kappa}_{\infty-} f(0) = g'(0) = 0, \quad \ldots \quad , {}^{\mathrm{RL}}\mathrm{D}^{\alpha-1}_{\infty-} f(0) = g^{(\kappa-1)}(0) = 0,$$

and

$$J^{\kappa-\alpha}_{\infty-} f(0) = g(0).$$

In the previous inequality, the only initial condition appearing is the one assigned in (11).

4 Proof of Theorem 1

We are now ready to prove Theorem 1.

Proof (Theorem 1) Let u be a global nontrivial weak solution to (2), with initial condition (12), in the sense of Definition 4. Assume that (14) holds.

For any $R \geq 1$ and $T \geq 1$, we fix

$$\varphi(t,x) = f(t/T)\,\psi(x/R),$$

where $f \in \mathscr{C}_c^{\max\{\kappa,m\}}([0,\infty))$, is a suitable, nonnegative, test function, constant in $[0,1/2]$, with $\operatorname{supp} f \subset [0,1]$, and $\psi \in \mathscr{C}_c^\infty(\mathbb{R}^n)$ is a suitable, nonnegative, test function, constant in $B_{1/2} = \{x \in \mathbb{R}^n : |x| \leq 1/2\}$, with $\operatorname{supp}\psi \subset B_1 = \{x \in \mathbb{R}^n : |x| \leq 1\}$.

According to Remark 7, it follows that

$$\begin{aligned}
I_{R,T} &= \int_0^\infty \int_{\mathbb{R}^n} |u(t,x)|^p\, f(t/T)\,\psi(x/R)\,dx\,dt \\
&\leq \int_0^\infty \int_{\mathbb{R}^n} u(t,x)\,({}^{\mathrm{C}}\mathrm{D}^\alpha_{\infty-} + A^*(x,\partial_x)) f(t/T)\,\psi(x/R)\,dx\,dt \\
&\quad - \int_{\mathbb{R}^n} u_{\alpha-1}(x)\,\varphi(0,x)\,dx \\
&= \int_0^\infty \int_{\mathbb{R}^n} u(t,x)\,({}^{\mathrm{C}}\mathrm{D}^\alpha_{\infty-} f(t/T))\,\psi(x/R)\,dx\,dt \\
&\quad + \int_0^\infty \int_{\mathbb{R}^n} u(t,x)\, f(t/T)\,(A^*(x,\partial_x)\psi(x/R))\,dx\,dt \\
&\quad - f(0)\int_{\mathbb{R}^n} u_{\alpha-1}(x)\,\psi(x/R)\,dx.
\end{aligned}$$

We notice that $I_{R,T}$ is nonnegative, since f and ψ are nonnegative. By homogeneity, we obtain

$${}^{\mathrm{C}}\mathrm{D}^\alpha_{\infty-} f(t/T) = T^{-\alpha}\,({}^{\mathrm{C}}\mathrm{D}^\alpha_{\infty-} f)(t/T),$$

so that, by Hölder's inequality, we derive

$$\begin{aligned}
&\int_0^\infty \int_{\mathbb{R}^n} |u(t,x)|\,|({}^{\mathrm{C}}\mathrm{D}^\alpha_{\infty-} f(t/T))|\,\psi(x/R)\,dx\,dt \\
&\quad = T^{-\alpha}\int_0^\infty \int_{\mathbb{R}^n} |u(t,x)|\,|({}^{\mathrm{C}}\mathrm{D}^\alpha_{\infty-} f)(t/T)|\,\psi(x/R)\,dx\,dt \\
&\quad \lesssim T^{-\alpha}\, I_{R,T}^{\frac{1}{p}} \Big(\int_0^\infty \int_{\mathbb{R}^n} |({}^{\mathrm{C}}\mathrm{D}^\alpha_{\infty-} f)(t/T)|^{p'}\, f(t/T)^{-\frac{p'}{p}}\,\psi(x/R)\,dx\,dt\Big)^{\frac{1}{p'}} \\
&\quad \lesssim T^{-\alpha}\, I_{R,T}^{\frac{1}{p}}\,(R^n\,T)^{\frac{1}{p'}},
\end{aligned}$$

provided that

$$|({}^{\mathrm{C}}\mathrm{D}^\alpha_{\infty-} f)|\, f^{-\frac{1}{p}} \leq C, \tag{37}$$

for some $C > 0$. We notice that, in the last inequality, we used that $f(t/T)$ is supported in $[0,T]$, that $\psi(x/R)$ is supported in $B_R(0)$. In order to obtain (37), it is sufficient to ask that $f(t) > 0$ in $[0, 1)$, with $f(t) = c_0(1-t)^\ell$, in a left neighborhood of $t = 1$, for some $c_0 > 0$, for a sufficiently large $\ell \geq \max\{\kappa, m\}$. Indeed, for any fixed $\varepsilon > 0$, condition (37) trivially holds in $[0, 1-\varepsilon]$, since f is continuous and positive. On the other hand, in $[1-\varepsilon, 1]$ it holds

$$|({}^{\mathrm{C}}\mathrm{D}^{\alpha}_{\infty-} f)(t)|\,(f(t))^{-\frac{1}{p}} \lesssim (1+t)^{\ell-\alpha-\frac{\ell}{p}},$$

so that it is sufficient to take $\ell \geq \alpha/(1-1/p)$ to derive (37).

For any $|\beta| \leq m$ and $\gamma \leq \beta$, let us set

$$c_{\beta,\gamma} = \partial_x^{\beta-\gamma} a_\beta(x).$$

We notice that $c_{\beta,\gamma}$ is homogeneous of degree $r_\beta - |\beta| + |\gamma|$, since a_β is homogeneous of degree r_β. Then it holds

$$\begin{aligned}
A_\beta^*(\psi(x/R)) &= \partial_x^\beta(a_\beta(x)\,\psi(x/R)) \\
&= \sum_{\gamma\leq\beta}\binom{\beta}{\gamma}(\partial_x^{\beta-\gamma}a_\beta(x))\,\partial_x^\gamma(\psi(x/R)) \\
&= \sum_{\gamma\leq\beta}\binom{\beta}{\gamma}R^{-|\gamma|}\,c_{\beta,\gamma}(x)\,(\partial_x^\gamma\psi)(x/R) \\
&= \sum_{\gamma\leq\beta}\binom{\beta}{\gamma}R^{r_\beta-|\beta|}\,c_{\beta,\gamma}(x/R)\,(\partial_x^\gamma\psi)(x/R) \\
&= R^{r_\beta-|\beta|}\sum_{\gamma\leq\beta}\binom{\beta}{\gamma}(c_{\beta,\gamma}\partial_x^\gamma\psi)(x/R) \\
&= R^{r_\beta-|\beta|}\,(A_\beta^*\psi)(x/R).
\end{aligned}$$

We also notice that, as a consequence of (5), $(A_\beta^*\psi)(x)$ vanishes in a neighborhood of the origin, since ψ is constant in a neighborhood of the origin.

By Hölder's inequality, and by using (13), we may now obtain

$$\begin{aligned}
&\int_0^\infty\int_{\mathbb{R}^n}|u(t,x)|\,f(t/T)\,|A^*(x,\partial_x)\psi(x/R)|\,dx\,dt \\
&\quad\lesssim I_{R,T}^{\frac{1}{p}}\Big(\int_0^\infty\int_{\mathbb{R}^n}f(t/T)\,|A^*(x,\partial_x)\psi(x/R)|^{p'}\,\psi(x/R)^{-\frac{p'}{p}}\,dx\,dt\Big)^{\frac{1}{p'}} \\
&\quad\lesssim I_{R,T}^{\frac{1}{p}}\,R^{-h}\,(R^n\,T)^{\frac{1}{p'}},
\end{aligned}$$

provided that

$$|A_\beta^* \psi(x)| \, \psi^{-\frac{1}{p}} \leq C,$$

for any $|\beta| \leq m$, for some $C > 0$. For a suitable function ψ, the above condition holds, as a consequence of the fact that the coefficients of A_β^* are smooth away from the origin. For instance, if $\psi = \tilde{\psi}^\ell$, with $\tilde{\psi} \in \mathscr{C}_c^\infty$ and sufficiently large ℓ.

As a consequence of (14), there exists a sufficiently large $\bar{R}$ such that

$$-f(0) \int_{\mathbb{R}^n} u_{\alpha-1}(x) \, \psi(x/R) \, dx \leq 0, \qquad \forall R \geq \bar{R}. \tag{38}$$

Summarizing, we proved that

$$I_{R,T} \lesssim I_{R,T}^{\frac{1}{p}} \, (T^{-\alpha} + R^{-h}) \, (R^n \, T)^{\frac{1}{p'}},$$

for $R \geq \bar{R}$. In the following we fix $R = R(T) = T^{\alpha/h}$, so that $T^{-\alpha} = R^{-h}$, and

$$I_{R(T),T} \lesssim I_{R(T),T}^{\frac{1}{p}} \, T^{-\alpha + \frac{1+n\alpha/h}{p'}}. \tag{39}$$

As a consequence,

$$I_{R(T),T} \lesssim T^{-\alpha p' + 1 + \frac{n\alpha}{h}}.$$

Assume, by contradiction, that

$$-\alpha p' + 1 + \frac{n\alpha}{h} < 0,$$

that is, $p < \tilde{p}(n, \alpha)$, where $\tilde{p}(n, \alpha)$ is as in (15). Due to

$$f(0) \, \psi(0) \int_0^{T/2} \int_{B_{R(T)/2}(0)} |u(t, x)|^p \, dx \, dt \leq I_{R(T),T} \lesssim T^{-\alpha p' + 1 + \frac{n\alpha}{h}},$$

taking the limit as $T \to \infty$, we deduce that

$$f(0) \, \psi(0) \int_0^\infty \int_{\mathbb{R}^n} |u(t, x)|^p \, dx \, dt \leq 0.$$

i.e., $u = 0$. Now let us consider the critical case $p = \tilde{p}(n, \alpha)$. In this case, we only get that $I_{R(T),T}$ is bounded by (39). That is, taking the limit as $T \to \infty$, we obtain that $u \in L^p$. On the other hand, taking into account that

$$\begin{aligned} ({}^{\mathrm{C}}\mathrm{D}^{\alpha}_{\infty-} f)(t/T) &= 0, \qquad && \forall t \in [0, T/2), \\ A^*(x, \partial_x)\psi(x/R) &= 0, \qquad && \forall x \in B_{R/2}, \end{aligned}$$

we may refine estimate (39) to derive

$$I_{R(T),T} \lesssim \Big(\int_{G_T^c} |u(t,x)|^p \, f(t/T)\, \psi(x/R)\, dx\, dt \Big)^{\frac{1}{p}},$$

where $G(T) = [0, T/2] \times B_{R/2}$ and $G_T^c = ([0,\infty) \times \mathbb{R}^n) \setminus G_T$. Due to the fact that $u \in L^p$ and $G_T \nearrow [0,\infty) \times \mathbb{R}^n$, we obtain that the right-hand side in the previous inequality vanishes as $T \to \infty$. That is, proceeding as before, we deduce that $u = 0$.

Therefore, u is trivial if $p \leq \tilde{p}(n, \alpha)$. This concludes the proof.

If we replace (14) by (16), then we may replace (38) by

$$\begin{aligned} - f(0) \int_{\mathbb{R}^n} & u_{\alpha-1}(x)\, \psi(x/R)\, dx \\ &\leq -f(0) \int_{B_{R/2}} u_{\alpha-1}(x)\, dx \\ &\leq -\varepsilon\, C\, R^{n-\theta}, \qquad \forall R \geq \bar{R}, \end{aligned}$$

for some $C > 0$. As a consequence, we may refine (39) into

$$I_{R(T),T} \lesssim I_{R,T}^{\frac{1}{p}}\, T^{-\alpha + \frac{1+n\alpha/h}{p'}} - \varepsilon T^{\frac{(n-\theta)\alpha}{h}}.$$

By Young inequality, we derive

$$I_{R(T),T} \lesssim T^{-\alpha p' + 1 + \frac{n\alpha}{h}} - \varepsilon T^{\frac{(n-\theta)\alpha}{h}}.$$

For any fixed $\varepsilon > 0$, the right-hand side is negative, for sufficiently large R, if

$$-\alpha p' + 1 + \frac{\theta\alpha}{h} < 0,$$

that is, if $p < \tilde{p}(\theta, \alpha)$, where $\tilde{p}(\theta, \alpha)$ is as in (17). This concludes the proof.

5 Proof of Theorem 2

We are now ready to prove Theorem 2.

Proof (Theorem 2) Let u be a global nontrivial weak solution to (1), with initial condition (11), in the sense of Definition 3. Assume that (18) holds.

For any $R \geq 1$ and $T \geq 1$, we fix

$$\varphi(t,x) = f(t/T)\,\psi(x/R),$$

where $\psi \in \mathscr{C}_c^\infty(\mathbb{R}^n)$ is a suitable, nonnegative, test function, constant in $B_{1/2} = \{x \in \mathbb{R}^n : |x| \leq 1/2\}$, with $\operatorname{supp}\psi \subset B_1 = \{x \in \mathbb{R}^n : |x| \leq 1\}$, as in the proof of Theorem 1. On the other hand, we choose f as in Lemma 2 and Remark 8. More precisely, we fix $g \in \mathscr{C}^\ell([0,\infty))$, positive, non-increasing, with $\operatorname{supp} g = [0,1]$, and constant in $[0,1/2]$, be such that (35) holds, that is,

$$g(t) = c_0(1-t)^{\ell+1}, \quad \text{in a left neighborhood of } t = 1,$$

for some $c_0 > 0$. Then we put

$$f(t) = {}^{\mathrm{RL}}\mathrm{D}_{\infty-}^{\kappa-\alpha} g(t).$$

According to Remark 8, it follows that

$$\begin{aligned}
I_{R,T} &= \int_0^\infty \int_{\mathbb{R}^n} |u(t,x)|^p\,\varphi(t,x)\,dx\,dt \\
&\leq \int_0^\infty \int_{\mathbb{R}^n} u(t,x)\,({}^{\mathrm{RL}}\mathrm{D}_{\infty-}^{\alpha} + A^*(x,\partial_x))\varphi(t,x)\,dx\,dt \\
&\quad - \int_{\mathbb{R}^n} u_{\kappa-1}(x)\,(J_{\infty-}^{\kappa-\alpha}\varphi)(0,x)\,dx \\
&= \int_0^\infty \int_{\mathbb{R}^n} u(t,x)\,{}^{\mathrm{RL}}\mathrm{D}_{\infty-}^{\alpha}(f(t/T))\psi(x/R)\,dx\,dt \\
&\quad + \int_0^\infty \int_{\mathbb{R}^n} u(t,x)\,f(t/T)A^*(x,\partial_x))(\psi(x/R))\,dx\,dt \\
&\quad - g(0)\,T^{\kappa-\alpha} \int_{\mathbb{R}^n} u_{\kappa-1}(x)\,\psi(x/R)\,dx.
\end{aligned}$$

Here we used that, by homogeneity, it holds

$$\begin{aligned}
(J_{\infty-}^{\kappa-\alpha}(f(t/T)))|_{t=0} &= T^{\kappa-\alpha}(J_{\infty-}^{\kappa-\alpha} f)(t/T)|_{t=0} \\
&= T^{\kappa-\alpha} g(t/T)|_{t=0}.
\end{aligned}$$

By homogeneity, we also obtain

$$ {}^{\mathrm{RL}}\mathrm{D}^{\alpha}_{\infty-}(f(t/T)) = T^{-\alpha}\,({}^{\mathrm{RL}}\mathrm{D}^{\alpha}_{\infty-}f)(t/T), $$

so that, by Hölder's inequality, we derive

$$ \begin{aligned} &\int_0^\infty \int_{\mathbb{R}^n} |u(t,x)|\,|({}^{\mathrm{RL}}\mathrm{D}^{\alpha}_{\infty-}f(t/T))|\,\psi(x/R)\,dx\,dt \\ &\quad = T^{-\alpha}\int_0^\infty \int_{\mathbb{R}^n} |u(t,x)|\,|({}^{\mathrm{RL}}\mathrm{D}^{\alpha}_{\infty-}f)(t/T)|\,\psi(x/R)\,dx\,dt \\ &\quad \lesssim T^{-\alpha}\,I_{R,T}^{\frac{1}{p}}\Big(\int_0^\infty \int_{\mathbb{R}^n} |({}^{\mathrm{RL}}\mathrm{D}^{\alpha}_{\infty-}f)(t/T)|^{p'}\,f(t/T)^{-\frac{p'}{p}}\,\psi(x/R)\,dx\,dt\Big)^{\frac{1}{p'}} \\ &\quad \lesssim T^{-\alpha}\,I_{R,T}^{\frac{1}{p}}\,(R^n\,T)^{\frac{1}{p'}}, \end{aligned} $$

provided that

$$ |({}^{\mathrm{RL}}\mathrm{D}^{\alpha}_{\infty-}f)|\,f^{-\frac{1}{p}} \leq C, \tag{40} $$

for some $C > 0$. We notice that, in the last inequality, we used that $f(t/T)$ is supported in $[0,T]$ and that $\psi(x/R)$ is supported in $B_R(0)$. In order to obtain (40), it is sufficient to take a sufficiently large $\ell \geq \max\{\kappa, m\}$. Indeed, for any sufficiently small, fixed $\varepsilon > 0$, condition (40) trivially holds in $[0, 1-\varepsilon]$, since f is continuous and positive. On the other hand, in $[1-\varepsilon, 1]$ it holds

$$ f(t) = c_0\,\frac{\Gamma(\ell+2)}{\Gamma(\ell+2-(\kappa-\alpha))}\,(1-t)^{\ell+1-(\kappa-\alpha)}, $$

so that

$$ |({}^{\mathrm{RL}}\mathrm{D}^{\alpha}_{\infty-}f)(t)|\,(f(t))^{-\frac{1}{p}} \lesssim (1+t)^{\ell+1-\kappa-\frac{\ell+\alpha+1-\kappa}{p}}, $$

and it is sufficient to take $\ell \geq \kappa - 1 + \alpha/(p-1)$ to derive (40).

Proceeding as in the proof of Theorem 1, we derive once again

$$ \begin{aligned} &\int_0^\infty \int_{\mathbb{R}^n} |u(t,x)|\,f(t/T)\,|A^*(x,\partial_x)\psi(x/R)|\,dx\,dt \\ &\quad \lesssim I_{R,T}^{\frac{1}{p}}\,R^{-h}\,(R^n\,T)^{\frac{1}{p'}}. \end{aligned} $$

We set $R = R(T) = T^{\alpha/h}$, so that $T^{-\alpha} = R^{-h}$, as in the proof of Theorem 1.

We now distinguish the three cases, according to which data assumption we take among (18), (19) or (21).

Let us assume (18). In this case, we can only deduce that

$$-g(0)\,T^{\kappa-\alpha}\int_{\mathbb{R}^n} u_{\kappa-1}(x)\,\psi(x/R)\,dx \leq 0,$$

so that we derive (39), as in the proof of Theorem 1. Therefore, repeating the steps in the proof of Theorem 1, we find again that u is trivial if $p \leq \tilde{p}(n,\alpha)$. This concludes the proof.

Now let us assume (19). In this case, there exists a sufficiently large $\bar{R}$, such that

$$-g(0)\,T^{\kappa-\alpha}\int_{\mathbb{R}^n} u_{\kappa-1}(x)\,\psi(x/R)\,dx \leq -\varepsilon\,T^{\kappa-\alpha}, \qquad \forall R \geq \bar{R},$$

where

$$\varepsilon = \frac{1}{2}\,g(0)\int_{\mathbb{R}^n} u_{\kappa-1}(x)\,dx.$$

As a consequence, we may refine (39) into

$$I_{R(T),T} \lesssim I_{R,T}^{\frac{1}{p}}\,T^{-\alpha+\frac{1+n\alpha/h}{p'}} - \varepsilon T^{\kappa-\alpha}.$$

By Young inequality, we derive

$$I_{R(T),T} \lesssim T^{-\alpha p'+1+\frac{n\alpha}{h}} - \varepsilon T^{\kappa-\alpha}.$$

For any fixed $\varepsilon > 0$, the right-hand side is negative, for sufficiently large R, if

$$-\alpha p' + 1 + \frac{n\alpha}{h} < \kappa - \alpha,$$

that is, if $p < \bar{p}(n,\alpha)$, where $\bar{p}(n,\alpha)$ is as in (20). This concludes the proof.

Finally, let us assume (21). In this case, we may estimate

$$\begin{aligned} &-g(0)\,T^{\kappa-\alpha}\int_{\mathbb{R}^n} u_{\kappa-1}(x)\,\psi(x/R)\,dx \\ &\quad\leq -g(0)\,T^{\kappa-\alpha}\int_{B_{R/2}} u_{\kappa-1}(x)\,dx \\ &\quad\leq -\varepsilon\,C\,T^{\kappa-\alpha}\,R^{n-\theta}, \qquad \forall R \geq \bar{R}, \end{aligned}$$

for some $C > 0$. As a consequence, we may refine (39) into

$$I_{R(T),T} \lesssim I_{R,T}^{\frac{1}{p}} T^{-\alpha+\frac{1+n\alpha/h}{p'}} - \varepsilon T^{\kappa-\alpha+\frac{(n-\theta)\alpha}{h}}.$$

By Young inequality, we derive

$$I_{R(T),T} \lesssim T^{-\alpha p'+1+\frac{n\alpha}{h}} - \varepsilon T^{\frac{\kappa-\alpha+(n-\theta)\alpha}{h}}.$$

For any fixed $\varepsilon > 0$, the right-hand side is negative, for sufficiently large R, if

$$-\alpha p' + 1 + \frac{\theta\alpha}{h} < \kappa - \alpha,$$

that is, if $p < \bar{p}(\theta, \alpha)$, where $\bar{p}(\theta, \alpha)$ is as in (22). This concludes the proof.

6 Decay Estimates for the Fractional Subdiffusive Equation

Let us consider the Cauchy-type problem of fractional order $\alpha \in (0, 1)$ with Riemann-Liouville derivative

$$\begin{cases} {}^{\mathrm{RL}}\mathrm{D}_{0+}^{\alpha} y - \lambda y = f(t), \quad t \geq 0, \\ J_{0+}^{1-\alpha} y(0) = y_{\alpha-1}, \end{cases} \tag{41}$$

and with Caputo derivative

$$\begin{cases} {}^{\mathrm{C}}\mathrm{D}_{0+}^{\alpha} y - \lambda y = f(t), \quad t \geq 0 \\ y(0) = y_0. \end{cases} \tag{42}$$

where $\lambda \in \mathbb{R}$. If f is sufficiently smooth, then the solutions to these problems are given (see Examples 4.1 and 4.9 in [11]), respectively, by:

$$y(t) = y_{\alpha-1}\, t^{\alpha-1}\, E_{\alpha,\alpha}(\lambda t^{\alpha}) + \int_0^t (t-s)^{\alpha-1}\, E_{\alpha,\alpha}(\lambda(t-s)^{\alpha}) f(s)\, ds, \tag{43}$$

$$y(t) = y_0\, E_{\alpha,1}(\lambda t^{\alpha}) + \int_0^t (t-s)^{\alpha-1}\, E_{\alpha,\alpha}(\lambda(t-s)^{\alpha}) f(s)\, ds. \tag{44}$$

Here $E_{\alpha,\beta}$ is Mittag-Leffler function [12], described by its analytic expression

$$E_{\alpha,\beta}(z) = \sum_{j=0}^{\infty} \frac{z^j}{\Gamma(\beta + j/\alpha)}.$$

In particular, if $f \in \mathscr{C}_{1-\alpha}$, that is, $t^{1-\alpha} f(t)$ is continuous, then the solution to (41) may be constructed in the space $\mathscr{C}^{\alpha}_{1-\alpha}$, that is, y and ${}^{\mathrm{RL}}\mathrm{D}^{\alpha}_{0+}y$ are both in $\mathscr{C}_{1-\alpha}$. On the other hand, if f is continuous, then the solution to (42) may be constructed in the space $\mathscr{C}^{\alpha,0}$, that is, y and ${}^{\mathrm{C}}\mathrm{D}^{\alpha}_{0+}y$ are both continuous.

For $\alpha \in (0, 1)$ and $|\arg z| \in (\alpha\pi, \pi]$, the Mittag-Leffler function $E_{\alpha,\beta}(z)$ may be represented by (see Theorem 1.1.2 in [13]):

$$E_{\alpha,\beta}(z) = \frac{1}{\alpha\pi} \int_0^\infty \frac{\tau \sin(\beta\pi) - z \sin(\pi(\beta-\alpha))}{g_\alpha(\tau, z)} \tau^{\frac{1-\beta}{\alpha}} e^{-\tau^{\frac{1}{\alpha}}} d\tau, \tag{45}$$

where

$$g_\alpha(\tau, z) = \tau^2 - 2\tau z \cos(\alpha\pi) + z^2 . \tag{46}$$

Let us define the convolution operator

$$K_{\alpha,\beta}(t, \cdot) *_{(x)} \varphi(x) = t^{\alpha-1} \mathscr{F}^{-1}\big(E_{\alpha,\beta}(-t^\alpha|\xi|^2)\hat{\varphi}(\xi)\big), \tag{47}$$

where $\mathscr{F}$ denotes the Fourier transform with respect to x, and $\hat{\varphi} = \mathscr{F}\varphi$. Then, as a consequence of (43) and (44), we derive that u is a solution to (28) or, respectively, to (25), if, and only if,

$$\begin{aligned} u(t,x) &= t^{\alpha-1} K_{\alpha,\alpha}(t, \cdot) *_{(x)} u_{\alpha-1}(x) \\ &\quad + \int_0^t (t-s)^{\alpha-1} K_{\alpha,\alpha}(t-s, \cdot) *_{(x)} |u(s,\cdot)|^p(x) ds, \end{aligned} \tag{48}$$

or, respectively,

$$\begin{aligned} u(t,x) &= K_{\alpha,1}(t, \cdot) *_{(x)} u_0(x) \\ &\quad + \int_0^t (t-s)^{\alpha-1} K_{\alpha,\alpha}(t-s, \cdot) *_{(x)} |u(s,\cdot)|^p(x) ds, \end{aligned} \tag{49}$$

for any $t \in (0, T)$, in a suitable space.

In order to prove Theorems 4 and 3, we will rely on the following.

Lemma 3 *Let $\alpha \in (0, 1)$. Then there exists $\delta \in (0, 1)$ such that we have the following pointwise estimate:*

$$|K_{\alpha,\alpha}(1,x)| \leq C \begin{cases} (1+|x|)^{-n-\delta} & \text{if } n \leq 3, \\ |x|^{-\delta}(1+|x|)^{-4} & \text{if } n = 4, \\ |x|^{-(n-4)}(1+|x|)^{-4-\delta} & \text{if } n \geq 5, \end{cases} \tag{50}$$

$$|K_{\alpha,1}(1,x)| \leq C \begin{cases} (1+|x|)^{-1} & \text{if } n=1, \\ |x|^{-\delta}(1+|x|)^{-2+\delta} & \text{if } n=2, \\ |x|^{-(n-2)}(1+|x|)^{-2} & \text{if } n\geq 3. \end{cases} \tag{51}$$

As a consequence of Lemma 3 and of the scale invariance of $K_{\alpha,\beta}$, we immediately derive the following.

Corollary 1 *Let $\alpha \in (0,1)$ and $t>0$. Then $K_{\alpha,\alpha}(t,\cdot) \in L^1 \cap L^\infty$ if $n \leq 3$ and*

$$K_{\alpha,\alpha}(t,\cdot) \in L^1 \cap L^p, \qquad \forall 1 \leq p < 1 + \frac{4}{n-4}, \tag{52}$$

if $n \geq 4$. On the other hand, $K_{\alpha,1}(t,\cdot) \in L^r \cap L^\infty$, for any $r>1$, if $n=1$ and

$$K_{\alpha,1}(t,\cdot) \in L^r \cap L^p, \qquad \forall 1 < r < p < 1 + \frac{2}{n-2}, \tag{53}$$

if $n \geq 2$. Moreover,

$$\|K_{\alpha,\beta}(t,\cdot)\|_{L^q} = t^{-\frac{n\alpha}{2}\left(1-\frac{1}{q}\right)} \|K_{\alpha,\beta}(1,\cdot)\|_{L^q},$$

for any $t>0$ and admissible q.

By Young inequality, we immediately obtain the following result.

Corollary 2 *Let $K_{\alpha,\beta}$ as in (47), with $\beta = \alpha$, 1, and $1 \leq p \leq q \leq \infty$. Assume that*

$$n\left(\frac{1}{p}-\frac{1}{q}\right) < 4,$$

if $\beta = \alpha$, or

$$0 < n\left(\frac{1}{p}-\frac{1}{q}\right) < 2,$$

if $\beta = 1$. Then

$$\|K_{\alpha,\beta}(t,\cdot) *_{(x)} \varphi\|_{L^q} \leq C\, t^{-\frac{n\alpha}{2}\left(\frac{1}{p}-\frac{1}{q}\right)} \|\varphi\|_{L^p}, \tag{54}$$

where $C>0$ is independent of φ.

Estimate (54) has been proved in [1, Lemma 1] for $\beta = 1$, without relying on the representation of Mittag-Leffler functions in (45).

In Corollary 1, we cannot prove that $K_{\alpha,\alpha}$ and $K_{\alpha,1}$ are in the endpoint spaces, as a consequence of the estimates in (50) and (51). That is, we cannot prove that $K_{\alpha,\alpha}(t,\cdot) \in L^{\frac{n}{n-4}}$, when $n \geq 4$, that $K_{\alpha,1}(t,\cdot) \in L^{\frac{n}{n-2}}$, when $n \geq 2$, and that $K_{\alpha,1}(t,\cdot) \in L^1$, for any $n \geq 1$.

However, by Hardy-Littlewood-Sobolev theorem, we may extend Corollary 2 to cover the endpoints, if we restrict to $1 < p \leq q < \infty$.

Corollary 3 *Let $K_{\alpha,\beta}$ as in (47), with $\beta = \alpha$, 1, and $1 < p \leq q < \infty$. Assume that*

$$n\left(\frac{1}{p} - \frac{1}{q}\right) \leq 4,$$

if $\beta = \alpha$, or

$$n\left(\frac{1}{p} - \frac{1}{q}\right) \leq 2,$$

if $\beta = 1$. Then (54) holds, where $C > 0$ is independent of φ.

6.1 *Proof of Lemma* 3

We are now ready to prove Lemma 3.

Proof (Lemma 3*)* Thanks to the integral representation (45), we may write

$$K_{\alpha,\alpha}(1,x) = \frac{\sin(\alpha\pi)}{\alpha(2\pi)^n\pi} \int_{\mathbb{R}^n} e^{ix\xi} \int_0^\infty \frac{\tau^{\frac{1}{\alpha}}}{g_\alpha(\tau,-|\xi|^2)}\, e^{-\tau^{\frac{1}{\alpha}}}\, d\tau\, d\xi$$

$$K_{\alpha,1}(1,x) = \frac{\sin(\pi(1-\alpha))}{\alpha(2\pi)^n\pi} \int_{\mathbb{R}^n} e^{ix\xi} \int_0^\infty \frac{|\xi|^2}{g_\alpha(\tau,-|\xi|^2)}\, e^{-\tau^{\frac{1}{\alpha}}}\, d\tau\, d\xi$$

As a consequence of

$$g_\alpha(\tau,-|\xi|^2) = \tau^2 + 2\tau|\xi|^2\cos(\alpha\pi) + |\xi|^4 \geq c(\tau^2 + |\xi|^4), \tag{55}$$

where $c = 1 - \cos(\alpha\pi)$, we immediately derive that

$$|K_{\alpha,\alpha}(1,x)| \lesssim \int_{\mathbb{R}^n} \int_0^\infty \frac{\tau^{\frac{1}{\alpha}}}{\tau^2 + |\xi|^4}\, e^{-\tau^{\frac{1}{\alpha}}}\, d\tau\, d\xi.$$

In particular, $K_{\alpha,\alpha} \in L^\infty$ if $n = 1, 2, 3$. Indeed, it is sufficient to estimate

$$\int_0^\infty \frac{\tau^{\frac{1}{\alpha}}}{\tau^2 + |\xi|^4}\, e^{-\tau^{\frac{1}{\alpha}}}\, d\tau \leq \int_0^\infty \tau^{\frac{1}{\alpha}-2}\, e^{-\tau^{\frac{1}{\alpha}}}\, d\tau < \infty,$$

for $|\xi| \leq 1$, and

$$\int_0^\infty \frac{\tau^{\frac{1}{\alpha}}}{\tau^2 + |\xi|^4}\, e^{-\tau^{\frac{1}{\alpha}}}\, d\tau \leq |\xi|^{-4} \int_0^\infty \tau^{\frac{1}{\alpha}}\, e^{-\tau^{\frac{1}{\alpha}}}\, d\tau = C|\xi|^{-4},$$

for $|\xi| \geq 1$. Similarly, we derive that

$$|K_{\alpha,1}(1,x)| \lesssim \int_{\mathbb{R}^n} \int_0^\infty \frac{|\xi|^2}{\tau^2 + |\xi|^4}\, e^{-\tau^{\frac{1}{\alpha}}}\, d\tau\, d\xi,$$

in particular, $K_{\alpha,1} \in L^\infty$ if $n = 1$. Indeed, it is sufficient to use the change of variable $\tau = |\xi|^2\sigma$ to estimate

$$\begin{aligned}\int_0^\infty \frac{|\xi|^2}{\tau^2 + |\xi|^4}\, e^{-\tau^{\frac{1}{\alpha}}}\, d\tau &\leq \int_0^\infty \frac{|\xi|^2}{\tau^2 + |\xi|^4}\, d\tau \\ &= \int_0^\infty \frac{1}{\sigma^2 + 1}\, d\sigma = \frac{\pi}{2},\end{aligned}$$

for $|\xi| \leq 1$, and

$$\int_0^\infty \frac{|\xi|^2}{\tau^2 + |\xi|^4}\, e^{-\tau^{\frac{1}{\alpha}}}\, d\tau \leq |\xi|^{-2} \int_0^\infty e^{-\tau^{\frac{1}{\alpha}}}\, d\tau = C|\xi|^{-2},$$

for $|\xi| \geq 1$.

It is easy to check that

$$\Big|\partial_\xi^\gamma \Big(\frac{1}{g_\alpha(\tau, -|\xi|^2)}\Big)\Big| \lesssim |\xi|^{-|\gamma|}(\tau + |\xi|^2)^{-2}, \quad \forall \gamma \in \mathbb{N}^n, \tag{56}$$

thanks to (55). If $\gamma \neq 0$, we may refine estimate (56) to

$$\Big|\partial_\xi^\gamma \Big(\frac{1}{g_\alpha(\tau, -|\xi|^2)}\Big)\Big| \lesssim |\xi|^{2-|\gamma|}(\tau + |\xi|^2)^{-3}, \quad \forall |\gamma| \geq 1. \tag{57}$$

More in general, if $2k + 1 \leq |\gamma|$ for some $k \in \mathbb{N}$, then we may refine estimate (57) to

$$\Big|\partial_\xi^\gamma \Big(\frac{1}{g_\alpha(\tau, -|\xi|^2)}\Big)\Big| \lesssim |\xi|^{2(k+1)-|\gamma|}(\tau + |\xi|^2)^{-(k+3)},$$

but estimates (56) and (57) will be sufficient for us. Indeed,

$$\begin{aligned}\partial_{\xi_j}\Big(\frac{1}{g_\alpha(\tau,-|\xi|^2)}\Big) &= -\frac{2\xi_j(\cos(\pi\alpha)\tau+2|\xi|^2)}{(g_\alpha(\tau,-|\xi|^2))^2},\\ \partial_{\xi_j}\partial_{\xi_k}\Big(\frac{1}{g_\alpha(\tau,-|\xi|^2)}\Big) &= -\frac{2\delta_j^k(\cos(\pi\alpha)\tau+2|\xi|^2)+4\xi_j\xi_k}{(g_\alpha(\tau,-|\xi|^2))^2}\\ &\quad+2\,\frac{4\xi_j\xi_k(\cos(\pi\alpha)\tau+2|\xi|^2)^2}{(g_\alpha(\tau,-|\xi|^2))^3},\end{aligned}$$

and so on. To derive the desired estimate, it is sufficient to notice that

$$\begin{aligned}|\cos(\pi\alpha)|\tau+2|\xi|^2 &\lesssim \tau+|\xi|^2,\\ (g_\alpha(\tau,-|\xi|^2))^{-1} &\lesssim (\tau+|\xi|^2)^{-2},\end{aligned}$$

as a consequence of (55). As a consequence of (56) and (55), we derive

$$\Big|\partial_\xi^\gamma\Big(\frac{|\xi|^2}{g_\alpha(\tau,-|\xi|^2)}\Big)\Big| \lesssim |\xi|^{2-|\gamma|}(\tau+|\xi|^2)^{-2},\quad \forall\gamma\in\mathbb{N}^n, \tag{58}$$

for any γ. Moreover, if $1+2k\leq|\gamma|$ for some $k\in\mathbb{N}$, then we may estimate

$$\Big|\partial_\xi^\gamma\Big(\frac{|\xi|^2}{g_\alpha(\tau,-|\xi|^2)}\Big)\Big| \lesssim |\xi|^{2(k+1)-|\gamma|}(\tau+|\xi|^2)^{-(k+2)},$$

but estimate (58) will be sufficient for us.

Thanks to the identity

$$e^{ix\xi} = \sum_{j=1}^n \frac{x_j}{i|x|^2}\frac{\partial}{\partial\xi_j}e^{ix\cdot\xi}, \tag{59}$$

for any $m\in\mathbb{N}$, we may integrate by parts m times $K_{\alpha,\alpha}$ and $K_{\alpha,1}$, obtaining:

$$\begin{aligned}|K_{\alpha,\alpha}(1,x)| &\lesssim |x|^{-m}\sum_{|\gamma|=m}\Big|\int_{\mathbb{R}^n}e^{ix\xi}\int_0^\infty\Big(\partial_\xi^\gamma\frac{1}{g_\alpha(\tau,-|\xi|^2)}\Big)\,\tau^{\frac{1}{\alpha}}\,e^{-\tau^{\frac{1}{\alpha}}}\,d\tau\,d\xi\Big|\\ |K_{\alpha,1}(1,x)| &\lesssim |x|^{-m}\sum_{|\gamma|=m}\Big|\int_{\mathbb{R}^n}e^{ix\xi}\int_0^\infty\Big(\partial_\xi^\gamma\frac{|\xi|^2}{g_\alpha(\tau,-|\xi|^2)}\Big)\,e^{-\tau^{\frac{1}{\alpha}}}\,d\tau\,d\xi\Big|\end{aligned}$$

Let us consider first $K_{\alpha,\alpha}$. For any fixed γ with $|\gamma|=m$, we split the integral into two parts, $\{|\xi|\leq|x|^{-1}\}$ and $\{|\xi|\geq|x|^{-1}\}$, and we perform one additional step of integration by parts in this latter:

$$
\begin{aligned}
&\int_{\mathbb{R}^n} e^{ix\xi} \int_0^\infty \Big(\partial_\xi^\gamma \frac{1}{g_\alpha(\tau, -|\xi|^2)}\Big)\, \tau^{\frac{1}{\alpha}}\, e^{-\tau^{\frac{1}{\alpha}}}\, d\tau\, d\xi \\
&\quad = \int_{|\xi|\leq |x|^{-1}} e^{ix\xi} \int_0^\infty \Big(\partial_\xi^\gamma \frac{1}{g_\alpha(\tau, -|\xi|^2)}\Big)\, \tau^{\frac{1}{\alpha}}\, e^{-\tau^{\frac{1}{\alpha}}}\, d\tau\, d\xi \\
&\qquad + \int_{|\xi|\geq |x|^{-1}} e^{ix\xi} \int_0^\infty \Big(\partial_\xi^\gamma \frac{1}{g_\alpha(\tau, -|\xi|^2)}\Big)\, \tau^{\frac{1}{\alpha}}\, e^{-\tau^{\frac{1}{\alpha}}}\, d\tau\, d\xi \\
&\quad = I_0 + I_1 + I_\infty,
\end{aligned}
$$

where

$$
\begin{aligned}
I_0 &= \int_{|\xi|\leq |x|^{-1}} e^{ix\xi} \int_0^\infty \Big(\partial_\xi^\gamma \frac{1}{g_\alpha(\tau, -|\xi|^2)}\Big)\, \tau^{\frac{1}{\alpha}}\, e^{-\tau^{\frac{1}{\alpha}}}\, d\tau\, d\xi, \\
I_1 &= \sum_{j=1}^n \frac{x_j}{i|x|^2} \int_{|\xi|= |x|^{-1}} e^{ix\xi} \int_0^\infty \Big(\partial_\xi^\gamma \frac{1}{g_\alpha(\tau, -|\xi|^2)}\Big)\, \tau^{\frac{1}{\alpha}}\, e^{-\tau^{\frac{1}{\alpha}}}\, d\tau\, dS, \\
I_\infty &= -\sum_{j=1}^n \frac{x_j}{i|x|^2} \int_{|\xi|\geq |x|^{-1}} e^{ix\xi} \int_0^\infty \Big(\partial_{\xi_j}\partial_\xi^\gamma \frac{1}{g_\alpha(\tau, -|\xi|^2)}\Big)\, \tau^{\frac{1}{\alpha}}\, e^{-\tau^{\frac{1}{\alpha}}}\, d\tau\, d\xi.
\end{aligned}
$$

Moreover, we may perform one additional step of integration by parts in I_∞, that is, $I_\infty = J_2 + J_\infty$, where

$$
\begin{aligned}
J_2 &= \sum_{j,k=1}^n \frac{x_j x_k}{|x|^4} \int_{|\xi|= |x|^{-1}} e^{ix\xi} \int_0^\infty \Big(\partial_{\xi_j}\partial_\xi^\gamma \frac{1}{g_\alpha(\tau, -|\xi|^2)}\Big)\, \tau^{\frac{1}{\alpha}}\, e^{-\tau^{\frac{1}{\alpha}}}\, d\tau\, dS, \\
J_\infty &= -\sum_{j,k=1}^n \frac{x_j x_k}{|x|^4} \int_{|\xi|\geq |x|^{-1}} e^{ix\xi} \int_0^\infty \Big(\partial_{\xi_k}\partial_{\xi_j}\partial_\xi^\gamma \frac{1}{g_\alpha(\tau, -|\xi|^2)}\Big)\, \tau^{\frac{1}{\alpha}}\, e^{-\tau^{\frac{1}{\alpha}}}\, d\tau\, d\xi.
\end{aligned}
$$

We claim that we may estimate

$$
|K_{\alpha,1}(1,x)| \lesssim |x|^{-(n-4)}, \qquad \forall |x|\leq 1,\ n\geq 5 \tag{60}
$$

and, for any $\delta\in(0,1)$, we may estimate

$$
|K_{\alpha,1}(1,x)| \lesssim |x|^{-\delta}, \qquad \forall |x|\leq 1,\ \text{if } n=4. \tag{61}
$$

By the fact that $K_{\alpha,\alpha}\in L^\infty$ if $n=1,2,3$, and by claims (60) and (61), we conclude estimate (50) for $|x|\leq 1$.

To prove claim (60), we fix $m = n-5$. Thanks to (56) with $|\gamma| = n-5$, we may use

$$\left|\partial_\xi^\gamma\Big(\frac{1}{g_\alpha(\tau, -|\xi|^2)}\Big)\right| \lesssim |\xi|^{-(n-5)}(\tau + |\xi|^2)^{-2} \leq |\xi|^{-n+1},$$

to estimate

$$|I_0| \lesssim \int_{|\xi|\leq|x|^{-1}} |\xi|^{-n+1} \int_0^\infty \tau^{\frac{1}{\alpha}}\, e^{-\tau^{\frac{1}{\alpha}}}\, d\tau\, d\xi \lesssim |x|^{-1},$$

and

$$|I_1| \lesssim |x|^{-1} \int_{|\xi|=|x|^{-1}} |\xi|^{-n+1} \int_0^\infty \tau^{\frac{1}{\alpha}}\, e^{-\tau^{\frac{1}{\alpha}}}\, d\tau\, dS = C|x|^{-1}.$$

Similarly, setting $\gamma_1 = \gamma + e_j$, so that $\partial_{\xi_j}\partial_\xi^\gamma = \partial_\xi^{\gamma_1}$, thanks to (56) with $|\gamma_1| = n-4$, we may use

$$\left|\partial_\xi^{\gamma_1}\Big(\frac{1}{g_\alpha(\tau, -|\xi|^2)}\Big)\right| \lesssim |\xi|^{-(n-4)}(\tau + |\xi|^2)^{-2} \leq |\xi|^{-n},$$

to estimate

$$|J_2| \lesssim |x|^{-2} \int_{|\xi|=|x|^{-1}} |\xi|^{-n} \int_0^\infty \tau^{\frac{1}{\alpha}}\, e^{-\tau^{\frac{1}{\alpha}}}\, d\tau\, dS = C|x|^{-1}.$$

Finally, setting $\gamma_2 = \gamma + e_j + e_k$, so that $\partial_{\xi_k}\partial_{\xi_j}\partial_\xi^\gamma = \partial_\xi^{\gamma_2}$, thanks to (56) with $|\gamma_2| = n-3$, we may use

$$\left|\partial_\xi^{\gamma_2}\Big(\frac{1}{g_\alpha(\tau, -|\xi|^2)}\Big)\right| \lesssim |\xi|^{-(n-3)}(\tau + |\xi|^2)^{-2} \leq |\xi|^{-n-1},$$

to estimate

$$|J_\infty| \lesssim |x|^{-2} \int_{|\xi|\geq|x|^{-1}} |\xi|^{-n-1} \int_0^\infty \tau^{\frac{1}{\alpha}}\, e^{-\tau^{\frac{1}{\alpha}}}\, d\tau\, dS \lesssim |x|^{-1}.$$

This proves (60). In space dimension $n = 4$, we shall fix $\gamma = 0$, since we cannot take $|\gamma| = n-5 = -1$. However, for any small $\delta \in (0,1)$, thanks to (55), we may use

$$\frac{1}{g_\alpha(\tau, -|\xi|^2)} \lesssim (\tau + |\xi|^2)^{-2} \lesssim \tau^{-\frac{\delta}{2}}\, |\xi|^{-4+\delta},$$

to estimate

$$|I_0| \lesssim \int_{|\xi|\le|x|^{-1}} |\xi|^{-4+\delta} \int_0^\infty \tau^{-\frac{\delta}{2}+\frac{1}{\alpha}}\, e^{-\tau^{\frac{1}{\alpha}}}\, d\tau\, d\xi \lesssim |x|^{-\delta},$$

and

$$|I_1| \lesssim |x|^{-1} \int_{|\xi|=|x|^{-1}} |\xi|^{-4+\delta} \int_0^\infty \tau^{-\frac{\delta}{2}+\frac{1}{\alpha}}\, e^{-\tau^{\frac{1}{\alpha}}}\, d\tau\, dS = C|x|^{-\delta}.$$

On the other hand, setting $\gamma_1 = e_j$, so that $\partial_{\xi_j} = \partial_\xi^{\gamma_1}$, thanks to (56) with $|\gamma_1| = 1$, we may use

$$\left|\partial_{\xi_j}\Big(\frac{1}{g_\alpha(\tau, -|\xi|^2)}\Big)\right| \lesssim |\xi|^{-1}(\tau + |\xi|^2)^{-2} \le |\xi|^{-5},$$

to estimate

$$|I_\infty| \lesssim |x|^{-1} \int_{|\xi|\ge|x|^{-1}} |\xi|^{-5} \int_0^\infty \tau^{\frac{1}{\alpha}}\, e^{-\tau^{\frac{1}{\alpha}}}\, d\tau\, dS \le C.$$

This proves (61).

We now claim that, for a sufficiently small $\delta \in (0, 1)$, we may estimate

$$|K_{\alpha,1}(1, x)| \lesssim |x|^{-n-\delta}, \qquad \forall |x| \ge 1,\ n \ge 1. \tag{62}$$

By claim (62), we conclude estimate (50) for $|x| \ge 1$.

To prove our claim, we fix $m = n$. By (57) with $|\gamma| = n$, we use

$$\left|\partial_\xi^\gamma\Big(\frac{1}{g_\alpha(\tau, -|\xi|^2)}\Big)\right| \lesssim |\xi|^{2-n}(\tau + |\xi|^2)^{-3} \lesssim \tau^{-2-\frac{\delta}{2}}\, |\xi|^{-n+\delta},$$

where we assume $\delta/2 < 1/\alpha - 1$, to estimate

$$|I_0| \lesssim \int_{|\xi|\le|x|^{-1}} |\xi|^{-n+\delta} \int_0^\infty \tau^{\frac{1}{\alpha}-2-\frac{\delta}{2}}\, e^{-\tau^{\frac{1}{\alpha}}}\, d\tau\, d\xi \lesssim |x|^{-\delta}.$$

and

$$|I_1| \lesssim |x|^{-1} \int_{|\xi|=|x|^{-1}} |\xi|^{-n+\delta} \int_0^\infty \tau^{\frac{1}{\alpha}-2-\frac{\delta}{2}}\, e^{-\tau^{\frac{1}{\alpha}}}\, d\tau\, dS = C|x|^{-\delta}.$$

On the other hand, setting $\gamma_1 = \gamma + e_j$, so that $\partial_{\xi_j}\partial_\xi^\gamma = \partial_\xi^{\gamma_1}$, by (57) with $|\gamma_1| = n+1$, we use

$$\Big|\partial_\xi^{\gamma_1}\Big(\frac{1}{g_\alpha(\tau,-|\xi|^2)}\Big)\Big| \lesssim |\xi|^{2-(n+1)}(\tau+|\xi|^2)^{-3} \lesssim \tau^{-2-\frac{\delta}{2}}|\xi|^{-n-1+\delta},$$

to estimate

$$|I_\infty| \lesssim |x|^{-1}\int_{|\xi|\geq|x|^{-1}} |\xi|^{-n-1+\delta}\int_0^\infty \tau^{\frac{1}{\alpha}-2-\frac{\delta}{2}}\,e^{-\tau^{\frac{1}{\alpha}}}\,d\tau\,dS \lesssim |x|^{-\delta}.$$

Summarizing, we proved estimate (50) for $K_{\alpha,\alpha}$.

We now consider $K_{\alpha,1}$. For any fixed γ with $|\gamma| = m$, we split the integral into two parts, $\{|\xi|\leq|x|^{-1}\}$ and $\{|\xi|\geq|x|^{-1}\}$, and we perform one additional step of integration by parts in this latter, so that

$$\int_{\mathbb{R}^n} e^{ix\xi}\int_0^\infty \Big(\partial_\xi^\gamma\frac{|\xi|^2}{g_\alpha(\tau,-|\xi|^2)}\Big)e^{-\tau^{\frac{1}{\alpha}}}\,d\tau\,d\xi = I_0+I_1+I_\infty,$$

where

$$I_0 = \int_{|\xi|\leq|x|^{-1}} e^{ix\xi}\int_0^\infty \Big(\partial_\xi^\gamma\frac{|\xi|^2}{g_\alpha(\tau,-|\xi|^2)}\Big)e^{-\tau^{\frac{1}{\alpha}}}\,d\tau\,d\xi,$$

$$I_1 = \sum_{j=1}^n \frac{x_j}{i|x|^2}\int_{|\xi|=|x|^{-1}} e^{ix\xi}\int_0^\infty \Big(\partial_\xi^\gamma\frac{|\xi|^2}{g_\alpha(\tau,-|\xi|^2)}\Big)e^{-\tau^{\frac{1}{\alpha}}}\,d\tau\,dS,$$

$$I_\infty = \sum_{j=1}^n \frac{x_j}{i|x|^2}\int_{|\xi|\geq|x|^{-1}} e^{ix\xi}\int_0^\infty \Big(\partial_{\xi_j}\partial_\xi^\gamma\frac{|\xi|^2}{g_\alpha(\tau,-|\xi|^2)}\Big)e^{-\tau^{\frac{1}{\alpha}}}\,d\tau\,d\xi.$$

Moreover, we may perform one additional step of integration by parts in I_∞, that is, $I_\infty = J_2 + J_\infty$, where

$$J_2 = \sum_{j,k=1}^n \frac{x_jx_k}{|x|^4}\int_{|\xi|=|x|^{-1}} e^{ix\xi}\int_0^\infty \Big(\partial_{\xi_j}\partial_\xi^\gamma\frac{|\xi|^2}{g_\alpha(\tau,-|\xi|^2)}\Big)e^{-\tau^{\frac{1}{\alpha}}}\,d\tau\,dS,$$

$$J_\infty = -\sum_{j,k=1}^n \frac{x_jx_k}{|x|^4}\int_{|\xi|\geq|x|^{-1}} e^{ix\xi}\int_0^\infty \Big(\partial_{\xi_k}\partial_{\xi_j}\partial_\xi^\gamma\frac{|\xi|^2}{g_\alpha(\tau,-|\xi|^2)}\Big)e^{-\tau^{\frac{1}{\alpha}}}\,d\tau\,d\xi.$$

We claim that we may estimate

$$|K_{\alpha,1}(1,x)| \lesssim |x|^{-(n-2)}, \qquad \forall|x|\leq 1,\ n\geq 3, \tag{63}$$

and that, for any small $\delta \in (0, 1)$, we may estimate

$$|K_{\alpha,1}(1,x)| \lesssim |x|^{-\delta}, \qquad \forall |x| \leq 1, \text{ if } n = 2. \tag{64}$$

By the fact that $K_{\alpha,1} \in L^\infty$ if $n = 1$, and by claims (63) and (64), we conclude estimate (51) for $|x| \leq 1$.

To prove claim (63), we fix $m = n - 3$. By (58) with $|\gamma| = n - 3$, we use

$$\left|\partial_\xi^\gamma \Big(\frac{|\xi|^2}{g_\alpha(\tau, -|\xi|^2)}\Big)\right| \lesssim |\xi|^{2-(n-3)}(\tau + |\xi|^2)^{-2} \lesssim |\xi|^{-n+1},$$

to estimate

$$|I_0| \lesssim \int_{|\xi|\leq|x|^{-1}} |\xi|^{-n+1} \int_0^\infty e^{-\tau^{\frac{1}{\alpha}}}\, d\tau\, d\xi \lesssim |x|^{-1}.$$

and

$$|I_1| \lesssim |x|^{-1} \int_{|\xi|=|x|^{-1}} |\xi|^{-n+1} \int_0^\infty e^{-\tau^{\frac{1}{\alpha}}}\, d\tau\, dS = C|x|^{-1}.$$

Similarly, setting $\gamma_1 = \gamma + e_j$, so that $\partial_{\xi_j}\partial_\xi^\gamma = \partial_\xi^{\gamma_1}$, thanks to (58) with $|\gamma_1| = n-2$, we may use

$$\left|\partial_\xi^{\gamma_1} \Big(\frac{1}{g_\alpha(\tau, -|\xi|^2)}\Big)\right| \lesssim |\xi|^{2-(n-2)}(\tau + |\xi|^2)^{-2} \leq |\xi|^{-n},$$

to estimate

$$|J_2| \lesssim |x|^{-2} \int_{|\xi|=|x|^{-1}} |\xi|^{-n} \int_0^\infty e^{-\tau^{\frac{1}{\alpha}}}\, d\tau\, dS = C|x|^{-1}.$$

Finally, setting $\gamma_2 = \gamma + e_j + e_k$, so that $\partial_{\xi_k}\partial_{\xi_j}\partial_\xi^\gamma = \partial_\xi^{\gamma_2}$, thanks to (58) with $|\gamma_2| = n - 1$, we may use

$$\left|\partial_\xi^{\gamma_2} \Big(\frac{1}{g_\alpha(\tau, -|\xi|^2)}\Big)\right| \lesssim |\xi|^{2-(n-1)}(\tau + |\xi|^2)^{-2} \leq |\xi|^{-n+1},$$

to estimate

$$|J_\infty| \lesssim |x|^{-2} \int_{|\xi|\geq|x|^{-1}} |\xi|^{-n+1} \int_0^\infty e^{-\tau^{\frac{1}{\alpha}}}\, d\tau\, dS \lesssim |x|^{-1}.$$

This proves (63). In space dimension $n = 2$, we shall fix $\gamma = 0$, since we cannot take $|\gamma| = n - 3 = -1$. However, for any small $\delta \in (0, 1)$, thanks to (55), we may use

$$\frac{|\xi|^2}{g_\alpha(\tau,-|\xi|^2)} \lesssim |\xi|^2(\tau+|\xi|^2)^{-2} \lesssim \tau^{-\frac{\delta}{2}}\,|\xi|^{-2+\delta},$$

to estimate

$$|I_0| \lesssim \int_{|\xi|\leq|x|^{-1}} |\xi|^{-2+\delta} \int_0^\infty \tau^{-\frac{\delta}{2}}\,e^{-\tau^{\frac{1}{\alpha}}}\,d\tau\,d\xi \lesssim |x|^{-\delta},$$

and

$$|I_1| \lesssim |x|^{-1} \int_{|\xi|=|x|^{-1}} |\xi|^{-2+\delta} \int_0^\infty \tau^{-\frac{\delta}{2}}\,e^{-\tau^{\frac{1}{\alpha}}}\,d\tau\,dS = C|x|^{-\delta}.$$

On the other hand, setting $\gamma_1 = e_j$, so that $\partial_{\xi_j} = \partial_\xi^{\gamma_1}$, thanks to (58) with $|\gamma_1| = 1$, we may use

$$\left|\partial_{\xi_j}\Big(\frac{|\xi|^2}{g_\alpha(\tau,-|\xi|^2)}\Big)\right| \lesssim |\xi|(\tau+|\xi|^2)^{-2} \leq |\xi|^{-3},$$

to estimate

$$|I_\infty| \lesssim |x|^{-1} \int_{|\xi|\geq|x|^{-1}} |\xi|^{-3} \int_0^\infty e^{-\tau^{\frac{1}{\alpha}}}\,d\tau\,dS \leq C.$$

This proves (64).

We now claim that we may estimate

$$|K_{\alpha,1}(1,x)| \lesssim |x|^{-n}, \qquad \forall |x| \geq 1. \tag{65}$$

By claim (65), we conclude estimate (51) for $|x| \geq 1$.

To prove our claim, we fix $m = n-1$. By (58) with $|\gamma| = n-1$, we get

$$\left|\partial_\xi^\gamma\Big(\frac{|\xi|^2}{g_\alpha(\tau,-|\xi|^2)}\Big)\right| \lesssim |\xi|^{2-(n-1)}(\tau+|\xi|^2)^{-2}.$$

We now use the change of variable $\tau = |\xi|^2\sigma$ to estimate

$$\begin{aligned}|I_0| &\lesssim \int_{|\xi|\leq|x|^{-1}} |\xi|^{-(n-1)} \int_0^\infty \frac{|\xi|^2}{(\tau+|\xi|^2)^2}\,d\tau\,d\xi \\ &= \int_{|\xi|\leq|x|^{-1}} |\xi|^{-(n-1)} \int_0^\infty \frac{1}{(\sigma+1)^2}\,d\sigma\,d\xi \lesssim |x|^{-1},\end{aligned}$$

and

$$|I_1| \lesssim |x|^{-1} \int_{|\xi|=|x|^{-1}} |\xi|^{-(n-1)} \int_0^\infty \frac{1}{(\sigma+1)^2} \, d\sigma \, dS = C|x|^{-1}.$$

Similarly, setting $\gamma_1 = \gamma + e_j$, so that $\partial_{\xi_j} \partial_\xi^\gamma = \partial_\xi^{\gamma_1}$, thanks to (58) with $|\gamma_1| = n$, we may use

$$\left| \partial_\xi^{\gamma_1} \Big(\frac{1}{g_\alpha(\tau, -|\xi|^2)} \Big) \right| \lesssim |\xi|^{2-n} (\tau + |\xi|^2)^{-2},$$

and the change of variable $\tau = |\xi|^2 \sigma$ to estimate

$$|J_2| \lesssim |x|^{-2} \int_{|\xi|=|x|^{-1}} |\xi|^{-n} \int_0^\infty \frac{1}{(\sigma+1)^2} \, d\sigma \, dS = C|x|^{-1}.$$

Finally, setting $\gamma_2 = \gamma + e_j + e_k$, so that $\partial_{\xi_k} \partial_{\xi_j} \partial_\xi^\gamma = \partial_\xi^{\gamma_2}$, thanks to (58) with $|\gamma_2| = n+1$, we may use

$$\left| \partial_\xi^{\gamma_2} \Big(\frac{1}{g_\alpha(\tau, -|\xi|^2)} \Big) \right| \lesssim |\xi|^{2-(n+1)} (\tau + |\xi|^2)^{-2},$$

and the change of variable $\tau = |\xi|^2 \sigma$ to estimate

$$|J_\infty| \lesssim |x|^{-2} \int_{|\xi| \geq |x|^{-1}} |\xi|^{-n+1} \int_0^\infty \frac{1}{(\sigma+1)^2} \, d\sigma \, dS \lesssim |x|^{-1}.$$

This proves (65).

Summarizing, we proved estimate (51) for $K_{\alpha,1}$. This concludes the proof.

6.2 Decay Estimates

Thanks to Corollary 2 and 3, we are in the position to prove the following.

Lemma 4 *Let* $\alpha \in (0, 1)$, $n \geq 1$ *and* $p \in (1, \infty]$, *with* $p < 1 + 2/(n-2)$ *if* $n \geq 2$. *Assume that* $u_0 \in L^1 \cap L^p$ *and* $f(t, \cdot) \in L^1$ *for any* $t > 0$, *and that*

$$\|f(t, \cdot)\|_{L^1} \leq A\,(1+t)^{-\eta},$$

for any $t \geq 0$, *for some* $A > 0$ *and* $\eta \in \mathbb{R}$. *Then the solution to*

$$\begin{cases} {}^{C}D_{0+}^{\alpha} u - \Delta u = f(t,x), & t > 0, \ x \in \mathbb{R}^n, \\ u(0,x) = u_0(x), \end{cases} \tag{66}$$

verifies the following estimate

$$\|u(t,\cdot)\|_{L^p} \leq C(1+t)^{-\frac{n\alpha}{2}\left(1-\frac{1}{p}\right)} \left(\|u_0\|_{L^1} + \|u_0\|_{L^p}\right)$$

$$+ CA\,(1+t)^{\alpha-1-\frac{n\alpha}{2}\left(1-\frac{1}{q}\right)} \times \begin{cases} 1 & \text{if } \eta > 1, \\ \log(e+t) & \text{if } \eta = 1, \\ (1+t)^{1-\eta} & \text{if } \eta < 1, \end{cases} \tag{67}$$

for any $t \geq 0$, for some $C > 0$, independent of u_0.

Lemma 4 allows us to use a contraction argument, which leads to the global existence of small data solutions to (25) stated in Theorem 3.

Proof The proof is a direct consequence of Corollary 2 and (48). On the one hand,

$$\|K_{\alpha,1}(t,\cdot) *_{(x)} u_0\|_{L^p} \lesssim \|u_0\|_{L^p},$$

for small $t \in (0,1]$. On the other hand, since $n(1-1/p) < 2$ as a consequence of $p < 1 + 2/(n-2)$ if $n \geq 2$, we get

$$\|K_{\alpha,1}(t,\cdot) *_{(x)} u_0\|_{L^p} \lesssim t^{-\frac{n\alpha}{2}\left(1-\frac{1}{p}\right)} \|u_0\|_{L^1},$$

for large $t \geq 1$. Moreover, we may estimate

$$\begin{aligned} &\int_0^t (t-s)^{\alpha-1} \|K_{\alpha,\alpha}(t-s,\cdot) *_{(x)} f(s,\cdot)\|_{L^p}\, ds \\ &\lesssim \int_0^t (t-s)^{\alpha\left(1-\frac{n}{2}\left(1-\frac{1}{p}\right)\right)-1} \|f(s,\cdot)\|_{L^1}\, ds \\ &\lesssim A \int_0^t (t-s)^{\alpha\left(1-\frac{n}{2}\left(1-\frac{1}{p}\right)\right)-1} (1+s)^{-\eta}\, ds. \end{aligned}$$

The proof follows by noticing that

$$\alpha - \frac{n\alpha}{2}\left(1-\frac{1}{p}\right) = \alpha\left(1 - \frac{n}{2}\left(1-\frac{1}{p}\right)\right) > 0,$$

thanks to $p < 1 + 2/(n-2)$ if $n \geq 2$, and by relying on the following well-known result:

$$\forall a \in (0,1): \quad \int_0^t (t-s)^{-a}(1+s)^{-b}\, ds \approx \begin{cases} (1+t)^{-a} & \text{if } b > 1, \\ (1+t)^{-a}\log(e+t) & \text{if } b = 1, \\ (1+t)^{1-a-b} & \text{if } b < 1. \end{cases} \tag{68}$$

The corresponding result for the problem with Riemann-Liouville fractional derivative has to take into account of the singularity $t^{\alpha-1}$ at $t = 0$.

Lemma 5 *Let $\alpha \in (0, 1)$, $n \geq 1$ and $p \in [1, \infty]$, with $p < 1 + 2/(n-2)$ if $n \geq 2$. Assume that $u_{\alpha-1} \in L^1 \cap L^p$ and $f(t, \cdot) \in L^1$ for any $t > 0$, and that*

$$\|f(t,\cdot)\|_{L^1} \leq A\, t^{-b}\, (1+t)^{-\eta},$$

for any $t > 0$, for some $A > 0$, $b \in [0, 1)$ and $\eta \in \mathbb{R}$. Then the solution to

$$\begin{cases} {}^{\mathrm{RL}}\mathrm{D}^{\alpha}_{0+} u - \Delta u = f(t,x), & t > 0,\ x \in \mathbb{R}^n, \\ J^{1-\alpha}_{0+} u(0,x) = u_{\alpha-1}(x), \end{cases} \tag{69}$$

verifies the following estimate

$$\begin{aligned} t^{1-\alpha}\|u(t,\cdot)\|_{L^p} &\leq C(1+t)^{-\frac{n\alpha}{2}\left(1-\frac{1}{p}\right)} \left(\|u_{\alpha-1}\|_{L^1} + \|u_{\alpha-1}\|_{L^p}\right) \\ &+ CA\, t^{1-b-\frac{n\alpha}{2}\left(1-\frac{1}{p}\right)} \times \begin{cases} (1+t)^{b-1} & \text{if } \eta > 1-b, \\ (1+t)^{-\eta}\log(e+t) & \text{if } \eta = 1-b, \\ (1+t)^{-\eta} & \text{if } \eta < 1-b, \end{cases} \end{aligned} \tag{70}$$

for any $t > 0$, for some $C > 0$, independent of $u_{\alpha-1}$.

Remark 9 For

$$b + \frac{n\alpha}{2}\left(1 - \frac{1}{p}\right) \leq 1, \tag{71}$$

if $p > 1$, or $b \in (0, 1)$ if $p = 1$, then the term $t^{1-b-\frac{n\alpha}{2}\left(1-\frac{1}{p}\right)}$ is not singular. In particular, if $\eta > 1 - b$, from (70) we derive:

$$t^{1-\alpha}\|u(t,\cdot)\|_{L^p} \leq C(1+t)^{-\frac{n\alpha}{2}\left(1-\frac{1}{p}\right)} \left(\|u_{\alpha-1}\|_{L^1} + \|u_{\alpha-1}\|_{L^p} + A\right) \tag{72}$$

Lemma 5 and, more precisely, Remark 9, allow us to use a contraction argument in the space $C_{1-\alpha}([0, \infty))$, which leads to the global existence of small data solutions to (28) stated in Theorem 4.

To prove Lemma 5, we need to modify the integral estimate given in (68).

Lemma 6 *Let $c \in \mathbb{R}$. Then we have the following:*

$$\forall a, b \in (0,1): \quad I_{a,b,c}(t) \approx t^{1-a-b} \times \begin{cases} (1+t)^{b-1} & \text{if } b+c>1, \\ (1+t)^{-c}\log(e+t) & \text{if } b+c=1, \\ (1+t)^{-c} & \text{if } b+c<1, \end{cases} \tag{73}$$

where we set

$$I_{a,b,c}(t) = \int_0^t (t-s)^{-a}\, s^{-b}\,(1+s)^{-c}\, ds.$$

Estimate (73) is the singular analogous of (68). Its proof is straight-forward, but we include it for the ease of reading.

Proof First of all, we notice that

$$\begin{aligned} I_{a,b,c}(t) &= \int_{t/2}^t (t-s)^{-a}\, s^{-b}\,(1+s)^{-c}\, ds + \int_0^{t/2} (t-s)^{-a}\, s^{-b}\,(1+s)^{-c}\, ds \\ &\approx t^{-b}\,(1+t)^{-c} \int_{t/2}^t (t-s)^{-a}\, ds + t^{-a} \int_0^{t/2} s^{-b}\,(1+s)^{-c}\, ds. \end{aligned}$$

It is immediate to obtain

$$t^{-b}\,(1+t)^{-c} \int_{t/2}^t (t-s)^{-a}\, ds \approx t^{1-a-b}\,(1+t)^{-c}.$$

In particular, this estimate is not worse than the estimate in (73).

When we consider the second integral, we distinguish two cases. If $t \geq 2$, then we may estimate

$$\begin{aligned} t^{-a} & \int_0^{t/2} s^{-b}\,(1+s)^{-c}\, ds \\ &= t^{-a} \int_0^1 s^{-b}\,(1+s)^{-c}\, ds + t^{-a} \int_1^{t/2} s^{-b}\,(1+s)^{-c}\, ds \\ &\approx t^{-a} \int_0^1 s^{-b}\, ds + t^{-a} \int_1^{t/2} (1+s)^{-b-c}\, ds. \end{aligned}$$

Due to $b < 1$, we obtain

$$t^{-a} \int_0^1 s^{-b}\, ds \approx t^{-a}.$$

On the other hand,

$$t^{-a}\int_1^{t/2}(1+s)^{-b-c}\,ds\approx t^{-a}\times\begin{cases}1 & \text{if } b+c>1,\\ \log(e+t) & \text{if } b+c=1,\\ (1+t)^{1-b-c} & \text{if } b+c<1.\end{cases}$$

This concludes the proof of (73), for any $t\geq 2$. Now let $t\in(0,2)$. Then we may estimate

$$t^{-a}\int_0^{t/2}s^{-b}\,(1+s)^{-c}\,ds\approx t^{-a}\int_0^{t/2}s^{-b}\,ds\approx t^{1-a-b}.$$

This concludes the proof of (73), for $t\in(0,2)$.

We may now immediately prove Lemma 5.

Proof (Lemma 5) The proof is a direct consequence of Corollary 2 and (48). On the one hand,

$$\|K_{\alpha,\alpha}(t,\cdot)*_{(x)}u_{\alpha-1}\|_{L^p}\lesssim\|u_{\alpha-1}\|_{L^p},$$

for small $t\in(0,1]$. On the other hand,

$$\|K_{\alpha,\alpha}(t,\cdot)*_{(x)}u_{\alpha-1}\|_{L^p}\lesssim t^{-\frac{n\alpha}{2}\left(1-\frac{1}{p}\right)}\|u_{\alpha-1}\|_{L^1},$$

for large $t\geq 1$. Moreover, we may estimate

$$\begin{aligned}&\int_0^t(t-s)^{\alpha-1}\|K_{\alpha,\alpha}(t-s,\cdot)*_{(x)}f(s,\cdot)\|_{L^p}\,ds\\ &\lesssim\int_0^t(t-s)^{\alpha\left(1-\frac{n}{2}\left(1-\frac{1}{p}\right)\right)-1}\|f(s,\cdot)\|_{L^1}\,ds\\ &\lesssim A\int_0^t(t-s)^{\alpha\left(1-\frac{n}{2}\left(1-\frac{1}{p}\right)\right)-1}s^{-b}(1+s)^{-\eta}\,ds.\end{aligned}$$

We notice, as in the proof of Lemma 4, that

$$\alpha-\frac{n\alpha}{2}\left(1-\frac{1}{p}\right)=\alpha\left(1-\frac{n}{2}\left(1-\frac{1}{p}\right)\right)>0,$$

as a consequence of $p<1+2/(n-2)$ if $n\geq 2$. Applying (73) with $c=\eta$ and

$$a=1-\alpha\left(1-\frac{n}{2}\left(1-\frac{1}{p}\right)\right),$$

we conclude the proof.

6.3 Proof of Theorems 3 and 4

We are now ready to prove Theorems 3 and 4.

To employ our contraction argument, for any $T > 0$ we define the Banach space $X_0 = \mathscr{C}([0,T], L^p)$ equipped with the norm

$$\|u\|_{X_0} = \sup_{t\geq 0} (1+t)^{\frac{n\alpha}{2}\left(1-\frac{1}{p}\right)} \|u(t,\cdot)\|_{L^p},$$

and the Banach space $X_{1-\alpha} = \mathscr{C}_{1-\alpha}([0,T], L^p)$ (see Theorem 3.1.1 in [11] for the contraction argument for ordinary fractional differential equations in the space $C_{1-\alpha}([0,T])$) equipped with the norm

$$\|u\|_{X_{1-\alpha}} = \sup_{t\geq 0} t^{1-\alpha} (1+t)^{\frac{n\alpha}{2}\left(1-\frac{1}{p}\right)} \|u(t,\cdot)\|_{L^p}.$$

A function u in X_0, is a solution to (25) if, and only if,

$$u(t,x) = \varphi_0(t,x) + Fu(t,x), \tag{74}$$

for any $t > 0$, a.e. in x, where

$$\begin{aligned} \varphi_0(t,x) &= K_{\alpha,1}(t,\cdot) *_{(x)} u_0(x), \\ Fu(t,x) &= \int_0^t (t-s)^{\alpha-1} K_{\alpha,\alpha}(t-s,\cdot) *_{(x)} |u(s,\cdot)|^p \, ds. \end{aligned}$$

Similarly, a function u in $X_{1-\alpha}$, is a solution to (28) if, and only if,

$$u(t,x) = \varphi_{1-\alpha}(t,x) + Fu(t,x), \tag{75}$$

for any $t > 0$, a.e. in x, where

$$\varphi_{1-\alpha}(t,x) = t^{\alpha-1} K_{\alpha,\alpha}(t,\cdot) *_{(x)} u_0(x).$$

As a consequence of Lemmas 4 and 5, we know that $\varphi_0 \in X_0$ and $\varphi_{1-\alpha} \in X_{1-\alpha}$, with

$$\|\varphi_0\|_{X_0} \leq C\,\|u_0\|_{\mathscr{A}}, \tag{76}$$

$$\|\varphi_{1-\alpha}\|_{X_{1-\alpha}} \leq C\,\|u_{\alpha-1}\|_{\mathscr{A}}, \tag{77}$$

where $C > 0$ is independent of $T > 0$. We also claim that

$$\|Fu\|_X \leq C\|u\|_X^p, \tag{78}$$

$$\|Fu - Fv\|_X \leq C\|u - v\|_X\big(\|u\|_X^{p-1} + \|v\|_X^{p-1}\big), \tag{79}$$

for any $u, v \in X$, where X stays for X_0 or $X_{1-\alpha}$, and $C > 0$ is independent of $T > 0$.

By standard fixed point arguments, (76), (77), (78) and (79) lead to the existence of a unique solution $u \in X_0$ to (74) or $u \in X_{1-\alpha}$ to (75), respectively, for sufficiently small data. Since the constants in (76), (77), (78) and (79) are independent of $T > 0$, the solution can be globally prolonged. By the definition of the norms of the spaces X_0 and $X_{1-\alpha}$, we derive decay estimates (26) and (29).

Therefore, to prove Theorems 3 and 4, it remains to prove (78) and (79).

Proof (Theorem 3) We prove (78). We apply (67) with $f(t,x) = |u(t,x)|^p$, so that

$$\|f(t,\cdot)\|_{L^1} = \|u(t,\cdot)\|_{L^p}^p \lesssim (1+t)^{-\frac{n\alpha}{2}(p-1)}\,\|u\|_{X_0}^p,$$

and we get:

$$\|Nu(t,\cdot)\|_{L^p} \lesssim A\,(1+t)^{\alpha-1-\frac{n\alpha}{2}\left(1-\frac{1}{p}\right)} \times \begin{cases} 1 & \text{if } n\alpha(p-1)/2 > 1, \\ \log(e+t) & \text{if } n\alpha(p-1)/2 > 1, \\ (1+t)^{1-\frac{n\alpha}{2}(p-1)} & \text{if } n\alpha(p-1)/2 < 1, \end{cases} \tag{80}$$

where $A = \|u\|_{X_0}^p$. Using $p \geq \bar{p}(n,\alpha) = 1 + 2/n$, that is, $n(p-1)/2 \geq 1$, we obtain

$$\|Nu(t,\cdot)\|_{L^p} \lesssim (1+t)^{-\frac{n\alpha}{2}\left(1-\frac{1}{p}\right)}\,\|u\|_{X_0}^p,$$

that is, we proved (78). Now we prove (79). We apply (67) with $f(t,x) = |u(t,x)|^p - |v(t,x)|^p$. Due to

$$|u|^p - |v|^p \lesssim |u-v|\big(|u|^{p-1} + |v|^{p-1}\big),$$

by Hölder inequality,

$$\begin{aligned} \|f(t,\cdot)\|_{L^1} &\leq \|(u-v)(t,\cdot)\|_{L^p}\,\||u(t,\cdot)|^{p-1} + |v(t,\cdot)|^{p-1}\|_{L^{\frac{p}{p-1}}} \\ &\lesssim \|(u-v)(t,\cdot)\|_{L^p}\left(\|u(t,\cdot)\|_{L^p}^{p-1} + \|v(t,\cdot)\|_{L^p}^{p-1}\right) \\ &\lesssim (1+t)^{-\frac{n\alpha}{2}\left(1-\frac{1}{p}\right)}\,\|u-v\|_{X_0}\left(\|u\|_{X_0}^{p-1} + \|v\|_{X_0}^{p-1}\right). \end{aligned}$$

Then we get (80) with

$$A = \|u - v\|_{X_0} \left(\|u\|_{X_0}^{p-1} + \|v\|_{X_0}^{p-1} \right),$$

and this proves (79), by virtue of $p \geq \bar{p}(n, \alpha) = 1 + 2/n$. This concludes the proof.

Proof (Theorem 4) We prove (78). We want to apply (70) with $f(t, x) = |u(t, x)|^p$, so that

$$\|f(t, \cdot)\|_{L^1} = \|u(t, \cdot)\|_{L^p}^p \lesssim t^{(\alpha-1)p} (1 + t)^{-\frac{n\alpha}{2}(p-1)} \|u\|_{X_0}^p.$$

Setting $b = (1 - \alpha)p$, we see that (71) holds, as a consequence of (27). Moreover, setting $\eta = n\alpha(p - 1)/2$, we can see that $b + \eta > 1$ if, and only if, $p > \tilde{p}(n, \alpha)$.

Therefore, by (72), we derive:

$$\|Nu(t, \cdot)\|_{L^p} \lesssim A\, t^{\alpha-1} (1 + t)^{-\frac{n\alpha}{2}\left(1-\frac{1}{p}\right)},$$

where $A = \|u\|_{X_0}^p$. That is, we proved (78). To prove (79), we proceed as in the proof of Theorem 3. This concludes the proof.

Acknowledgements The author has been supported by INdAM – GNAMPA Project 2017 *Equazioni di tipo dispersivo e proprietà asintotiche.*

References

1. B. Andrade, A. Viana, On a fractional reaction-diffusion equation. Z. Angew. Math. Phys. **68**, 59 (2017)
2. I. Capuzzo Dolcetta, A. Cutrì, On the Liouville property for the sublaplacians. Ann. Scuola Norm. Sup. Pisa Cl. Sci. **25**, 239–256 (1997)
3. T. Cazenave, F. Dickstein, F.B. Weissler, An equation whose Fujita critical exponent is not given by scaling. Nonlinear Anal. **68**, 862–874 (2008). https://doi.org/10.1016/j.na.2006.11.042
4. M. D'Abbicco, The influence of a nonlinear memory on the damped wave equation. Nonlinear Anal. **95**, 130–145 (2014). https://doi.org/10.1016/j.na.2013.09.006
5. M. D'Abbicco, A wave equation with structural damping and nonlinear memory. Nonlinear Differ. Equ. Appl. **21**, 751–773 (2014). https://doi.org/10.1007/s00030-014-0265-2
6. M. D'Abbicco, S. Lucente, A modified test function method for damped wave equations. Adv. Nonlinear Stud. **13**, 867–892 (2013)
7. M. D'Abbicco, M.R. Ebert, T. Picon, Global existence of small data solutions to the semilinear fractional wave equation, in *New Trends in Analysis and Interdisciplinary Applications*. Trends in Mathematics, ed. by P. Dang, M. Ku, T. Qian, L. Rodino (Birkhäuser, Cham, 2017). https://doi.org/10.1007/978-3-319-48812-7_59
8. M. D'Abbicco, M.R. Ebert, T.H. Picon, The critical exponent(s) for the fractional diffusive equation. J. Fourier Anal. Appl. (2018). https://doi.org/10.1007/s00041-018-9627-1
9. L. D'Ambrosio, S. Lucente, Nonlinear Liouville theorems for Grushin and Tricomi operators. J. Differ. Equ. **123**, 511–541 (2003)

10. H. Fujita, On the blowing-up of solutions of the Cauchy problem for $u_t = \Delta u + u^{1+\alpha}$. J. Fac. Sci. Univ. Tokyo Sect. 1, **13**, 109–124 (1966)
11. A.A. Kilbas, H.M. Srivastava, J.J. Trujillo, *Theory and Applications of Fractional Differential Equations*. North-Holland Mathematics Studies, vol. 204 (Elsevier Science B.V., Amsterdam, 2006)
12. M.G. Mittag–Leffler, Sur la fonction nouvelle $E_{1/\alpha}(x)$. C. R. Acad. Sci. Paris **137**(2), 554–558 (1903)
13. A.Yu. Popov, A.M. Sedletskii, Distribution of roots of Mittag-Leffler functions. J. Math. Sci. **190**, 209–409 (2013)

Weakly Coupled Systems of Semilinear Effectively Damped Waves with Different Time-Dependent Coefficients in the Dissipation Terms and Different Power Nonlinearities

Abdelhamid Mohammed Djaouti and Michael Reissig

Abstract We study the global existence of small data solutions to the Cauchy problem for the coupled system of semilinear damped wave equations with different effective dissipation terms and different exponents of power nonlinearities. The data are supposed to belong to different classes of regularity. We will show the interaction of the exponents p and q on the one hand and on the other hand the interaction of the dissipation terms $b_1(t)u_t$ and $b_2(t)v_t$.

1 Introduction

Let us consider the following Cauchy problem for a weakly coupled system of classical semilinear damped wave equations

$$\begin{aligned} u_{tt} - \Delta u + u_t &= |v|^p, \qquad u(0,x) = u_0(x), \quad u_t(0,x) = u_1(x), \\ v_{tt} - \Delta v + v_t &= |u|^q, \qquad v(0,x) = v_0(x), \quad v_t(0,x) = v_1(x), \end{aligned} \tag{1}$$

where $t \in [0, \infty)$, $x \in \mathbb{R}^n$. Recently, K. Nishihara and Y. Wakasugi studied in [11] the Cauchy problem (1). Using the weighted energy method they found the critical exponents for any space dimension. In particular, if the inequality

$$\frac{\max\{p; q\} + 1}{pq - 1} < \frac{n}{2}$$

A. Mohammed Djaouti
Faculty of Exact and Computer Sciences, Hassiba Ben Bouali University, Chlef, Algeria
e-mail: djaouti_abdelhamid@yahoo.fr

M. Reissig (✉)
Faculty for Mathematics and Computer Science, Technical University Bergakademie, Freiberg, Germany
e-mail: reissig@math.tu-freiberg.de

M. D'Abbicco et al. (eds.), *New Tools for Nonlinear PDEs and Application*, Trends in Mathematics, https://doi.org/10.1007/978-3-030-10937-0_3

is satisfied, then for suitable given small data there exists a global (in time) solution which satisfies some decay estimates. On the contrary, if this condition is not satisfied, then every Sobolev solution to given initial data having positive average value does not exist globally (in time).

In [10], our main concern is to study the Cauchy problem for a weakly coupled system of semilinear effectively damped waves with time-dependent coefficient in the dissipation term. The model we have in mind is

$$\begin{aligned} u_{tt} - \Delta u + b(t)u_t &= |v|^p, \qquad u(0,x) = u_0(x), \quad u_t(0,x) = u_1(x), \\ v_{tt} - \Delta v + b(t)v_t &= |u|^q, \qquad v(0,x) = v_0(x), \quad v_t(0,x) = v_1(x). \end{aligned}$$

In this paper we explained the interaction of the exponents p and q, the influence of the additional regularity parameter $m \in [1,2)$, and the regularity parameter s on the global (in time) existence of small data solutions.

In the present paper we allow different coefficients in the dissipation terms. We restrict ourselves to the special structure $b_1(t)u_t$ and $b_2(t)v_t$, where

$$b_1(t) = \frac{1}{(1+t)^{r_1}} \quad \text{and} \quad b_2(t) = \frac{1}{(1+t)^{r_2}}$$

with exponents

$$r_1,\, r_2 \in (-1, 1).$$

Therefore, we are concerned with the following model:

$$\begin{aligned} u_{tt} - \Delta u + \tfrac{1}{(1+t)^{r_1}} u_t &= |v|^p, \qquad u(0,x) = u_0(x), \quad u_t(0,x) = u_1(x), \\ v_{tt} - \Delta v + \tfrac{1}{(1+t)^{r_2}} v_t &= |u|^q, \qquad v(0,x) = v_0(x), v_t(0,x) = v_1(x). \end{aligned} \tag{2}$$

Additionally to the influence of the parameters which we have described above, we take into account the interaction of the parameters r_1 and r_2 on the global (in time) existence of small data solutions. It turns out that instead of the exponents p and q we introduce (hint of M. D'Abbicco) the parameters $\tilde{p}$ and $\tilde{q}$ depending on r_1, r_2 and m which allow to describe the interaction in an effective way.

1.1 *Notations*

We introduce for $s > 0$ and $m \in [1,2)$ the function space

$$\mathcal{A}_{m,s} := (H^s \cap L^m) \times (H^{\max\{s-1;0\}} \cap L^m)$$

with the norm

$$\|(u,v)\|_{\mathcal{A}_{m,s}} := \|u\|_{H^s} + \|u\|_{L^m} + \|v\|_{H^{\max\{s-1;0\}}} + \|v\|_{L^m}.$$

We denote by $B(t, \tau)$ the primitive of $1/b(t)$ which vanishes at $t = \tau$, that is,

$$B(t, \tau) = \int_\tau^t \frac{1}{b(r)} dr = B(t, 0) - B(\tau, 0).$$

The dissipation term $b(t)u_t$ is called effective if $b = b(t)$ satisfies the following properties introduced by J. Wirth in [15] and [16]:

- b is a positive and monotonic function with $tb(t) \to \infty$ as $t \to \infty$,
- $((1+t)^2 b(t))^{-1} \in L^1(0, \infty)$,
- $b \in \mathscr{C}^3[0, \infty)$ and $|b^{(k)}(t)| \lesssim \frac{b(t)}{(1+t)^k}$ for $k = 1, 2, 3$,
- $\frac{1}{b} \notin L^1(0, \infty)$ and there exists a constant $a \in [0, 1)$ such that

$$tb'(t) \leq ab(t).$$

In [3], the authors showed that both coefficients $b_1(t) = \frac{1}{(1+t)^{r_1}}$ and $b_2(t) = \frac{1}{(1+t)^{r_2}}$ generate for $r_1, r_2 \in (-1, 1)$ effective dissipation terms $b_1(t)u_t$ and $b_2(t)v_t$. The corresponding primitives $B_1 = B_1(t, \tau)$ and $B_2 = B_2(t, \tau)$ satisfy the following lemma (definition as above with $b_1 = b_1(r)$ and $b_2 = b_2(r)$).

Lemma 1.1 *The primitive $B(t, \tau)$ satisfies the following properties:*

$$B(t, \tau) \approx B(t, 0) \text{ for all } \tau \in \Big[0, \frac{t}{2}\Big],$$

$$B(\tau, 0) \approx B(t, 0) \text{ for all } \tau \in \Big[\frac{t}{2}, t\Big],$$

$$\int_{\frac{t}{2}}^t \frac{1}{b(\tau)} \big(1 + B(t, \tau)\big)^{-\frac{j}{2} - l} d\tau \lesssim (1 + B(t, 0))^{1 - \frac{j}{2} - l} \log{(1 + B(t, 0))^l},$$

where $j + l = 0, 1$

In order to use Duhamels principle we need the following results in the proofs of our main results.

Theorem 1.2 *The Sobolev solutions to the Cauchy problem*

$$u_{tt} - \Delta u + b(t)u_t = 0, \quad u(0, x) = u_0(x), \quad u_t(0, x) = u_1(x)$$

satisfy the following estimates:

for low regular data ($0 < s < 1$)*:*

$$\|u(t, \cdot)\|_{L^2} \lesssim \big(1 + B(t, 0)\big)^{-\frac{n}{2}\left(\frac{1}{m} - \frac{1}{2}\right)} \|(u_0, u_1)\|_{\mathscr{A}_{m,s}},$$

$$\||D|^s u(t, \cdot)\|_{L^2} \lesssim \big(1 + B(t, 0)\big)^{-\frac{n}{2}\left(\frac{1}{m} - \frac{1}{2}\right) - \frac{s}{2}} \|(u_0, u_1)\|_{\mathscr{A}_{m,s}};$$

for data from the energy space ($s = 1$)*:*

$$\|\nabla^j \partial_t^l u(t,\cdot)\|_{L^2} \lesssim (b(t))^{-l}\big(1+B(t,0)\big)^{-\frac{n}{2}\left(\frac{1}{m}-\frac{1}{2}\right)-\frac{j}{2}-l}\|(u_0,u_1)\|_{\mathscr{A}_{m,1}},$$

where $j+l \le 1$*;*

for high regular data ($s > 1$)*:*

$$\|u(t,\cdot)\|_{L^2} \lesssim \big(1+B(t,0)\big)^{-\frac{n}{2}\left(\frac{1}{m}-\frac{1}{2}\right)}\|(u_0,u_1)\|_{\mathscr{A}_{m,s}},$$

$$\|u_t(t,\cdot)\|_{L^2} \lesssim b(t)^{-1}\big(1+B(t,0)\big)^{-\frac{n}{2}\left(\frac{1}{m}-\frac{1}{2}\right)-1}\|(u_0,u_1)\|_{\mathscr{A}_{m,s}},$$

$$\||D|^s u(t,\cdot)\|_{L^2} \lesssim \big(1+B(t,0)\big)^{-\frac{n}{2}\left(\frac{1}{m}-\frac{1}{2}\right)-\frac{s}{2}}\|(u_0,u_1)\|_{\mathscr{A}_{m,s}},$$

$$\||D|^{s-1} u_t(t,\cdot)\|_{L^2} \lesssim b(t)^{-1}\big(1+B(t,0)\big)^{-\frac{n}{2}\left(\frac{1}{m}-\frac{1}{2}\right)-\frac{s-1}{2}-1}\|(u_0,u_1)\|_{\mathscr{A}_{m,s}}.$$

Proof The proof of this theorem can be concluded from [15] and [16].

Theorem 1.3 *The Sobolev solutions to the parameter-dependent family of Cauchy problems*

$$v_{tt} - \Delta v + b(t)v_t = 0, \quad v(\tau,x) = 0, \quad v_t(\tau,x) = v_1(x)$$

satisfy the following estimates:

for low regular data ($0 < s < 1$)*:*

$$\|v(t,\cdot)\|_{L^2} \lesssim b(\tau)^{-1}\big(1+B(t,\tau)\big)^{-\frac{n}{2}\left(\frac{1}{m}-\frac{1}{2}\right)}\|v_1\|_{L^2\cap L^m},$$

$$\||D|^s v(t,\cdot)\|_{L^2} \lesssim b(\tau)^{-1}\big(1+B(t,\tau)\big)^{-\frac{n}{2}\left(\frac{1}{m}-\frac{1}{2}\right)-\frac{s}{2}}\|v_1\|_{L^2\cap L^m}; \tag{3}$$

for data from the energy space ($s = 1$)*:*

$$\|\nabla^j \partial_t v(t,\cdot)\|_{L^2} \lesssim b(t)^{-1}b(\tau)^{-l}\big(1+B(t,\tau)\big)^{-\frac{n}{2}\left(\frac{1}{m}-\frac{1}{2}\right)-\frac{j}{2}-l}\|v_1\|_{L^2\cap L^m},$$

where $j+l \le 1$*;*

for high regular data ($s > 1$)*:*

$$\|v(t,\cdot)\|_{L^2} \lesssim b(\tau)^{-1}\big(1+B(t,\tau)\big)^{-\frac{n}{2}\left(\frac{1}{m}-\frac{1}{2}\right)}\|v_1\|_{H^{s-1}\cap L^m},$$

$$\|v_t(t,\cdot)\|_{L^2} \lesssim b(\tau)^{-1}b(t)^{-1}\big(1+B(t,\tau)\big)^{-\frac{n}{2}\left(\frac{1}{m}-\frac{1}{2}\right)-1}\|v_1\|_{H^{s-1}\cap L^m},$$

$$\||D|^s v(t,\cdot)\|_{L^2} \lesssim b(\tau)^{-1}\big(1+B(t,\tau)\big)^{-\frac{n}{2}\left(\frac{1}{m}-\frac{1}{2}\right)-\frac{s}{2}}\|v_1\|_{H^{s-1}\cap L^m},$$

$$\||D|^{s-1} v_t(t,\cdot)\|_{L^2} \lesssim b(\tau)^{-1}b(t)^{-1}\big(1+B(t,\tau)\big)^{-\frac{n}{2}\left(\frac{1}{m}-\frac{1}{2}\right)-\frac{s-1}{2}-1}\|v_1\|_{H^{s-1}\cap L^m}. \quad (4)$$

Proof The proof of this theorem can be concluded from [3].

2 Main Results

We study the Cauchy problem (2) in several cases with respect to the regularity of the data. Therefore, we introduce the following classification of regularity:

low regular data, data from energy space, data from Sobolev spaces with suitable regularity and, finally, large regular data.

2.1 Low Regular Data

In this section we are interested in the system (2), where the data are taken from the Sobolev space $H^s(\mathbb{R}^n)$ for $s \in [0,1)$, with the same additional regularity $L^m(\mathbb{R}^n)$. We remark immediately that $p_{Fuj,m}(n) := 1 + \frac{2m}{n} > s$ which means that the regularity parameter has a weak influence on the admissible range for $\tilde{p} = \tilde{p}_{r_1,r_2}$ and $\tilde{q} = \tilde{q}_{r_1,r_2}$. Therefore, we compare in our statements the modified exponents $\tilde{p}_{r_1,r_2}$ and $\tilde{q}_{r_1,r_2}$ with the modified Fujita exponent $p_{Fuj,m}(n)$.

Theorem 2.1 *Let $n \le \frac{4s}{2-m}$, $n < \max\left\{\frac{2sm}{m-s}; \frac{2m(2-s)}{2-m}\right\}$, $r_1, r_2 \in (-1,1)$, $m \in [1,2)$ and $s \in (0,1)$. The data (u_0,u_1), (v_0,v_1) are assumed to belong to $\mathcal{A}_{m,s} \times \mathcal{A}_{m,s}$. Moreover, let the modified exponents satisfy*

$$\min\{\tilde{p}_{r_1,r_2}; \tilde{q}_{r_1,r_2}\} < p_{Fuj,m}(n) < \max\{\tilde{p}_{r_1,r_2}; \tilde{q}_{r_1,r_2}\}, \quad (5)$$

$$\frac{n}{2} > m\Big(\frac{\max\{\tilde{p}_{r_1,r_2}; \tilde{q}_{r_1,r_2}\} + \gamma}{\min\{\tilde{p}_{r_1,r_2}; \tilde{q}_{r_1,r_2}\} \times \max\{\tilde{p}_{r_1,r_2}; \tilde{q}_{r_1,r_2}\} - 1 + (\min\{\tilde{p}_{r_1,r_2}; \tilde{q}_{r_1,r_2}\} - 1)\delta}\Big), \quad (6)$$

where

$$\tilde{q}_{r_1,r_2} = \frac{1+r_1}{1+r_2}(q-1)+1, \quad \tilde{p}_{r_1,r_2} = \frac{1+r_2}{1+r_1}(p-1)+1,$$

and

$$\gamma = \frac{1+r_1}{1+r_2}, \quad \delta = \frac{r_1 - r_2}{1+r_2} \text{ if } \tilde{p}_{r_1,r_2} < \tilde{q}_{r_1,r_2},$$

$$\gamma = \frac{1+r_2}{1+r_1}, \quad \delta = \frac{r_2 - r_1}{1+r_1} \text{ if } \tilde{q}_{r_1,r_2} < \tilde{p}_{r_1,r_1}.$$

The exponents p and q of the power nonlinearities satisfy

$$\begin{cases} \frac{2}{m} \leq \min\{p;q\} \leq \max\{p;q\} < \infty & \text{if } n=1 \quad \text{and} \quad s \in [\frac{1}{2}, 1), \\ \frac{2}{m} < \min\{p;q\} \leq \max\{p;q\} \leq p_{GN,s}(1) & \text{if } n=1 \quad \text{and} \quad s \in (0, \frac{1}{2}), \\ \frac{2}{m} < \min\{p;q\} \leq \max\{p;q\} \leq p_{GN,s}(n) & \text{if} \quad n \geq 2, \end{cases} \tag{7}$$

where $p_{GN,s}(n) := \frac{n}{n-2s}$.

Then, there exists a constant ϵ_0 such that if

$$\|(u_0, u_1)\|_{\mathscr{A}_{m,s}} + \|(v_0, v_1)\|_{\mathscr{A}_{m,s}} \leq \epsilon_0,$$

then there exists a uniquely determined global (in time) Sobolev solution to (2) *in* $(\mathscr{C}([0,\infty), H^s(\mathbb{R}^n)))^2$. *Furthermore, the solution satisfies the following estimates:*

$$\|u(t,\cdot)\|_{L^2(\mathbb{R}^n)} \lesssim \big(1 + B_1(t,0)\big)^{-\frac{n}{2}\left(\frac{1}{m}-\frac{1}{2}\right)+[\gamma_{n,m}(\tilde{p}_{r_1,r_2})]^+} \big(\|(u_0,u_1)\|_{\mathscr{A}_{m,s}} + \|(v_0,v_1)\|_{\mathscr{A}_{m,s}}\big),$$

$$\||D|^s u(t,\cdot)\|_{L^2(\mathbb{R}^n)} \lesssim \big(1 + B_1(t,0)\big)^{-\frac{n}{2}\left(\frac{1}{m}-\frac{1}{2}\right)-\frac{s}{2}+[\gamma_{n,m}(\tilde{p}_{r_1,r_2})]^+} \big(\|(u_0,u_1)\|_{\mathscr{A}_{m,s}} + \|(v_0,v_1)\|_{\mathscr{A}_{m,s}}\big),$$

$$\|v(t,\cdot)\|_{L^2(\mathbb{R}^n)} \lesssim \big(1 + B_2(t,0)\big)^{-\frac{n}{2}\left(\frac{1}{m}-\frac{1}{2}\right)+[\gamma_{n,m}(\tilde{q}_{r_1,r_2})]^+} \big(\|(u_0,u_1)\|_{\mathscr{A}_{m,s}} + \|(v_0,v_1)\|_{\mathscr{A}_{m,s}}\big),$$

$$\||D|^s v(t,\cdot)\|_{L^2(\mathbb{R}^n)} \lesssim \big(1 + B_2(t,0)\big)^{-\frac{n}{2}\left(\frac{1}{m}-\frac{1}{2}\right)-\frac{s}{2}+[\gamma_{n,m}(\tilde{p}_{r_1,r_2})]^+} \big(\|(u_0,u_1)\|_{\mathscr{A}_{m,s}} + \|(v_0,v_1)\|_{\mathscr{A}_{m,s}}\big),$$

where

$$\gamma_{n,m}(\tilde{p}_{r_1,r_2}) = -\frac{n}{2m}(\tilde{p}_{r_1,r_2} - 1) + 1 \quad \text{or} \quad \gamma_{n,m}(\tilde{q}_{r_1,r_2}) = -\frac{n}{2m}(\tilde{q}_{r_1,r_2} - 1) + 1$$

represents the loss of decay in comparison with the corresponding decay estimates for the solution u or v to the linear Cauchy problem with vanishing right hand-side.

Remark 2.2 If we have $\tilde{p}_{r_1,r_2} = p_{Fuj,m}(n) < \tilde{q}_{r_1,r_2}$ in condition (5), then we obtain a small loss of decay $\gamma_{n,m}(\tilde{p}_{r_1,r_2}) = \varepsilon$ for arbitrarily small positive ε. This small loss of decay is generated by a log term appearing in the step when we control the nonlinear term of u, in particular, the integral over $[\frac{t}{2}, t]$.

Example 2.3 Let us assume $n = 2$. We choose the additional regularity $m = \frac{3}{2}$ and the regularity parameter $s = \frac{9}{10}$. Then we get $\frac{2}{m} = \frac{4}{3}$, $p_{Fuj,\frac{3}{2}}(2) = \frac{5}{2}$ and $p_{GN,\frac{9}{10}} = 10$. If we take $p = \frac{19}{10} \in \left[\frac{4}{3}, 10\right]$ and $q = \frac{49}{9} \in \left[\frac{4}{3}, 10\right]$, then for $\frac{1+r_1}{1+r_2} = \frac{9}{10}$ we get

$$\tilde{p} = 2 < p_{Fuj,\frac{3}{2}}(2) < 5 = \tilde{q}.$$

The modified exponents $\tilde{p}$ and $\tilde{q}$ satisfy condition (6). Moreover, the solution satisfies the following estimates:

$$\begin{aligned}
\|u(t,\cdot)\|_{L^2(\mathbb{R}^n)} &\lesssim (1+t)^{\frac{1}{6}}\big(\|(u_0,u_1)\|_{\mathcal{A}_{\frac{3}{2},\frac{9}{10}}} + \|(v_0,v_1)\|_{\mathcal{A}_{\frac{3}{2},\frac{9}{10}}}\big),\\
\||D|^s u(t,\cdot)\|_{L^2(\mathbb{R}^n)} &\lesssim (1+t)^{-\frac{17}{60}}\big(\|(u_0,u_1)\|_{\mathcal{A}_{\frac{3}{2},\frac{9}{10}}} + \|(v_0,v_1)\|_{\mathcal{A}_{\frac{3}{2},\frac{9}{10}}}\big),\\
\|v(t,\cdot)\|_{L^2(\mathbb{R}^n)} &\lesssim (1+t)^{-\frac{1}{6}}\big(\|(u_0,u_1)\|_{\mathcal{A}_{\frac{3}{2},\frac{9}{10}}} + \|(v_0,v_1)\|_{\mathcal{A}_{\frac{3}{2},\frac{9}{10}}}\big),\\
\||D|^s v(t,\cdot)\|_{L^2(\mathbb{R}^n)} &\lesssim (1+t)^{-\frac{37}{60}}\big(\|(u_0,u_1)\|_{\mathcal{A}_{\frac{3}{2},\frac{9}{10}}} + \|(v_0,v_1)\|_{\mathcal{A}_{\frac{3}{2},\frac{9}{10}}}\big).
\end{aligned}$$

Remark 2.4 If we assume instead of (5) that both modified exponents are larger than $p_{Fuj,m}(n)$, i.e.,

$$\max\{\tilde{p}_{r_1,r_2}; \tilde{q}_{r_1,r_2,m}\} > p_{Fuj,m}(n), \tag{8}$$

then we can prove a similar result to Theorem 2.1 but without a loss of decay. Moreover, it is possible to assume that one of the exponents p or q is smaller than $p_{Fuj,m}(n)$ without any loss of decay in the solution, which is impossible in the case that the coefficients of the dissipation terms coincide. The following example explains this effect.

Example 2.5 Let us choose the dimension $n = 1$. The coefficients of the dissipation terms are $b_1(t) = (1+t)^{-\frac{1}{2}}$ and $b_2(t) = (1+t)^{\frac{1}{2}}$. The data (u_0, u_1), (v_0, v_1) are supposed to belong to $\mathcal{A}_{\frac{3}{2},\frac{1}{2}} \times \mathcal{A}_{\frac{3}{2},\frac{1}{2}}$. Then admissible values for the exponents p and q to guarantee the global (in time) existence of small data solutions can be chosen as follows:

$$\frac{4}{3} \le p = \frac{8}{3} < p_{Fuj,\frac{3}{2}}(1) = 4 \qquad \text{and} \qquad q = 13 \ge \frac{4}{3}.$$

Here we obtain as modified exponents $\tilde{p} = 6$ and $\tilde{q} = 5$ which satisfy the condition (8). The solution satisfies the following decay estimates:

$$\|u(t,\cdot)\|_{L^2(\mathbb{R}^n)} \lesssim (1+t)^{-\frac{1}{24}}\big(\|(u_0,u_1)\|_{\mathscr{A}_{\frac{3}{2},\frac{1}{2}}} + \|(v_0,v_1)\|_{\mathscr{A}_{\frac{3}{2},\frac{1}{2}}}\big),$$

$$\||D|^s u(t,\cdot)\|_{L^2(\mathbb{R}^n)} \lesssim (1+t)^{-\frac{1}{6}}\big(\|(u_0,u_1)\|_{\mathscr{A}_{\frac{3}{2},\frac{1}{2}}} + \|(v_0,v_1)\|_{\mathscr{A}_{\frac{3}{2},\frac{1}{2}}}\big),$$

$$\|v(t,\cdot)\|_{L^2(\mathbb{R}^n)} \lesssim (1+t)^{-\frac{1}{8}}\big(\|(u_0,u_1)\|_{\mathscr{A}_{\frac{3}{2},\frac{1}{2}}} + \|(v_0,v_1)\|_{\mathscr{A}_{\frac{3}{2},\frac{1}{2}}}\big),$$

$$\||D|^s v(t,\cdot)\|_{L^2(\mathbb{R}^n)} \lesssim (1+t)^{-\frac{1}{2}}\big(\|(u_0,u_1)\|_{\mathscr{A}_{\frac{3}{2},\frac{1}{2}}} + \|(v_0,v_1)\|_{\mathscr{A}_{\frac{3}{2},\frac{1}{2}}}\big).$$

2.2 *Data from Energy Space*

Similarly to the case of low regular data we may treat the limit case of Theorem 2.1, that is, the data (u_0,u_1), (v_0,v_1) are supposed to belong to $\mathscr{A}_{m,1} \times \mathscr{A}_{m,1}$. Therefore, we can prove a global (in time) existence result of small data energy solutions. But, now the data has a larger regularity which allows to define energy solutions and to introduce the norms $\|u_t(\tau,\cdot)\|_{L^2(\mathbb{R}^n)}$ and $\|v_t(\tau,\cdot)\|_{L^2(\mathbb{R}^n)}$ in the solution space $X(t)$. We include also in this case the additional regularity parameter m in the definition of the modified exponents $\tilde{p} = \tilde{p}_{r_1,r_2,m}$ or $\tilde{q} = \tilde{q}_{r_1,r_2,m}$ of the power nonlinearities. We distinguish between several cases with respect to the values of r_1 and r_2 on the one hand and the order relation between $\tilde{p} = \tilde{p}_{r_1,r_2,m}$ and $\tilde{q} = \tilde{q}_{r_1,r_2,m}$ on the other hand. The cases we have in mind are the followings:

- $r_1 < r_2$ and $\tilde{p} < \tilde{q}$,
- $r_1 > r_2$ and $\tilde{p} < \tilde{q}$,
- $r_1 < r_2$ and $\tilde{p} > \tilde{q}$,
- $r_1 > r_2$ and $\tilde{p} > \tilde{q}$.

In the following theorem we will present the results for the first case.

Theorem 2.6 *Let* $n < \frac{2m^2}{2-m}$, $n \leq \frac{2m}{m-1}$, $0 < r_1 < r_2 < 1$, *and* $m \in [1,2)$. *The data* (u_0,u_1), (v_0,v_1) *are assumed to belong to* $\mathscr{A}_{m,1} \times \mathscr{A}_{m,1}$. *Moreover, the modified exponents satisfy*

$$\tilde{p}_{r_1,r_2} < p_{Fuj,m}(n) < \tilde{q}_{r_1,r_2,m}$$

and

$$\frac{n}{2} > m\Big(\frac{\tilde{q}_{r_1,r_2,m} + 1 + \frac{m}{2}\big(\frac{r_1-r_2}{1+r_2}\big)}{\tilde{p}_{r_1,r_2}\tilde{q}_{r_1,r_2,m} - 1 + \frac{m}{2}(\tilde{p}-1)\big(\frac{r_1-r_2}{1+r_2}\big)}\Big), \tag{9}$$

where

$$\tilde{q} = \tilde{q}_{r_1,r_2,m} = \frac{1+r_1}{1+r_2}\Big(q - \frac{m}{2}\Big) + \frac{m}{2}, \quad \tilde{p} = \tilde{p}_{r_1,r_2} = \frac{1+r_2}{1+r_1}(p-1)+1.$$

The exponents p and q of the power nonlinearities satisfy

$$\begin{aligned} \tfrac{2}{m} \leq \min\{p;q\} \leq \max\{p;q\} < \infty \quad & \text{if } n \leq 2, \\ \tfrac{2}{m} \leq \min\{p;q\} \leq \max\{p;q\} \leq p_{GN}(n) \ & \text{if } n > 2. \end{aligned}$$

Then, there exists a constant ϵ_0 such that if

$$\|(u_0,u_1)\|_{\mathscr{A}_{m,1}} + \|(v_0,v_1)\|_{\mathscr{A}_{m,1}} \leq \epsilon_0,$$

then there exists a uniquely determined global (in time) energy solution to (2) *in*

$$\Big(\mathscr{C}([0,\infty),H^1(\mathbb{R}^n)) \cap \mathscr{C}^1([0,\infty),L^2(\mathbb{R}^n))\Big)^2.$$

Furthermore, the solution satisfies the following estimates:

$$\begin{aligned}
\|u(t,\cdot)\|_{L^2(\mathbb{R}^n)} &\lesssim \big(1+B_1(t,0)\big)^{-\frac{n}{2}\left(\frac{1}{m}-\frac{1}{2}\right)+\gamma_{n,m}(\tilde{p}_{r_1,r_2})} \\
&\qquad \times\big(\|(u_0,u_1)\|_{\mathscr{A}_{m,1}} + \|(v_0,v_1)\|_{\mathscr{A}_{m,1}}\big), \\
\|\nabla u(t,\cdot)\|_{L^2(\mathbb{R}^n)} &\lesssim \big(1+B_1(t,0)\big)^{-\frac{n}{2}\left(\frac{1}{m}-\frac{1}{2}\right)-\frac{1}{2}+\gamma_{n,m}(\tilde{p}_{r_1,r_2})} \\
&\qquad \times\big(\|(u_0,u_1)\|_{\mathscr{A}_{m,1}} + \|(v_0,v_1)\|_{\mathscr{A}_{m,1}}\big), \\
\|u_t(t,\cdot)\|_{L^2(\mathbb{R}^n)} &\lesssim b_1(t)^{-1}\big(1+B_1(t,0)\big)^{-\frac{n}{2}\left(\frac{1}{m}-\frac{1}{2}\right)-1+\gamma_{n,m}(\tilde{p}_{r_1,r_2})} \\
&\qquad \times\big(\|(u_0,u_1)\|_{\mathscr{A}_{m,1}} + \|(v_0,v_1)\|_{\mathscr{A}_{m,1}}\big), \\
\|v(t,\cdot)\|_{L^2(\mathbb{R}^n)} &\lesssim \big(1+B_2(t,0)\big)^{-\frac{n}{2}\left(\frac{1}{m}-\frac{1}{2}\right)} \\
&\qquad \times\big(\|(u_0,u_1)\|_{\mathscr{A}_{m,1}} + \|(v_0,v_1)\|_{\mathscr{A}_{m,1}}\big), \\
\|\nabla v(t,\cdot)\|_{L^2(\mathbb{R}^n)} &\lesssim \big(1+B_2(t,0)\big)^{-\frac{n}{2}\left(\frac{1}{m}-\frac{1}{2}\right)-\frac{1}{2}} \\
&\qquad \times\big(\|(u_0,u_1)\|_{\mathscr{A}_{m,1}} + \|(v_0,v_1)\|_{\mathscr{A}_{m,1}}\big).
\end{aligned}$$

$$\begin{aligned}
\|v_t(t,\cdot)\|_{L^2(\mathbb{R}^n)} &\lesssim b_2(t)^{-1}\big(1+B_2(t,0)\big)^{-\frac{n}{2}\left(\frac{1}{m}-\frac{1}{2}\right)-1} \\
&\qquad \times\big(\|(u_0,u_1)\|_{\mathscr{A}_{m,1}} + \|(v_0,v_1)\|_{\mathscr{A}_{m,1}}\big),
\end{aligned}$$

where

$$\gamma_{n,m}(\tilde{p}_{r_1,r_2}) = -\frac{n}{2m}(\tilde{p} - 1) + 1$$

represent the loss of decay in comparison with the corresponding decay estimates for the solution u of the linear Cauchy problem with vanishing right hand-side.

Example 2.7 Let us choose $n = 2$ in Theorem 2.6. If we choose the additional regularity $m = \frac{7}{4}$, then we obtain $\frac{2}{m} = \frac{8}{7}$ and $p_{Fuj,\frac{7}{4}}(2) = \frac{11}{4}$. Finally, for $p = \frac{3}{2} \in \left[\frac{8}{7}, \infty\right)$, $q = 20 \in \left[\frac{8}{7}, \infty\right)$ and $\frac{1+r_1}{1+r_2} = \frac{1}{2}$ we get

$$\tilde{p} = 2 < p_{Fuj,\frac{7}{4}}(2) < \frac{167}{16} = \tilde{q}.$$

Moreover, the modified exponents $\tilde{p}$ and $\tilde{q}$ satisfy condition (9).

We summarize our results for all cases with respect to $r_1, r_2, \tilde{p}$ and $\tilde{q}$ in the following table:

	$\tilde{p} < p_{Fuj,m}(n) < \tilde{q}$	$\tilde{q} < p_{Fuj,m}(n) < \tilde{p}$
$r_1 < r_2$ $\tilde{q} = \tilde{q}_{r_1,r_2,m}$ $\tilde{p} = \tilde{p}_{r_1,r_2}$	Loss of decay in the estimate for u $\gamma_{n,m}(\tilde{p}) = -\frac{n}{2m}(\tilde{p} - 1) + 1$ interaction condition $\frac{n}{2} > m\left(\frac{\tilde{q}+1+\frac{m}{2}\left(\frac{r_1-r_2}{1+r_2}\right)}{\tilde{p}\tilde{q}-1+\frac{m}{2}(\tilde{p}-1)\left(\frac{r_1-r_2}{1+r_2}\right)}\right)$	Loss of decay in the estimate for v $\gamma_{n,m}(\tilde{q}) = -\frac{n}{2m}(\tilde{q} - 1) + 1 + \varepsilon$ interaction condition $\frac{n}{2} > m\left(\frac{\tilde{p}+\frac{1+r_2}{1+r_1}}{\tilde{p}\tilde{q}-\frac{1+r_2}{1+r_1}+\tilde{q}\left(\frac{r_2-r_1}{1+r_1}\right)}\right)$
$r_2 < r_1$ $\tilde{q} = \tilde{q}_{r_1,r_2}$ $\tilde{p} = \tilde{p}_{r_1,r_2,m}$	Loss of decay in the estimate for u $\gamma_{n,m}(\tilde{p}) = -\frac{n}{2m}(\tilde{p} - 1) + 1 + \varepsilon$ interaction condition $\frac{n}{2} > m\left(\frac{\tilde{q}+\frac{1+r_1}{1+r_2}}{\tilde{p}\tilde{q}-\frac{1+r_1}{1+r_2}+\tilde{p}\left(\frac{r_1-r_2}{1+r_2}\right)}\right)$	Loss of decay in the estimate for v $\gamma_{n,m}(\tilde{q}) = -\frac{n}{2m}(\tilde{q} - 1) + 1$ interaction condition $\frac{n}{2} > m\left(\frac{\tilde{p}+1+\frac{m}{2}\left(\frac{r_2-r_1}{1+r_1}\right)}{\tilde{p}\tilde{q}-1+\frac{m}{2}(\tilde{q}-1)\left(\frac{r_2-r_1}{1+r_1}\right)}\right)$

Remark 2.4 remains still true in the case that the data are chosen from the energy space.

2.3 *Data from Sobolev Spaces with Suitable Regularity*

In this case we assume that the data have a suitable larger regularity with an additional regularity $L^m(\mathbb{R}^n)$, $m \in [1, 2)$. In this section we shall use a generalized (fractional) Gagliardo-Nirenberg inequality used in the papers [7] and [13]. Furthermore, we shall use a fractional Leibniz rule and a fractional chain rule which are explained in Propositions A.2 and A.3 from the Appendix.

Theorem 2.8 *Let $n \geq 4$, $s_1, s_2 \in [3, \frac{n}{2}+1]$, $0 < s_2 - s_1 < 1$, $\lceil s_1 \rceil \neq \lceil s_2 \rceil$, and $-1 < r_1 < r_2 < 1$. The data $(u_0, u_1), (v_0, v_1)$ are supposed to belong to $\mathscr{A}_{m,s_1} \times \mathscr{A}_{m,s_2}$ with $m \in [1, 2)$. Furthermore, we require*

$$\tilde{q} > \frac{2m}{n}\Big(\frac{s_2+1}{2}\Big) + 1, \tag{10}$$

where $\tilde{q} = \tilde{q}_{r_1,r_2,m} = \frac{1+r_1}{1+r_2}(q - \frac{m}{2}) + \frac{m}{2}$. The exponents p and q of the power nonlinearities satisfy the conditions

$$\begin{array}{llll} \lceil s_1 \rceil < p, & \lceil s_2 \rceil < q & \text{if} & n \leq 2s_1, \\ \lceil s_1 \rceil < p, & \lceil s_2 \rceil < q \leq 1 + \frac{2}{n-2s_1} & \text{if } 2s_1 < & n \leq 2s_2, \\ \lceil s_1 \rceil < p \leq 1 + \frac{2}{n-2s_2}, & \lceil s_2 \rceil < q \leq 1 + \frac{2}{n-2s_1} & \text{if} & n > 2s_2. \end{array} \tag{11}$$

Then, there exists a constant ϵ_0 such that if

$$\|(u_0, u_1)\|_{\mathscr{A}_{m,s_1}} + \|(v_0, v_1)\|_{\mathscr{A}_{m,s_2}} \leq \epsilon_0,$$

then there exists a uniquely determined globally (in time) energy solution to (2) *in*

$$\Big(\mathscr{C}\big([0,\infty), H^{s_1}(\mathbb{R}^n)\big) \cap \mathscr{C}^1\big([0,\infty), H^{s_1-1}(\mathbb{R}^n)\big)\Big) \\ \times \Big(\mathscr{C}\big([0,\infty), H^{s_2}(\mathbb{R}^n)\big) \cap \mathscr{C}^1\big([0,\infty), H^{s_2-1}(\mathbb{R}^n)\big)\Big).$$

Furthermore, the solution satisfies the estimates

$$\|u(t,\cdot)\|_{L^2(\mathbb{R}^n)} \lesssim \big(1 + B_1(t,0)\big)^{-\frac{n}{2}\left(\frac{1}{m}-\frac{1}{2}\right)} \\ \times \Big(\|(u_0,u_1)\|_{\mathscr{A}_{m,s_1}} + \|(v_0,v_1)\|_{\mathscr{A}_{m,s_2}}\Big),$$

$$\||D|^{s_1} u(t,\cdot)\|_{L^2(\mathbb{R}^n)} \lesssim \big(1 + B_1(t,0)\big)^{-\frac{n}{2}\left(\frac{1}{m}-\frac{1}{2}\right)-\frac{s_1}{2}} \\ \times \Big(\|(u_0,u_1)\|_{\mathscr{A}_{m,s_1}} + \|(v_0,v_1)\|_{\mathscr{A}_{m,s_2}}\Big),$$

$$\|u_t(t,\cdot)\|_{L^2(\mathbb{R}^n)} \lesssim b_1(t)^{-1}\big(1 + B_1(t,0)\big)^{-\frac{n}{2}\left(\frac{1}{m}-\frac{1}{2}\right)-1} \\ \times \Big(\|(u_0,u_1)\|_{\mathscr{A}_{m,s_1}} + \|(v_0,v_1)\|_{\mathscr{A}_{m,s_2}}\Big),$$

$$\||D|^{s_1-1} u_t(t,\cdot)\|_{L^2(\mathbb{R}^n)} \lesssim b_1(t)^{-1}\big(1 + B_1(t,0)\big)^{-\frac{n}{2}\left(\frac{1}{m}-\frac{1}{2}\right)-1-\frac{s_1-1}{2}} \\ \times \Big(\|(u_0,u_1)\|_{\mathscr{A}_{m,s_1}} + \|(v_0,v_1)\|_{\mathscr{A}_{m,s_2}}\Big),$$

$$\|v(t,\cdot)\|_{L^2(\mathbb{R}^n)} \lesssim \big(1+B_2(t,0)\big)^{-\frac{n}{2}\left(\frac{1}{m}-\frac{1}{2}\right)} \times \Big(\|(u_0,u_1)\|_{\mathscr{A}_{m,s_1}} + \|(v_0,v_1)\|_{\mathscr{A}_{m,s_2}}\Big),$$

$$\||D|^{s_2}v(t,\cdot)\|_{L^2(\mathbb{R}^n)} \lesssim \big(1+B_2(t,0)\big)^{-\frac{n}{2}\left(\frac{1}{m}-\frac{1}{2}\right)-\frac{s_2}{2}} \times \Big(\|(u_0,u_1)\|_{\mathscr{A}_{m,s_1}} + \|(v_0,v_1)\|_{\mathscr{A}_{m,s_2}}\Big),$$

$$\|v_t(t,\cdot)\|_{L^2(\mathbb{R}^n)} \lesssim b_2(t)^{-1}\big(1+B_2(t,0)\big)^{-\frac{n}{2}\left(\frac{1}{m}-\frac{1}{2}\right)-1} \times \Big(\|(u_0,u_1)\|_{\mathscr{A}_{m,s_1}} + \|(v_0,v_1)\|_{\mathscr{A}_{m,s_2}}\Big),$$

$$\||D|^{s_2-1}v_t(t,\cdot)\|_{L^2(\mathbb{R}^n)} \lesssim b_2(t)^{-1}\big(1+B_2(t,0)\big)^{-\frac{n}{2}\left(\frac{1}{m}-\frac{1}{2}\right)-1-\frac{s_2-1}{2}} \times \Big(\|(u_0,u_1)\|_{\mathscr{A}_{m,s_1}} + \|(v_0,v_1)\|_{\mathscr{A}_{m,s_2}}\Big).$$

Example 2.9 Let us choose the dimension $n=6$. The coefficients of the dissipation terms are $b_1(t)=(1+t)^{-\frac{1}{2}}$ and $b_2(t)=(1+t)^{\frac{1}{2}}$. The data (u_0,u_1), (v_0,v_1) are supposed to belong to $\mathscr{A}_{1,3}\times\mathscr{A}_{1,4}$. Then admissible values for the exponents p and q to guarantee the global (in time) existence of small data solutions can be chosen as follows:

$$p=4>\lceil 3\rceil=3, \qquad \text{and} \qquad q=5>\lceil 4\rceil=4.$$

Here we obtain as modified exponent $\tilde{q}=2$ which satisfies the condition (10).

Remark 2.10 If we suppose in Theorem 2.8 the assumption $-1<r_2<r_1<1$, then we replace condition (10) by the following condition:

$$\tilde{p} > \frac{2m}{n}\Big(\frac{s_1+1}{2}\Big)+1,$$

where $\tilde{p}=\tilde{p}_{r_1,r_2,m}=\frac{1+r_1}{1+r_2}\big(p-\frac{m}{2}\big)+\frac{m}{2}$.

2.4 Large Regular Data

This case has been classified to benefit from embedding in $L^\infty(\mathbb{R}^n)$, where the data are supposed to have a high regularity.

Theorem 2.11 *Let $n \geq 4$, $(u_0, u_1), (v_0, v_1) \in \mathscr{A}_{m,s_1} \times \mathscr{A}_{m,s_2}$, $m \in [1, 2)$, $s_2 > s_1 > \frac{n}{2} + 1$, and $-1 < r_1 < r_2 < 1$. Moreover, let*

$$p > s_1,\ q > \tilde{s}_2,\ \tilde{q} \geq \frac{2m}{n}\Big(\frac{s_2+1}{2}\Big) + 1,$$

where

$$\tilde{s}_2 \in (s_1, s_1+1),\ \ \tilde{s}_2 \leq s_2,\ \ \tilde{q} = \tilde{q}_{r_1,r_2,m} = \frac{1+r_1}{1+r_2}\Big(q - \frac{m}{2}\Big) + \frac{m}{2}.$$

Then, there exists a constant ϵ_0 such that if

$$\|(u_0, u_1)\|_{\mathscr{A}_{m,s_1}} + \|(v_0, v_1)\|_{\mathscr{A}_{m,s_2}} \leq \epsilon_0,$$

then there exists a uniquely determined globally (in time) energy solution to (2) *in*

$$\Big(\mathscr{C}\big([0,\infty), H^{s_1}(\mathbb{R}^n)\big) \cap \mathscr{C}^1\big([0,\infty), H^{s_1-1}(\mathbb{R}^n)\big)\Big)$$
$$\times\Big(\mathscr{C}\big([0,\infty), H^{\tilde{s}_2}(\mathbb{R}^n)\big) \cap \mathscr{C}^1\big([0,t], H^{\tilde{s}_2-1}(\mathbb{R}^n)\big)\Big).$$

Furthermore, the solution satisfies the following estimates:

$$\|u(t,\cdot)\|_{L^2(\mathbb{R}^n)} \lesssim \big(1 + B_1(t,0)\big)^{-\frac{n}{2}\left(\frac{1}{m}-\frac{1}{2}\right)}$$
$$\times\Big(\|(u_0, u_1)\|_{\mathscr{A}_{m,s_1}} + \|(v_0, v_1)\|_{\mathscr{A}_{m,s_2}}\Big),$$

$$\||D|^{s_1} u(t,\cdot)\|_{L^2(\mathbb{R}^n)} \lesssim \big(1 + B_1(t,0)\big)^{-\frac{n}{2}\left(\frac{1}{m}-\frac{1}{2}\right)-\frac{s_1}{2}}$$
$$\times\Big(\|(u_0, u_1)\|_{\mathscr{A}_{m,s_1}} + \|(v_0, v_1)\|_{\mathscr{A}_{m,s_2}}\Big),$$

$$\|u_t(t,\cdot)\|_{L^2(\mathbb{R}^n)} \lesssim b_1(t)^{-1}\big(1 + B_1(t,0)\big)^{-\frac{n}{2}\left(\frac{1}{m}-\frac{1}{2}\right)-1}$$
$$\times\Big(\|(u_0, u_1)\|_{\mathscr{A}_{m,s_1}} + \|(v_0, v_1)\|_{\mathscr{A}_{m,s_2}}\Big),$$

$$\||D|^{s_1-1} u_t(t,\cdot)\|_{L^2(\mathbb{R}^n)} \lesssim b_1(t)^{-1}\big(1 + B_1(t,0)\big)^{-\frac{n}{2}\left(\frac{1}{m}-\frac{1}{2}\right)-1-\frac{s_1-1}{2}}$$
$$\times\Big(\|(u_0, u_1)\|_{\mathscr{A}_{m,s_1}} + \|(v_0, v_1)\|_{\mathscr{A}_{m,s_2}}\Big),$$

$$\|v(t,\cdot)\|_{L^2(\mathbb{R}^n)} \lesssim \big(1 + B_2(t,0)\big)^{-\frac{n}{2}\left(\frac{1}{m}-\frac{1}{2}\right)}$$
$$\times\Big(\|(u_0, u_1)\|_{\mathscr{A}_{m,s_1}} + \|(v_0, v_1)\|_{\mathscr{A}_{m,s_2}}\Big),$$

$$\||D|^{\tilde{s}_2} v(t,\cdot)\|_{L^2(\mathbb{R}^n)} \lesssim \big(1+B_2(t,0)\big)^{-\frac{n}{2}\left(\frac{1}{m}-\frac{1}{2}\right)-\frac{\tilde{s}_2}{2}} \times \Big(\|(u_0,u_1)\|_{\mathscr{A}_{m,s_1}} + \|(v_0,v_1)\|_{\mathscr{A}_{m,s_2}}\Big),$$

$$\|v_t(t,\cdot)\|_{L^2(\mathbb{R}^n)} \lesssim b_2(t)^{-1}\big(1+B_2(t,0)\big)^{-\frac{n}{2}\left(\frac{1}{m}-\frac{1}{2}\right)-1} \times \Big(\|(u_0,u_1)\|_{\mathscr{A}_{m,s_1}} + \|(v_0,v_1)\|_{\mathscr{A}_{m,s_2}}\Big),$$

$$\||D|^{\tilde{s}_2-1} v_t(t,\cdot)\|_{L^2(\mathbb{R}^n)} \lesssim b_2(t)^{-1}\big(1+B_2(t,0)\big)^{-\frac{n}{2}\left(\frac{1}{m}-\frac{1}{2}\right)-1-\frac{\tilde{s}_2-1}{2}} \times \Big(\|(u_0,u_1)\|_{\mathscr{A}_{m,s_1}} + \|(v_0,v_1)\|_{\mathscr{A}_{m,s_2}}\Big).$$

Example 2.12 Let us consider the model from Example 2.9. If we choose data (u_0,u_1), (v_0,v_1) from $\mathscr{A}_{1,5}\times\mathscr{A}_{1,\frac{11}{2}}$, then the admissible range for the exponents p, q and $\tilde{q}$ to guarantee the global (in time) existence of small data solutions can be chosen as follows:

$$p>5,\quad q>\frac{11}{2},\quad \tilde{q}>\frac{25}{22}.$$

3 Philosophy of Our Approach

We define the norm of the solution space $X(t)$ by

$$\|(u,v)\|_{X(t)} = \sup_{\tau\in[0,t]}\big\{M_1(\tau,u)+M_2(\tau,v)\big\},$$

where we shall choose $M(\tau,u)$ and $M(\tau,v)$ with respect to the goals of each theorem.

Let N be the mapping on $X(t)$ which is defined by

$$N:(u,v)\in X(t)\to N(u,v)=\big(u^{ln}+u^{nl}, v^{ln}+v^{nl}\big),$$

where

$$u^{ln}(t,x) := E_{1,0}(t,0,x) *_{(x)} u_0(x) + E_{1,1}(t,0,x) *_{(x)} u_1(x),$$

$$u^{nl}(t,x) := \int_0^t E_{1,1}(t,\tau,x) *_{(x)} |v(\tau,x)|^p d\tau,$$

$$v^{ln}(t,x) := E_{2,0}(t,0,x) *_{(x)} v_0(x) + E_{2,1}(t,0,x) *_{(x)} v_1(x),$$

$$v^{nl}(t,x) := \int_0^t E_{2,1}(t,\tau,x) *_{(x)} |u(\tau,x)|^q d\tau.$$

Our aim is to prove the following inequalities which imply among other things the global (in time) existence of small data solutions:

$$\|N(u,v)\|_{X(t)} \lesssim \|(u_0,u_1)\|_{\mathscr{A}_{m,s_1}} + \|(v_0,v_1)\|_{\mathscr{A}_{m,s_2}} + \|(u,v)\|_{X(t)}^p + \|(u,v)\|_{X(t)}^q, \tag{12}$$

$$\begin{aligned}\|N(u,v) - N(\tilde{u},\tilde{v})\|_{X(t)} &\lesssim \|(u,v)-(\tilde{u},\tilde{v})\|_{X(t)} \\ &\times\big(\|(u,v)\|_{X(t)}^{p-1} + \|(\tilde{u},\tilde{v})\|_{X(t)}^{p-1} + \|(u,v)\|_{X(t)}^{q-1} + \|(\tilde{u},\tilde{v})\|_{X(t)}^{q-1}\big).\end{aligned} \tag{13}$$

We can immediately obtain from the introduced norm of the solution space $X(t)$ the following inequality:

$$\|(u^{ln},v^{ln})\|_{X(t)} \lesssim \|(u_0,u_1)\|_{\mathscr{A}_{m,s_1}} + \|(v_0,v_1)\|_{\mathscr{A}_{m,s_2}}.$$

This inequality implies (12) with the following estimate which we shall prove separately for each case:

$$\|(u^{nl},v^{nl})\|_{X(t)} \lesssim \|(u,v)\|_{X(t)}^p + \|(u,v)\|_{X(t)}^q. \tag{14}$$

Summarizing, we will prove for each case the inequalities (14) and (13).

3.1 Proof of Theorem 2.1

For $s_1 = s_2 = s$ and without loss of generality we assume $\tilde{p}_{r_1,r_2} < p_{Fuj,m}(n) < \tilde{q}_{r_1,r_2}$. Let us choose

$$\begin{aligned}M_1(\tau,u) &= (1+B_1(\tau,0))^{\frac{n}{2}\left(\frac{1}{m}-\frac{1}{2}\right)-\gamma_{n,m}(\tilde{p}_{r_1,r_2})}\|u(\tau,\cdot)\|_{L^2(\mathbb{R}^n)} \\ &\quad +(1+B_1(\tau,0))^{\frac{n}{2}\left(\frac{1}{m}-\frac{1}{2}\right)+\frac{s}{2}-\gamma_{n,m}(\tilde{p}_{r_1,r_2})}\||D|^s u(\tau,\cdot)\|_{L^2(\mathbb{R}^n)}, \\ M_2(\tau,v) &= (1+B_2(\tau,0))^{\frac{n}{2}\left(\frac{1}{m}-\frac{1}{2}\right)}\|v(\tau,\cdot)\|_{L^2(\mathbb{R}^n)} \\ &\quad +(1+B_2(\tau,0))^{\frac{n}{2}\left(\frac{1}{m}-\frac{1}{2}\right)+\frac{s}{2}}\||D|^s v(\tau,\cdot)\|_{L^2(\mathbb{R}^n)}.\end{aligned}$$

Lemma 3.1 *Using the Gagliardo-Nirenberg inequality we get for* $0 \le \tau \le t$ *the estimates*

$$\||v(\tau,\cdot)|^p\|_{L^2} \lesssim (1+B_2(\tau,0))^{-\frac{n}{2m}p+\frac{n}{4}}\|(u,v)\|_{X(t)}^p, \tag{15}$$

$$\||v(\tau,\cdot)|^p\|_{L^m} \lesssim (1+B_2(\tau,0))^{-\frac{n}{2m}p+\frac{n}{2m}}\|(u,v)\|_{X(t)}^p, \tag{16}$$

$$\||u(\tau,\cdot)|^q\|_{L^2} \lesssim (1+B_1(\tau,0))^{-\frac{n}{2m}q+\frac{n}{4}+\gamma_{n,m}(\tilde{p}_{r_1,r_2})q}\|(u,v)\|_{X(t)}^q, \tag{17}$$

$$\||u(\tau,\cdot)|^q\|_{L^m} \lesssim (1+B_1(\tau,0))^{-\frac{n}{2m}q+\frac{n}{2m}+\gamma_{n,m}(\tilde{p}_{r_1,r_2})q}\|(u,v)\|^q_{X(t)}, \tag{18}$$

$$\begin{aligned}\big\||v(\tau,\cdot)|^p-|\tilde{v}(\tau,\cdot)|^p\big\|_{L^2} &\lesssim (1+B_2(\tau,0))^{-\frac{n}{2m}p+\frac{n}{4}} \\ &\quad\times\|v-\tilde{v}\|_{X(t)}\big(\|v\|^{p-1}_{X(t)}+\|\tilde{v}\|^{p-1}_{X(t)}\big),\end{aligned} \tag{19}$$

$$\begin{aligned}\big\||v(\tau,\cdot)|^p-|\tilde{v}(\tau,\cdot)|^p\big\|_{L^m} &\lesssim (1+B_2(\tau,0))^{-\frac{n}{2m}p+\frac{n}{2m}} \\ &\quad\times\|v-\tilde{v}\|_{X(t)}\big(\|v\|^{p-1}_{X(t)}+\|\tilde{v}\|^{p-1}_{X(t)}\big),\end{aligned} \tag{20}$$

$$\begin{aligned}\big\||u(\tau,\cdot)|^q-|\tilde{u}(\tau,\cdot)|^q\big\|_{L^2} &\lesssim (1+B_1(\tau,0))^{-\frac{n}{2m}q+\frac{n}{4}+\gamma_{n,m}(\tilde{p}_{r_1,r_2})q} \\ &\quad\times\|u-\tilde{u}\|_{X(t)}\big(\|u\|^{q-1}_{X(t)}+\|\tilde{u}\|^{q-1}_{X(t)}\big),\end{aligned} \tag{21}$$

$$\begin{aligned}\big\||u(\tau,\cdot)|^q-|\tilde{u}(\tau,\cdot)|^q\big\|_{L^m} &\lesssim (1+B_1(\tau,0))^{-\frac{n}{2m}q+\frac{n}{2m}+\gamma_{n,m}(\tilde{p}_{r_1,r_2})q} \\ &\quad\times\|u-\tilde{u}\|_{X(t)}\big(\|u\|^{q-1}_{X(t)}+\|\tilde{u}\|^{q-1}_{X(t)}\big),\end{aligned} \tag{22}$$

provided that the condition (7) *is satisfied.*

Proof Let us prove (16). Using the Gagliardo-Nirenberg inequality we obtain

$$\begin{aligned}&\big\||v(\tau,x)|^p\big\|_{L^m(\mathbb{R}^n)} \\ &= \Big(\int_{\mathbb{R}^n}|v(\tau,x)|^{mp}dx\Big)^{\frac{1}{mp}p} = \|v(\tau,\cdot)\|^p_{L^{mp}(\mathbb{R}^n)} \lesssim \|v(\tau,\cdot)\|^{(1-\theta)p}_{L^2(\mathbb{R}^n)}\||D|^s v(\tau,\cdot)\|^{\theta p}_{L^2(\mathbb{R}^n)},\end{aligned}$$

where we choose

$$\theta = \frac{n}{s}\Big(\frac{1}{2}-\frac{1}{mp}\Big)\in[0,1]$$

due to condition (7) for p. By using for $0\le\tau\le t$ the definition of the norm of the solution space $X(t)$ we get

$$\|v(\tau,\cdot)\|^p_{L^{mp}(\mathbb{R}^n)} \lesssim (1+B_2(\tau,0))^{(1-\theta)p\left(-\frac{n}{2}\left(\frac{1}{m}-\frac{1}{2}\right)\right)+\theta p\left(-\frac{n}{2}\left(\frac{1}{m}-\frac{1}{2}\right)-\frac{s}{2}\right)}\|(u,v)\|^p_{X(t)}.$$

Then

$$\|v(\tau,\cdot)\|^p_{L^{mp}(\mathbb{R}^n)} \lesssim (1+B_2(\tau,0))^{-\frac{n}{2m}p+\frac{n}{2m}}\|(u,v)\|^p_{X(t)}.$$

In the same way we can prove (18). After setting $m=2$ we conclude (15) and (17).

For (20), we get after using Hölder's inequality

$$\begin{aligned}&\big\||v(\tau,x)|^p-|\tilde{v}(\tau,x)|^p\big\|_{L^m(\mathbb{R}^n)} \\ &\quad\lesssim \big\|v(\tau,\cdot)-\tilde{v}(\tau,\cdot)\big\|_{L^{mp}(\mathbb{R}^n)}\big(\|v(\tau,\cdot)\|^{p-1}_{L^{mp}(\mathbb{R}^n)}+\|\tilde{v}(\tau,\cdot)\|^{p-1}_{L^{mp}(\mathbb{R}^n)}\big).\end{aligned}$$

By using the definition of the norm of the solution space $X(t)$ and after applying the classical Gagliardo-Nirenberg inequality to estimate the norms $\big\|v(\tau,\cdot)-\tilde{v}(\tau,\cdot)\big\|_{L^{mp}(\mathbb{R}^n)}$, $\|v(\tau,\cdot)\|^{p-1}_{L^{mp}(\mathbb{R}^n)}$ and $\|\tilde{v}(\tau,\cdot)\|^{p-1}_{L^{mp}(\mathbb{R}^n)}$ as we did for (16) we obtain for $0\le\tau\le t$ the desired estimates. Following the same ideas we obtain (22). Finally, by setting $m=2$ one may conclude (19) and (21).

We come back to the proof of Theorem 2.1. To estimate the nonlinear part we begin with u^{nl}. Let us begin to estimate the norm $\big\||D|^s u^{nl}(t,\cdot)\big\|_{L^2}$. Using the estimates (15), (16) and the estimate (3) of Theorem 1.3 we get

$$\begin{aligned}
&\||D|^s u^{nl}(t,\cdot)\|_{L^2(\mathbb{R}^n)}\\
&\lesssim \int_0^t b_1(\tau)^{-1}(1+B_1(t,\tau))^{-\frac{n}{2}\left(\frac{1}{m}-\frac{1}{2}\right)-\frac{s}{2}}\||v(\tau,\cdot)|^p\|_{L^m(\mathbb{R}^n)\cap L^2(\mathbb{R}^n)}d\tau\\
&\lesssim \|(u,v)\|^p_{X(t)}\int_0^{\frac{t}{2}} b_1(\tau)^{-1}(1+B_1(t,\tau))^{-\frac{n}{2}\left(\frac{1}{m}-\frac{1}{2}\right)-\frac{s}{2}}\\
&\qquad\qquad\times(1+B_1(\tau,0))^{-\frac{n}{2m}(\tilde{p}_{r_1,r_2}-1)}d\tau\\
&\quad+\|(u,v)\|^p_{X(t)}\int_{\frac{t}{2}}^t b_1(\tau)^{-1}(1+B_1(t,\tau))^{-\frac{n}{2}\left(\frac{1}{m}-\frac{1}{2}\right)-\frac{s}{2}}\\
&\qquad\qquad\times(1+B_1(\tau,0))^{-\frac{n}{2m}(\tilde{p}_{r_1,r_2}-1)}d\tau\\
&\lesssim \|(u,v)\|^p_{X(t)}(1+B_1(t,0))^{-\frac{n}{2}\left(\frac{1}{m}-\frac{1}{2}\right)-\frac{s}{2}}\\
&\qquad\qquad\times\int_0^{\frac{t}{2}} b_1(\tau)^{-1}(1+B_1(\tau,0))^{-\frac{n}{2m}(\tilde{p}_{r_1,r_2}-1)}d\tau\\
&\quad+\|(u,v)\|^p_{X(t)}(1+B_1(t,0))^{-\frac{n}{2m}(\tilde{p}_{r_1,r_2}-1)}\\
&\qquad\qquad\times\int_{\frac{t}{2}}^t b_1(\tau)^{-1}(1+B_1(t,\tau))^{-\frac{n}{2}\left(\frac{1}{m}-\frac{1}{2}\right)-\frac{s}{2}}d\tau.
\end{aligned}$$

If $\tau\in[0,\frac{t}{2}]$, then

$$\int_0^{\frac{t}{2}} b_1(\tau)^{-1}(1+B_1(\tau,0))^{-\frac{n}{2m}(\tilde{p}_{r_1,r_2}-1)}d\tau\lesssim(1+B_1(t,0))^{\gamma_{n,m}(\tilde{p}_{r_1,r_2})}.$$

If $\tau\in[\frac{t}{2},t]$ and $-\frac{n}{2}\left(\frac{1}{m}-\frac{1}{2}\right)-\frac{s}{2}>-1$ which is equivalent to $n<\frac{2m(2-s)}{2-m}$, then

$$\int_{\frac{t}{2}}^t b_1(\tau)^{-1}(1+B_1(t,\tau))^{-\frac{n}{2}\left(\frac{1}{m}-\frac{1}{2}\right)-\frac{s}{2}}d\tau\lesssim(1+B_1(t,0))^{-\frac{n}{2}\left(\frac{1}{m}-\frac{1}{2}\right)-\frac{s}{2}+1}.$$

Consequently, we get

$$\||D|^s u^{nl}(t,\cdot)\|_{L^2(\mathbb{R}^n)} \lesssim \|(u,v)\|^p_{X(t)}(1+B_1(t,0))^{-\frac{n}{2}\left(\frac{1}{m}-\frac{1}{2}\right)-\frac{s}{2}+\gamma_{n,m}(\tilde{p}_{r_1,r_2})}. \quad (23)$$

In the same way we obtain

$$\|u^{nl}(t,\cdot)\|_{L^2(\mathbb{R}^n)} \lesssim \|(u,v)\|^p_{X(t)}(1+B_1(t,0))^{-\frac{n}{2}\left(\frac{1}{m}-\frac{1}{2}\right)+\gamma_{n,m}(\tilde{p}_{r_1,r_2})}. \quad (24)$$

On the other hand, for v^{nl} by using the estimates (17), (18) and the estimate (3) of Theorem 1.3 we obtain

$$\begin{aligned}
&\||D|^s v^{nl}(t,\cdot)\|_{L^2(\mathbb{R}^n)}\\
&\lesssim \int_0^t b_2(\tau)^{-1}(1+B_2(t,\tau))^{-\frac{n}{2}\left(\frac{1}{m}-\frac{1}{2}\right)-\frac{s}{2}}\||u(\tau,\cdot)|^q\|_{L^m(\mathbb{R}^n)\cap L^2(\mathbb{R}^n)}d\tau\\
&\lesssim \|(u,v)\|^q_{X(t)}\int_0^{\frac{t}{2}} b_2(\tau)^{-1}(1+B_2(t,\tau))^{-\frac{n}{2}\left(\frac{1}{m}-\frac{1}{2}\right)-\frac{s}{2}}\\
&\qquad\times(1+B_2(\tau,0))^{-\frac{n}{2m}(\tilde{q}_{r_1,r_2}-1)+\gamma_{n,m}(\tilde{p}_{r_1,r_2})q\left(\frac{1+r_1}{1+r_2}\right)}d\tau\\
&\quad+\|(u,v)\|^q_{X(t)}\int_{\frac{t}{2}}^t b_2(\tau)^{-1}(1+B_2(t,\tau))^{-\frac{n}{2}\left(\frac{1}{m}-\frac{1}{2}\right)-\frac{s}{2}}\\
&\qquad\times(1+B_2(\tau,0))^{-\frac{n}{2m}(\tilde{q}_{r_1,r_2}-1)+\gamma_{n,m}(\tilde{p}_{r_1,r_2})q\left(\frac{1+r_1}{1+r_2}\right)}d\tau.
\end{aligned}$$

If $\tau\in[0,\frac{t}{2}]$, then we have

$$\begin{aligned}
&\int_0^{\frac{t}{2}} b_2(\tau)^{-1}(1+B_2(t,\tau))^{-\frac{n}{2}\left(\frac{1}{m}-\frac{1}{2}\right)-\frac{s}{2}}\\
&\quad\times(1+B_2(\tau,0))^{-\frac{n}{2m}(\tilde{q}_{r_1,r_2}-1)+\gamma_{n,m}(\tilde{p}_{r_1,r_2})q\left(\frac{1+r_1}{1+r_2}\right)}d\tau \lesssim (1+B_2(t,0))^{-\frac{n}{2}\left(\frac{1}{m}-\frac{1}{2}\right)-\frac{s}{2}},
\end{aligned}$$

where we used

$$-\frac{n}{2m}(\tilde{q}_{r_1,r_2}-1)+\gamma_{n,m}(\tilde{p}_{r_1,r_2})q\left(\frac{1+r_1}{1+r_2}\right)+1<0$$

which is equivalent to (6) after taking account of $\tilde{p}_{r_1,r_2} < \tilde{q}_{r_1,r_2}$. If $\tau \in [\frac{t}{2}, t]$, then we have

$$\begin{aligned}\int_{\frac{t}{2}}^{t} & b_2(\tau)^{-1}(1 + B_2(t,\tau))^{-\frac{n}{2}\left(\frac{1}{m}-\frac{1}{2}\right)-\frac{s}{2}} \\ & \qquad \times (1 + B_2(\tau,0))^{-\frac{n}{2m}(\tilde{q}_{r_1,r_2}-1)+\gamma_{n,m}(\tilde{p}_{r_1,r_2})q\left(\frac{1+r_1}{1+r_2}\right)} d\tau \\ & \lesssim (1 + B_2(t,0))^{-\frac{n}{2m}(\tilde{q}_{r_1,r_2}-1)+\gamma_{n,m}(\tilde{p}_{r_1,r_2})q\left(\frac{1+r_1}{1+r_2}\right)} \\ & \qquad \times \int_{\frac{t}{2}}^{t} b_2(\tau)^{-1}(1 + B_2(t,\tau))^{-\frac{n}{2}\left(\frac{1}{m}-\frac{1}{2}\right)-\frac{s}{2}} d\tau \\ & \lesssim (1 + B_2(t,0))^{-\frac{n}{2}\left(\frac{1}{m}-\frac{1}{2}\right)-\frac{s}{2}},\end{aligned}$$

where we used again

$$-\frac{n}{2m}(\tilde{q}_{r_1,r_2} - 1) + \gamma_{n,m}(\tilde{p})q\left(\frac{1+r_1}{1+r_2}\right) + 1 < 0.$$

Consequently, we have

$$\| |D|^s v^{nl}(t,\cdot)\|_{L^2(\mathbb{R}^n)} \lesssim \|(u,v)\|_{X(t)}^q (1 + B_2(t,0))^{-\frac{n}{2}\left(\frac{1}{m}-\frac{1}{2}\right)-\frac{s}{2}}. \tag{25}$$

In the same way we may conclude

$$\|v^{nl}(t,\cdot)\|_{L^2(\mathbb{R}^n)} \lesssim \|(u,v)\|_{X(t)}^q (1 + B_2(t,0))^{-\frac{n}{2}\left(\frac{1}{m}-\frac{1}{2}\right)}. \tag{26}$$

Finally, from (23) to (26) we get (14).

Now we prove (13). Let us assume that (u, v) and $(\tilde{u}, \tilde{u})$ are two vector-functions belonging to $X(t)$. Then we have

$$\begin{aligned}N(u,v) - N(\tilde{u},\tilde{v}) = \Big(& \textstyle\int_0^t E_{1,1}(t,\tau,x) *_{(x)} \left(|v(\tau,x)|^p - |\tilde{v}(\tau,x)|^p\right) d\tau, \\ & \textstyle\int_0^t E_{2,1}(t,\tau,x) *_{(x)} \left(|u(\tau,x)|^p - |\tilde{u}(\tau,x)|^p\right) d\tau\Big).\end{aligned} \tag{27}$$

Analogously to (23)–(26) by using (19)–(22) we may conclude

$$\begin{aligned}\Big\| |D|^s & \textstyle\int_0^t E_{1,1}(t,\tau,x) *_{(x)} \left(|v(\tau,x)|^p - |\tilde{v}(\tau,x)|^p\right) d\tau \Big\|_{L^2(\mathbb{R}^n)} \\ & \lesssim (1 + B_1(t,0))^{-\frac{n}{2}\left(\frac{1}{m}-\frac{1}{2}\right)-\frac{s}{2}+\gamma_{n,m}(\tilde{p}_{r_1,r_2})} \\ & \qquad \times \|(u,v) - (\tilde{u},\tilde{v})\|_{X(t)} \left(\|(u,v)\|_{X(t)}^{p-1} + \|(\tilde{u},\tilde{v})\|_{X(t)}^{p-1}\right),\end{aligned} \tag{28}$$

$$\begin{aligned}
&\Big\| \int_0^t E_{1,1}(t,\tau,x) *_{(x)} \big(|v(\tau,x)|^p - |\tilde{v}(\tau,x)|^p\big)\, d\tau \Big\|_{L^2(\mathbb{R}^n)} \\
&\quad \lesssim (1+B_1(t,0))^{-\frac{n}{2}\left(\frac{1}{m}-\frac{1}{2}\right)+\gamma_{n,m}(\tilde{p}_{r_1,r_2})} \\
&\qquad \times \|(u,v)-(\tilde{u},\tilde{v})\|_{X(t)}\big(\|(u,v)\|_{X(t)}^{p-1} + \|(\tilde{u},\tilde{v})\|_{X(t)}^{p-1}\big),
\end{aligned} \tag{29}$$

$$\begin{aligned}
&\Big\| |D|^s \int_0^t E_{2,1}(t,\tau,x) *_{(x)} \big(|u(\tau,x)|^q - |\tilde{u}(\tau,x)|^q\big)\, d\tau \Big\|_{L^2(\mathbb{R}^n)} \\
&\quad \lesssim (1+B_2(t,0))^{-\frac{n}{2}\left(\frac{1}{m}-\frac{1}{2}\right)-\frac{s}{2}} \\
&\qquad \times \|(u,v)-(\tilde{u},\tilde{v})\|_{X(t)}\big(\|(u,v)\|_{X(t)}^{q-1} + \|(\tilde{u},\tilde{v})\|_{X(t)}^{q-1}\big),
\end{aligned} \tag{30}$$

$$\begin{aligned}
&\Big\| \int_0^t E_{2,1}(t,\tau,x) *_{(x)} \big(|u(\tau,x)|^q - |\tilde{u}(\tau,x)|^q\big)\, d\tau \Big\|_{L^2(\mathbb{R}^n)} \\
&\quad \lesssim (1+B_2(t,0))^{-\frac{n}{2}\left(\frac{1}{m}-\frac{1}{2}\right)} \\
&\qquad \times \|(u,v)-(\tilde{u},\tilde{v})\|_{X(t)}\big(\|(u,v)\|_{X(t)}^{q-1} + \|(\tilde{u},\tilde{v})\|_{X(t)}^{q-1}\big).
\end{aligned} \tag{31}$$

In this way we complete the proof.

3.2 *Proof of Theorem 2.6*

We can prove this theorem by following the same steps used in the proof of Theorem 2.1, after setting $s_1 = s_2 = 1$. But in this case we define a modified solution space $X(t)$, with additional terms formed by suitable norms of u_t and v_t, namely

$$X(t) = \Big\{(u,v) \in \big(\mathscr{C}([0,t], H^1(\mathbb{R}^n)) \cap \mathscr{C}^1([0,t], L^2(\mathbb{R}^n))\big)^2\Big\}$$

with the norm

$$\begin{aligned}
M_1(\tau,u) &= (1+B_1(\tau,0))^{\frac{n}{2}\left(\frac{1}{m}-\frac{1}{2}\right)-\gamma_{n,m}(\tilde{p}_{r_1,r_2})}\|u(\tau,\cdot)\|_{L^2(\mathbb{R}^n)} \\
&\quad + (1+B_1(\tau,0))^{\frac{n}{2}\left(\frac{1}{m}-\frac{1}{2}\right)+\frac{1}{2}-\gamma_{n,m}(\tilde{p}_{r_1,r_2})}\|\nabla u(\tau,\cdot)\|_{L^2(\mathbb{R}^n)} \\
&\quad + b_1(\tau)(1+B_1(\tau,0))^{\frac{n}{2}\left(\frac{1}{m}-\frac{1}{2}\right)+1-\gamma_{n,m}(\tilde{p}_{r_1,r_2})}\|u_t(\tau,\cdot)\|_{L^2(\mathbb{R}^n)}, \\
M_2(\tau,v) &= (1+B_2(\tau,0))^{\frac{n}{2}\left(\frac{1}{m}-\frac{1}{2}\right)}\|v(\tau,\cdot)\|_{L^2(\mathbb{R}^n)} \\
&\quad + (1+B_2(\tau,0))^{\frac{n}{2}\left(\frac{1}{m}-\frac{1}{2}\right)+\frac{1}{2}}\|\nabla v(\tau,\cdot)\|_{L^2(\mathbb{R}^n)} \\
&\quad + b_2(\tau)(1+B_2(\tau,0))^{\frac{n}{2}\left(\frac{1}{m}-\frac{1}{2}\right)+1}\|v_t(\tau,\cdot)\|_{L^2(\mathbb{R}^n)}.
\end{aligned}$$

From the proof of Theorem 2.1 we can conclude (23)–(26) for $s = 1$. Analogously, we can prove the following estimates for $\|u_t^{nl}(t,\cdot)\|_{L^2(\mathbb{R}^n)}$ and $\|v_t^{nl}(t,\cdot)\|_{L^2(\mathbb{R}^n)}$:

$$\|u_t^{nl}(t,\cdot)\|_{L^2(\mathbb{R}^n)} \lesssim \|(u,v)\|_{X(t)}^p b_1(t)^{-1}(1+B_1(t,0))^{-\frac{n}{2}\left(\frac{1}{m}-\frac{1}{2}\right)-1+\gamma_{n,m}(\tilde{p}_{r_1,r_2})}, \quad (32)$$

$$\|v_t^{nl}(t,\cdot)\|_{L^2(\mathbb{R}^n)} \lesssim \|(u,v)\|_{X(t)}^p b_2(t)^{-1}(1+B_2(t,0))^{-\frac{n}{2}\left(\frac{1}{m}-\frac{1}{2}\right)-1}, \quad (33)$$

where we use condition (9). Finally, from (32) and (33) together with (23)–(26) for $s = 1$ we conclude (14).

To prove (13), we conclude the estimates (28)–(31) for $s = 1$ and, similarly, to (32) and (33) we prove

$$\begin{aligned}
&\left\|\partial_t \int_0^t E_{1,1}(t,\tau,x) *_{(x)} \left(|v(\tau,x)|^p - |\tilde{v}(\tau,x)|^p\right) d\tau\right\|_{L^2(\mathbb{R}^n)} \\
&\qquad \lesssim b_1(t)^{-1}(1+B_1(t,0))^{-\frac{n}{2}\left(\frac{1}{m}-\frac{1}{2}\right)-1+\gamma_{n,m}(\tilde{p}_{r_1,r_2})} \\
&\qquad\qquad \times \|(u,v)-(\tilde{u},\tilde{v})\|_{X(t)}\left(\|(u,v)\|_{X(t)}^{p-1} + \|(\tilde{u},\tilde{v})\|_{X(t)}^{p-1}\right), \\
&\left\|\partial_t \int_0^t E_{2,1}(t,\tau,x) *_{(x)} \left(|u(\tau,x)|^p - |\tilde{u}(\tau,x)|^p\right) d\tau\right\|_{L^2(\mathbb{R}^n)} \\
&\qquad \lesssim b_2(t)^{-1}(1+B_2(t,0))^{-\frac{n}{2}\left(\frac{1}{m}-\frac{1}{2}\right)-1} \\
&\qquad\qquad \times \|(u,v)-(\tilde{u},\tilde{v})\|_{X(t)}\left(\|(u,v)\|_{X(t)}^{p-1} + \|(\tilde{u},\tilde{v})\|_{X(t)}^{p-1}\right).
\end{aligned}$$

The proof is completed.

3.3 *Proof of Theorem 2.8*

Let us define the space of solutions $X(t)$ by

$$\begin{aligned}
X(t) = \Big\{(u,v) \in &\left[\mathscr{C}([0,t], H^{s_1}(\mathbb{R}^n)) \cap \mathscr{C}^1([0,t], H^{s_1-1}(\mathbb{R}^n))\right] \\
&\times \left[\mathscr{C}([0,t], H^{s_2}(\mathbb{R}^n)) \cap \mathscr{C}^1([0,t], H^{s_2-1}(\mathbb{R}^n))\right]\Big\}
\end{aligned}$$

with the norm

$$
\begin{aligned}
M_1(\tau,u) &= \big(1+B_1(\tau,0)\big)^{\frac{n}{2}\left(\frac{1}{m}-\frac{1}{2}\right)}\|u(\tau,\cdot)\|_{L^2(\mathbb{R}^n)} \\
&+b_1(\tau)\big(1+B_1(\tau,0)\big)^{\frac{n}{2}\left(\frac{1}{m}-\frac{1}{2}\right)+1}\|u_t(\tau,\cdot)\|_{L^2(\mathbb{R}^n)} \\
&+b_1(\tau)\big(1+B_1(\tau,0)\big)^{\frac{n}{2}\left(\frac{1}{m}-\frac{1}{2}\right)+\frac{s_1-1}{2}+1}\||D|^{s_1-1}u_t(\tau,\cdot)\|_{L^2(\mathbb{R}^n)} \\
&+\big(1+B_1(\tau,0)\big)^{\frac{n}{2}\left(\frac{1}{m}-\frac{1}{2}\right)+\frac{s_1}{2}}\||D|^{s_1}u(\tau,\cdot)\|_{L^2(\mathbb{R}^n)},
\end{aligned}
$$

and

$$
\begin{aligned}
M_2(\tau,v) &= \big(1+B_2(\tau,0)\big)^{\frac{n}{2}\left(\frac{1}{m}-\frac{1}{2}\right)}\|v(\tau,\cdot)\|_{L^2(\mathbb{R}^n)} \\
&+b_2(\tau)\big(1+B_2(\tau,0)\big)^{\frac{n}{2}\left(\frac{1}{m}-\frac{1}{2}\right)+1}\|v_t(\tau,\cdot)\|_{L^2(\mathbb{R}^n)} \\
&+b_2(\tau)\big(1+B_2(\tau,0)\big)^{\frac{n}{2}\left(\frac{1}{m}-\frac{1}{2}\right)+\frac{s_2-1}{2}+1}\||D|^{s_2-1}v_t(\tau,\cdot)\|_{L^2(\mathbb{R}^n)} \\
&+\big(1+B_2(\tau,0)\big)^{\frac{n}{2}\left(\frac{1}{m}-\frac{1}{2}\right)+\frac{s_2}{2}}\||D|^{s_2}v(\tau,\cdot)\|_{L^2(\mathbb{R}^n)}.
\end{aligned}
$$

Lemma 3.2 *Under the assumptions of Theorem* 2.8 *and the choice of the above introduced norms, the following inequalities hold for* $0 \le \tau \le t$*:*

$$
\||v(\tau,\cdot)|^p\|_{\dot H^{s_1-1}} \lesssim (1+B_2(\tau,0))^{-\frac{n}{2m}p+\frac{n}{4}-\frac{s_2-1}{2}}\|(u,v)\|^p_{X(t)}, \tag{34}
$$

$$
\||u(\tau,\cdot)|^q\|_{\dot H^{s_2-1}} \lesssim (1+B_1(\tau,0))^{-\frac{n}{2m}q+\frac{n}{4}-\frac{s_1-1}{2}}\|(u,v)\|^q_{X(t)}, \tag{35}
$$

$$
\begin{aligned}
\||v(\tau,\cdot)|^p-|\tilde v(\tau,\cdot)|^p\|_{\dot H^{s_1-1}} &\lesssim (1+B_2(\tau,0))^{-\frac{n}{2m}p+\frac{n}{4}-\frac{s_2-1}{2}} \\
&\times M_2(t,v-\tilde v)\Big(\|(u,v)\|^{p-1}_{X(t)}+\|(\tilde u,\tilde v)\|^{p-1}_{X(t)}\Big),
\end{aligned} \tag{36}
$$

$$
\begin{aligned}
\||u(\tau,\cdot)|^q-|\tilde u(\tau,\cdot)|^q\|_{\dot H^{s_2-1}} &\lesssim (1+B_1(\tau,0))^{-\frac{n}{2m}q+\frac{n}{4}-\frac{s_1-1}{2}} \\
&\times M_1(t,u-\tilde u)\Big(\|(u,v)\|^{q-1}_{X(t)}+\|(\tilde u,\tilde v)\|^{q-1}_{X(t)}\Big).
\end{aligned} \tag{37}
$$

Proof Let us begin with (34). Taking into consideration the Propositions A.1 and A.3, in particular, formula (48) we may conclude for $p > \lceil s_1-1\rceil$ and $0 \le \tau \le t$ the following estimate:

$$
\begin{aligned}
&\||v(\tau,\cdot)|^p\|_{\dot H^{s_1-1}} \\
&\quad\lesssim \big\|v(\tau,\cdot)\big\|^{p-1}_{L^{q_1}}\big\||D|^{s_1-1}(\tau,\cdot)\big\|_{L^{q_2}}
\end{aligned}
$$

$$\lesssim \|v(\tau,\cdot)\|_{L^2}^{(p-1)(1-\theta_1)} \||D|^{s_2}v(\tau,\cdot)\|_{L^2}^{(p-1)\theta_1} \|v(\tau,\cdot)\|_{L^2}^{1-\theta_2} \||D|^{s_2}v(\tau,\cdot)\|_{L^2}^{\theta_2}$$
$$\lesssim (1+B_2(\tau,0))^{-\frac{n}{2m}p+\frac{n}{4}-\frac{s_2-1}{2}} \|(u,v)\|_{X(t)}^p,$$

where

$$\frac{p-1}{q_1}+\frac{1}{q_2}=\frac{1}{2},\ \theta_1=\frac{n}{s}\Big(\frac{1}{2}-\frac{1}{q_1}\Big)\in[0,1],\ \theta_2=\frac{n}{s_2}\Big(\frac{1}{2}-\frac{1}{q_2}\Big)+\frac{s_1-1}{s_2}\in\Big[\frac{s_1-1}{s_2},1\Big].$$

To satisfy the last conditions for the parameters θ_1 and θ_2 we choose $q_2=\frac{2n}{n-2}$ and $q_1=n(p-1)$. This choice implies the condition

$$1+\frac{2}{n}\le p\le 1+\frac{2}{n-2s_2}.$$

Consequently, we obtain (34). Analogously, we can prove (35).

Now we prove (36). Using the fractional Leibniz rule from Proposition A.2 we get

$$\big\||v(\tau,\cdot)|^p-|\tilde{v}(\tau,\cdot)|^p\big\|_{\dot{H}^{s_1-1}}$$
$$\lesssim \int_0^1 \Big\||D|^{s_1-1}\Big\{(v-\tilde{v})(v-r(v-\tilde{v}))|v-r(v-\tilde{v})|^{p-2}\Big\}\Big\|_{L^2}dr$$
$$\lesssim \int_0^1 \big\||D|^{s_1-1}(v-\tilde{v})\big\|_{L^{q_1}} \big\|(v-r(v-\tilde{v}))|v-r(v-\tilde{v})|^{p-2}\big\|_{L^{q_2}}dr$$
$$+\int_0^1 \big\|v-\tilde{v}\big\|_{L^{q_3}} \big\||D|^{s_1-1}\big[(v-r(v-\tilde{v}))|v-r(v-\tilde{v})|^{p-2}\big]\big\|_{L^{q_4}}dr,$$

where

$$\frac{1}{2}=\frac{1}{q_1}+\frac{1}{q_2}=\frac{1}{q_3}+\frac{1}{q_4}.$$

For the first integral we use the classical Gagliardo-Nirenberg inequality and obtain for $0\le\tau\le t$ the estimates

$$\big\||D|^{s_1-1}(v-\tilde{v})\big\|_{L^{q_1}}$$
$$\lesssim \big\|v-\tilde{v}\big\|_{L^2}^{1-\theta_1} \big\||D|^{s_2}(v-\tilde{v})\big\|_{L^2}^{\theta_1} \lesssim (1+B_2(\tau,0))^{-\frac{n}{2m}-\frac{s_2-1}{2}+\frac{n}{2q_1}} \big\|v-\tilde{v}\big\|_{X(t)},$$

and

$$
\begin{aligned}
&\big\|(v-r(v-\tilde{v}))|v-r(v-\tilde{v})|^{p-2}\big\|_{L^{q_2}}\\
&\quad\lesssim \big\|v-r(v-\tilde{v})\big\|_{L^2}^{(1-\theta_2)(p-1)}\big\||D|^{s_2}\left(v-r(v-\tilde{v})\right)\big\|_{L^2}^{\theta_2(p-1)}\\
&\quad\lesssim (1+B_2(\tau,0))^{-\frac{n}{2m}(p-1)+\frac{n}{2q_2}}\big\|v-r(v-\tilde{v})\big\|_{X(t)}^{p-1}
\end{aligned}
$$

for

$$
\theta_1=\frac{n}{s_2}\Big(\frac{1}{2}-\frac{1}{q_1}+\frac{s_1-1}{n}\Big)\in\Big[\frac{s_1-1}{s_2},1\Big],\ \ \theta_2=\frac{n}{s_2}\Big(\frac{1}{2}-\frac{1}{q_2(p-1)}\Big)\in[0,1]
$$

which is satisfied from condition (11) after choosing $q_1=\frac{2n}{n-2}$ and $q_2=n$.

To estimate the first norm in the second integral we use again the Gagliardo-Nirenberg inequality. In this way we obtain

$$
\big\|v-\tilde{v}\big\|_{L^{q_3}}\lesssim\big\|v-\tilde{v}\big\|_{L^2}^{1-\theta_3}\big\||D|^{s_2}(v-\tilde{v})\big\|_{L^2}^{\theta_3}\lesssim(1+B_2(\tau,0))^{-\frac{n}{2m}+\frac{n}{2q_3}}\big\|v-\tilde{v}\big\|_{X(t)},
$$

where

$$
\theta_3=\frac{n}{s_2}\Big(\frac{1}{2}-\frac{1}{q_3}\Big)\in[0,1].
$$

To estimate the second norm we use the fractional chain rule from Proposition A.3 for $p-1>\lceil s_1-1\rceil$. Then we get

$$
\begin{aligned}
&\big\||D|^{s_1-1}\big[(v-r(v-\tilde{v}))|v-r(v-\tilde{v})|^{p-2}\big]\big\|_{L^{q_4}}\\
&\qquad\lesssim\|v-r(v-\tilde{v})\|_{L^{q_5}}^{p-2}\||D|^{s_1-1}(v-r(v-\tilde{v}))\|_{L^{q_6}},
\end{aligned}
$$

where

$$
\frac{1}{q_4}=\frac{p-2}{q_5}+\frac{1}{q_6}.
$$

Using the Gagliardo-Nirenberg inequality to estimate the last two norms we get

$$
\begin{aligned}
\|v-r(v-\tilde{v})\|_{L^{q_5}}^{p-2}&\lesssim\|v-r(v-\tilde{v})\|_{L^2}^{(p-2)(1-\theta_5)}\||D|^{s_2}(v-r(v-\tilde{v}))\|_{L^2}^{(p-2)\theta_5}\\
&\lesssim(1+B_2(\tau,0))^{-\frac{n}{2m}(p-2)+\frac{n}{2q_5}(p-2)}\big\|v-r(v-\tilde{v})\big\|_{X(t)}^{p-2},
\end{aligned}
$$

and

$$\||D|^{s_1-1}(v-r(v-\tilde{v}))\|_{L^{q_6}} \lesssim \|v-r(v-\tilde{v})\|_{L^2}^{1-\theta_6}\||D|^{s_2}(v-r(v-\tilde{v}))\|_{L^2}^{\theta_6}$$
$$\lesssim (1+B_2(\tau,0))^{-\frac{n}{2m}+\frac{n}{2q_6}-\frac{s_2-1}{2}}\big\|v-r(v-\tilde{v})\big\|_{X(t)}$$

for

$$\theta_5=\frac{n}{s_2}\Big(\frac{1}{2}-\frac{1}{q_5}\Big)\in[0,1],\ \ \theta_6=\frac{n}{s_2}\Big(\frac{1}{2}-\frac{1}{q_6}\Big)+\frac{s_1-1}{s_2}\in\Big[\frac{s_1-1}{s_2},1\Big].$$

One possibility to choose the parameters q_3, q_4, q_5 and q_6 satisfying the last conditions is

$$q_3=n(p-1),\ \ q_4=\frac{2n(p-1)}{n(p-1)-2},\ \ q_5=n(p-1),\ \ q_6=\frac{2n}{n-2}.$$

This choice implies the condition

$$1+\frac{2}{n}\le p\le 1+\frac{2}{n-2s_2}$$

which follows from (11). Consequently, we get (36). Analogously, we can prove (37).

Let us come back to the proof of Theorem 2.8. We have from the estimate (4) of Theorem 1.3

$$\||D|^{s_1-1}u_t^{nl}(t,\cdot)\|_{L^2(\mathbb{R}^n)}$$
$$\lesssim \int_0^t b_1(\tau)^{-1}b_1(t)^{-1}(1+B_1(t,\tau))^{-\frac{n}{2}\left(\frac{1}{m}-\frac{1}{2}\right)-\frac{s_1-1}{2}-1}$$
$$\times\||v(\tau,\cdot)|^p\|_{L^m(\mathbb{R}^n)\cap L^2(\mathbb{R}^n)\cap\dot{H}^{s_1-1}(\mathbb{R}^n)}d\tau$$
$$\lesssim \|(u,v)\|_{X(t)}^p b_1(t)^{-1}(1+B_1(t,0))^{-\frac{n}{2}\left(\frac{1}{m}-\frac{1}{2}\right)-\frac{s_1-1}{2}-1},$$

where we use the estimates (15), (16), (34) and

$$\tilde{p}:=\tilde{p}_{r_1,r_2}=\frac{1+r_2}{1+r_1}(p-1)+1>\frac{2m}{n}\Big(\frac{s_1+1}{2}\Big)+1$$

from (11), in particular,

$$\tilde{p}>p>\lceil s_1\rceil>\frac{2m}{n}\Big(\frac{s_1+1}{2}\Big)+1\ \text{ for } n\ge 4,\ s_1>3 \text{ and } r_1<r_2.$$

Consequently, we obtain

$$\||D|^{s_1-1}u_t^{nl}(t,\cdot)\|_{L^2(\mathbb{R}^n)} \lesssim \|(u,v)\|_{X(t)}^p b_1(t)^{-1}(1+B_1(t,0))^{-\frac{n}{2}\left(\frac{1}{m}-\frac{1}{2}\right)-\frac{s_1-1}{2}-1}. \quad (38)$$

In a similar way we derive the estimates

$$\|u^{nl}(t,\cdot)\|_{L^2(\mathbb{R}^n)} \lesssim \|(u,v)\|_{X(t)}^p (1+B_1(t,0))^{-\frac{n}{2}\left(\frac{1}{m}-\frac{1}{2}\right)}, \quad (39)$$

$$\|u_t^{nl}(t,\cdot)\|_{L^2(\mathbb{R}^n)} \lesssim \|(u,v)\|_{X(t)}^p b_1(t)^{-1}(1+B_1(t,0))^{-\frac{n}{2}\left(\frac{1}{m}-\frac{1}{2}\right)-1}, \quad (40)$$

$$\||D|^{s_1}u^{nl}(t,\cdot)\|_{L^2(\mathbb{R}^n)} \lesssim \|(u,v)\|_{X(t)}^p (1+B_1(t,0))^{-\frac{n}{2}\left(\frac{1}{m}-\frac{1}{2}\right)-\frac{s_1}{2}}. \quad (41)$$

Following the same ideas to estimate $\||D|^{s_1-1}u_t^{nl}(t,\cdot)\|_{L^2(\mathbb{R}^n)}$, one can arrive at the following estimates:

$$\begin{aligned}&\||D|^{s_2-1}v_t^{nl}(t,\cdot)\|_{L^2(\mathbb{R}^n)}\\ &\quad\lesssim \|(u,v)\|_{X(t)}^q b_2(t)^{-1}(1+B_2(t,0))^{-\frac{n}{2}\left(\frac{1}{m}-\frac{1}{2}\right)-\frac{s_2-1}{2}-1},\end{aligned} \quad (42)$$

$$\|v^{nl}(t,\cdot)\|_{L^2(\mathbb{R}^n)} \lesssim \|(u,v)\|_{X(t)}^q (1+B_2(t,0))^{-\frac{n}{2}\left(\frac{1}{m}-\frac{1}{2}\right)}, \quad (43)$$

$$\|v_t^{nl}(t,\cdot)\|_{L^2(\mathbb{R}^n)} \lesssim \|(u,v)\|_{X(t)}^q b_2(t)^{-1}(1+B_2(t,0))^{-\frac{n}{2}\left(\frac{1}{m}-\frac{1}{2}\right)-1}, \quad (44)$$

$$\||D|^{s_2}v^{nl}(t,\cdot)\|_{L^2(\mathbb{R}^n)} \lesssim \|(u,v)\|_{X(t)}^q (1+B_2(t,0))^{-\frac{n}{2}\left(\frac{1}{m}-\frac{1}{2}\right)-\frac{s_2}{2}}. \quad (45)$$

But now we need the condition (10). This condition is not included in the condition $q > \lceil s_2 \rceil$ because of $q > \tilde{q}$ for $r_1 < r_2$.

Finally, from (38) to (45) we conclude (14).

To prove (13), we use (19)–(22) with (36), (37) in (27), to get in a similar way to (38)–(45) the following estimates:

$$\begin{aligned}&\Big\||D|^{s_1-1}\partial_t \int_0^t E_{1,1}(t,\tau,x) *_{(x)} \big(|v(\tau,x)|^p - |\tilde{v}(\tau,x)|^p\big)d\tau\Big\|_{L^2(\mathbb{R}^n)}\\ &\qquad\lesssim b_1(t)^{-1}(1+B_1(t,0))^{-\frac{n}{2}\left(\frac{1}{m}-\frac{1}{2}\right)-\frac{s_1-1}{2}-1}\\ &\qquad\qquad\times\|(u,v)-(\tilde{u},\tilde{v})\|_{X(t)}\big(\|(u,v)\|_{X(t)}^{p-1} + \|(\tilde{u},\tilde{v})\|_{X(t)}^{p-1}\big),\end{aligned}$$

$$\begin{aligned}&\Big\|\int_0^t E_{1,1}(t,\tau,x) *_{(x)} \big(|v(\tau,x)|^p - |\tilde{v}(\tau,x)|^p\big)d\tau\Big\|_{L^2(\mathbb{R}^n)}\\ &\qquad\lesssim (1+B_1(t,0))^{-\frac{n}{2}\left(\frac{1}{m}-\frac{1}{2}\right)}\\ &\qquad\qquad\times\|(u,v)-(\tilde{u},\tilde{v})\|_{X(t)}\big(\|(u,v)\|_{X(t)}^{p-1} + \|(\tilde{u},\tilde{v})\|_{X(t)}^{p-1}\big),\end{aligned}$$

$$\Big\|\partial_t\int_0^t E_{1,1}(t,\tau,x)*_{(x)}\big(|v(\tau,x)|^p-|\tilde{v}(\tau,x)|^p\big)d\tau\Big\|_{L^2(\mathbb{R}^n)}$$
$$\lesssim b_1(t)^{-1}(1+B_1(t,0))^{-\frac{n}{2}\left(\frac{1}{m}-\frac{1}{2}\right)-1}$$
$$\times\|(u,v)-(\tilde{u},\tilde{v})\|_{X(t)}\big(\|(u,v)\|_{X(t)}^{p-1}+\|(\tilde{u},\tilde{v})\|_{X(t)}^{p-1}\big),$$

$$\Big\||D|^{s_1}\int_0^t E_{1,1}(t,\tau,x)*_{(x)}\big(|v(\tau,x)|^p-|\tilde{v}(\tau,x)|^p\big)d\tau\Big\|_{L^2(\mathbb{R}^n)}$$
$$\lesssim (1+B_1(t,0))^{-\frac{n}{2}\left(\frac{1}{m}-\frac{1}{2}\right)-\frac{s_1}{2}}$$
$$\times\|(u,v)-(\tilde{u},\tilde{v})\|_{X(t)}\big(\|(u,v)\|_{X(t)}^{p-1}+\|(\tilde{u},\tilde{v})\|_{X(t)}^{p-1}\big),$$

$$\Big\|\int_0^t E_{2,1}(t,\tau,x)*_{(x)}\big(|u(\tau,x)|^q-|\tilde{u}(\tau,x)|^q\big)d\tau\Big\|_{L^2(\mathbb{R}^n)}$$
$$\lesssim (1+B_2(t,0))^{-\frac{n}{2}\left(\frac{1}{m}-\frac{1}{2}\right)}$$
$$\times\|(u,v)-(\tilde{u},\tilde{v})\|_{X(t)}\big(\|(u,v)\|_{X(t)}^{q-1}+\|(\tilde{u},\tilde{v})\|_{X(t)}^{q-1}\big),$$

$$\Big\|\partial_t\int_0^t E_{2,1}(t,\tau,x)*_{(x)}\big(|u(\tau,x)|^q-|\tilde{u}(\tau,x)|^q\big)d\tau\Big\|_{L^2(\mathbb{R}^n)}$$
$$\lesssim b_2(t)^{-1}(1+B_2(t,0))^{-\frac{n}{2}\left(\frac{1}{m}-\frac{1}{2}\right)-1}$$
$$\times\|(u,v)-(\tilde{u},\tilde{v})\|_{X(t)}\big(\|(u,v)\|_{X(t)}^{q-1}+\|(\tilde{u},\tilde{v})\|_{X(t)}^{q-1}\big),$$

$$\Big\||D|^{s_2-1}\partial_t\int_0^t E_{2,1}(t,\tau,x)*_{(x)}\big(|u(\tau,x)|^q-|\tilde{u}(\tau,x)|^q\big)d\tau\Big\|_{L^2(\mathbb{R}^n)}$$
$$\lesssim b_2(t)^{-1}(1+B_2(t,0))^{-\frac{n}{2}\left(\frac{1}{m}-\frac{1}{2}\right)-\frac{s_2-1}{2}-1}$$
$$\times\|(u,v)-(\tilde{u},\tilde{v})\|_{X(t)}\big(\|(u,v)\|_{X(t)}^{q-1}+\|(\tilde{u},\tilde{v})\|_{X(t)}^{q-1}\big),$$

$$\Big\||D|^{s_2}\int_0^t E_{2,1}(t,\tau,x)*_{(x)}\big(|u(\tau,x)|^q-|\tilde{u}(\tau,x)|^q\big)d\tau\Big\|_{L^2(\mathbb{R}^n)}$$
$$\lesssim (1+B_2(t,0))^{-\frac{n}{2}\left(\frac{1}{m}-\frac{1}{2}\right)-\frac{s_2}{2}}$$
$$\times\|(u,v)-(\tilde{u},\tilde{v})\|_{X(t)}\big(\|(u,v)\|_{X(t)}^{q-1}+\|(\tilde{u},\tilde{v})\|_{X(t)}^{q-1}\big).$$

In this way the proof is completed.

3.4 Proof of Theorem 2.11

To prove this theorem we choose the same norm of the solution space which is used in the proof of Theorem 2.8 with $\tilde{s}_2$ instead of s_2. Then we obtain the following lemma.

Lemma 3.3 *The following inequalities hold for* $0 \le \tau \le t$*:*

$$\||v(\tau,\cdot)|^p\|_{\dot{H}^{s_1-1}(\mathbb{R}^n)} \lesssim (1+B_2(\tau,0))^{-\frac{n}{2}\left(\frac{1}{m}-\frac{1}{2}\right)p-\frac{s_1-1}{2}-\frac{s^*}{2}(p-1)}\|(u,v)\|^p_{X(t)},$$

$$\||u(\tau,\cdot)|^q\|_{\dot{H}^{\tilde{s}_2-1}(\mathbb{R}^n)} \lesssim (1+B_1(\tau,0))^{-\frac{n}{2}\left(\frac{1}{m}-\frac{1}{2}\right)q-\frac{\tilde{s}_2-1}{2}-\frac{s^*}{2}(q-1)}\|(u,v)\|^q_{X(t)},$$

$$\begin{aligned}\Big\||v(\tau,\cdot)|^p-|\tilde{v}(\tau,\cdot)|^p\Big\|_{\dot{H}^{s_1-1}(\mathbb{R}^n)} &\lesssim (1+B_2(\tau,0))^{-\frac{n}{2m}p+\frac{n}{4}(p-1)-\frac{s_1-1}{2}-\frac{s^*}{2}(p-1)}\\ &\quad\times\|(u,v)-(\tilde{u},\tilde{v})\|_{X(t)}\left(\|(u,v)\|^{p-1}_{X(t)}+\|(\tilde{u},\tilde{v})\|^{p-1}_{X(t)}\right),\end{aligned}$$

$$\begin{aligned}\Big\||u(\tau,\cdot)|^q-|\tilde{u}(\tau,\cdot)|^q\Big\|_{\dot{H}^{\tilde{s}_2-1}(\mathbb{R}^n)} &\lesssim (1+B_1(\tau,0))^{-\frac{n}{2m}q+\frac{n}{4}(q-1)-\frac{\tilde{s}_2-1}{2}-\frac{s^*}{2}(q-1)}\\ &\quad\times\|(u,v)-(\tilde{u},\tilde{v})\|_{X(t)}\left(\|(u,v)\|^{p-1}_{X(t)}+\|(\tilde{u},\tilde{v})\|^{p-1}_{X(t)}\right),\end{aligned}$$

where we used $p > s_1$, $q > \tilde{s}_2$ *and* s^* *from Lemma* A.6.

The proof of this lemma can be obtained by using the introduced norm of solution space $X(t)$ and the rules for fractional powers from Corollary A.5 and Lemma A.6. We follow the same steps of the proof of Theorem 2.8, but now by using the estimates from Lemma 3.3. Taking into consideration these estimates, we get (38) to (45), for $\tilde{s}_2$ instead of s_2 under the following conditions:

1. $-\frac{n}{2m}\tilde{p}+\frac{n}{4}-\left(1-\frac{m}{2}\right)\frac{r_2-r_1}{1+r_1} < -\frac{n}{2}\left(\frac{1}{m}-\frac{1}{2}\right)-\frac{s_1-1}{2}-1$ which is satisfied due to the condition $p > s_1$,
2. $-\frac{n}{2m}\tilde{q}+\frac{n}{4} < -\frac{n}{2}\left(\frac{1}{m}-\frac{1}{2}\right)-\frac{s_1-1}{2}-1$ which is equivalent to $\tilde{q} \ge \frac{2m}{n}\left(\frac{s_2+1}{2}\right)+1$ supposed in the statement of the theorem.

In this way we can complete the proof.

4 Concluding Remarks

In this section we sketch possible generalizations of the results of this paper. Let us choose the time-dependent coefficients $b_1 = b_1(t)$ and $b_2 = b_2(t)$ in such a way that the dissipation terms $b_1(t)u_t$ and $b_2(t)v_t$ become effective and the following condition is satisfied:

$$B_2(t,0) \approx B_1(t,0)^{\alpha}, \tag{46}$$

where α is a positive real number. It is clear that (46) covers a larger class of effective dissipation terms comparing with those are treated in the previous sections of this paper.

Example 4.1 The following coefficients are effective and satisfy condition (46):

1. $b_1(t) = \frac{\mu_1}{(1+t)^{r_1}}, b_2(t) = \frac{\mu_2}{(1+t)^{r_2}}$ for some $\mu_1, \mu_2 > 0$ and $r_1, r_2 \in (-1, 1)$, where $\alpha = \frac{1+r_2}{1+r_1}$,
2. $b_1(t) = \frac{\mu_1}{(1+t)^{r_1}}(\log(c_{r_1,\gamma_1} + t))^{\gamma_1}, b_2(t) = \frac{\mu_2}{(1+t)^{r_2}}(\log(c_{r_2,\gamma_2} + t))^{\gamma_2}$ for some $\mu_1, \mu_2 > 0$, $\gamma_1, \gamma_2 > 0$ and $\alpha = \frac{1+r_2}{1+r_1} = \frac{\gamma_2}{\gamma_1}$,
3. $b_1(t) = \frac{\mu_1}{(1+t)^{r_1}(\log(c_{r_1,\gamma_1}+t))^{\gamma_1}}, b_2(t) = \frac{\mu_2}{(1+t)^{r_2}(\log(c_{r_2,\gamma_2}+t))^{\gamma_2}}$ for some $\mu_1, \mu_1 > 0$, $\gamma_1, \gamma_1 > 0$ and $\alpha = \frac{1+r_2}{1+r_1} = \frac{\gamma_2}{\gamma_1}$,

where c_{r_1,γ_1} and c_{r_1,γ_1} are sufficiently large positive constants.

Using the new class of dissipation terms we summarize generalizations of the first two cases of regularity of data in the following tables:

Low regular data

$\tilde{p} < p_{Fuj,m}(n) < \tilde{q}$	$\tilde{q} < p_{Fuj,m}(n) < \tilde{p}$
Loss of decay in the estimate for u $\gamma_{n,m}(\tilde{p}) = -\frac{n}{2m}(\tilde{p}-1)+1$ interaction condition $\frac{n}{2} > m\Big(\frac{\tilde{q}+\frac{1}{\alpha}}{\tilde{p}\tilde{q}-1+(\tilde{p}-1)\frac{1-\alpha}{\alpha}}\Big)$	Loss of decay in the estimate for v $\gamma_{n,m}(\tilde{q}) = -\frac{n}{2m}(\tilde{q}-1)+1$ interaction condition $\frac{n}{2} > m\Big(\frac{\tilde{p}+\alpha}{\tilde{p}\tilde{q}-1+(\tilde{q}-1)(\alpha-1)}\Big)$

Data from the energy space

$s = 1$	$\tilde{p} < p_{Fuj,m}(n) < \tilde{q}$	$\tilde{q} < p_{Fuj,m}(n) < \tilde{p}$
$r_1 < r_2$ $\tilde{q} = \tilde{q}_m$ $\tilde{p} = \tilde{p}$	Loss of decay in the estimate for u $\gamma_{n,m}(\tilde{p}) = -\frac{n}{2m}(\tilde{p}-1)+1$ interaction condition $\frac{n}{2} > m\Big(\frac{\tilde{q}+1+\frac{m}{2}\left(\frac{\alpha-1}{\alpha}\right)}{\tilde{p}\tilde{q}-1+\frac{m}{2}(\tilde{p}-1)\left(\frac{\alpha-1}{\alpha}\right)}\Big)$	Loss of decay in the estimate for v $\gamma_{n,m}(\tilde{q}) = -\frac{n}{2m}(\tilde{q}-1)+1+\varepsilon$ interaction condition $\frac{n}{2} > m\Big(\frac{\tilde{p}+\alpha}{\tilde{p}\tilde{q}-\alpha+\tilde{q}(\alpha-1)}\Big)$
$r_2 < r_1$ $\tilde{q} = \tilde{q}$ $\tilde{p} = \tilde{p}_m$	Loss of decay in the estimate for u $\gamma_{n,m}(\tilde{p}) = -\frac{n}{2m}(\tilde{p}-1)+1+\varepsilon$ interaction condition $\frac{n}{2} > m\Big(\frac{\tilde{q}+\frac{1}{\alpha}}{\tilde{p}\tilde{q}-\frac{1}{\alpha}+\tilde{p}\left(\frac{\alpha-1}{\alpha}\right)}\Big)$	Loss of decay in the estimate for v $\gamma_{n,m}(\tilde{q}) = -\frac{n}{2m}(\tilde{q}-1)+1$ interaction condition $\frac{n}{2} > m\Big(\frac{\tilde{p}+1+\frac{m}{2}(\alpha-1)}{\tilde{p}\tilde{q}-1+\frac{m}{2}(\tilde{q}-1)(\alpha-1)}\Big)$

where

$$\begin{aligned} \tilde{q}_m &= \tfrac{1}{\alpha}\left(q-\tfrac{m}{2}\right)+\tfrac{m}{2},\ \tilde{q}_{m=2} = \tilde{q} = \tfrac{1}{\alpha}(q-1)+1, \\ \tilde{p}_m &= \alpha\left(p-\tfrac{m}{2}\right)+\tfrac{m}{2},\ \tilde{p}_{m=2} = \tilde{p} = \alpha(p-1)+1. \end{aligned} \tag{47}$$

We can present possible generalizations of the last two theorems, too, by using the modified exponents of power nonlinearities (47) and dissipation terms satisfying condition (46).

Appendix

In the Appendix we collect some background material which is helpful and important for our approach. Most of these tools are from the theory of harmonic analysis and function spaces. In particular, these tools allow us to estimate power nonlinearities in different scales of function spaces.

Proposition A.1 *Let* $1 < p, p_0, p_1 < \infty, \sigma > 0$ *and* $s \in [0, \sigma)$. *Then it holds the following fractional Gagliardo-Nirenberg inequality for all* $u \in L^{p_0}(\mathbb{R}^n) \cap \dot{H}^{\sigma}_{p_1}(\mathbb{R}^n)$:

$$\|u\|_{\dot{H}^s_p(\mathbb{R}^n)} \lesssim \|u\|^{(1-\theta)}_{L^{p_0}(\mathbb{R}^n)} \|u\|^{\theta}_{\dot{H}^{\sigma}_{p_1}(\mathbb{R}^n)}, \tag{48}$$

where $\theta = \theta_{s,\sigma} := \frac{\frac{1}{p_0} - \frac{1}{p} + \frac{s}{n}}{\frac{1}{p_0} - \frac{1}{p_1} + \frac{\sigma}{n}}$ *and* $\frac{s}{\sigma} \le \theta \le 1$.

Proof For the proof see [7] and [1, 4–6, 8, 9].

Proposition A.2 *Let us assume* $s > 0$ *and* $1 \le r \le \infty, 1 < p_1, p_2, q_1, q_2 \le \infty$ *satisfying the following relation*

$$\frac{1}{r} = \frac{1}{p_1} + \frac{1}{p_2} = \frac{1}{q_1} + \frac{1}{q_2}.$$

Then the following fractional Leibniz rule holds:

$$\||D|^s(fg)\|_{L^r(\mathbb{R}^n)} \lesssim \||D|^s f\|_{L^{p_1}(\mathbb{R}^n)} \|g\|_{L^{p_2}(\mathbb{R}^n)} + \|f\|_{L^{q_1}(\mathbb{R}^n)} \||D|^s g\|_{L^{q_2}(\mathbb{R}^n)} \tag{49}$$

for all $f \in \dot{H}^s_{p_1}(\mathbb{R}^n) \cap L^{q_1}(\mathbb{R}^n)$ *and* $g \in \dot{H}^s_{q_2}(\mathbb{R}^n) \cap L^{p_2}(\mathbb{R}^n)$.

For more details concerning fractional Leibniz rules see [4].

Proposition A.3 *Let us choose* $s > 0$, $p > \lceil s \rceil$ *and* $1 < r, r_1, r_2 < \infty$ *satisfying*

$$\frac{1}{r} = \frac{p-1}{r_1} + \frac{1}{r_2}.$$

Let us denote by $F(u)$ one of the functions $|u|^p, \pm|u|^{p-1}u$. Then it holds the following fractional chain rule:

$$\| |D|^s F(u)\|_{L^r(\mathbb{R}^n)} \lesssim \|u\|^{p-1}_{L^{r_1}(\mathbb{R}^n)} \| |D|^s u\|_{L^{r_2}(\mathbb{R}^n)}. \tag{50}$$

Proof For the proof see [12].

Proposition A.4 *Let $p > 1$ and $u \in H^s_m(\mathbb{R}^n)$, where $s \in \left(\frac{n}{m}, p\right)$. Then the following estimates hold:*

$$\| |u|^p \|_{H^s_m(\mathbb{R}^n)} \lesssim \|u\|_{H^s_m(\mathbb{R}^n)} \|u\|^{p-1}_{L^\infty(\mathbb{R}^n)},$$

$$\| u|u|^{p-1} \|_{H^s_m(\mathbb{R}^n)} \lesssim \|u\|_{H^s_m(\mathbb{R}^n)} \|u\|^{p-1}_{L^\infty(\mathbb{R}^n)}.$$

Proof For the proof see [14].

We can derive from Proposition A.4 the following corollary.

Corollary A.5 *Under the assumptions of Proposition* A.4 *it holds:*

$$\| |u|^p \|_{\dot{H}^s_m(\mathbb{R}^n)} \lesssim \|u\|_{\dot{H}^s_m(\mathbb{R}^n)} \|u\|^{p-1}_{L^\infty(\mathbb{R}^n)},$$

$$\| u|u|^{p-1} \|_{\dot{H}^s_m(\mathbb{R}^n)} \lesssim \|u\|_{\dot{H}^s_m(\mathbb{R}^n)} \|u\|^{p-1}_{L^\infty(\mathbb{R}^n)}.$$

Proof For the proof see [13].

Lemma A.6 *Let $0 < 2s^* < n < 2s$. Then for any function $f \in \dot{H}^{s^*}(\mathbb{R}^n) \cap \dot{H}^s(\mathbb{R}^n)$ one has*

$$\|f\|_{L^\infty(\mathbb{R}^n)} \leq \|f\|_{\dot{H}^{s^*}(\mathbb{R}^n)} + \|f\|_{\dot{H}^s(\mathbb{R}^n)}.$$

Proof For the proof see [2].

References

1. F. Christ, M. Weinstein, Dispersion of small-amplitude solutions of the generalized Korteweg-de Vries equation. J. Funct. Anal. **100**, 87–109 (1991)
2. M. D'Abbicco, The threshold of effective damping for semilinear wave equations. Math. Meth. Appl. Sci. **38**, 1032–1045 (2015)
3. M. D'Abbicco, S. Lucente, M. Reissig, Semi-linear wave equations with effective damping. Chin. Ann. Math. **34B**(3), 345–380 (2013)
4. L. Grafakos, *Classical and Modern Fourier Analysis* (Prentice Hall, Upper Saddle River, 2004)
5. L. Grafakos, S. Oh, The Kato Ponce inequality. Commun. Partial Differ. Equ. **39**(6), 1128–1157 (2014)

6. A. Gulisashvili, M. Kon, Exact smoothing properties of Schrödinger semigroups. Am. J. Math. **118**, 1215–1248 (1996)
7. H. Hajaiej, L. Molinet, T. Ozawa, B. Wang, Necessary and sufficient conditions for the fractional Gagliardo-Nirenberg inequalities and applications to Navier-Stokes and generalized boson equations. Harmon. Anal. Nonlinear Partial Differ. Equ. **B26**, 159–175 (2011)
8. T. Kato, G. Ponce, Well-posedness and scattering results for the generalized Korteweg-de Vries equation via the contraction principle. Commun. Pure Appl. Math. **46**(4), 527–620 (1993)
9. C.E. Kenig, G. Ponce, L. Vega, Commutator estimates and the Euler and Navier Stokes equations. Commun. Pure Appl. Math. **41**, 891–907 (1988)
10. A. Mohammed Djaouti, M. Reissig, Weakly coupled systems of semilinear effectively damped waves with time-dependent coefficient, different power nonlinearities and different regularity of the data. Nonlinear Anal. **175**, 28–55 (2018)
11. K. Nishihara, Y. Wakasugi, Critical exponant for the Cauchy problem to the weakly coupled wave system. Nonlinear Anal. **108**, 249–259 (2014)
12. A. Palmieri, M. Reissig, Semi-linear wave models with power non-linearity and scale invariant time-dependent mass and dissipation, II. Math. Nachr. **291**(11–12), 1859–1892 (2018)
13. D.T. Pham, M. Kainane Mezadek, M. Reissig, Global existence for semilinear structurally damped $\sigma-$evolution models. J. Math. Anal. Appl. **431**, 569–596 (2015)
14. T. Runst, W. Sickel, *Sobolev Spaces of Fractional Order, Nemytskij Operators, and Nonlinear Partial Differential Equations*. De Gruyter Series in Nonlinear Analysis and Applications (Walter de Gruyter & Co., Berlin, 1996)
15. J. Wirth, Asymptotic properties of solutions to wave equations with time-dependent dissipation. Ph.D. thesis, TU Bergakademie Freiberg (2004)
16. J. Wirth, Wave equations with time-dependent dissipation II, Effective dissipation. J. Differ. Equ. **232**, 74–103 (2007)

Incompressible Limits for Generalisations to Symmetrisable Systems

Michael Dreher

Abstract We shortly review the incompressible limit of the barotropic Euler system of gas dynamics, also known as low Mach number limit, and the quasineutral limit of a simplified Euler–Poisson system. Then we develop a general pseudodifferential framework which is able to cover both examples, called generalised symmetrisable systems. This framework can also handle incompressible limits. As an application, we then discuss a barotropic Euler–Poisson system.

1 Introduction

Let us recall some standard results on symmetric hyperbolic systems: we consider

$$A_0(t,x)\partial_t U(t,x) + A(t,x,D_x)U(t,x) = F(t,x), \quad (t,x) \in (0,\infty) \times \mathbb{R}^n, \tag{1}$$

$$U(0,x) = U_0(x), \quad x \in \mathbb{R}^n, \tag{2}$$

where $U(t,x) \in \mathbb{C}^N$ is the vector-valued unknown function and

$$A(t,x,D_x) = \sum_{k=1}^{n} A_k(t,x) D_{x_{le}}, \qquad D = \frac{1}{\mathrm{i}}\nabla, \qquad \mathrm{i}^2 = -1,$$

with $A_k(t,x)$ being matrices from $\mathbb{C}^{N\times N}$, for $k = 0, 1, \dots, n$. The matrix A_0 is assumed hermitian, $A_0 = A_0^*$, and bounded positive definite: there is some $\delta > 0$ such that for all (t,x), we have $\delta I_N \le A_0(t,x) \le \delta^{-1} I_N$. Then the expression

$$\langle A_0 V, W\rangle := \int_{\mathbb{R}^n} (A_0(t,x)V(t,x)) \cdot \overline{W(t,x)}\,\mathrm{d}x$$

is (for all t) a symmetric positive definite bilinear form in $L^2(\mathbb{R}^n, \mathbb{C}^N)$.

M. Dreher (✉)
Institute of Mathematics, University of Rostock, Rostock, Germany
e-mail: michael.dreher@uni-rostock.de

M. D'Abbicco et al. (eds.), *New Tools for Nonlinear PDEs and Application*, Trends in Mathematics, https://doi.org/10.1007/978-3-030-10937-0_4

We write A^* for the L^2 adjoint differential operator to A and obtain (assuming U to be a smooth solution to (1)) the estimate

$$\begin{aligned} \partial_t \langle A_0 U, U \rangle &= \big\langle ((\partial_t A_0) - (A + A^*))U, U \big\rangle + 2\Re \langle F, U \rangle \\ &\leq \text{const}\ \langle A_0 U, U \rangle + \langle A_0 F, F \rangle \end{aligned} \tag{3}$$

if $A + A^*$ is a zeroth order differential operator with bounded coefficients (which is the definition of (1) being a symmetric hyperbolic system), with some constant that only depends on δ, $\|\partial_t A_0\|_{L^\infty}$ and $\|\partial_j A_j\|_{L^\infty}$. Having found such an a priori estimate in differential form, the Gronwall lemma can be applied, leading to a theory of well-posedness of the Cauchy problem in the usual way, even in a quasilinear setting, compare the classical results in [5].

The purpose of this article is to pursue the following questions:

Question 1 what happens if A_0 (and possibly A) are replaced by matrix pseudodifferential operators of unspecified order, for instance

$$A_0 = \begin{pmatrix} -\triangle & 0 \\ 0 & I_{N-1} \end{pmatrix}.$$

Question 2 is there a general framework for performing incompressible limits, assuming well-prepared initial data?

For first order pseudodifferential systems of symmetrisable hyperbolic type, singular perturbations that lead to incompressible limits have been studied in [3], and we will extend those results to matrix pseudodifferential operators of unspecified order.

1.1 An Example: The Incompressible Limit for the Euler System

The barotropic Euler system of gas dynamics reads (after a suitable scaling)

$$\begin{aligned} \partial_t \rho + \operatorname{div}(\rho \mathbf{u}) &= 0, \\ \partial_t \mathbf{u} + (\mathbf{u} \cdot \nabla)\mathbf{u} + \frac{\Lambda^2}{\rho} \nabla p(\rho) &= 0, \end{aligned}$$

for $(t, x) \in (0, \infty) \times \mathbb{R}^3$, with gas density $\rho(t, x)$, vectorial gas velocity $\mathbf{u}(t, x)$, and pressure $p(\rho) = c\rho^\gamma$ for some constants $c > 0$ and $\gamma \geq 1$. The first equation describes the conservation of mass, the second equation describes the balance of momentum. This system becomes incompressible when we perform the limit $\Lambda \to \infty$ which can be seen when we follow the approach of [6, 7]. We also mention [8].

Let $\hat{\rho} > 0$ be a constant reference value for the density and set $\hat{p} := p(\hat{\rho})$. We introduce

$$U := \begin{pmatrix} \Lambda(p - \hat{p}) \\ \mathbf{u} \end{pmatrix}$$

and let ρ_0, $\mathbf{u}_0$, U_0 denote the initial values at time $t = 0$. Then we have the symmetric hyperbolic system

$$\begin{pmatrix} \frac{1}{\gamma p} & 0 \\ 0 & \rho I_3 \end{pmatrix} \left(\partial_t + \sum_{j=1}^{3} u_j \partial_j \right) U + \Lambda \begin{pmatrix} 0 & \text{div} \\ \nabla & 0_{3\times 3} \end{pmatrix} U = 0, \tag{4}$$

$$A_0 = \begin{pmatrix} \frac{1}{\gamma p} & 0 \\ 0 & \rho I_3 \end{pmatrix}, \qquad A(t, x, D_x) = A_0(\mathbf{u} \cdot \nabla) + \Lambda \begin{pmatrix} 0 & \text{div} \\ \nabla & 0 \end{pmatrix},$$

$$A + A^* = -\sum_{j=1}^{3} \partial_j (A_0 u_j),$$

and in particular, $A + A^*$ does not depend on the huge singular parameter Λ. Assuming that ρ (and hence p) is bounded from above and below by positive constants, we then obtain the estimate (3) with some constant independent of Λ (and $F \equiv 0$ obviously). We require the initial data to be well-prepared, which means (as a first condition) that the initial energy $\langle A_0 U, U\rangle(t = 0)$ is bounded uniformly for $\Lambda \to \infty$, which yields $p_0 = \hat{p} + \mathcal{O}(\Lambda^{-1})$, and then we get uniform estimates for $\|U(t, \cdot)\|_{L^2(\mathbb{R}^3)}$ on some time interval. Higher order estimates can be deduced after differentiating (4):

$$A_0(\partial_t + \mathbf{u} \cdot \nabla)\partial^\alpha U + \Lambda \begin{pmatrix} 0 & \text{div} \\ \nabla & 0 \end{pmatrix} \partial^\alpha U$$

$$= F_\alpha \left(\{\partial^\beta U,\ \Lambda^{-1} \partial^\gamma \partial_t U : \quad |\beta| \le |\alpha|, \quad |\gamma| \le |\alpha| - 1\} \right),$$

with $\alpha \in \mathbb{N}_0^3$, and the factor Λ^{-1} originates when we apply at least one spatial derivative to A_0. Note that F_α depends on the highest order derivatives $\partial^\beta U$ with $|\beta| = |\alpha|$ only linearly. Then we substitute (4) into F_α and find uniform in Λ estimates of higher order derivatives, leading to a time interval $[0, T]$ of existence that does not depend on Λ. To perform the incompressible limit, we take one time derivative of (4) and assume $\langle A_0 \partial_t U, \partial_t U\rangle(t = 0)$ to be uniformly bounded (second condition of the initial data being well-prepared), which amounts to $p_0 = \hat{p} + \mathcal{O}(\Lambda^{-2})$ and $\text{div}(\mathbf{u}_0) = \mathcal{O}(\Lambda^{-1})$. Then uniform in Λ estimates of $\partial_t U$ in $L^2(\mathbb{R}^3)$ follow, and compactness arguments give us a sub-sequence that converges in $C([0, T], L^2(B_R))$ with B_R being a ball of radius R. A Cantor diagonal trick for increasing R then ensures the existence of a sub-sequence with convergence in

$C([0,T], L^2_{loc}(\mathbb{R}^3))$, and the limit $(\rho_\infty, \mathbf{u}_\infty)$ then is a sufficiently regular solution to the incompressible Euler equations

$$\rho_\infty = \hat{\rho}, \qquad \operatorname{div}\mathbf{u}_\infty = 0, \qquad \partial_t \mathbf{u}_\infty + (\mathbf{u}_\infty \cdot \nabla)\mathbf{u}_\infty + \nabla \pi = 0,$$

for some unknown scalar pressure π. Classical solutions to this system are unique, and therefore the complete sequence $(\rho, \mathbf{u})$ converges in a local Sobolev space $C([0,T], H^s_{loc}(\mathbb{R}^3))$ to the limit $(\rho_\infty, \mathbf{u}_\infty)$, not just a sub-sequence.

1.2 An Example: The Quasineutral Limit for the Euler–Poisson System

Under certain assumptions, the electron transport in a crystal lattice can be described by the system

$$\begin{aligned} \partial_t \rho + \operatorname{div}(\rho \mathbf{u}) &= 0, \\ \partial_t \mathbf{u} + (\mathbf{u} \cdot \nabla)\mathbf{u} + \frac{1}{\rho} \nabla p + \nabla \Phi &= 0, \\ -\lambda^2 \triangle \Phi &= \rho - 1, \end{aligned}$$

for $(t,x) \in (0,\infty) \times \mathbb{R}^3$, with electron density $\rho(t,x)$, electron velocity $\mathbf{u}(t,x)$, a certain pressure p of the electron gas, and an unknown electric potential $\Phi(t,x)$. The Poisson equation tells us that this electric potential is being generated by the electrons (with negative charge and density ρ) and the immobile ions in the lattice (with positive charge and density 1). The parameter λ is called *Debye length*, and we are interested in the limit $\lambda \to 0$. Then formally $\rho \equiv 1$, hence the device is locally of neutral charge, explaining the name of the limit. Initial data ρ_0 and $\mathbf{u}_0$ are being prescribed at $t = 0$, which are bounded in certain norms, and ρ_0 is in a neighbourhood of 1. The existence of smooth solutions has been established for instance in [1] and [4]. Concerning the quasineutral limit, we refer to [10] and [12].

In this paper we will re-establish the quasineutral limit in a more general way. For sake of simplicity, assume a constant pressure p. Then we introduce

$$U := \begin{pmatrix} q \\ \mathbf{u} \end{pmatrix}, \qquad q := \lambda \Phi$$

and obtain the system

$$\begin{pmatrix} -\triangle & 0 \\ 0 & I_3 \end{pmatrix} \partial_t U + \begin{pmatrix} -\operatorname{div}(\mathbf{u} \triangle \cdot) & 0 \\ 0 & \mathbf{u} \cdot \nabla \end{pmatrix} U + \frac{1}{\lambda} \begin{pmatrix} 0 & \operatorname{div} \\ \nabla & 0_{3\times 3} \end{pmatrix} U = 0. \tag{5}$$

The claim of this article is that this system can be called a *generalisation to a symmetrisable system*. To see this, we put

$$A_0 := \begin{pmatrix} -\triangle & 0 \\ 0 & I_3 \end{pmatrix}, \qquad A(t,x,D_x) = \begin{pmatrix} -\operatorname{div}(\mathbf{u}\,\triangle\,\cdot) & 0 \\ 0 & \mathbf{u}\cdot\nabla \end{pmatrix} + \frac{1}{\lambda}\begin{pmatrix} 0 & \operatorname{div} \\ \nabla & 0 \end{pmatrix}$$

and observe that A_0 is positive semidefinite,

$$\langle A_0 U, U\rangle = \|\nabla q\|_{L^2}^2 + \|\mathbf{u}\|_{L^2}^2,$$

with some kernel consisting of constant functions q. The matrix operator A has a third order term in the upper left corner, but $A + A^*$ almost cancels in the sense of

$$2\Re\langle AW, W\rangle = -2\Re\int(\nabla\mathbf{u})\colon(\nabla w_0 \otimes \nabla\overline{w_0})\,\mathrm{d}x + \int(\operatorname{div}\mathbf{u})(|\nabla w_0|^2 - |\mathbf{w}|^2)\,\mathrm{d}x,$$

for $W = (w_0, \mathbf{w})^\top$ and real-valued $\mathbf{u}$. Note that w_0 always appears as ∇w_0, and therefore $\Re\langle AW, W\rangle$ can be estimated by $\langle A_0 W, W\rangle$, assuming natural bounds of derivatives of $\mathbf{u}$. The existence of solutions U can be established using the methods of [1] and [4], we only need uniformly in λ estimates of various norms of U in order to perform the limit of $\lambda \to 0$, and these estimates can be obtained in the same way as in Sect. 1.1.

2 Assumptions and Main Results

We consider the problem

$$A_0(U)\partial_t U(t,x) + A_1(U)U(t,x) + \frac{1}{\epsilon}LU(t,x) = 0, \quad (t,x) \in (0,\infty)\times\mathbb{R}^n, \tag{6}$$

$$U(0,x) = U_0(x), \quad x \in \mathbb{R}^n, \tag{7}$$

with A_0, A_1 and L being matrix pseudodifferential operators of size $N \times N$. The coefficients of A_0 and A_1 may depend on the solution U in a certain way specified below. Currently, we make no assumptions on the orders of these operators.

Let $\mathscr{S}(\mathbb{R}^n, \mathbb{C}^N)$ be the Schwartz space of smooth functions with rapid decay, and $\mathscr{S}'(\mathbb{R}^n, \mathbb{C}^N)$ be its topological dual space, consisting of temperate distributions.

Property 1 (Dependence of A_0 and A_1 on U) There is a pseudodifferential operator $\pi_0\colon \mathscr{S}'(\mathbb{R}^n, \mathbb{C}^N) \to \mathscr{S}'(\mathbb{R}^n, \mathbb{C}^M)$ with pseudodifferential symbol only depending on ξ such that A_0 and A_1 (which are possibly in divergence form) depend smoothly on $\pi_0 U$, with values $\pi_0 U \in \mathscr{B}$, where $\mathscr{B}$ is an open subset of $\mathbb{R}^M$. There is no other dependence of A_0 and A_1 on U. The operators A_0 and A_1 can depend on ϵ, but all estimates of their mapping properties are uniform in $\epsilon \in (0, 1)$.

When we wish to emphasise the dependence of A_0 or A_1 on ϵ, we write $A_0(U;\epsilon)$ or $A_1(U;\epsilon)$, otherwise we just write $A_0(U)$ and $A_1(U)$.

Remark 1 For (4), we have $n = 3$, $N = M = n+1$ and $\pi_0 = I_N$. The set $\mathscr{B}$ is given by $U \in (-\frac{1}{2}\hat{p}, \frac{1}{2}\hat{p}) \times \mathbb{R}^n$, and $\Lambda^{-1} \in (0,1)$ figures as the singular parameter ϵ. The operator A_0 depends on ϵ via $p = \hat{p} + \epsilon u_1$.

Remark 2 For (5), we have $n = 3$, $N = n+1$, $M = 2n$, and, with ∇ as column,

$$\pi_0 = \begin{pmatrix} \nabla & 0 \\ 0 & I_n \end{pmatrix}. \tag{8}$$

The operator A_0 does not depend on U at all, and A_1 only depends on U via $\mathbf{u}$,

$$A_1(U) = \begin{pmatrix} -\operatorname{div}(\mathbf{u} \,\triangle\, \cdot) & 0 \\ 0 & \mathbf{u}\cdot\nabla \end{pmatrix} = \pi_0^* \circ \begin{pmatrix} \mathbf{u} \otimes \operatorname{div} & 0 \\ 0 & \mathbf{u}\cdot\nabla \end{pmatrix} \circ \pi_0, \quad U = \begin{pmatrix} q \\ \mathbf{u} \end{pmatrix}. \tag{9}$$

For sake of completeness, we mention that

$$L = \begin{pmatrix} 0 & \operatorname{div} \\ \nabla & 0 \end{pmatrix} = \pi_0^* \circ \begin{pmatrix} 0 & -I_n \\ I_n & 0 \end{pmatrix} \circ \pi_0. \tag{10}$$

Property 2 (The order reducing operator π_0) The pseudodifferential operator π_0 has the following property: For every $s \in \mathbb{N}$, the set

$$H^s_\pi := \left\{ V \in \mathscr{S}'(\mathbb{R}^n, \mathbb{C}^N) \colon \ \pi_0 V \in H^s(\mathbb{R}^n, \mathbb{C}^M) \right\}$$

is a Hilbert space when we equip it with the norm $\|V\|_{H^s_\pi} := \|\pi_0 V\|_{H^s(\mathbb{R}^n)}$. Here and in the sequel we tacitly identify $V \in \mathscr{S}'$ with its equivalence class $V + \ker \pi_0$.

The set H^s_π characterises to which functions V the operators A_0 and A_1 can be meaningfully applied. We need to define some more sets of functions: $\mathscr{U}$ describes which functions U can be substituted into $A_0(U)$ and $A_1(U)$. And finally, $\mathscr{M}_{s_0,s_1,s_*}(T)$ is a set in which the solution U to (6) is expected to live.

Definition 1 We choose $s_\infty := \lceil \frac{n+1}{2} \rceil$, take $\mathscr{B}$ from Property 1 and define a set

$$\mathscr{U} = \left\{ U \in H^{s_\infty}_\pi \colon \ (\pi_0 U)(x) \in \mathscr{B} \quad \text{for all} \quad x \in \mathbb{R}^n \right\}.$$

For some numbers $s_0, s_1, s_* \in \mathbb{N}$ with $s_0 \geq s_1 > s_* > s_\infty$, and for positive real numbers T, M_0, we define

$$\mathscr{M}_{s_0,s_1,s_*}(T) := \Big\{ U \in C([0,T], \mathscr{U}) \colon \ \partial_t^j U \in C([0,T], H^{s_j}_\pi), \quad j = 0, 1,$$

$$\sup_{[0,T]} \|(\pi_0 U)(t,\cdot)\|_{H^{s_*}} \leq M_0 \Big\}.$$

Having introduced these sets, we now list the assumptions on the operators A_0, A_1, L. The constants C with subscripts appearing below do neither depend on $\epsilon \in (0, 1)$ nor on T, but on M_0. The constants δ_+, δ_- only depend on the set $\mathscr{B} \subset \mathbb{R}^M$, nothing else.

Property 3 (Mapping properties) The operators A_0, A_1, L map between Sobolev spaces as follows. There is some $d \in \mathbb{N}$ and some C_m (depending on s) such that

$$\left\| A_j(U)V \right\|_{H^s} \leq C_m(1 + \|\pi_0 U\|_{H^{s+d}}) \|\pi_0 V\|_{H^{s+d}}, \quad j = 0, 1, \quad s \geq s_*,$$
$$\|LV\|_{H^s} \leq C_m \|\pi_0 V\|_{H^{s+d}}, \quad s \geq 0,$$

for all $V \in H_\pi^{s+d}$ and all $U \in \mathscr{M}_{s_0,s_1,s_*}(T)$ with $U(t, \cdot) \in H_\pi^{s+d}$. We also have

$$\left|\left\langle D^\alpha(A_1(U)V), W\right\rangle\right| \leq C_m \|\pi_0 V\|_{H^{s_\infty+d}} \|\pi_0 W\|_{L^2}, \quad |\alpha| \leq s_\infty,$$

for all $(V, W) \in H_\pi^{s_*} \times H_\pi^0$.

Property 4 (Bilinear forms) For each $U \in \mathscr{U}$, the operator $A_0(U)$ and the real part of $A_1(U)$ generate bilinear forms $a_0(\cdot, \cdot)$, $a_1(\cdot, \cdot)$ in the sense of

$$\langle A_0(U)V, W\rangle = a_0(\pi_0 V, \pi_0 W) = a_0(\pi_0 W, \pi_0 V), \quad \text{for all} \quad V, W \in H_\pi^0,$$
$$2\Re\langle A_1(U)V, V\rangle = a_1(\pi_0 V, \pi_0 V), \quad \text{for all} \quad V \in H_\pi^d.$$

There are positive δ_-, δ_+ such that, for all $U \in \mathscr{M}_{s_0,s_1,s_*}(T)$ and all $V, W \in L^2$,

$$\delta_- \|V\|_{L^2}^2 \leq a_0(V, V), \qquad |a_0(V, W)| \leq \frac{1}{\delta_+} \|V\|_{L^2} \|W\|_{L^2}.$$

Moreover, the form a_1 and a similar form generated by L are bounded by some constant C_f in the following sense: for all $U \in \mathscr{M}_{s_0,s_1,s_*}(T)$, we have the estimates

$$|a_1(V, V)| \leq C_f \|V\|_{L^2}^2, \quad \text{for all} \quad V \in L^2,$$
$$|\langle A_1(U)V, W\rangle| \leq C_f \|\pi_0 V\|_{H^d} \|\pi_0 W\|_{L^2}, \quad \text{for all} \quad (V, W) \in H_\pi^d \times H_\pi^0,$$
$$|\langle LV, W\rangle| \leq C_f \|\pi_0 V\|_{H^d} \|\pi_0 W\|_{L^2}, \quad \text{for all} \quad (V, W) \in H_\pi^d \times H_\pi^0.$$

Property 5 The pseudodifferential symbol of the operator L does neither depend on t nor on x, hence L commutes with every scalar constant coefficient pseudodifferential operator. Moreover, L is anti-selfadjoint: $L + L^* = 0$.

Remark 3 The operators π_0 and π_0^* can be seen as order-reducing operators, and then A_0 is selfadjoint, when conjugated with π_0. Moreover, A_1 and L are "anti-selfadjoint plus bounded, when conjugated with π_0". This concept can be called *generalised hyperbolicity*. It should be noted though that the form estimate of a_1 in case of (5) follows from the representation (9) of A_1 not immediately, but after repeated integration by parts.

Property 6 (Regularity of the dependence of A_0 and A_1 on U and ϵ) For each $U \in \mathcal{M}_{s_0,s_1,s_*}(T)$, the operator $\partial_t A_0(U)$ generates a hermitian bilinear form $a_{0,t}$ in the sense of

$$\langle (\partial_t A_0(U))V, W\rangle = a_{0,t}(\pi_0 V, \pi_0 W) = a_{0,t}(\pi_0 W, \pi_0 V), \quad \text{for all} \quad V, W \in H^0_\pi.$$

This form $a_{0,t}$ is bounded: there is some $C_{0,t}$ such that, for all $U \in \mathcal{M}_{s_0,s_1,s_*}(T)$,

$$\left|a_{0,t}(V, W)\right| \leq \epsilon C_{0,t} \left\|\pi_0 \partial_t U\right\|_{H^{s_\infty}} \left\|V\right\|_{L^2} \left\|W\right\|_{L^2}, \quad \text{for all} \quad V, W \in L^2. \tag{11}$$

We also have

$$\begin{aligned} &\left|\langle A_0(U;\epsilon)V, W\rangle - \left\langle A_0(\check{U};\check{\epsilon})V, W\right\rangle\right| \\ &\quad \leq C_{0,t}\Big(\max(\epsilon,\check{\epsilon}) \left\|\pi_0(U-\check{U})\right\|_{H^{s_\infty}} + |\epsilon - \check{\epsilon}|\Big) \left\|\pi_0 V\right\|_{L^2} \left\|\pi_0 W\right\|_{L^2}, \end{aligned}$$

for all $\epsilon, \check{\epsilon} \in [0, 1]$, all $U, \check{U} \in \mathcal{M}_{s_0,s_1,s_*}(T)$ and all $V, W \in H^0_\pi$.

There is some $C_{1,t}$ and some C_μ such that we have, for all $U \in \mathcal{M}_{s_0,s_1,s_*}(T)$,

$$\begin{aligned} \left|\langle (\partial_t A_1(U))V, W\rangle\right| &\leq C_{1,t} \left\|\pi_0 \partial_t U\right\|_{L^2} \left\|\pi_0 V\right\|_{H^{s_\infty+d}} \left\|\pi_0 W\right\|_{L^2} \\ &\quad + C_\mu \epsilon \left\|\pi_0 \partial_t U\right\|_{L^2} \left\|\pi_0 V\right\|_{H^{s_\infty+d}} \left\|\pi_0 W\right\|_{H^\mu}, \end{aligned} \tag{12}$$

for all $(V, W) \in H^{s_\infty+d}_\pi \times H^\mu_\pi$. We also require

$$\begin{aligned} &\left|\langle A_1(U;\epsilon)V, W\rangle - \left\langle A_1(\check{U};\check{\epsilon})V, W\right\rangle\right| \\ &\quad \leq C_{1,t}\Big(\left\|\pi_0(U-\check{U})\right\|_{H^{s_\infty+d-1}(\mathbb{R}^n)} + |\epsilon - \check{\epsilon}|\Big) \left\|\pi_0 V\right\|_{H^{s_\infty+d}} \left\|\pi_0 W\right\|_{H^\mu}, \end{aligned} \tag{13}$$

for all $\epsilon, \check{\epsilon} \in [0, 1]$, all $U, \check{U} \in \mathcal{M}_{s_0,s_1,s_*}(T)$ and all $(V, W) \in H^d_\pi \times H^\mu_\pi$.

The dependence of $A_1(U)$ on U can be estimated locally, which means: there is some $\mu \in \mathbb{N}_0$ (with $\mu < s_\infty$) and a dense subset $\mathcal{H}$ of $L^2((0,T), H^\mu_\pi)$ such that for each $W \in \mathcal{H}$ there is some positive R with

$$\begin{aligned} &\int_{t=0}^T \left|\langle A_1(U;\epsilon)V, W\rangle(t) - \left\langle A_1(\check{U};\check{\epsilon})V, W\right\rangle(t)\right| \,\mathrm{d}t \\ &\quad \leq C_{1,t}\Big(\left\|\pi_0(U-\check{U})\right\|_{L^\infty((0,T),C^{d-1}(B_R(0)))} + |\epsilon - \check{\epsilon}|\Big) \times \\ &\qquad \times \left\|\pi_0 V\right\|_{L^2((0,T),H^{s_\infty+d})} \left\|\pi_0 W\right\|_{L^2((0,T),H^\mu)} \\ &\quad + C_\mu |\epsilon - \check{\epsilon}|^{1/2} \max(\epsilon,\check{\epsilon})^{1/2} \left\|\pi_0 V\right\|_{L^2((0,T),H^{s_\infty+d})} \left\|\pi_0 W\right\|_{L^2((0,T),H^\mu)} \end{aligned} \tag{14}$$

for all $\epsilon, \check{\epsilon} \in [0, 1]$, all $U, \check{U} \in \mathcal{M}_{s_0,s_1,s_*}(T)$ and all $V \in L^2((0,T), H^{s_\infty+d}_\pi)$.

Remark 4 Some comments are in order. The mysterious μ in (12) is not needed when all the spatial derivatives in $\langle(\partial_t A_1(U))V, W\rangle$ can by arranged (using integration by parts) in such a way that no derivative lands on $\pi_0\partial_t U$ or $\pi_0 W$. This would be the optimal situation, which we enjoy for (4) and (5), but we will not be so lucky for the application in Sect. 5, when a perturbing higher order operator of small norm appears, which will come from the requirement of having A_1 *and* L in symmetric/antisymmetric form.

We also note that the condition of a localisable dependence will be necessary because A_0 and A_1 *will* be pseudodifferential operators, hence non-local.

Next we make assumptions on how A_0 and A_1 commute with spatial derivatives.

Property 7 (Commutation relations for A_0) There is some $C_{0,c,0}$ such that, for all $\alpha \in \mathbb{N}_0$ with $|\alpha| \leq s_0$ and all $U \in \mathscr{M}_{s_0,s_1,s_*}(T)$, we have

$$\begin{aligned}&\left|\left\langle[A_0(U), D^\alpha]V, W\right\rangle\right|\\ &\quad\leq \epsilon C_{0,c,0}\left(\|\pi_0 U\|_{H^{s_\infty+1}}\|\pi_0 V\|_{H^{|\alpha|-1}} + \|\pi_0 U\|_{H^{|\alpha|}}\|\pi_0 V\|_{H^{s_\infty}}\right)\|\pi_0 W\|_{L^2},\end{aligned}$$

for all $V \in H_\pi^{|\alpha|-1} \cap H_\pi^{s_\infty}$ and all $W \in H_\pi^0$.

For all $\alpha \in \mathbb{N}_0^n$, there are zero-th order operators $B_{0\alpha\beta}$ with

$$\left[A_0(U), D^\alpha\right] = \sum_{\beta<\alpha} B_{0\alpha\beta}(U)D^\beta \circ A_0(U),$$

and they are (jointly with A_1 and L) bounded as follows: there are constants $C_{0,c,1}$ and $C_{0,c,L}$ such that for all α with $|\alpha| \leq s_*$ and all $U \in \mathscr{M}_{s_0,s_1,s_*}(T)$, we have

$$\begin{aligned}\left|\left\langle B_{0\alpha\beta}(U)D^\beta(A_1(U)V), D^\alpha V\right\rangle\right| &\leq \epsilon C_{0,c,1}\|\pi_0 V\|_{H^{s_*}}^2, \quad \text{for all} \quad V \in H_\pi^{s_*},\\ \left|\left\langle B_{0\alpha\beta}(U)D^\beta LV, D^\alpha V\right\rangle\right| &\leq \epsilon C_{0,c,L}\|\pi_0 V\|_{H^{s_*}}^2, \quad \text{for all} \quad V \in H_\pi^{s_*}.\end{aligned}$$

Remark 5 For (4), A_0 is a strictly positive definite matrix. Then $B_{0\alpha\beta}$ is being obtained when at least one derivative lands on $A_0(U)$, and $\frac{\partial A_0}{\partial U} = \mathscr{O}(\epsilon)$. And for (5), A_0 has constant coefficients, hence all its commutators vanish anyway.

Property 8 (Commutation relation for A_1) There is a positive constant $C_{1,c}$ such that for all $U \in \mathscr{M}_{s_0,s_1,s_*}(T)$ and $V \in H_\pi^{s_*}$, we have

$$\left|\left\langle[A_1(U), D^\alpha]V, D^\alpha V\right\rangle\right| \leq C_{1,c}\|\pi_0 V\|_{H^{s_*}}^2, \quad |\alpha| \leq s_*.$$

Remark 6 It is easy to check that these inequalities hold for (4). And in case of (5), we recall (9) and that the operators π_0^*, π_0 commute with every derivative ∂_t and D^α.

The following is a common feature of first-order quasilinear symmetric and symmetrisable hyperbolic systems: the solution can only blow up if its C^1 norm explodes, and we will make a very similar assumption in this paper.

Property 9 (Local existence and persistence of solutions) There are integers s_0, s_1 and s_* with $s_0 \geq s_1 \geq s_* + d$, $s_* > \frac{n}{2} + 1$, which have the following property: for every compact set $\mathcal{K} \subset \mathcal{B}$ and for each $U_0 \in \mathcal{U}$ with $(\pi_0 U_0)(x) \in \mathcal{K}$ for all x, there is some positive number T and a unique solution $U \in C([0,T), \mathcal{U})$ with $\pi_0 \partial_t^j U \in C([0,T), H^{s_j}(\mathbb{R}^n))$ $(j = 0, 1)$ to the Cauchy problem (6), (7). And if, for one such initial function U_0, the norm $\sup_{[0,T)} \|\pi_0 U(t,\cdot)\|_{H^{s_*}}$ remains bounded, then this solution U can be extended beyond T in the sense of $\pi_0 \partial_t^j U \in C([0, T+\gamma], H^{s_j})$ for some positive γ and $j = 0, 1$.

Theorem 1 (Uniform existence interval) *We assume Property* 1 *through Property* 9 *and suppose additionally that* $s_* \geq s_\infty + \max(1, d)$. *Then for each compact subset* $\mathcal{K}$ *of* $\mathcal{B}$ *and for each positive real number* M_0*, some positive real number* T *exists with the following property: for each* $\epsilon \in (0,1)$ *and for each* $U_0 \in H^{s_0}_\pi$ *with* $\|\pi_0 U_0\|_{H^{s_*}} \leq \frac{\delta_+ \delta_-}{2} M_0$ *and* $(\pi_0 U_0)(x) \in \mathcal{K}$ *for all* $x \in \mathbb{R}^n$*, there is a solution* $U \in \mathcal{M}_{s_0,s_1,s_*}(T)$ *to* (6), (7).

In particular, the life-span T of the solution U has a lower bound that does not depend on ϵ.

Theorem 2 (Incompressible Limit) *Under the conditions of Theorem* 1*, let the family* $\{U_0^\epsilon : 0 < \epsilon < 1\}$ *be initial data with* $U_0^\epsilon \in H^{s_0}_\pi$, $\|\pi_0 U_0^\epsilon\|_{H^{s_*}} \leq \frac{\delta_+ \delta_-}{2} M_0$ *and* $(\pi_0 U_0^\epsilon)(x) \in \mathcal{K}$ *for all* x *and all* ϵ. *Let* $U^\epsilon \in \mathcal{M}_{s_0,s_1,s_*}(T)$ *be the unique solution to* (6) *with initial data* U_0^ϵ*, provided by Theorem* 1. *Suppose the uniform bound*

$$\left\|\pi_0 \partial_t U^\epsilon(0,\cdot)\right\|_{L^2} \leq C, \quad 0 < \epsilon < 1. \tag{15}$$

Then the following holds.

(a) *The item* LU^ϵ*, when interpreted as an element of the dual space* $(H^0_\pi)'$*, converges to zero for* $\epsilon \to 0$ *in the sense of*

$$\sup\left\{|\langle LU^\epsilon, \varphi\rangle| : \varphi \in H^0_\pi,\ \|\pi_0\varphi\|_{L^2} \leq 1\right\} \leq C\epsilon, \quad 0 \leq t \leq T, \quad 0 < \epsilon < 1. \tag{16}$$

Further, as ϵ *goes to zero, the sequence* $(U^\epsilon)_{\epsilon\to 0}$ *contains a subsequence (which we do not relabel) which converges to a limit* U^* *in the following sense:*

$$\pi_0 U^\epsilon \rightharpoonup \pi_0 U^* \quad \text{in} \quad L^2((0,T), H^{s_*}(\mathbb{R}^n)), \tag{17}$$

$$\pi_0 \partial_t U^\epsilon \rightharpoonup \pi_0 \partial_t U^* \quad \text{in} \quad L^2((0,T) \times \mathbb{R}^n), \tag{18}$$

$$\pi_0 U^\epsilon \to \pi_0 U^* \quad \text{in} \quad C([0,T], C^{d-1}(B_R(0))), \tag{19}$$

for all balls $B_R(0) \subset \mathbb{R}^n$.

(b) *The limit* U^* *solves*

$$A_0(U^*;0)\partial_t U^* + A_1(U^*;0)U^* + R = 0 \tag{20}$$

as an identity in $L^2((0,T),(H_\pi^\mu)')$, *with some* $R \in L^2((0,T),(H_\pi^\mu)' \cap \text{range}\,(L))$, *and* μ *from Property* 6.

(c) *If the operator* L *can be decomposed as* $L = \pi_0^* \circ \tilde{L}$ *with some pseudodifferential operator* $\tilde{L}$, *then a sufficient condition for* (15) *is*

$$\left\| \tilde{L} U_0^\epsilon \right\|_{L^2} \leq C\epsilon, \quad 0 < \epsilon < 1.$$

Remark 7 The solution U^ϵ has a certain life span T_ϵ before its regularity breaks down. One may wonder what can be said about the limit of T_ϵ for ϵ approaching zero. Assuming we had shown a strong convergence of $\pi_0 U^\epsilon$ to $\pi_0 U^*$ in $C([0,T], L^2(\mathbb{R}^n))$, a natural conjecture would be that T_ϵ converges to the life span of strong solutions to (20). We know that usually strong solutions to the incompressible Euler equation blow up in finite time, hence we expect the same for (20).

3 The Uniform Existence Interval

In this section, we prove Theorem 1.

Let U_0 be given. By Property 9, there is some positive T_ϵ and a solution $U \in C([0,T_\epsilon),\mathscr{U})$ to (6), (7) with $\pi_0 \partial_t^j U \in C([0,T_\epsilon), H^{s_j})$, and we can assume that $\sup_{[0,T_\epsilon)} \|\pi_0 U(t,\cdot)\|_{H^{s_*}} \leq M_0$ by decreasing T_ϵ if needed.

In the following, we will construct some time interval (independent of ϵ) on which the H^{s_*} norm of $\pi_0 U$ stays bounded (uniformly in ϵ). By the persistence criterion, the H^{s_0} life-span of the solution then has a uniform lower bound, proving the claim.

We begin with an estimate of $\|\pi_0 \partial_t U(t,\cdot)\|_{H^{s_\infty}}$, for $t \in (0,T_\epsilon)$. We choose α with $|\alpha| \leq s_\infty$ and apply D^α to (6):

$$A_0(U)\partial_t D^\alpha U + D^\alpha(A_1(U)U) + \frac{1}{\epsilon} L D^\alpha U = \left[A_0(U), D^\alpha\right]\partial_t U,$$

and then we can estimate

$$\begin{aligned}
\delta_- \left\|\pi_0 \partial_t D^\alpha U\right\|_{L^2}^2 &\leq \left\langle A_0(U)\partial_t D^\alpha U, \partial_t D^\alpha U\right\rangle \\
&= -\left\langle D^\alpha(A_1(U)U), \partial_t D^\alpha U\right\rangle - \frac{1}{\epsilon}\left\langle L D^\alpha U, \partial_t D^\alpha U\right\rangle + \left\langle [A_0(U), D^\alpha]\partial_t U, \partial_t D^\alpha U\right\rangle
\end{aligned}$$

$$
\begin{aligned}
&\le C_m \|\pi_0 U\|_{H^{s_\infty+d}} \left\|\pi_0 \partial_t D^\alpha U\right\|_{L^2} + \frac{C_f}{\epsilon} \left\|\pi_0 D^\alpha U\right\|_{H^d} \left\|\pi_0 \partial_t D^\alpha U\right\|_{L^2} \\
&\quad + \epsilon C_{0,c,0} \|\pi_0 \partial_t U\|_{H^{s_\infty}} \left\|\pi_0 \partial_t D^\alpha U\right\|_{L^2} \\
&\le \frac{C}{\epsilon} \|\pi_0 U\|_{H^{s_\infty+d}} \|\pi_0 \partial_t U\|_{H^{s_\infty}} + \epsilon C \|\pi_0 \partial_t U\|_{H^{s_\infty}}^2 .
\end{aligned}
$$

We now sum up over all α with $|\alpha| \le s_\infty$ and get, at least for small ϵ,

$$
\|\pi_0 \partial_t U(t,\cdot)\|_{H^{s_\infty}} \le \frac{C M_0}{\epsilon}, \qquad 0 < t < T_\epsilon. \tag{21}
$$

Then the estimate (11) of the bilinear form $a_{0,t}$ from Property 6 becomes

$$
\left|a_{0,t}(V,W)\right| \le \tilde{C}_{0,t} \|V\|_{L^2} \|W\|_{L^2}, \quad V, W \in L^2.
$$

Next we choose some α with $|\alpha| \le s_*$. Recalling Property 3 and that $s_* \le s_1 - d \le s_0 - d$ we observe that D^α can be applied to each of the items $A_0 \partial_t U$, $A_1 U$, LU of (6), hence we find

$$
\begin{aligned}
&A_0(U)\partial_t D^\alpha U + A_1(U) D^\alpha U + \frac{1}{\epsilon} L D^\alpha U \\
&\quad = \left[A_0(U), D^\alpha\right] \partial_t U + \left[A_1(U), D^\alpha\right] U =: F_{\alpha,0} + F_{\alpha,1},
\end{aligned}
$$

and we deduce that

$$
\begin{aligned}
\partial_t \langle A_0(U) D^\alpha U, D^\alpha U\rangle &= \langle(\partial_t A_0(U)) D^\alpha U, D^\alpha U\rangle + 2\Re\langle A_0(U)\partial_t D^\alpha U, D^\alpha U\rangle \\
&= a_{0,t}(\pi_0 D^\alpha U, \pi_0 D^\alpha U) - a_1(\pi_0 D^\alpha U, \pi_0 D^\alpha U) - \frac{2}{\epsilon}\Re\langle L D^\alpha U, D^\alpha U\rangle \\
&\quad + 2\Re\langle F_{\alpha,0} + F_{\alpha,1}, D^\alpha U\rangle \\
&\le \frac{\tilde{C}_{0,t} + C_f}{\delta_-}\langle A_0(U) D^\alpha U, D^\alpha U\rangle + 2\Re\langle F_{\alpha,0} + F_{\alpha,1}, D^\alpha U\rangle.
\end{aligned}
$$

Now we can estimate the right-hand side like this:

$$
\begin{aligned}
\left|\langle F_{\alpha,0}, D^\alpha U\rangle\right| &\le \sum_{\beta<\alpha} \left|\langle B_{0\alpha\beta} D^\beta (A_0(U)\partial_t U), D^\alpha U\rangle\right| \\
&\le \sum_{\beta<\alpha} \left|\langle B_{0\alpha\beta} D^\beta (A_1(U) U), D^\alpha U\rangle\right| + \frac{2}{\epsilon} \sum_{\beta<\alpha} \left|\langle B_{0\alpha\beta} D^\beta L U, D^\alpha U\rangle\right| \\
&\le C\epsilon C_{0,c,1} \|\pi_0 U\|_{H^{s_*}}^2 + C C_{0,c,L} \|\pi_0 U\|_{H^{s_*}}^2 , \\
\left|\langle F_{\alpha,1}, D^\alpha U\rangle\right| &= \left|\langle [A_1(U), D^\alpha] U, D^\alpha U\rangle\right| \le C_{1,c} \|\pi_0 U\|_{H^{|\alpha|}}^2 .
\end{aligned}
$$

By a Plancherel argument, this results in

$$\partial_t \langle A_0(U) D^\alpha U, D^\alpha U \rangle \le C \, \|\pi_0 U\|_{H^{s*}}^2 \le C \sum_{|\beta| \le s_*} \|\pi_0 D^\beta U\|_{L^2}^2$$
$$\le C \sum_{|\beta| \le s_*} \langle A_0(U) D^\beta U, D^\beta U \rangle,$$

and now Gronwall's Lemma can be applied (after summing over α),

$$\sum_{|\alpha| \le s_*} \langle A_0(U) D^\alpha U, D^\alpha U \rangle(t) \le \exp(Ct) \sum_{|\alpha| \le s_*} \langle A_0(U) D^\alpha U, D^\alpha U \rangle(0),$$

for all $0 \le t < T_\epsilon$, with a constant C that does not depend on ϵ. We estimate the form $a_0(\cdot, \cdot)$ from above and below,

$$\sum_{|\alpha| \le s_*} \|\pi_0 D^\alpha U(t)\|_{L^2}^2 \le \frac{\exp(Ct)}{\delta_+ \delta_-} \sum_{|\alpha| \le s_*} \|\pi_0 D^\alpha U(0)\|_{L^2}^2 \le \frac{\exp(Ct)}{\delta_+ \delta_-} \cdot \frac{\delta_+ \delta_-}{2} M_0,$$

for all $0 \le t < T_\epsilon$. That is a (uniform in ϵ) growth estimate of the solution, which concludes the proof of Theorem 1.

4 The Incompressible Limit

Now we present the proof of Theorem 2, beginning with part (a). The idea is to take a time derivative of (6), but we have some trouble to explain what $\partial_t^2 U^\epsilon$ should be (remember that A_0 need not be invertible). To this end, we define the forward shift $\tau_h W(t) := W(t + h)$, for any real positive number h and any function $W(t)$. The time t is now restricted to the interval $(0, T - h)$. Subtracting equation (6) from τ_h (6) then yields

$$A_0(U^\epsilon) \partial_t (\tau_h U^\epsilon - U^\epsilon) + A_1(U^\epsilon)(\tau_h U^\epsilon - U^\epsilon) + \frac{1}{\epsilon} L(\tau_h U^\epsilon - U^\epsilon)$$
$$= -(\tau_h A_0(U^\epsilon) - A_0(U^\epsilon)) \partial_t \tau_h U^\epsilon - (\tau_h A_1(U^\epsilon) - A_1(U^\epsilon)) \tau_h U^\epsilon.$$

We introduce the abbreviation $V_h := h^{-1}(\tau_h U^\epsilon - U^\epsilon)$, and it follows that

$$\begin{aligned} \partial_t \langle A_0(U^\epsilon) V_h, V_h \rangle &= a_{0,t}(\pi_0 V_h, \pi_0 V_h) - a_1(\pi_0 V_h, \pi_0 V_h) \\ &\quad -2\Re \Big\langle h^{-1}(\tau_h A_0(U^\epsilon) - A_0(U^\epsilon)) \partial_t \tau_h U^\epsilon, V_h \Big\rangle \\ &\quad -2\Re \Big\langle h^{-1}(\tau_h A_1(U^\epsilon) - A_1(U^\epsilon)) \tau_h U^\epsilon, V_h \Big\rangle. \end{aligned}$$

We have assumed $U^\epsilon \in \mathscr{M}_{s_0,s_1,s_*}(T)$, which yields $\partial_t \pi_0 U^\epsilon \in C([0,T], H^{s_1})$, and therefore the right-hand side has a limit for $h \to 0$, and then it follows that

$$\begin{aligned}
\partial_t \big\langle A_0(U^\epsilon)\partial_t U^\epsilon, \partial_t U^\epsilon\big\rangle &= a_{0,t}(\pi_0\partial_t U^\epsilon, \pi_0\partial_t U^\epsilon) - a_1(\pi_0\partial_t U^\epsilon, \pi_0\partial_t U^\epsilon) \\
&\quad -2\Re\big\langle(\partial_t A_0(U^\epsilon))\partial_t U^\epsilon, \partial_t U^\epsilon\big\rangle - 2\Re\big\langle(\partial_t A_1(U^\epsilon))U^\epsilon, \partial_t U^\epsilon\big\rangle \\
&\le C\big\langle A_0(U^\epsilon)\partial_t U^\epsilon, \partial_t U^\epsilon\big\rangle + 2\epsilon C_{0,t} \left\|\pi_0\partial_t U^\epsilon\right\|_{H^{s_\infty}} \left\|\pi_0\partial_t U^\epsilon\right\|_{L^2} \left\|\pi_0\partial_t U^\epsilon\right\|_{L^2} \\
&\quad +2C_{1,t} \left\|\pi_0\partial_t U^\epsilon\right\|_{L^2} \left\|\pi_0 U^\epsilon\right\|_{H^{s_\infty+d}} \left\|\pi_0\partial_t U^\epsilon\right\|_{L^2} \\
&\quad +2C_\mu\epsilon \left\|\pi_0\partial_t U^\epsilon\right\|_{L^2} \left\|\pi_0 U^\epsilon\right\|_{H^{s_\infty+d}} \left\|\pi_0\partial_t U^\epsilon\right\|_{H^\mu}
\end{aligned}$$

owing to Property 6, in particular (12). We recall (21), $\|\pi_0 U^\epsilon\|_{H^{s_\infty+d}} \le \|\pi_0 U^\epsilon\|_{H^{s_*}} \le M_0$, and complex interpolation:

$$\left\|\pi_0\partial_t U^\epsilon\right\|_{H^\mu} \le C \left\|\pi_0\partial_t U^\epsilon\right\|_{L^2}^{1-\frac{\mu}{s_\infty}} \left\|\pi_0\partial_t U^\epsilon\right\|_{H^{s_\infty}}^{\frac{\mu}{s_\infty}} \le C \left\|\pi_0\partial_t U^\epsilon\right\|_{L^2}^{\theta} \epsilon^{\theta-1},$$

with $\theta = 1 - \frac{\mu}{s_\infty} \in (0,1]$. From this we observe that $y(t) := \langle A_0(U^\epsilon)\partial_t U^\epsilon, \partial_t U^\epsilon\rangle(t)$ satisfies the inequality

$$y'(t) \le C(y(t) + \epsilon^\theta y^\theta(t)), \quad 0 < t < T.$$

We may replace y^θ by $y^{1/p}$ with $p \in \mathbb{N}$ and $\frac{1}{p} \le \theta$, changing C if needed. By integration we then have

$$0 \le y(t) \le y(0) + \int_{s=0}^{t} c_1 \left(y(s) + \epsilon^\theta y^{\frac{1}{p}}(s)\right) \,\mathrm{d}s, \quad 0 \le t \le T,$$

for some positive c_1. Then the Inequality of Bihari [2] and LaSalle [9] reveals

$$y(t) \le G^{-1}\big(G(y(0)) + c_1 t\big),$$

with G^{-1} being the inverse function to G, and

$$G(z) := \int_{\zeta=0}^{z} \frac{\mathrm{d}\zeta}{\zeta + \epsilon^\theta \zeta^{1/p}} = \frac{p}{p-1} \ln\left(1 + \frac{z^{\frac{p-1}{p}}}{\epsilon^\theta}\right).$$

We find $G^{-1}(Z) \le \epsilon^{\theta p/(p-1)} \exp(Z)$, hence

$$y(t) \le \left(\epsilon^\theta + (y(0))^{\frac{p-1}{p}}\right)^{\frac{p}{p-1}} e^{c_1 t}.$$

We now apply (15) and get the uniform estimates

$$\left\|\pi_0 U^\epsilon\right\|_{C([0,T],H^{s_*})} \le C, \qquad \left\|\partial_t \pi_0 U^\epsilon\right\|_{C([0,T],L^2)} \le C, \tag{22}$$

and the weak convergences (17) and (18) then follow from boundedness in Hilbert spaces. Choose some positive R. Then, by Lions-Aubin-Dubinskii compactness arguments, a subsequence of $\{U^\epsilon\}_{\epsilon\to 0}$ converges in $C([0,T], C^{d-1}(B_R(0)))$ to U^*. We choose some larger R and get a smaller subsequence. We repeat this step indefinitely and apply a Cantor diagonal argument. This gives (19).

And the strong $(H^0_\pi)'$ convergence (16) of the full sequence $\{LU^\epsilon\}_{\epsilon\to 0}$ to zero follows when we test (6) with $\varphi \in H^0_\pi$:

$$\begin{aligned}\left|\langle LU^\epsilon, \varphi\rangle\right| &\le \epsilon \left|\langle A_0(U^\epsilon)\partial_t U^\epsilon, \varphi\rangle\right| + \epsilon \left|\langle A_1(U^\epsilon)U^\epsilon, \varphi\rangle\right| \\ &\le \epsilon\delta_+^{-1} \left\|\pi_0\partial_t U^\epsilon\right\|_{L^2} \|\pi_0\varphi\|_{L^2} + \epsilon C_f \left\|\pi_0 U^\epsilon\right\|_{H^d} \|\pi_0\varphi\|_{L^2}.\end{aligned}$$

Part (b) is demonstrated as follows.

Take some $W \in \mathscr{H}$, as in Property 6. We begin with

$$\begin{aligned}\left|\int_{t=0}^T \langle A_0(U^\epsilon;\epsilon)\partial_t U^\epsilon, W\rangle(t) - \langle A_0(U^*;0)\partial_t U^*, W\rangle(t)\,\mathrm{d}t\right| &\le I_1 + |I_2|,\\ I_1 := \int_{t=0}^T \left|\langle A_0(U^\epsilon;\epsilon)\partial_t U^\epsilon, W\rangle(t) - \langle A_0(U^*;0)\partial_t U^\epsilon, W\rangle(t)\right|\,\mathrm{d}t&,\\ I_2 := \int_{t=0}^T \langle A_0(U^*;0)\partial_t U^\epsilon, W\rangle(t) - \langle A_0(U^*;0)\partial_t U^*, W\rangle(t)\,\mathrm{d}t&,\end{aligned}$$

and by Property 6, Property 4 we can estimate

$$\begin{aligned}I_1 &\le C_{0,t}T\epsilon \left\|\pi_0(U^\epsilon - U^*)\right\|_{L^\infty([0,T],H^{s_\infty})} \left\|\pi_0\partial_t U^\epsilon\right\|_{L^2((0,T)\times\mathbb{R}^n)} \|\pi_0 W\|_{L^2((0,T)\times\mathbb{R}^n)},\\ |I_2| &\le \sqrt{T}C_f \left\|\pi_0(\partial_t U^\epsilon - \partial_t U^*)\right\|_{L^2((0,T)\times\mathbb{R}^n)} \|\pi_0 W\|_{L^2((0,T)\times\mathbb{R}^n)}.\end{aligned}$$

We immediately see $\lim_{\epsilon\to 0} I_1 = 0$, and I_2 (understood as a mapping that sends $\pi_0(\partial_t U^\epsilon - \partial_t U^*)$ to $I_2 \in \mathbb{C}$) is a bounded linear functional on $L^2((0,T)\times\mathbb{R}^n)$. By the Riesz representation theorem and (18), we then also have $\lim_{\epsilon\to 0} I_2 = 0$. Now we put

$$\mathscr{T}_\epsilon(W) := \int_{t=0}^T \langle A_0(U^\epsilon;\epsilon)\partial_t U^\epsilon, W\rangle(t)\,\mathrm{d}t, \quad 0 < \epsilon < 1,$$

as a mapping from $L^2((0,T), H^\mu_\pi)$ into $\mathbb{C}$. Owing to Property 4 and (22), we have estimates $|\mathscr{T}_\epsilon(W)| \le C\|W\|_{L^2((0,T),H^\mu_\pi)}$ that are uniform in ϵ. The above bounds on I_1 and I_2 tell us

$$\lim_{\epsilon\to 0} \mathscr{T}_\epsilon(W) = \int_{t=0}^T \langle A_0(U^*;0)\partial_t U^*, W\rangle(t)\,\mathrm{d}t, \quad \text{for all} \quad W \in \mathscr{H}. \tag{23}$$

From Property 2 we know H^μ_π to be a Hilbert space, and then also $L^2((0,T), H^\mu_\pi)$ is a Hilbert space. The Banach–Steinhaus Theorem then guarantees the convergence in (23) for all $W \in L^2((0,T), H^\mu_\pi)$.

With very similar arguments, we can also show, for all $W \in L^2((0,T), H^\mu_\pi)$,

$$\lim_{\epsilon\to 0} \int_{t=0}^{T} \big\langle A_1(U^\epsilon;\epsilon)U^\epsilon, W\big\rangle(t)\,\mathrm{d}t = \int_{t=0}^{T} \big\langle A_1(U^*;0)U^*, W\big\rangle(t)\,\mathrm{d}t,$$

which completes the proof of part (b).

To prove part (c), we estimate

$$\begin{aligned}\delta_- \left\|\pi_0\partial_t U^\epsilon\right\|_{L^2}^2 &\le \big\langle A_0(U^\epsilon)\partial_t U^\epsilon, \partial_t U^\epsilon\big\rangle \\ &= -\big\langle A_1(U^\epsilon)U^\epsilon, \partial_t U^\epsilon\big\rangle - \frac{1}{\epsilon}\Big\langle \pi_0^*\tilde{L}U^\epsilon, \partial_t U^\epsilon\Big\rangle \\ &\le C_f \left\|\pi_0 U^\epsilon\right\|_{H^d} \left\|\pi_0\partial_t U^\epsilon\right\|_{L^2} + \frac{1}{\epsilon}\left\|\tilde{L}U^\epsilon\right\|_{L^2}\left\|\pi_0\partial_t U^\epsilon\right\|_{L^2},\end{aligned}$$

and now it suffices to evaluate this inequality at $t = 0$. The proof of Theorem 2 is complete.

5 An Application

Now we consider the Euler–Poisson system

$$\begin{aligned}\partial_t\rho + \mathrm{div}(\rho\mathbf{u}) &= 0,\\ \partial_t\mathbf{u} + (\mathbf{u}\cdot\nabla)\mathbf{u} + \frac{1}{\rho}\nabla p(\rho) + \nabla\Phi &= 0,\\ -\lambda^2 \triangle \Phi &= \rho - 1\end{aligned}$$

with some non-trivial pressure $p(\rho) = \rho^2$. The claim is that the general framework above, which culminates in Theorems 1 and 2, can be applied to this situation. We have chosen this version of the pressure $p(\rho)$ instead of some more general $p(\rho) = \rho^\gamma$ in order to avoid heavy use of pseudodifferential techniques. In the current situation, we can write

$$\frac{1}{\rho}\nabla p(\rho) + \nabla\Phi = \nabla\left(1 - 2\lambda^2\triangle\right)\Phi,$$

and $1 - 2\lambda^2\triangle$ is a nice invertible elliptic operator with constant coefficients. We introduce

$$\tilde{U} = \begin{pmatrix} q \\ \mathbf{u}\end{pmatrix}, \qquad q = \lambda\Phi,$$

and then it follows that

$$\begin{pmatrix} -\triangle & 0 \\ 0 & I \end{pmatrix} \partial_t \tilde{U} + \begin{pmatrix} -\mathrm{div}(\mathbf{u} \triangle \cdot) & 0 \\ 0 & \mathbf{u} \cdot \nabla \end{pmatrix} \tilde{U} + \frac{1}{\lambda} \begin{pmatrix} 0 & \mathrm{div} \\ (1 - 2\lambda^2 \triangle)\nabla & 0 \end{pmatrix} \tilde{U} = 0,$$

and the third matrix in this equation is the singular perturbation, which is not yet anti-selfadjoint. To repair this, we define an operator

$$m := \sqrt{1 - 2\lambda^2 \triangle}$$

as a Fourier multiplier, and put $M := \mathrm{diag}(m, I)$, and $U := M\tilde{U}$. Noting that M commutes with all derivatives ∂_t and D^α, we find (after applying M from the left)

$$\begin{pmatrix} -\triangle & 0 \\ 0 & I \end{pmatrix} \partial_t U + M \begin{pmatrix} -\mathrm{div}(\mathbf{u} \triangle \cdot) & 0 \\ 0 & \mathbf{u} \cdot \nabla \end{pmatrix} M^{-1} U + \frac{1}{\lambda} \begin{pmatrix} 0 & m\mathrm{div} \\ m\nabla & 0 \end{pmatrix} U = 0,$$

and we will write this equation as $A_0 \partial_t U + M\tilde{A}_1(U)M^{-1}U + \lambda^{-1}LU = 0$. We also put $A_1 := M\tilde{A}_1 M^{-1}$.

Before we begin to check the various assumptions of the general framework, we have a look at the main idea—to estimate the time derivative of the expression $\langle A_0 U, U\rangle$—from the perspective of physics. The mentioned expression equals

$$\begin{aligned} \langle A_0 U, U\rangle &= \left\langle A_0 M\tilde{U}, M\tilde{U}\right\rangle = \left\langle A_0 \tilde{U}, M^2 \tilde{U}\right\rangle \\ &= \int_{\mathbb{R}^n} (-\triangle q)(1 - 2\lambda^2 \triangle) q + |\mathbf{u}|^2 \,\mathrm{d}x = \int_{\mathbb{R}^n} |\nabla q|^2 + 2\lambda^2 (\triangle q)^2 + |\mathbf{u}|^2 \,\mathrm{d}x \\ &= \lambda^2 \int_{\mathbb{R}^n} |\nabla \Phi|^2 \,\mathrm{d}x + 2\int_{\mathbb{R}^n} (\rho - 1)^2 \,\mathrm{d}x + \int_{\mathbb{R}^n} |\mathbf{u}|^2 \,\mathrm{d}x, \end{aligned}$$

and we recall that the physical energy is the conserved quantity

$$\mathscr{E}(t) = \frac{\lambda^2}{2} \int_{\mathbb{R}^n} |\nabla \Phi|^2 \,\mathrm{d}x + \int_{\mathbb{R}^n} H(\rho) \,\mathrm{d}x + \int_{\mathbb{R}^n} \frac{\rho}{2} |\mathbf{u}|^2 \,\mathrm{d}x,$$

with $H(\rho)$ as the enthalpy, defined by $H'(\rho) = h(\rho)$ and $h'(\rho) = \frac{1}{\rho} p'(\rho)$. Then we see that the expression $\langle A_0 U, U\rangle$ almost equals the physical energy, the only difference appearing in the contribution of the kinetic energy (with 1 instead of ρ). From this observation we conclude that our approach of massively symmetrising the system (at the expense of introducing pseudodifferential operators) has its motivation on the grounds of physics.

Now we verify the assumptions. Concerning Property 1, we have π_0 as in (8). The operator A_0 does not depend itself on $\pi_0 U$ or $\epsilon = \lambda$, and A_1 depends on $\mathbf{u}$ which are the lower components of $\pi_0 U$. Further $\mathscr{B} = \mathbb{R}^n$, and the operators M and div appearing in A_1 are "outer derivatives". In that sense, A_1 is a pseudodifferential operator in divergence form, and A_1 does not in itself depend on derivatives of $\mathbf{u}$. More on that below.

Property 2 requires

$$H_\pi^s := \{(v_0, \mathbf{v}) \in \mathscr{S}'(\mathbb{R}^n, \mathbb{C}^{n+1}) \colon \nabla v_0 \in H^s(\mathbb{R}^n, \mathbb{C}^n), \mathbf{v} \in H^s(\mathbb{R}^n, \mathbb{C}^n)\}$$

to be Hilbert spaces, which follows from Chapter 5 of [11]. The kernel of π_0 contains constant functions in the upper components and zero in the lower components.

To check Property 3, we first calculate

$$\frac{1 + \sqrt{2}\lambda|\xi|}{\sqrt{2}} \leq m(\xi) = \sqrt{1 + 2\lambda^2|\xi|^2} \leq 1 + \sqrt{2}\lambda|\xi|,$$

and therefore

$$C^{-1}\left(\|\phi\|_{H^s} + \lambda\,\|\nabla\phi\|_{H^s}\right) \leq \|m\phi\|_{H^s} \leq C\left(\|\phi\|_{H^s} + \lambda\,\|\nabla\phi\|_{H^s}\right)$$

After recalling (10), we then immediately get $\|LV\|_{H^s} \leq C\,\|\pi_0 V\|_{H^{s+2}}$, hence $d = 2$. We also see $\|A_0 V\|_{H^s} \leq C\,\|\pi_0 V\|_{H^{s+1}}$, because of $A_0 = \pi_0^* \circ \pi_0$. And concerning A_1, we have $s \geq s_* > s_\infty$, which turns all appearing Sobolev spaces into algebras, hence

$$\begin{aligned}
\|A_1(U)V\|_{H^s} &= \left\|M\tilde{A}_1(U)M^{-1}V\right\|_{H^s} \\
&\leq C\left(\left\|m\mathrm{div}(\mathbf{u}\,\triangle\, m^{-1}v_0)\right\|_{H^s} + \|(\mathbf{u}\cdot\nabla)\mathbf{v}\|_{H^s}\right) \\
&\leq C\left(\left\|\mathbf{u}m^{-1}\,\triangle\, v_0\right\|_{H^{s+1}} + \lambda\left\|\mathbf{u}m^{-1}\,\triangle\, v_0\right\|_{H^{s+2}} + \|\mathbf{u}\|_{H^s}\,\|\mathbf{v}\|_{H^{s+1}}\right) \\
&\leq C\,\|\mathbf{u}\|_{H^{s+2}}\left(\|\nabla v_0\|_{H^{s+2}} + \|\mathbf{v}\|_{H^{s+1}}\right).
\end{aligned}$$

We come to Property 4. The bilinear form a_0 on $L^2(\mathbb{R}^n, \mathbb{C}^M)$ is obviously

$$a_0(V, W) = \int_{\mathbb{R}^n} V\overline{W}\,\mathrm{d}x,$$

hence $\delta_- = \delta_+ = 1$. We also have $|\langle LV, W\rangle| \leq \|\pi_0 V\|_{L^2}\,\|\pi_0 W\|_{L^2}$, by (10). And to calculate $\langle A_1(U)V, W\rangle$ for $V, W \in H_\pi^d$, we put $\tilde{V} = M^{-1}V$, $\tilde{W} = M^{-1}W$ and get

$$\begin{aligned}
\langle A_1(U)V, W\rangle &= \left\langle \tilde{A}_1(U)\tilde{V}, M^2\tilde{W}\right\rangle \\
&= \int_{\mathbb{R}^n} (-\mathrm{div}(\mathbf{u}\,\triangle\,\tilde{v}_0))(1 - 2\lambda^2\triangle)\overline{\tilde{w}_0} + ((\mathbf{u}\cdot\nabla)\mathbf{v})\overline{\mathbf{w}}\,\mathrm{d}x
\end{aligned}$$

$$= \int_{\mathbb{R}^n} (\mathbf{u} \vartriangle \tilde{v}_0)\nabla\overline{\tilde{w}_0}\,\mathrm{d}x + 2\lambda^2 \int_{\mathbb{R}^n} (\operatorname{div}(\mathbf{u} \vartriangle \tilde{v}_0)) \vartriangle \overline{\tilde{w}_0}\,\mathrm{d}x + \int_{\mathbb{R}^n} ((\mathbf{u}\cdot\nabla)\mathbf{v})\overline{\mathbf{w}}\,\mathrm{d}x$$

$$= \int_{\mathbb{R}^n} \left(\mathbf{u}\frac{\vartriangle}{m}v_0\right)\frac{\nabla}{m}\overline{w_0} + 2\left(\operatorname{div}\left(\mathbf{u}\frac{\lambda\vartriangle}{m}v_0\right)\right)\frac{\lambda\vartriangle}{m}\overline{w_0} + ((\mathbf{u}\cdot\nabla)\mathbf{v})\overline{\mathbf{w}}\,\mathrm{d}x.$$

Note that $U \in \mathscr{M}_{s_0,s_1,s_*}(T)$ enforces $\mathbf{u}$ to have real values. Then

$$\begin{aligned}
a_1(\pi_0 V, \pi_0 V) &= \langle A_1(U)V, V\rangle + \langle V, A_1(U)V\rangle \\
&= \int_{\mathbb{R}^n} (\mathbf{u} \vartriangle \tilde{v}_0)\nabla\overline{\tilde{v}_0} + (\mathbf{u} \vartriangle \overline{\tilde{v}_0})\nabla\tilde{v}_0 + ((\mathbf{u}\cdot\nabla)\mathbf{v})\overline{\mathbf{v}} + ((\mathbf{u}\cdot\nabla)\overline{\mathbf{v}})\mathbf{v} \\
&\qquad + 2\lambda^2(\operatorname{div}(\mathbf{u} \vartriangle \tilde{v}_0)) \vartriangle \overline{\tilde{v}_0} + 2\lambda^2(\operatorname{div}(\mathbf{u} \vartriangle \overline{\tilde{v}_0})) \vartriangle \tilde{v}_0\,\mathrm{d}x \\
&= -\sum_{j,k}\int_{\mathbb{R}^n} (\partial_k u_j)\cdot 2\Re((\partial_k\tilde{v}_0)(\partial_j\overline{\tilde{v}_0})) + u_j\partial_j|\partial_k\tilde{v}_0|^2 + (\partial_j u_j)|v_k|^2 \\
&\qquad + 2\lambda^2 u_j\partial_j|\vartriangle \tilde{v}_0|^2\,\mathrm{d}x \\
&= -2\int_{\mathbb{R}^n} (\nabla\mathbf{u}) : \Re((\nabla\tilde{v}_0)\otimes(\nabla\overline{\tilde{v}_0}))\,\mathrm{d}x \\
&\qquad + \int_{\mathbb{R}^n} (\operatorname{div}\mathbf{u})\left(|\nabla\tilde{v}_0|^2 - |\mathbf{v}|^2 + 2|\lambda \vartriangle \tilde{v}_0|^2\right)\mathrm{d}x,
\end{aligned}$$

and now it suffices to add $\|\lambda \vartriangle \tilde{v}_0\|_{L^2} = \left\|\lambda m^{-1} \vartriangle v_0\right\|_{L^2} \le C\,\|\nabla v_0\|_{L^2}$.

Property 5 and the statements of Property 6 that relate to A_0 are obviously satisfied. To show (12), we write

$$\begin{aligned}
|\langle\partial_t A_1(U)V, W\rangle| \le \int_{\mathbb{R}^n} |\mathbf{u}_t|\left|\frac{\vartriangle}{m}v_0\right|\left|\frac{1}{m}\nabla w_0\right|\,\mathrm{d}x + 2\int_{\mathbb{R}^n} |\mathbf{u}_t|\left|\frac{\vartriangle}{m}v_0\right|\left|\frac{\lambda^2\vartriangle}{m}\nabla w_0\right|\,\mathrm{d}x \\
+ C\int_{\mathbb{R}^n} |\mathbf{u}_t|\,|\nabla\mathbf{v}|\,|\mathbf{w}|\,\mathrm{d}x,
\end{aligned}$$

and estimate $\left\|\Delta m^{-1} v_0\right\|_{L^\infty} \le C\,\|\nabla v_0\|_{H^{s_\infty+1}}$, $\|\nabla\mathbf{v}\|_{L^\infty} \le C\,\|\mathbf{v}\|_{H^{s_\infty+1}}$, $\left\|m^{-1}\nabla w_0\right\|_{L^2} \le C\,\|\nabla w_0\|_{L^2}$. By Plancherel, we have

$$\begin{aligned}
\left\|\frac{\lambda^2\vartriangle}{m}\nabla w_0\right\|_{L^2}^2 &= \int_{\mathbb{R}^n}\left(\frac{\lambda^2|\xi|^3}{\sqrt{1+2\lambda^2|\xi|^2}}|\hat{w}_0(\xi)|\right)^2\mathrm{d}\xi \le \int_{\mathbb{R}^n}\left(\lambda|\xi|^2|\hat{w}_0(\xi)|\right)^2\mathrm{d}\xi \\
&\le C\lambda^2\,\|\nabla w_0\|_{H^1}^2,
\end{aligned}$$

from which we get (12) with $\mu = 1$.

To prove (13) and (14), we begin with

$$\langle A_1(U;\lambda)V, W\rangle - \left\langle A_1(\check{U};\check{\lambda})V, W\right\rangle = \int_{\mathbb{R}^n} \left((\mathbf{u}-\check{\mathbf{u}})\frac{\triangle}{m} v_0 \right) \frac{\nabla}{m} \overline{w_0}\, \mathrm{d}x \tag{24}$$
$$+ \int_{\mathbb{R}^n} \left(\mathbf{u}\left(\frac{\triangle}{m} - \frac{\triangle}{\check{m}}\right) v_0 \right) \frac{\nabla}{m}\overline{w_0} - \left(\operatorname{div}\left(\check{\mathbf{u}}\frac{\triangle}{\check{m}} v_0\right)\right)\left(\frac{1}{m} - \frac{1}{\check{m}}\right)\overline{w_0}\, \mathrm{d}x$$
$$+2\int_{\mathbb{R}^n} \left(\operatorname{div}\left((\mathbf{u}-\check{\mathbf{u}})\frac{\triangle}{m} v_0\right)\right)\frac{\lambda^2\triangle}{m}\overline{w_0}\, \mathrm{d}x$$
$$+2\int_{\mathbb{R}^n} \left(\operatorname{div}\left(\check{\mathbf{u}}\lambda\left(\frac{\triangle}{m} - \frac{\triangle}{\check{m}}\right) v_0\right)\right)\frac{\lambda\triangle}{m}\overline{w_0}$$
$$+2\int_{\mathbb{R}^n} \left(\operatorname{div}\left(\check{\mathbf{u}}\frac{\triangle}{\check{m}} v_0\right)\right)\left(\frac{\lambda^2}{m} - \frac{\check{\lambda}^2}{\check{m}}\right)\triangle\,\overline{w_0}\, \mathrm{d}x + \int_{\mathbb{R}^n} \left(\left((\mathbf{u}-\check{\mathbf{u}})\cdot\nabla\right)\mathbf{v}\right)\overline{\mathbf{w}}\, \mathrm{d}x,$$

and now we estimate like this for the proof of (13): the terms $|\mathbf{u}-\check{\mathbf{u}}|$, $|\operatorname{div}(\mathbf{u}-\check{\mathbf{u}})|$, $|\check{\mathbf{u}}|$ and $|\operatorname{div}(\check{\mathbf{u}})|$ always receive the $L^\infty(\mathbb{R}^n)$ norm. Further,

$$\left\| \nabla \frac{\triangle}{m} v_0 \right\|_{L^2} \le C \left\|\nabla v_0\right\|_{H^d},$$
$$\left| \frac{\lambda}{m(\xi)} - \frac{\lambda}{\check{m}(\xi)} \right| = \frac{2\left|\check{\lambda}^2 - \lambda^2\right|\lambda|\xi|^2}{(\check{m}(\xi)+m(\xi))m(\xi)\check{m}(\xi)} \le C|\lambda-\check{\lambda}|,$$
$$\left| \frac{1}{m(\xi)} - \frac{1}{\check{m}(\xi)} \right| \le C|\lambda-\check{\lambda}||\xi|,$$
$$\left\| \nabla\lambda\left(\frac{\triangle}{m} - \frac{\triangle}{\check{m}}\right) v_0 \right\|_{L^2} \le C|\lambda-\check{\lambda}| \left\|\nabla v_0\right\|_{H^d},$$
$$\left\| \frac{\lambda\triangle}{m} w_0 \right\|_{L^2} \le C\left\|\nabla w_0\right\|_{L^2},$$
$$\left| \frac{\lambda^2|\xi|^2}{m(\xi)} - \frac{\check{\lambda}^2|\xi|^2}{\check{m}(\xi)} \right| \le \left| \frac{\lambda^2}{m(\xi)} - \frac{\lambda^2}{\check{m}(\xi)} \right| |\xi|^2 + \frac{|\lambda^2-\check{\lambda}^2|}{\check{m}(\xi)}|\xi|^2$$
$$\le C|\lambda-\check{\lambda}||\xi|, \quad \text{provided} \quad \check{\lambda} > \lambda,$$

and then it can be concluded that

$$\left| \langle A_1(U;\lambda)V, W\rangle - \left\langle A_1(\check{U};\check{\lambda})V, W\right\rangle \right|$$
$$\le C\left(\left\|\mathbf{u}-\check{\mathbf{u}}\right\|_{C^{d-1}(\mathbb{R}^n)} + |\lambda-\check{\lambda}| \right) \left\|\nabla v_0\right\|_{H^d} \left\|\nabla w_0\right\|_{L^2},$$

as desired. And now it remains to show (14). We choose

$$\mathscr{H} = \left\{ W \in C([0,T], C_0^\infty(\mathbb{R}^n)) \colon \ \exists R > 1 \text{ with } \operatorname{supp} W \subset [0,T] \times B_{R-1}(0) \right\},$$

and we note that R may depend on W, but is uniform in $t \in [0,T]$. Then we split all terms $\frac{1}{m}\nabla w_0$ appearing in (24) as

$$\frac{1}{m}\nabla w_0 = \nabla w_0 + \frac{1-m}{m}\nabla w_0, \qquad \left\| \frac{1-m}{m}\nabla w_0 \right\|_{H^s} \le C\lambda \left\| \nabla w_0 \right\|_{H^{s+1}},$$

and we also make use of

$$\left| \frac{1}{m(\chi)} - \frac{1}{\check{m}(\xi)} \right| \le C|\lambda - \check{\lambda}||\xi| \le C\sqrt{|\lambda - \check{\lambda}|}\sqrt{\max(\lambda, \check{\lambda})}|\xi|,$$

and so on. The appearing products of the type "$(\mathbf{u} - \check{\mathbf{u}})\nabla w_0$" are being estimated as $\left\|\mathbf{u} - \check{\mathbf{u}}\right\|_{C([0,T],B_R(0))} \left\|\nabla w_0\right\|_{L^2}$. In contrast, the products of the type "$(\mathbf{u} - \check{\mathbf{u}})\frac{1-m}{m}\nabla w_0$" then are estimated as $(\left\|\mathbf{u}\right\|_{L^\infty(\mathbb{R}^n)} + \left\|\check{\mathbf{u}}\right\|_{L^\infty})\lambda \left\|\nabla w_0\right\|_{H^\mu}$ with $\mu = 1$.

Property 7 is empty. The first estimate of Property 8 follows from

$$\begin{aligned}
\left|\left\langle D^\alpha(A_1(U)V), W\right\rangle\right| &\le \int_{\mathbb{R}^n} \left| D^\alpha\left(\mathbf{u}\frac{\Delta m}{v_0}\right)\right| \left|\frac{\nabla}{m} w_0\right| \,\mathrm{d}x \\
&+ 2\int_{\mathbb{R}^n} \left| \operatorname{div} D^\alpha\left(\mathbf{u}\frac{\lambda\Delta}{m} v_0\right)\right| \left|\frac{\lambda\Delta}{m} w_0\right| \,\mathrm{d}x + \int_{\mathbb{R}^n} \left| D^\alpha(\mathbf{u}\cdot\nabla)\mathbf{v}\right| |\mathbf{w}| \,\mathrm{d}x,
\end{aligned}$$

and noting that $|\alpha| \le s_\infty \le s_* - d$, $\|\mathbf{u}\|_{H^{s_*}} \le M_0$.

And the second estimate of Property 8 can be obtained from

$$\begin{aligned}
\left\langle [A_1(U), D^\alpha]V, D^\alpha V\right\rangle &= \int_{\mathbb{R}^n} \left(\left[\mathbf{u}, D^\alpha\right] \frac{\Delta}{m} v_0 \right) \frac{\nabla}{m} D^\alpha \overline{v_0} \,\mathrm{d}x \\
&+ 2\int_{\mathbb{R}^n} \left(\operatorname{div}\left([\mathbf{u}, D^\alpha] \frac{\lambda\Delta}{m} v_0 \right)\right) \frac{\lambda\Delta}{m} D^\alpha \overline{v_0} \,\mathrm{d}x + \int_{\mathbb{R}^n} \left(\left([\mathbf{u}, D^\alpha]\cdot\nabla\right)\mathbf{v}\right) D^\alpha \overline{\mathbf{v}} \,\mathrm{d}x.
\end{aligned}$$

And the Euler–Poisson system also has Property 9, because this is a symmetrisable hyperbolic system for the unknown functions $(\rho, \mathbf{u})$, with a nonlocal lower order term hidden in $\nabla\Phi$. The local existence of solutions to symmetrisable hyperbolic systems is a classical fact, and we also know that they persist as long as the C^1 norm of $(\rho, \mathbf{u})$ remains bounded, compare [5]. The charge density ρ being bounded in C^1 corresponds to $\nabla\Phi$ being bounded in C^2, because of $\rho = 1 - \lambda^2\Delta\Phi$, and therefore we have the boundedness of $\nabla\Phi$ in $H^{s_\infty+2}$ as a part of the sufficient conditions for a persistence of a smooth solution. Note that $d = 2$ in this Section, and $s_* \ge s_\infty + d$ has been our assumption.

Therefore we have shown that the barotropic Euler–Poisson system possesses all the Properties 1–9, and then the Theorems 1 and 2 can be applied, ensuring the local existence of solutions U^ϵ with uniform existence interval, and their incompressible limit for $\epsilon \to 0$, which in our case is a quasineutral limit.

Next we determine the limit system. We start with the uniform (in t and λ) estimates for $\|\pi_0 U^\lambda\|_{H^1}$:

$$C \geq \|\pi_0 U^\lambda\|_{H^1}^2 \geq \|m\lambda\nabla\Phi^\lambda\|_{H^1}^2 = \int_{\mathbb{R}^n_\xi} \left|(1+|\xi|^2)^{1/2} m(\xi)\lambda\xi\hat{\Phi}^\lambda(t,\xi)\right|^2 \mathrm{d}\xi$$

$$= \int_{\mathbb{R}^n_\xi} (1+|\xi|^2)\left(1+2\lambda^2|\xi|^2\right)\lambda^2|\xi|^2 \left|\hat{\Phi}^\lambda(t,\xi)\right|^2 \mathrm{d}\xi$$

$$\geq \int_{\mathbb{R}^n_\xi} \left|\lambda|\xi|^2\hat{\Phi}^\lambda(t,\xi)\right|^2 \mathrm{d}\xi = \|\lambda \triangle \Phi^\lambda(t,\cdot)\|_{L^2}^2,$$

and then the Poisson equation implies $\|\rho^\lambda - 1\|_{L^2} \leq C\lambda$. Hence $\rho^* \equiv 1$, and the limit velocity $\mathbf{u}^*$ belongs to $L^2((0,T), H^{s_*}(\mathbb{R}^n))$, with $\partial_t \mathbf{u}^* \in L^2((0,T)\times\mathbb{R}^n)$. From the limit of the mass conservation equations we then find $\operatorname{div}\mathbf{u}^* = 0$ in distributional sense. To determine the differential equation for $\mathbf{u}^*$, we recall the identity

$$A_0(U^*;0)\partial_t U^* + A_1(U^*;0)U^* + R = 0,$$

in the sense of $L^2((0,T),(H^1_\pi)')$, with some $R \in L^2((0,T),(H^1_\pi)' \cap \operatorname{range}(L))$. We drop the top component of this equation and obtain

$$\partial_t \mathbf{u}^* + (\mathbf{u}^* \cdot \nabla)\mathbf{u}^* + r = 0,$$

with $r \in L^2((0,T),(H^1)' \cap \operatorname{range}(\nabla))$, which then can be improved to $r = \nabla P \in L^2((0,T)\times\mathbb{R}^n)$, from the already known regularity of $\mathbf{u}^*$.

6 Concluding Remarks

We have presented a general pseudodifferential framework that allows to handle the low Mach number limit for Euler systems of gas dynamics as well as the quasineutral limit for Euler–Poisson systems. The approach is based on the newly introduced concept of generalised symmetrisable systems, which have as main novel feature matrix (pseudo)differentiable operators of a priori unspecified orders. Generalisations to symmetrisable systems with a viscous part can be done quite naturally. The incompressible limit is being shown for sufficiently strong solutions with well-prepared initial data. Further generalisations to other systems (for instance of quantum hydrodynamic type) or to less regular solutions will be part of forthcoming publications.

Acknowledgements The author wishes to thank the referees for constructive remarks which helped to improve an earlier version of the paper.

References

1. G. Alì, A. Jüngel, Global smooth solutions to the multi-dimensional hydrodynamic model for two-carrier plasmas. J. Differ. Equ. **190**(2), 663–685 (2003)
2. I. Bihari, A generalization of a lemma of Bellman and its application to uniqueness problems of differential equations. Acta Math. Acad. Sci. Hungar. **7**, 81–94 (1956)
3. E. Grenier, Pseudo-differential energy estimates of singular perturbations. Commun. Pure Appl. Math. **50**(9), 821–865 (1997)
4. L. Hsiao, P.A. Markowich, S. Wang, The asymptotic behavior of globally smooth solutions of the multidimensional isentropic hydrodynamic model for semiconductors. J. Differ. Equ. **192**(1), 111–133 (2003)
5. T. Kato, The Cauchy problem for quasi-linear symmetric hyperbolic systems. Arch. Ration. Mech. Anal. **58**(3), 181–205 (1975)
6. S. Klainerman, A. Majda, Singular limits of quasilinear hyperbolic systems with large parameters and the incompressible limit of compressible fluids. Commun. Pure Appl. Math. **34**(4), 481–524 (1981)
7. S. Klainerman, A. Majda, Compressible and incompressible fluids. Commun. Pure Appl. Math. **35**(5), 629–651 (1982)
8. H.-O. Kreiss, Problems with different time scales for partial differential equations. Commun. Pure Appl. Math. **33**(3), 399–439 (1980)
9. J. LaSalle, Uniqueness theorems and successive approximations. Ann. Math. (2) **50**, 722–730 (1949)
10. Y.-J. Peng, Y.-G. Wang, Convergence of compressible Euler-Poisson equations to incompressible type Euler equations. Asymptot. Anal. **41**(2), 141–160 (2005)
11. H. Triebel, *Theory of Function Spaces*. Volume 78 of Monographs in Mathematics (Birkhäuser Verlag, Basel, 1983)
12. S. Wang, Quasineutral limit of Euler-Poisson system with and without viscosity. Commun. Partial Differ. Equ. **29**(3–4), 419–456 (2004)

The Critical Exponent for Evolution Models with Power Non-linearity

Marcelo Rempel Ebert and Linniker Monteiro Lourenço

Abstract In this note we derive $L^r - L^q$ estimates for the solutions to the Cauchy problem

$$u_{tt} + (-\Delta)^\sigma u = 0\,, \qquad t \geq 0,\ x \in \mathbb{R}^n, \qquad u(0,x) = 0,\ \ u_t(0,x) = g(x),$$

with $\sigma > 1$. Moreover, we derived the critical index $p_c(n)$ for the existence of global in time small data solutions to the associated semilinear Cauchy problem with power nonlinearity $|u|^p$, $p > 1$.

1 Introduction

Let us consider the Cauchy problem for the evolution equation,

$$\begin{cases} u_{tt} + (-\Delta)^\sigma u = 0\,, & t \geq 0,\ x \in \mathbb{R}^n \\ u(0,x) = f(x), \qquad u_t(0,x) = g(x). & \end{cases} \tag{1}$$

It is a σ-evolution operator in the sense of Petrowsky, since its principal symbol $\tau^2 - |\xi|^{2\sigma}$ has only real and distinct roots $\tau = \pm|\xi|^\sigma$ for all $\xi \neq 0$. The Cauchy problem (1) is $H^s(\mathbb{R}^n)$ well-posed, that is, for $s \geq \sigma$, to given data $f \in H^s(\mathbb{R}^n)$ and $g \in H^{s-\sigma}(\mathbb{R}^n)$, there exists a uniquely determined energy solution $u \in C([0,T], H^s(\mathbb{R}^n)) \cap C^1([0,T], H^{s-\sigma}(\mathbb{R}^n))$.

M. R. Ebert (✉)
Departamento de Computação e Matemática, FFCLRP, Universidade de São Paulo (USP), Ribeirão Preto, SP, Brazil
e-mail: ebert@ffclrp.usp.br

L. M. Lourenço
Departamento de Matemática, Universidade Federal de São Carlos, São Carlos, SP, Brazil

M. D'Abbicco et al. (eds.), *New Tools for Nonlinear PDEs and Application*, Trends in Mathematics, https://doi.org/10.1007/978-3-030-10937-0_5

The study of the long-time asymptotics of the solution and, more in general, the study of long-time behavior of suitable energies, has been a topic of interest in the recent years. For the wave equation, the estimates in [9] and [12] imply that the solution to the Cauchy problem (1), with $\sigma = 1$ and $f \equiv 0$, satisfies the $L^r - L^q$ estimates

$$\|u(t,\cdot)\|_{L^q} \leq C\, t^{1-n\left(\frac{1}{r}-\frac{1}{q}\right)} \|g\|_{L^r}$$

uniformly for any $t > 0$, if, and only if, the point $(\frac{1}{r}, \frac{1}{q})$ belongs to the closed triangle with vertices

$$P_1 = \left(\tfrac{1}{2} + \tfrac{1}{n+1}, \tfrac{1}{2} - \tfrac{1}{n+1}\right),\ P_2 = \left(\tfrac{1}{2} - \tfrac{1}{n-1}, \tfrac{1}{2} - \tfrac{1}{n-1}\right),$$
$$\text{and } P_3 = \left(\tfrac{1}{2} + \tfrac{1}{n-1}, \tfrac{1}{2} + \tfrac{1}{n-1}\right).$$

In the case $n = 1$ or $n = 2$ we define $P_2 = (0, 0)$ and $P_3 = (1, 1)$.

The case $\sigma = 2$ in (1) is an important model in the literature, it is known as Germain-Lagrange operator, as well as beam operator and plate operator in the case of space dimension $n = 1$ and $n = 2$, respectively. The linear beam operator inherits some but not all properties from Schrödinger operator. In particular, we do not have the mass conservation law, since in the beam equation the coefficients are real. On the other hand, the functional representation of the solution contains oscillations like it happens for the wave operator. However, the Germain-Lagrange operator is not Kovalesvkian and we do not have the finite speed of propagation [3]. In textbook [5] the reader may found non-singular $L^r - L^q$ estimates in the dual line for the solutions to (1) in the case $\sigma = 2$, and also the influence of lower order terms in these estimates was considered.

If $f \neq 0$ in (1), one may not expect $L^q - L^q$ estimates for $q \neq 2$ for solutions to the wave and Germain-Lagrange equation, neither for solutions to the Cauchy problem for the Schrödinger equation.

The first goal in this paper is to determine (Theorem 2.1), for $\sigma > 1$ and $f \equiv 0$ in (1), the range for $1 \leq r \leq q \leq \infty$ for which $L^r - L^q$ estimates holds. Then, by homogeneity, it follows

$$\|u(t,\cdot)\|_{L^q} \leq C\, t^{1-\frac{n}{\sigma}\left(\frac{1}{r}-\frac{1}{q}\right)} \|g\|_{L^r(\mathbb{R}^n)}, \qquad \forall t > 0.$$

One may find a partial answer to this problem in [7] for a more general class of multipliers $m_{a,b}$, localized at high frequencies in the phase space, where it was proved a characterization for $m_{a,b}$ be a multiplier in Hardy spaces. Hence, to prove Theorem 2.1 it would be sufficient to prove estimates at low frequencies for the associated multipliers, however for the completeness of paper, in Proposition 3.4 we present an elementary proof also at high frequencies.

The application of linear estimates is a very useful tool to the study global existence of small data solutions for semilinear problems. Much has been devoted to the case of the equation

$$\begin{cases} u_{tt} + (-\Delta)^\sigma u = |u|^p\,, & t \geq 0,\ x \in \mathbb{R}^n \\ u(0,x) = f(x),\quad u_t(0,x) = g(x), \end{cases} \tag{2}$$

with $p > 1$ and $\sigma > 1$. If the data in the initial condition $u(0,x) = f(x)$ is small, then $|u|^p$ becomes small for large p. For this reason one is often able to prove such a global (in time) existence result only for some $p > p_c(n)$.

For the wave equation, the critical exponent $p_c(n)$ for the Cauchy problem (2) with $\sigma = 1$ is the positive root of

$$(n-1)p^2 - (n+1)p - 2 = 0.$$

This critical exponent is called the Strauss exponent [11]. By critical exponent we mean that suitable global small data solutions exist in the supercritical case, whereas global solutions cannot exist, under suitable sign assumption on the data, in the critical and subcritical cases.

If we consider the Cauchy problem (2) with $\sigma = 2$, under a standard scaling argument used for nonlinear dispersive equations, we derive that $s_p \doteq \frac{n}{2} - \frac{4}{p-1}$ is the critical index in the Sobolev space $\dot{H}^s(R^n)$, namely, one may expect that if $s > s_p$, then the Cauchy problem (2) is well-posed in $H^s(R^n)$, at least locally, whereas it is not well-posed for $s < s_p$. In [1] the authors derived Strichartz estimates for the solutions to the linear problem and proved the existence of small data global solutions to the Cauchy problem (2) in Besov spaces $\dot{B}^{s_p}_{2,q}(R^n)(1 \leq q < \infty)$ for any $n \geq 1$ and $p > 1 + \frac{8}{n}$, or in Besov spaces $B^s_{2,q}(R^n)$, $s > s_p$, for $p > 1 + \frac{8}{n}$ for $n \geq 4$, and for $p > 5$ in the case $n = 3$. We refer to [13] for an ill-posedness result in the case $0 < s < s_p$. If we look for energy solutions $C([0,T], H^2(\mathbb{R}^n)) \cap C^1([0,T], L^2(\mathbb{R}^n))$, then we have to assume $\frac{n}{2} - \frac{4}{p-1} < 2$ and it appears an upper bounded for p, that is, $p < \frac{n+4}{n-4}$ for $n > 4$ (see Remark 10).

According to Duhamel's principle, the restriction on p to derive the existence of small data global solutions to (2) is related to an integrability condition depending only on the kernel of the second data, whereas the regularity condition is related to the first initial data f. Since the question about regularity condition follows by scaling argument, for simplicity, throughout this paper we assume $f \equiv 0$ in (2), but under additional regularity on f (see Remark 5) one can still have the same results.

The second goal of this paper is to show that, using the derived linear estimates (Theorem 2.1) in the $L^q(R^n)$ basis, with $q \neq 2$, the critical exponent $p_c(n)$ for the existence of global in time small data solutions to

$$\begin{cases} u_{tt} + (-\Delta)^\sigma u = |u|^p\,, & t \geq 0,\ x \in \mathbb{R}^n \\ u(0,x) = 0,\quad u_t(0,x) = g(x), \end{cases} \tag{3}$$

with $p > 1$ and $\sigma > 1$, is given by

$$p_c(n) \doteq 1 + \frac{2\sigma}{[n-\sigma]_+}. \tag{4}$$

From Theorems 2.2 and 2.4 we conclude that $p_c(n)$ is really the critical exponent for integer $\sigma > 1$, with $\sigma < n \leq 2\sigma$, however, in the case $n > 2\sigma$, we may have a gap between the non-existence result Theorem 2.4 and the existence result given by Theorem 2.3.

It is interesting to compare the Cauchy problem (3) with the structurally damped semi-linear evolution equations

$$\begin{cases} u_{tt} + 2\mu(-\Delta)^\delta u_t + (-\Delta)^\sigma u = |u|^p, & t \geq 0, x \in \mathbb{R}^n, \\ (u, u_t)(0, x) = (f, g)(x). \end{cases} \tag{5}$$

In [2] the authors proved that for $\sigma \geq 2\delta$

$$p_0 \doteq 1 + \frac{2\sigma}{[n-2\delta]_+}, \tag{6}$$

is the critical exponent for global in time small data energy solutions. It is clear that (6) coincide with (4) for $\sigma = 2\delta$ and it is better for $\sigma > 2\delta$, i.e., in this case the dissipative term $(-\Delta)^\delta u_t$ improves the critical exponent. For $\sigma < 2\delta$ the dissipation in (5) is less effective than for $\sigma = 2\delta$, so we may conjecture that in this case the critical exponent is also given by (4).

1.1 Notation

Throughout this paper, we use the following.

Notation 1 *Let $f, g : \Omega \subset \mathbb{R}^n \to \mathbb{R}$ be two strictly positive functions. If there exists a constant $C > 0$ such that $f(y) \leq Cg(y)$ (resp. $f(y) \geq Cg(y)$) for all $y \in \Omega$, then we write $f \lesssim g$ (resp. $f \gtrsim g$).*

Notation 2 *Let $\chi \in C_c^\infty(\mathbb{R}^n)$ be a cut-off nonnegative functions satisfying*

$$\chi(\xi) = 1 \text{ if } |\xi| \leq \frac{1}{2}, \qquad \chi(\xi) = 0 \text{ if } |\xi| \geq 1 \qquad \text{and} \qquad \chi(\xi) \in [0, 1].$$

Notation 3 *We denote $\hat{f} = \mathfrak{F} f$, the Fourier transform of a function f with respect to the x variable. For $b \geq 0$, we denote by $(-\Delta)^b f = \mathfrak{F}^{-1}(|\xi|^{2b}\hat{f})$, the possibly fractional Laplace operator.*

Notation 4 *By $L_p^q = L_p^q(\mathbb{R}^n)$ we denote the space of tempered distributions T satisfying the estimate*

$$\|T * f\|_{L^q} \leq C\|f\|_{L^p}$$

for all f in the Schwartz space $S(\mathbb{R}^n)$ with a constant C which is independent of f.

The set of Fourier transforms $\hat{T}$ of distributions $T \in L_p^q$ is denoted by $M_p^q = M_p^q(\mathbb{R}^n)$. The elements in M_p^q are called multipliers of type (p, q). We define in $M_p^q(\mathbb{R}^n)$ the following norm

$$\|m\|_{M_p^q} := \sup\{\|\mathfrak{F}^{-1}(m\mathfrak{F}(f))\|_q : f \in \mathscr{S}, \|f\|_p = 1\}. \tag{7}$$

In the case $p = q$, we denote M_p^p by M_p.

2 Main Results

The following linear estimates plays a fundamental role in order to prove local and global in time existence results to the Cauchy problem (3).

Theorem 2.1 *Let $\sigma > 1$. If $f \equiv 0$ and $g \in L^r$, then the solution u to the Cauchy problem* (1) *satisfies the following estimate*

$$\|u(t,\cdot)\|_{L^q} \lesssim t^{1-\frac{n}{\sigma}\left(\frac{1}{r}-\frac{1}{q}\right)} \|g\|_{L^r(\mathbb{R}^n)}, \qquad \forall t > 0, \tag{8}$$

for all $1 < r \leq q < \infty$, with $\frac{1}{r} + \frac{1}{q} \leq 1$ and $\frac{1-\sigma}{r} - \frac{1}{q} \leq \sigma\left(\frac{1}{n} - \frac{1}{2}\right)$ or $\frac{1}{r} + \frac{1}{q} \geq 1$ and $\frac{1}{r} + \frac{\sigma-1}{q} \leq \sigma\left(\frac{1}{n} + \frac{1}{2}\right)$.

Remark 1 If $\sigma > \frac{2n}{n+2}$, for $r = 1 + \delta$, with $\delta > 0$ sufficiently small, (8) is true for all $\tilde{q} \leq q < \infty$, with $\tilde{q}$ given by

$$\frac{1}{\tilde{q}} \doteq \frac{1}{\sigma - 1}\left(\sigma\left(\frac{1}{n} + \frac{1}{2}\right) - 1\right). \tag{9}$$

For $2\sigma = n$ we get $\tilde{q} = 2$, whereas $\tilde{q} < 2$ for $2\sigma > n$ and $\tilde{q} > 2$ for $2\sigma < n$. Moreover, if $\sqrt{2n} < 2\sigma < n$, then $\tilde{q} < \frac{2n}{n-2\sigma}$

Remark 2 The restriction $r > 1$ in Theorem 2.1 is due to the applications of Mikhlin-Hörmander multiplier theorem in the proof at the low frequencies estimates. However, thanks to Theorem 4.1 in [7] we have $\frac{(1-\chi(\xi))\sin(|\xi|^\sigma)}{|\xi|^\sigma} \in M_1^q$ for $\tilde{q} < q \leq \infty$, whereas using Hausdorff - Young's inequality we conclude $\frac{\chi(\xi)\sin(|\xi|^\sigma)}{|\xi|^\sigma} \in M_1^q$ for $2 \leq q \leq \infty$. Therefore, if $\sigma > \frac{2n}{n+2}$, (8) is also true for $r = 1$ and $\tilde{q} < q \leq \infty$ if $n \geq 2\sigma$ or for $2 \leq q \leq \infty$ if $2\sigma > n$ (Fig. 1).

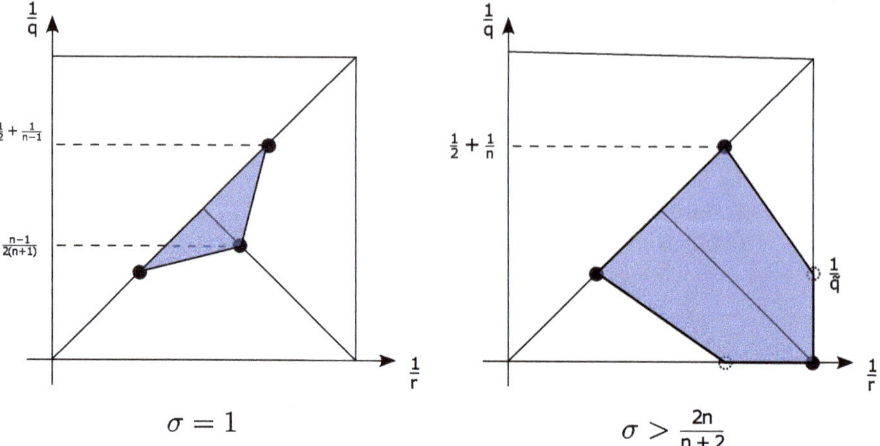

Fig. 1 $L^r - L^q$ estimates

It remains as an open problem to decide if (8) is true in the case $2\sigma > n$ for $r = 1$ and $\tilde{q} < q < 2$.

Remark 3 From Theorem A.4 of Appendix we conclude that (8) is not true for $\frac{1-\sigma}{r} - \frac{1}{q} > \sigma\left(\frac{1}{n} - \frac{1}{2}\right)$ and for $\frac{1}{r} + \frac{\sigma-1}{q} > \sigma\left(\frac{1}{n} + \frac{1}{2}\right)$.

Remark 4 If $1 - \frac{n}{\sigma}\left(\frac{1}{r} - \frac{1}{q}\right) \geq 0$ the singularity in (8) disappears. The intersection between the lines

$$\frac{1}{q} = \frac{1}{r} - \frac{\sigma}{n} \qquad \text{and} \qquad \frac{1-\sigma}{r} - \frac{1}{q} = \sigma\left(\frac{1}{n} - \frac{1}{2}\right)$$

is the point $(r, q) = (2, \frac{2n}{n-2\sigma})$. Let $\mathscr{A} \doteq]1, 2] \times [\tilde{q}, \infty)$ for $2\sigma \geq n$ and $\mathscr{A} \doteq]1, 2] \times \left[\tilde{q}, \frac{2n}{n-2\sigma}\right]$ for $\sqrt{2n} < 2\sigma < n$ (see Fig. 2). Hence, if $\sigma > \frac{2n}{n+2}$, $g \in L^1(\mathbb{R}^n) \cap L^2(\mathbb{R}^n)$ and $q \geq r$, for all $(r, q) \in \mathscr{A}$, there exists $\bar{r} \in (1, 2]$ such that (8) implies in

$$\|u(t, \cdot)\|_{L^q} \lesssim \begin{cases} \|g\|_{L^{\bar{r}}} \leq \|g\|_{L^1 \cap L^2}, & t \in [0, 1) \\ (1+t)^{1-\frac{n}{\sigma}\left(\frac{1}{r} - \frac{1}{q}\right)} \|g\|_{L^r \cap L^2}, & t \in [1, \infty). \end{cases}$$

In particular, for any $\epsilon > 0$, there exists $\delta > 0$ such that

$$\|u(t, \cdot)\|_{L^q} \lesssim (1+t)^{1-\frac{n}{\sigma}\left(1-\frac{1}{q}\right)+\epsilon} \|g\|_{L^{1+\delta} \cap L^2}, \qquad t \in [0, \infty), \tag{10}$$

for all $q \geq \tilde{q}$ if $2\sigma \geq n$ or for all $\tilde{q} \leq q \leq \frac{2n}{n-2\sigma}$ if $2\sigma < n$.

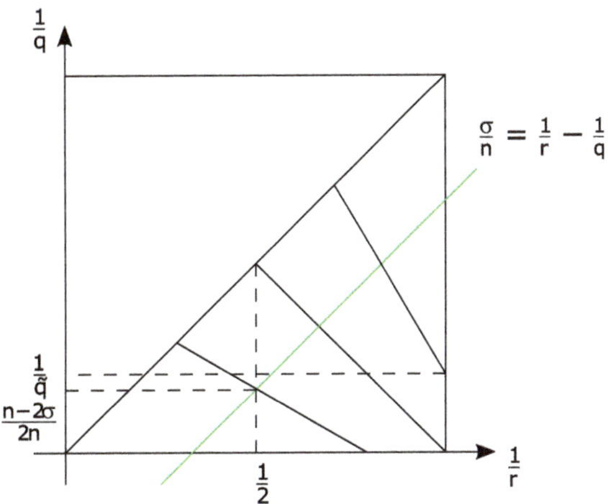

Fig. 2 $n > 2\sigma > \sqrt{2n}$

Remark 5 As it was explained in the introduction, if $f \neq 0$ and $q \neq 2$, one may not expect $L^q - L^q$ estimates for solutions to the Cauchy problem (1). However, following the proof of Theorem 2.1, under additional regularity for f we can conclude the estimate

$$\|u(t,\cdot)\|_{L^q} \leq ct^{\frac{n}{\sigma}(\frac{1}{q}-\frac{1}{r})}\Big(\|f\|_{L^r} + t\||D|^\sigma f\|_{L^r} + t\|g\|_{L^r}\Big), \tag{11}$$

for all $1 < r \leq q < \infty$, with $\frac{1}{r}+\frac{1}{q} \geq 1$ and $\frac{1}{r}+\frac{\sigma-1}{q} \leq \sigma\Big(\frac{1}{2}+\frac{1}{n}\Big)$ or $\frac{1}{r}+\frac{1}{q} \leq 1$ and $\frac{1-\sigma}{r}-\frac{1}{q} \leq \sigma\Big(\frac{1}{n}-\frac{1}{2}\Big)$. Here $|D|^\sigma$ denotes the pseudodifferential operator having the symbol $|\xi|^\sigma$. Moreover, if $g \equiv 0$, in order to avoid singular estimates at $t = 0$ in (11), we may assume that $f \in H^{\sigma,q}(\mathbb{R}^n)$(the Sobolev space of fractional order), to conclude

$$\|u(t,\cdot)\|_{L^q} \leq C(1+t)\|f\|_{H^{\sigma,q}}$$

for all $t \geq 0$ and $q > 1$ such that $\frac{1}{2}-\frac{1}{n} \leq \frac{1}{q} \leq \frac{1}{2}+\frac{1}{n}$. Therefore, for any $\epsilon > 0$, there exists $\delta > 0$ such that if

$$(f,g) \in \mathscr{D} \doteq (H^{\sigma,1+\delta}(\mathbb{R}^n) \cap H^{\sigma,\frac{2n}{[n-2]_+}}(\mathbb{R}^n)) \times (L^1(\mathbb{R}^n) \cap L^2(\mathbb{R}^n)),$$

thanks to $\tilde{q} \geq \frac{2n}{n+2}$ and using Remark 4, we may conclude that the solution to (1) satisfies

$$\|u(t,\cdot)\|_{L^q} \leq C\,(1+t)^{1-\frac{n}{\sigma}\left(1-\frac{1}{q}\right)+\epsilon}\,\|(f,g)\|_{\mathscr{D}}, \qquad \forall t \geq 0,$$

for all $\tilde{q} \leq q \leq \frac{2n}{[n-2]_+}$.

Due to Remark 4, it will be convenient to split the analysis of the semilinear problem (3) in two cases, for $2\sigma \geq n$ and for $2\sigma < n$.

We first give a definition of weak solution to (3).

Definition 1 *Let $q > 1$. We say that $u \in L^q_{\text{loc}}([0,\infty) \times \mathbb{R}^n)$ is a global weak solution to* (3)*, if, for any test function $F \in C_c^\infty([0,\infty) \times \mathbb{R}^n)$, it holds:*

$$I = \int_0^\infty \int_{\mathbb{R}^n} u(t,x)\big(F_{tt}(t,x) + (-\Delta)^\sigma F(t,x)\big)\,dxdt - \int_{\mathbb{R}^n} g(x)\,F(0,x)\,dx, \tag{12}$$

where

$$I = \int_0^\infty \int_{\mathbb{R}^n} |u(t,x)|^p F(t,x)\,dxdt.$$

Let $T > 0$. We say that $u \in L^q_{\text{loc}}([0,T] \times \mathbb{R}^n)$ is a local weak solution to (3)*, if* (12) *is verified for the test functions as above, under the additional assumption that* $\operatorname{supp} F \subset [0,T] \times \mathbb{R}^n$.

Theorem 2.2 *Let $n \geq 2$, $1 < \sigma < n \leq 2\sigma$ and $p_c(n) < p < \infty$. Then there exists a constant $\epsilon > 0$ such that for all $g \in \mathscr{D} \doteq L^1(\mathbb{R}^n) \cap L^2(\mathbb{R}^n)$ with*

$$||g||_{\mathscr{D}} \doteq \|g\|_{L^1} + \|g\|_{L^2} < \epsilon,$$

there exists a unique weak solution $u \in L^\infty([0,\infty), L^{\tilde{q}}(\mathbb{R}^n) \cap L^{q_1}(\mathbb{R}^n))$ to (3) *for any $q_1 > 2p$. Moreover, there exists a $\bar{\delta} > 0$ such that for any $\delta \in (0,\bar{\delta})$ the solution satisfies the following estimates*

$$\|u(t,\cdot)\|_{L^q} \leq C\,(1+t)^{1-\frac{n}{\sigma}\left(\frac{1}{1+\delta}-\frac{1}{q}\right)}\,||g||_{\mathscr{D}}, \qquad \forall t \geq 0, \tag{13}$$

for all $\tilde{q} \leq q \leq q_1$.

Remark 6 In the statement of Theorem 2.2 it is sufficient to assume $g \in L^{1+\delta}(\mathbb{R}^n) \cap L^2(\mathbb{R}^n)$, with $\delta > 0$ sufficiently small. If $g \in L^1(\mathbb{R}^n)$, taking into account Remark 2, we may use $L^1 - L^q$ estimates, with $q \geq 2$, and derive that the solution $u \in L^\infty([0,\infty), L^2(\mathbb{R}^n) \cap L^{q_1}(\mathbb{R}^n))$, with $\delta = 0$ in (13). Moreover, using the embedding of $H^1(\mathbb{R}^n) \hookrightarrow L^q(\mathbb{R}^n)$ for $2 \leq q \leq \frac{2n}{n-2}$, one may conclude that $u \in C^0([0,\infty), L^2(\mathbb{R}^n) \cap L^q(\mathbb{R}^n))$ for $2 \leq q \leq \frac{2n}{n-2}$.

Remark 7 If $(f,g) \in \mathscr{D} \doteq (H^{\sigma,1+\delta}(\mathbb{R}^n) \cap H^{\sigma,\frac{2n}{n-2}}(\mathbb{R}^n)) \times (L^1(\mathbb{R}^n) \cap L^2(\mathbb{R}^n))$, under the assumption of Theorem 2.2, by replacing $||g||_{\mathscr{D}}$ by $\|(f,g)||_{\mathscr{D}}$ and applying the linear estimates stated in Remark 5, one may conclude the existence of a unique global weak solution $u \in L^\infty([0,\infty), L^{\tilde{q}}(\mathbb{R}^n) \cap L^{\frac{2n}{n-2}}(\mathbb{R}^n))$ to the Cauchy problem (3) and it satisfies (13) for all $\tilde{q} \leq q \leq \frac{2n}{n-2}$.

In the case $n > 2\sigma$, in general we can not arrive in the critical index. However, we are able to derive a result where the lower order bound for p depends on a parameter $\bar{r}$ and we arrive at $p_c(n)$ if we can take $\bar{r} = 1$. Now, instead to use $L^{1+\delta} - L^{\tilde{q}}$ estimates as in the proof of Theorem 2.2, we shall use $L^{r+\delta} - L^q$ estimates, with (r, q) on the line segment with end points $\left(1, \frac{1}{\tilde{q}}\right)$ and $\left(\frac{n+2}{2n}, \frac{n+2}{2n}\right)$ given by Theorem 2.1, i.e.

$$\frac{1}{q(r)} = \frac{1}{\sigma - 1}\left(\sigma\left(\frac{1}{n} + \frac{1}{2}\right) - \frac{1}{r}\right). \tag{14}$$

Remark 8 If $n > 2\sigma$, the assumption $1 \le r \le r^\sharp \doteq \frac{2n}{n+2\sigma}$ implies that $q(r)$ given by (14) satisfies $q(r^\sharp) = 2 \le q \le \tilde{q} = q(1)$. Note that $\frac{n+r\sigma}{n-r\sigma}$ is an increasing function on r and $\frac{q(r^\sharp)}{r^\sharp} \le p_c(n) \le \frac{n+r^\sharp\sigma}{n-r^\sharp\sigma}$, hence there exists $0 < r_0 \le r^\sharp$ such that

$$\frac{q(r_0)}{r_0} = \frac{n + r_0\sigma}{n - r_0\sigma}. \tag{15}$$

Remark 9 Thanks to $\frac{q(r)}{r} \le \frac{n+r\sigma}{n-r\sigma} < \frac{n+2\sigma}{n-2\sigma}$ for $r \ge \bar{r} \doteq \max\{1, r_0\}$, we conclude that $q(r) < r\frac{n+2\sigma}{n-2\sigma} \le r^\sharp\frac{n+2\sigma}{n-2\sigma} = \frac{2n}{n-2\sigma}$. Hence, if $g \in L^1(\mathbb{R}^n) \cap L^2(\mathbb{R}^n)$, for all $(r, q) \in \mathscr{A} \doteq]\bar{r}, 2] \times \left[q(\bar{r}), \frac{2n}{n-2\sigma}\right)$ (Fig. 3), there exists $\tilde{r} \in (1, 2]$ such that (8) implies in

$$\|u(t,\cdot)\|_{L^q} \lesssim \begin{cases} \|g\|_{L^{\tilde{r}}} \le \|g\|_{L^1 \cap L^2}, & t \in [0, 1) \\ (1+t)^{1-\frac{n}{\sigma}\left(\frac{1}{r}-\frac{1}{q}\right)}\|g\|_{L^r \cap L^2}, & t \in [1, \infty). \end{cases}$$

In particular, for any $\epsilon > 0$, there exists $\delta > 0$ such that

$$\|u(t,\cdot)\|_{L^q} \lesssim (1+t)^{1-\frac{n}{\sigma}\left(1-\frac{1}{q}\right)+\epsilon}\|g\|_{L^{1+\delta} \cap L^2}, \qquad t \in [0, \infty). \tag{16}$$

Due to the last remark we are able to derive the next result without any restriction on the space dimension.

Theorem 2.3 *Let* $\frac{2n}{n+2} < \sigma < \frac{n}{2}$, $\frac{n+\bar{r}\sigma}{n-\bar{r}\sigma} < p < \frac{n+2\sigma}{n-2\sigma}$, *with* $\bar{r} \doteq \max\{1, r_0\}$ *and* r_0 *satisfying* (15)*. Then there exists* $\epsilon > 0$ *such that for all* $g \in \mathscr{D} \doteq L^1(\mathbb{R}^n) \cap L^2(\mathbb{R}^n)$ *with*

$$\|g\|_{\mathscr{D}} \doteq \|g\|_{L^1} + \|g\|_{L^2} < \epsilon,$$

there exists a unique weak solution $u \in L^\infty([0,\infty), L^{q(\bar{r})}(\mathbb{R}^n) \cap L^{\frac{2n}{n-2\sigma}}(\mathbb{R}^n))$ *to* (3)*, with* $q(\bar{r})$ *satisfying* (14)*. Moreover, there exists a* $\bar{\delta} > 0$ *such that for any* $\delta \in (0, \bar{\delta})$

Fig. 3 Region $\mathscr{A}$

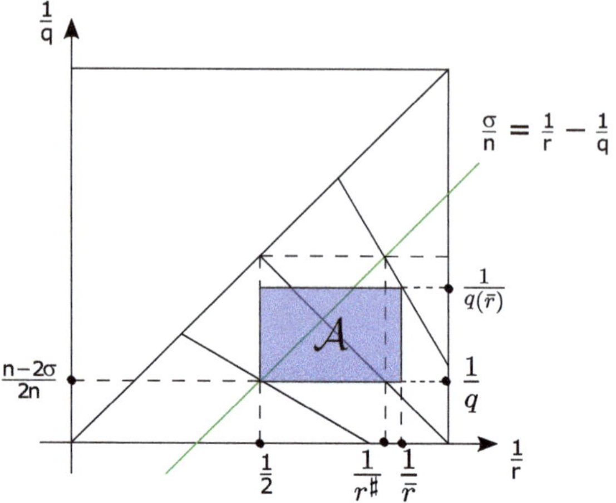

the solution satisfies the following estimates

$$\|u(t,\cdot)\|_{L^q} \leq C\,(1+t)^{1-\frac{n}{\sigma}\left(\frac{1}{\bar{r}+\delta}-\frac{1}{q}\right)}\,\|g\|_{\mathscr{D}}, \qquad \forall t \geq 0,$$

for all $q(\bar{r}) \leq q \leq \frac{2n}{n-2\sigma}$.

Remark 10 The conclusions of Theorem 2.3 is still true if we take $\max\{1, r_0\} \leq \bar{r} \leq r^\sharp$. For instance, if we chose $\bar{r} = r^\sharp$, then $q(\bar{r}) = 2$ and a lower bounded for p is given by $p > 1 + \frac{4\sigma}{n}$. In particular, for $\sigma = 2$ we recover the condition $1 + \frac{8}{n} < p < \frac{n+4}{n-4}$ obtained by [1] for $n > 4$.

Remark 11 From the scale argument, smallness of $\|g\|_{L^{\frac{n(p-1)}{\sigma(p+1)}}}$ is a necessary condition to have global existence to (3). Indeed, if u is a solution to (3), then

$$\lambda^h u(\lambda t, \lambda^{\frac{1}{\sigma}} x), \qquad \text{with} \quad h \doteq \frac{2}{p-1},$$

is a solution to the equation in (3) for any $\lambda > 0$, with initial velocity $\lambda^{h+1} g(\lambda^{\frac{1}{\sigma}} x)$. We have

$$\lambda^{h+1}\|g(\lambda^{\frac{1}{\sigma}}\cdot)\|_{L^q} = \lambda^{h+1-\frac{n}{\sigma q}}\,\|g\|_{L^q}\,,$$

so that the L^q norm is invariant if, and only if, $h + 1 = \frac{n}{\sigma q}$. Hence, we find $q = \frac{n(p-1)}{\sigma(p+1)}$. The assumption $g \in \mathscr{D}$ implies $g \in L^{\frac{n(p-1)}{\sigma(p+1)}}$, thanks to

$$1 < \frac{n(p-1)}{\sigma(p+1)} \leq 2 \Leftrightarrow p_c(n) < p \leq 1 + \frac{4\sigma}{[n-2\sigma]_+}.$$

Remark 12 If $(f, g) \in \mathscr{D} \doteq (H^{\sigma,1+\delta}(\mathbb{R}^n) \cap H^{\sigma,\frac{2n}{n-2}}(\mathbb{R}^n)) \times (L^1(\mathbb{R}^n) \cap L^2(\mathbb{R}^n))$, under the assumption of Theorem 2.3, by replacing $\|g\|_{\mathscr{D}}$ by $\|(f, g)\|_{\mathscr{D}}$ and applying the linear estimates stated in Remark 5, one may conclude the existence of a unique global weak solution $u \in L^\infty([0, \infty), L^{\tilde{q}}(\mathbb{R}^n) \cap L^{\frac{2n}{n-2}}(\mathbb{R}^n))$ to the Cauchy problem (2) and it satisfies (13) for all $\tilde{q} \leq q \leq \frac{2n}{n-2}$. However, in order that the range for q is not empty we have to assume $2\sigma \geq \frac{n+2}{2}$.

According to Definition 2.2 in [4], an operator $L(t, x, \partial_t, \partial_x)$ is quasi-homogeneous of type (h, d_1, d_2), if, for any $\lambda > 0$, (t, x), $(\tau, \xi) \in R^{1+n}$ it holds

$$L(\lambda^{-d_1}t, \lambda^{-d_2}x, \lambda^{d_1}\tau, \lambda^{d_2}\xi) = \lambda^h L(t, x, \tau, \xi).$$

If $\sigma > 1$ is an integer, then the operator $L = \partial_t^2 + (-\Delta)^\sigma$ is quasi-homogeneous of type $(2, 1, \sigma^{-1})$. Hence, the application of Theorem 2.1 in [4] gives the following nonexistence of global weak solution for any $1 < p \leq \frac{n+\sigma}{[n-\sigma]_+}$:

Theorem 2.4 (Theorem 2.1 **of [4])** *Let $\sigma > 1$ be an integer. If $1 < p \leq \frac{n+\sigma}{[n-\sigma]_+}$, then there exists no global in time nontrivial weak solution to* (3).

Now we present two sample models, to which our results applies:

Example 1 Let us consider the semilinear plate equation

$$u_{tt} + \Delta^2 u = |u|^p.$$

We find global existence of small data solutions to space dimension $n = 3, 4$ if $p > \frac{n+2}{n-2} = p_c(n)$ (Theorem 2.2). For space dimension $n > 5$ we can not arrive in the critical index, but we can derive a global existence result for all $1+\frac{8}{n} < p < \frac{n+4}{n-4}$ (Theorem 2.3 and Remark 10).

In general, global weak solutions cannot exist for space dimension $n = 1, 2$ and for space dimension $n \geq 3$ and $1 < p \leq \frac{n+2}{n-2}$ (Theorem 2.4).

Example 2 Let us consider the third-order evolution equation

$$u_{tt} - \Delta^3 u = |u|^p.$$

We find global existence of small data solutions in space dimension $n = 4, 5, 6$ if $p > \frac{n+3}{n-3} = p_c(n)$ (Theorem 2.4), for space dimension $n = 7$ if $p > p_c(7)$ (Theorem 2.3). For space dimension $n \geq 8$ we can not arrive in the critical index, but we can derive a global existence result for all $1 + \frac{12}{n} < p < \frac{n+6}{n-6}$ (Theorem 2.3 and Remark 10).

In general, global weak solutions cannot exist for space dimension $n = 1, 2, 3$ and for space dimension $n \geq 4$ and $1 < p \leq \frac{n+3}{n-3}$(Theorem 2.4).

3 The Linear Estimates

After applying the Fourier transform, the solution to (1) can be written as

$$u(t,x) = \mathfrak{F}^{-1}_{\xi\to x}\Big(\frac{e^{i|\xi|^\sigma t} + e^{-i|\xi|^\sigma t}}{2}\mathfrak{F}(f)(\xi)\Big) + \mathfrak{F}^{-1}_{\xi\to x}\Big(\Big(e^{i|\xi|^\sigma t} - e^{-i|\xi|^\sigma t}\Big)\frac{1}{2i|\xi|^\sigma}\mathfrak{F}(g)(\xi)\Big)$$

or

$$u(t,x) = \mathfrak{F}^{-1}_{\xi\to x}\Big(\cos(t|\xi|^\sigma)\mathfrak{F}(f)(\xi)\Big) + \mathfrak{F}^{-1}_{\xi\to x}\Big(\sin(t|\xi|^\sigma)\frac{1}{|\xi|^\sigma}\mathfrak{F}(g)(\xi)\Big).$$

In the following we will only consider the case $f \equiv 0$, but under additional regularity on f we may write

$$\begin{aligned}
&\mathfrak{F}^{-1}_{\xi\to x}\Big(\cos(t|\xi|^\sigma)\mathfrak{F}(f)(\xi)\Big)\\
=&\mathfrak{F}^{-1}_{\xi\to x}\Big(\chi(t|\xi|^\sigma)\cos(t|\xi|^\sigma)\mathfrak{F}(f)(\xi)\Big) + \mathfrak{F}^{-1}_{\xi\to x}\Big((1-\chi(t|\xi|^\sigma))\frac{e^{i|\xi|^\sigma t} + e^{-i|\xi|^\sigma t}}{2|\xi|^\sigma}|\xi|^\sigma\mathfrak{F}(f)(\xi)\Big)\\
=&\mathfrak{F}^{-1}_{\xi\to x}\Big(\chi(t|\xi|^\sigma)\cos(t|\xi|^\sigma)\mathfrak{F}(f)(\xi)\Big) + \mathfrak{F}^{-1}_{\xi\to x}\Big((1-\chi(t|\xi|^\sigma))\frac{e^{i|\xi|^\sigma t} + e^{-i|\xi|^\sigma t}}{2|\xi|^\sigma}\mathfrak{F}(|D|^\sigma f)(\xi)\Big),
\end{aligned}$$

and following as in the case $f \equiv 0$, one may derive $L^r - L^q$ estimates for solutions to (1) (see Remark 5).

By homogeneity, it is sufficient to prove (8) for $t = 1$. In order to do it, we will divide the phase space at low and high-frequencies. Let us first derive low-frequencies estimates:

Proposition 3.1 *For* $1 < p \le 2 \le q < \infty$ *and* $0 \le r \le n\Big(\frac{1}{p} - \frac{1}{q}\Big)$, *we have*

$$m(\xi) := \frac{\chi(\xi)e^{\pm i|\xi|^\sigma}}{|\xi|^r} \in M_p^q. \tag{17}$$

Proof Let us first consider the case $r > 0$. In this case we have

$$\text{meas}\{\xi \in \mathbb{R}^n : |m(\xi)| \ge l\} \le \text{meas}\{\xi \in \mathbb{R}^n : |\xi| \le l^{-\frac{1}{r}}\} \le Cl^{-\frac{n}{r}} \tag{18}$$

and $l^{-\frac{n}{r}} \le l^{-b}$ for $l \ge 1$, where $(\frac{1}{p} - \frac{1}{q}) = \frac{1}{b}$. If $0 < l < 1$, $\{\xi \in \mathbb{R}^n : |m(\xi)| \ge l\} \subset \{\xi \in \mathbb{R}^n : |\xi| \le 1\}$, then

$$\text{meas}\{\xi \in \mathbb{R}^n : |m(\xi)| \ge l\} \le Cl^{-b}, \tag{19}$$

Therefore, applying Theorem A.2 the result is concluded for $r > 0$.

For $r = 0$ and $l > 1$ we have that $\text{meas}\{\xi \in \mathbb{R}^n : |g(\xi)| \ge l\} = 0$, hence (19) holds for all $b > 0$, and again the conclusion follows from Theorem A.2.

By using that $\frac{\sin(|\xi|^\sigma)}{|\xi|^\sigma}$ is a bounded function, similar to the case $r = 0$ in Proposition 3.1, one may conclude the following result:

Proposition 3.2 *For* $1 < p \leq 2 \leq q < \infty$, *we have*

$$\frac{\chi(\xi)\sin(|\xi|^\sigma)}{|\xi|^\sigma} \in M_p^q. \tag{20}$$

As a consequence of the Mikhlin-Hörmander multiplier theorem we have:

Proposition 3.3 *For all* $1 < q < \infty$ *we have*

$$m(\xi) := \frac{\chi(\xi)\sin(|\xi|^\sigma)}{|\xi|^\sigma} \in M_q^q. \tag{21}$$

Proof For all multi-indice α and $\xi \neq 0$ we have

$$|\partial^\alpha |\xi|^{-\sigma}| \leq C_\alpha |\xi|^{-\sigma-|\alpha|}$$

and for all $0 < |\xi| \leq R$

$$|\partial^\alpha \sin(|\xi|^\sigma)| \leq C_{\alpha,R}|\xi|^{\sigma-|\alpha|}.$$

Hence, for all $0 < |\xi| \leq R$

$$\left|\partial^\alpha \frac{\sin(|\xi|^\sigma)}{|\xi|^\sigma}\right| \leq C_\alpha \sum_{\beta+\gamma=\alpha} |\partial^\beta \sin(|\xi|^\sigma)||\partial^\gamma |\xi|^{-\sigma}| \leq C_{\alpha,R}|\xi|^{-|\alpha|}.$$

Thanks to $m(\xi) = \frac{\sin(|\xi|^\sigma)}{|\xi|^\sigma}$ for small frequencies, we conclude that

$$|\partial^\alpha m(\xi)| \leq C_{\alpha,R}|\xi|^{-|\alpha|},$$

for all multi-indice α and the proof follows by using Theorem A.1.

Now, by following an idea used in [10] to derive $L^q - L^q$ estimates for a class of multipliers including the one from Schrödinger equation, but now we interpolate the point $p = q = 1$ with points from the conjugate line (see Fig. 4) we present an elementary proof of high frequencies estimates for the class of symbols considered in [7], on a region of the phase space described by Theorem 2.1, excluding the lines

$$\frac{1}{p} + \frac{a-1}{q} = \frac{b}{n} + \frac{a}{2}, \qquad \frac{1-a}{p} - \frac{1}{q} = \frac{b}{n} - \frac{a}{2}, \qquad p \leq q \tag{22}$$

for which we refer to [7], where the author derived such estimates in Hardy spaces $H^p(\mathbb{R}), 0 < p < \infty$.

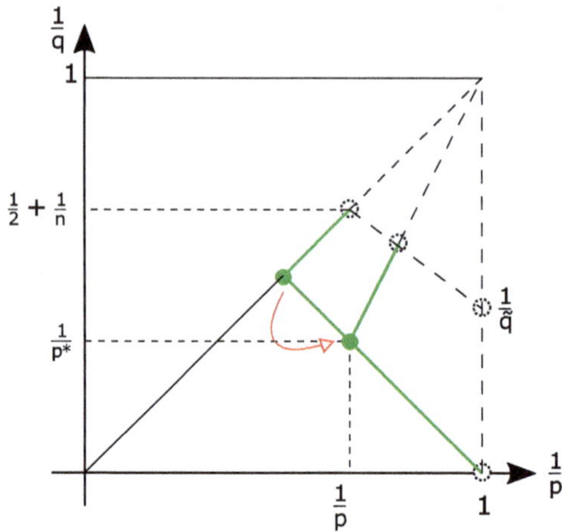

Fig. 4 High frequencies estimates

Proposition 3.4 *Let us consider Fourier multiplier*

$$m_{a,b}(\xi) = \frac{(1-\chi(\xi))e^{i|\xi|^a}}{|\xi|^b}, \qquad \xi \in \mathbb{R}^n, \qquad a > 0, a \neq 1, \qquad b \in \mathbb{R},$$

where χ *is as in Notation 2. Let* $1 < p \leq q$. *If* $\frac{1}{p} + \frac{1}{q} \geq 1$ *and* $\frac{1}{p} + \frac{a-1}{q} < \frac{b}{n} + \frac{a}{2}$, *then* $m_{a,b} \in M_p^q$. *The same conclusion is true if* $\frac{1}{p} + \frac{1}{q} \leq 1$ *and* $\frac{1-a}{p} - \frac{1}{q} < \frac{b}{n} - \frac{a}{2}$.

Proof By using duality argument, it is sufficient to prove Proposition 3.4 for $\frac{1}{p} + \frac{1}{q} \geq 1$. Let $\phi \in C_c^\infty(\mathbb{R}^n)$ supported $\{\xi : \frac{1}{2} \leq |\xi| \leq 2\}$ and $\phi_k(\xi) \doteq \phi(2^{-k}|\xi|)$, with k an integer. By using Plancherel's theorem we have

$$\|\phi_k \cdot m_{a,b}\|_{M_2^2} \leq C2^{-kb}. \tag{23}$$

Now, put $2^{-k}\xi \doteq \eta$, and since $(1-\chi(\xi))\frac{\phi_k(\xi)}{|\xi|^b} = \frac{\phi_k(\xi)}{|\xi|^b}$ for $|\xi| \geq 1$ and $k \geq 1$, by using Littman's lemma we conclude

$$\begin{aligned} \left\|\mathfrak{F}^{-1}_{\xi\to x}\left(e^{\pm i|\xi|^a}\frac{\phi_k(\xi)}{|\xi|^b}\right)\right\|_{L^\infty(\mathbb{R}^n)} &= 2^{k(n-b)}\left\|\mathfrak{F}^{-1}_{\eta\to x}\left(e^{\pm i2^{ka}|\eta|^\sigma}\frac{\phi(\eta)}{|\eta|^b}\right)\right\|_{L^\infty(\mathbb{R}^n)} \\ &\leq C2^{k(n-b)}(1+2^{ka})^{-\frac{n}{2}}, \end{aligned}$$

for all $k \geq 1$. Hence, take into account that $\chi(\xi)\phi_k(\xi) = 0$ for $k \leq 0$, Young's Inequality implies

$$\left\|\mathfrak{F}^{-1}_{\xi\to x}\Big(m_{a,b}(\xi)\phi_k(\xi)\mathfrak{F}(f)\Big)\right\|_{L^\infty(\mathbb{R}^n)} \leq C2^{k(n-b-a\frac{n}{2})}\|f\|_{L^1(\mathbb{R}^n)}, \tag{24}$$

for all integer k, or equivalent,

$$\|\phi_k \cdot m_{a,b}\|_{M_1^\infty} \leq C 2^{k(n-b-a\frac{n}{2})}. \tag{25}$$

As a consequence of (23), (25) and the Riesz-Thorin interpolation theorem we get

$$\|\phi_k \cdot m_{a,b}\|_{M_{p_0}^{q_0}} \leq C 2^{k\left(-b+(\frac{1}{p_0}-\frac{1}{2})(n(2-a))\right)} \tag{26}$$

for $\frac{1}{p_0} + \frac{1}{q_0} = 1$.

In order to derive an estimate for $\|\phi_k \cdot m_{a,b}\|_{M_1^1}$, one may prove the following estimates

$$\|D^M \phi_k \cdot m_{a,b}\|_{L^2} \leq C_M 2^{k(-b+M(a-1)+\frac{n}{2})} \tag{27}$$

and applying the Berstein's inequality (Proposition A.1) for $M > \frac{n}{2}$ we get

$$\|\phi_k \cdot m_{a,b}\|_{M_1} \leq \|\phi_k \cdot m_{a,b}\|_{L^2}^{(1-\frac{n}{2M})} \|D^M \left(\phi_k \cdot m_{a,b}\right)\|_{L^2}^{\frac{n}{2M}} \leq C 2^{ka(\frac{n}{2}-\frac{b}{a})}. \tag{28}$$

Using (26) and (28) and Riesz-Thorin interpolation theorem we conclude that

$$\|\phi_k \cdot m_{a,b}\|_{M_p^q} \leq C 2^{k\left(-b+(\frac{1}{p_0}-\frac{1}{2})(n(2-a))\right)(1-\theta)} 2^{ka(\frac{n}{2}-\frac{b}{\sigma})\theta} = C 2^{kn\left(\frac{1}{p}+\frac{a-1}{q}-\left(\frac{a}{2}+\frac{b}{n}\right)\right)},$$

where $0 < \theta < 1$, with $\frac{1}{p} = \frac{1-\theta}{p_0} + \theta$ and $\frac{1}{q} = \frac{1-\theta}{q_0} + \theta$.

Therefore, for large frequencies and for fixed k_0, using the dyadic decomposition we conclude the estimate

$$\|m_{a,b}\|_{M_p^q} \leq C \sum_{k \geq k_0} \|\phi_k \cdot m_{a,b}\|_{M_p^q},$$

which is convergent if $\frac{1}{p} + \frac{a-1}{q} - \left(\frac{a}{2} + \frac{b}{n}\right) < 0$ and the proof is concluded.

Proof (Theorem 2.1) Interpolating Propositions 3.2 and 3.3, for small frequencies we conclude that $\frac{\chi(\xi)\sin(|\xi|^\sigma)}{|\xi|^\sigma} \in M_r^q$ for all $1 < r \leq q < \infty$.

For large frequencies, applying Proposition 3.4 for $a = b = \sigma$, we conclude that $\frac{(1-\chi(\xi))\sin(|\xi|^\sigma)}{|\xi|^\sigma} \in M_r^q$, with $1 < r \leq q$, if $\frac{1}{r} + \frac{1}{q} \leq 1$ and $\frac{1-\sigma}{r} - \frac{1}{q} < \sigma\left(\frac{1}{n} - \frac{1}{2}\right)$ or if $\frac{1}{r} + \frac{1}{q} \geq 1$ and $\frac{1}{r} + \frac{\sigma-1}{q} < \sigma\left(\frac{1}{n} + \frac{1}{2}\right)$. The estimates in the segment lines given by (22) are a consequence of Theorem A.4.

4 Applications to Semilinear Evolution Equations

According to Duhamel's principle, a solution to (3) satisfies the nonlinear integral equation

$$u(t,x) = K_1(t,x) *_{(x)} g(x) + Nu(t,x),$$

where $K_1(t,x) *_{(x)} g(x)$ is the solutions of the Cauchy problem (1) with $f \equiv 0$ and

$$Nu(t,x) = \int_0^t K_1(t-s,x) * |u(s,x)|^p ds.$$

Motivated by the derived linear estimates, we choose the spaces for solutions $X(T)$ and data $\mathscr{D}$. For any $u \in X(T)$ we define

$$P : u \in X(T) \to Pu(t,x) := K_1(t,x) *_{(x)} g(x) + Nu(t,x). \tag{29}$$

Then we show that the following estimates are satisfied:

$$\begin{aligned}
&\|Pu\|_{X(T)} \leq C_0 \, \|g\|_{\mathscr{D}} + C_1(t)\|u\|^p_{X(T)},\\
&\|Pu - Pv\|_{X(T)} \leq C_2(t)\|u - v\|_{X(T)}\big(\|u\|^{p-1}_{X(T)} + \|v\|^{p-1}_{X(T)}\big),
\end{aligned}$$

for $t \in [0,\infty)$ with nonnegative constants C_0, $C_1(t)$ and $C_2(t)$. The estimates for the image Pu allow us to apply Banach's fixed point theorem. In this way we get simultaneously a unique solution to $Pu = u$ locally in time for large data and globally in time for small data.

In this section we will use the following well-known result (see for instance [2]):

Lemma 1 *Let $\alpha < 1 < \beta$. Then it holds*

$$\int_0^t (t-s)^{-\alpha} \, (1+s)^{-\beta} \, ds \lesssim (1+t)^{-\alpha}. \tag{30}$$

4.1 Local and Global Existence in the Case $2\sigma \geq n$

Let us first prove the following local in time existence result for large data and for all $p \geq 1$:

Proposition 4.1 *Let $\sigma > 1$, $2\sigma \geq n$, $1 \leq p < \infty$ and $q_1 > 2p$. For $R > 0$, there exists $0 < T < 1$ such that for all $g \in \mathscr{D} \doteq L^{\tilde{q}}(\mathbb{R}^n) \cap L^2(\mathbb{R}^n)$, with*

$$||g||_{\mathscr{D}} \doteq \|g\|_{L^{\tilde{q}}} + \|g\|_{L^2} < R,$$

there exists a unique local in time weak solution $u \in L^\infty([0,T], L^{\tilde q}(\mathbb{R}^n) \cap L^{q_1}(\mathbb{R}^n))$ *to* (3).

Proof Let us define for $q_1 \geq \tilde q$ the Banach space

$$X(T) = \{u \in L^\infty([0,T], L^{\tilde q}(\mathbb{R}^n)) \cap L^{q_1}(\mathbb{R}^n)) : \|u\|_{X(T)} < \infty\},$$

with the norm

$$\|u\|_{X(T)} = \operatorname*{ess\,sup}_{0\leq t\leq T} \Big(\|u(t,\cdot)\|_{L^{q_1}} + \|u(t,\cdot)\|_{L^{\tilde q}} \Big).$$

Thanks to the derived estimate in Theorem 2.1, the linear part $K_1(t,x) *_{(x)} g(x)$ of the solution is in $X(T)$, and for all $0 \leq t \leq T < 1$ we have

$$\|K_1(t,\cdot) *_{(x)} g\|_{X(T)} \leq C T^{1-\frac{n}{2\sigma}} \|g\|_{\mathscr{D}}.$$

By using Minkowski integral inequality and Theorem 2.1 for admissible pairs $(q,\tilde q)$, with $q = \tilde q$ and $q = q_1$, it holds

$$\begin{aligned}\|Nu(t,\cdot)\|_{L^q} &\lesssim \int_0^t (t-s)^{1-\frac{n}{\sigma}\left(\frac{1}{\tilde q}-\frac{1}{q}\right)} \||u(s,\cdot)|^p\|_{L^{\tilde q}} ds \lesssim \int_0^t (t-s)^{1-\frac{n}{\sigma}\left(\frac{1}{\tilde q}-\frac{1}{q}\right)} \|u(s,\cdot)\|^p_{L^{p\tilde q}} ds \\ &\lesssim \int_0^t (t-s)^{1-\frac{n}{\sigma}\left(\frac{1}{\tilde q}-\frac{1}{q}\right)} ds \|u\|^p_{X(T)} \lesssim T^{2-\frac{n}{\sigma}\left(\frac{1}{\tilde q}-\frac{1}{q}\right)} \|u\|^p_{X(T)},\end{aligned}$$

for any $u \in X(T)$, where in the third inequality we used that $\tilde q \leq p\tilde q \leq q_1$ and in the last one that $2 - \frac{n}{\sigma}\left(\frac{1}{\tilde q} - \frac{1}{q}\right) > 0$ for $q = \tilde q$ e $q = q_1$.

Hence, it follows

$$\|Pu\|_{X(T)} \leq C T^{1-\frac{n}{2\sigma}} \|g\|_{\mathscr{D}} + C_1 T^{2-\frac{n}{\sigma}\left(\frac{1}{\tilde q}-\frac{1}{q_1}\right)} \|u\|^p_{X(T)}.$$

This leads to $Pu \in L^\infty([0,T], L^{\tilde q}(\mathbb{R}^n) \cap L^{q_1}(\mathbb{R}^n))$ and, if $0 < T < 1$ is sufficiently small, P maps balls of $X(T)$ into balls of $X(T)$.

Moreover, thanks to the Mean Value theorem we have

$$||u|^p - |v|^p| \leq C_0 |u-v| \big(|u|^{p-1} + |v|^{p-1}\big),$$

and using Hölder's inequality we get

$$\||u|^p - |v|^p\|_{L^r} \leq C_0 \|u-v\|_{L^{rp}} \big(\|u\|^{p-1}_{L^{rp}} + \|v\|^{p-1}_{L^{rp}}\big).$$

Now, following as done to estimate $\|Nu(t,\cdot)\|_{L^q}$, we conclude that

$$\|Pu-Pv\|_{X(T)} = \|Nu-Nv\|_{X(T)} \lesssim T^{2-\frac{n}{\sigma}\left(\frac{1}{\tilde q}-\frac{1}{q}\right)} \|u-v\|_{X(T)} \big(\|u\|^{p-1}_{X(T)} + \|v\|^{p-1}_{X(T)}\big).$$

Therefore, if $0 < T < 1$ is sufficiently small, the existence of a unique local weak solution follows by contraction argument.

Now, under the additional assumption that $g \in L^1(\mathbb{R}^n)$ and $||g||_{\mathscr{D}}$ is small, for $p > p_c(n)$ we are able to prove that the local solution derived in Proposition 4.1 may be extended globally in time:

Proof (of Theorem 2.2*)* Let $\delta > 0$ sufficiently small such that $1+\delta \leq \tilde{q} < q_1$ and define

$$X(T) = \{u \in L^\infty([0,T], L^{\tilde{q}}(\mathbb{R}^n) \cap L^{q_1}(\mathbb{R}^n)) : \ \|u\|_{X(T)} < \infty\},$$

with the norm

$$\|u\|_{X(T)} = \operatorname*{ess\,sup}_{0\leq t\leq T} \left((1+t)^{\frac{n}{\sigma}\left(\frac{1}{1+\delta}-\frac{1}{\tilde{q}}\right)-1} \|u(t,\cdot)\|_{L^{\tilde{q}}} + (1+t)^{\frac{n}{\sigma}\left(\frac{1}{1+\delta}-\frac{1}{q_1}\right)-1} \|u(t,\cdot)\|_{L^{q_1}} \right).$$

For any $u, v \in X(T)$, the operator P given by (29) satisfies

$$\|Pu\|_{X(T)} \leq C_0 \, \|g\|_{\mathscr{D}} + C\|u\|^p_{X(T)} \tag{31}$$

$$\|Pu - Pv\|_{X(T)} \leq C_1 \|u-v\|_{X(T)} \big(\|u\|^{p-1}_{X(T)} + \|v\|^{p-1}_{X(T)}\big). \tag{32}$$

By using the derived linear estimates, we prove (31), but we omit the proof of (32), since it is analogous to the proof of (31).

Thanks to estimate (16), we have

$$\|K_1(t,x) *_{(x)} g\|_{X(T)} \leq C_0 \|g\|_{\mathscr{D}}.$$

By using Minkowski integral inequality and Theorem 2.1 it holds for $q \geq \tilde{q}$

$$\|Nu(t,\cdot)\|_{L^q} \lesssim \int_0^t \|K_1(t-s,\cdot) * |u(s,\cdot)|^p\|_{L^q} ds \lesssim \int_0^t (t-s)^{1-\frac{n}{\sigma}\left(\frac{1}{1+\delta}-\frac{1}{q}\right)} \||u(s,\cdot)|^p\|_{L^{1+\delta}} ds.$$

Moreover, thanks to the assumption $p_c(n) < p < \frac{q_1}{2}$ and $\tilde{q} \leq 2 < p_c(n)$, there exists $\bar{\delta} > 0$ such that $\tilde{q} < p(1+\delta) < q_1$ and $\beta \doteq \beta(\delta) > 1$ for all $\delta \in (0, \bar{\delta})$, where

$$\beta \doteq p\left(\frac{n}{\sigma(1+\delta)}\left(1-\frac{1}{p}\right) - 1\right). \tag{33}$$

Hence

$$\||u(s,\cdot)|^p\|_{L^{1+\delta}} = \|u(s,\cdot)\|^p_{L^{p(1+\delta)}} \lesssim (1+s)^{-\beta} \|u\|^p_{X(T)},$$

and using that $1-\frac{n}{\sigma}\left(\frac{1}{1+\delta}-\frac{1}{q}\right)>-1$ for $2\sigma\geq n$, Lemma 1 yields

$$\|Nu(t,\cdot)\|_{L^q}\lesssim\int_0^t(t-s)^{1-\frac{n}{\sigma}\left(\frac{1}{1+\delta}-\frac{1}{q}\right)}(1+s)^{-\beta}ds\|u\|_{X(T)}^p$$
$$\lesssim(1+t)^{1-\frac{n}{\sigma}\left(\frac{1}{1+\delta}-\frac{1}{q}\right)}\|u\|_{X(T)}^p.$$

We conclude (31) by choosing $q=\tilde{q}$ and $q=q_1$ in the last estimate.

This leads to $Pu\in L^\infty([0,T],L^{\tilde{q}}(\mathbb{R}^n)\cap L^{q_1}(\mathbb{R}^n))$. Moreover, the operator P maps balls of $X(T)$ into balls of $X(T)$ for small data in $\mathscr{D}$ and, the existence of a unique global in time weak solution u follows from estimates (31) and (32). Indeed, taking the recurrence sequence $u^{(-1)}=0,\ u^{(j)}=P(u^{(j-1)})$ for $j=0,1,2,\cdots$, we apply (31) with small $\|g\|_{\mathscr{D}}<\epsilon$, and we inductively see that

$$\|u^{(j)}\|_{X(T)}\leq 2C_0\epsilon,\qquad j=0,1,2,\cdots. \tag{34}$$

Here, we remark that we have $u^{(0)}=P(u^{(-1)})=K_1(t,x)*_{(x)}g(x)\in X(T)$ thanks to (16).

After checking the uniform estimate (34), we use (32) to find

$$\|u^{(j+1)}-u^{(j)}\|_{X(T)}\leq 2^{-1}\|u^{(j)}-u^{(j-1)}\|_{X(T)}. \tag{35}$$

From (35), we inductively obtain $\|u^{(j)}-u^{(j-1)}\|_{X(T)}\leq C2^{-j}$, so that $\{u^{(j)}\}$ is a Cauchy sequence in $X(T)$ and the limit function u satisfies $\|u\|_{X(T)}\leq 2C_0\epsilon$, $P(u)=u$ and, using again (31), we conclude that $\|u\|_{X(T)}\leq C_0\,\|g\|_{\mathscr{D}}$.

Moreover, u also satisfies (13). Indeed, for any $\tilde{q}\leq q\leq q_1$, by interpolation we get

$$\|u(t,\cdot)\|_{L^q}\lesssim\|u(t,\cdot)\|_{L^{\tilde{q}}}^{\theta}\|u(t,\cdot)\|_{L^{q_1}}^{(1-\theta)}$$
$$\lesssim(1+t)^{-\alpha}\|u\|_X\lesssim(1+t)^{-\alpha}\|g\|_{\mathscr{D}},$$

with $\frac{1}{q}=\frac{\theta}{\tilde{q}}+\frac{(1-\theta)}{q_1}$ and

$$\alpha\doteq\left(\frac{n}{\sigma}\left(\frac{1}{(1+\delta)}-\frac{1}{q}\right)-1\right).$$

This concludes the proof.

4.2 Local and Global Existence in the Case $n > 2\sigma$

From Remark 4 we understood that in the case $n > 2\sigma$, in order to have non-singular $L^r - L^q$ estimates for the solutions to (1), an upper bound for q appears and this produce some additional difficulties to derive existence results to the semilinear problem (3).

In the case $n > 2\sigma$, we have $\tilde{q} > 2$, so we will replace in Proposition 4.1 $\tilde{q}$ by the smaller q such that $L^q - L^q$ linear estimate holds, i.e., $q = \frac{2n}{n+2}$ and, in order to control the singular behaviour of the linear estimates it appears an upper bounded for q_1.

Proposition 4.2 *Let* $2 < 2\sigma < n$ *and* $\frac{2n}{n+2} \leq q_0 \leq 2 \leq q_1 < \frac{2n}{n-2\sigma}$. *Then, for* $R > 0$ *and for all* $1 \leq p \leq \frac{q_1}{q_0}$, *there exists* $0 < T < 1$ *such that for all* $g \in \mathscr{D} \doteq L^{q_0}(\mathbb{R}^n) \cap L^2(\mathbb{R}^n)$, *with*

$$||g||_{\mathscr{D}} \doteq \|g\|_{L^{q_0}} + \|g\|_{L^2} < R,$$

there exists a unique local in time weak solution $u \in L^\infty([0,T], L^{q_0}(\mathbb{R}^n) \cap L^{q_1}(\mathbb{R}^n))$ *to* (3).

Proof Let us define for $q_1 \geq \tilde{q}$ the Banach space

$$Z(T) = \{u \in L^\infty([0,T], L^{q_0}(\mathbb{R}^n)) \cap L^{q_1}(\mathbb{R}^n)) : \|u\|_{Z(T)} < \infty\},$$

with the norm

$$\|u\|_{Z(T)} = \operatorname*{ess\,sup}_{0\leq t\leq T} \big(\|u(t,\cdot)\|_{L^{q_0}} + \|u(t,\cdot)\|_{L^{q_1}}\big).$$

Thanks to the derived estimate in Theorem 2.1, the linear part $K_1(t,x) *_{(x)} g(x)$ of the solution is in $Z(T)$, and for any $0 \leq t \leq T < 1$ we have

$$\begin{aligned}\|K_1(t,\cdot) *_{(x)} g\|_{Z(T)} &= \operatorname*{ess\,sup}_{0\leq t\leq T} \big(\|K_1(t,\cdot) *_{(x)} g\|_{L^{q_0}} + \|K_1(t,\cdot) *_{(x)} g\|_{L^{q_1}}\big)\\ &\leq C_1 T\|g\|_{L^{q_0}} + C_2 T^{1-\frac{n}{\sigma}\left(\frac{1}{2}-\frac{1}{q_1}\right)}\|g\|_{L^2} \leq C T^{1-\frac{n}{\sigma}\left(\frac{1}{2}-\frac{1}{q_1}\right)}\|g\|_{\mathscr{D}}.\end{aligned}$$

Here, $1 - \frac{n}{\sigma}\left(\frac{1}{2} - \frac{1}{q_1}\right) > 0$ thanks to $q_1 < \frac{2n}{n-2\sigma}$.

By using Minkowski integral inequality and Theorem 2.1 for admissible pairs (q, q_0), with $q = q_0$ and $q = q_1$, it holds

$$\begin{aligned}\|Nu(t,\cdot)\|_{L^q} &\lesssim \int_0^t (t-s)^{1-\frac{n}{\sigma}\left(\frac{1}{q_0}-\frac{1}{q}\right)} \||u(s,\cdot)|^p\|_{L^{q_0}} ds \lesssim \int_0^t (t-s)^{1-\frac{n}{\sigma}\left(\frac{1}{q_0}-\frac{1}{q}\right)} \|u(s,\cdot)\|^p_{L^{pq_0}} ds\\ &\lesssim \int_0^t (t-s)^{1-\frac{n}{\sigma}\left(\frac{1}{q_0}-\frac{1}{q}\right)} ds \|u\|^p_{Z(T)} \lesssim T^{2-\frac{n}{\sigma}\left(\frac{1}{q_0}-\frac{1}{q}\right)} \|u\|^p_{Z(T)},\end{aligned}$$

for any $u \in Z(T)$, where in the third inequality we used that $q_0 \leq pq_0 \leq q_1$ and in the last one that $2 - \frac{n}{\sigma}\left(\frac{1}{q_0} - \frac{1}{q_1}\right) > 0$ thanks to $q_1 \leq \frac{2n}{n-2\sigma} < \frac{nq_0}{[n-2\sigma q_0]_+}$.

Hence, it follows

$$\|Pu\|_{Z(T)} \leq CT^{1-\frac{n}{\sigma}\left(\frac{1}{2}-\frac{1}{q_1}\right)}\|g\|_{\mathscr{D}} + C_1 T^{2-\frac{n}{\sigma}\left(\frac{1}{q_0}-\frac{1}{q_1}\right)}\|u\|^p_{Z(T)}.$$

This leads to $Pu \in L^\infty([0,T], L^{q_0}(\mathbb{R}^n) \cap L^{q_1}(\mathbb{R}^n))$ and, if $0 < T < 1$ is sufficiently small, P maps balls of $Z(T)$ into balls of $Z(T)$.

Now, following as done to estimate $\|Nu(t,\cdot)\|_q$, we conclude that

$$\|Pu - Pv\|_{Z(T)} = \|Nu - Nv\|_{Z(T)} \lesssim T^{2-\frac{n}{\sigma}\left(\frac{1}{q_0}-\frac{1}{q}\right)}\|u-v\|_{X(T)}\left(\|u\|^{p-1}_{Z(T)} + \|v\|^{p-1}_{Z(T)}\right).$$

Therefore, if $0 < T < 1$ is sufficiently small, the existence of a unique local weak solution follows by contraction argument.

If we fix $q_0 = 2$ in Proposition 4.2, by changing the argument in the proof of Proposition 4.2, we can relax the bounded from above to p, to obtain in the next result the existence of a unique local weak solution to (3), that under additional hypotheses, can be extended globally (see Theorem 2.3 with $\bar{q} = 2$):

Proposition 4.3 *Let $2 < 2\sigma < n$, $2 \leq q_1 < \frac{2n}{n-2\sigma}$ and $1 \leq p \leq 1 + \frac{4\sigma}{n-2\sigma}$. Then for $R > 0$, there exists $0 < T < 1$ such that for all $g \in L^2(\mathbb{R}^n)$, with $\|g\|_{L^2} < R$, there exists a unique local in time weak solution $u \in L^\infty([0,T], L^2(\mathbb{R}^n) \cap L^{q_1}(\mathbb{R}^n))$ to* (3).

Proof If $1 \leq p < 1 + \frac{2\sigma}{n}$, we may apply Proposition 4.2 with $q_0 = 2$, but if $1 + \frac{2\sigma}{n} \leq p \leq 1 + \frac{4\sigma}{n-2\sigma}$, to estimate $\|Nu(t,\cdot)\|_{L^q}$ we shall use $L^{r^\sharp} - L^q$ linear estimates, with $r^\sharp = \frac{2n}{n+2\sigma} + \varepsilon \in (1,2)$, with $\varepsilon > 0$ sufficiently small. Applying Theorem 2.1 for admissible pairs $(q, r^\sharp)$, with $q = 2$ and $q = \frac{2n}{n-2\sigma}$, it holds

$$\begin{aligned}\|Nu(t,\cdot)\|_{L^q} &\lesssim \int_0^t (t-s)^{1-\frac{n}{\sigma}\left(\frac{1}{r^\sharp}-\frac{1}{q}\right)}\||u(s,\cdot)|^p\|_{L^{r^\sharp}}ds \lesssim \int_0^t (t-s)^{1-\frac{n}{\sigma}\left(\frac{1}{r^\sharp}-\frac{1}{q}\right)}\|u(s,\cdot)\|^p_{L^{pr^\sharp}}ds \\ &\lesssim \int_0^t (t-s)^{1-\frac{n}{\sigma}\left(\frac{1}{r^\sharp}-\frac{1}{q}\right)}ds\|u\|^p_{Z(T)} \lesssim T^{2-\frac{n}{\sigma}\left(\frac{1}{r^\sharp}-\frac{1}{q}\right)}\|u\|^p_{Z(T)},\end{aligned}$$

for any $u \in Z(T)$, where we use that $2 \leq pr^\sharp \leq \frac{2n}{n-2\sigma}$ and $2 - \frac{n}{\sigma}\left(\frac{1}{r^\sharp} - \frac{1}{q}\right) > 0$.

Proof (of Theorem 2.3*)* Let $\delta > 0$ sufficiently small such that $1 + \delta \leq \bar{q} < q_1 \doteq \frac{2n}{n-2\sigma}$ and define

$$X(T) = \{u \in L^\infty([0,T], L^{\bar{q}}(\mathbb{R}^n) \cap L^{q_1}(\mathbb{R}^n)) : \|u\|_{X(T)} < \infty\},$$

with the norm

$$\|u\|_{X(T)} = \operatorname*{ess\,sup}_{0\le t\le T} \left((1+t)^{\frac{n}{\sigma}\left(\frac{1}{\bar r+\delta}-\frac{1}{\bar q}\right)-1} \|u(t,\cdot)\|_{L^{\bar q}} + (1+t)^{\frac{n}{\sigma}\left(\frac{1}{\bar r+\delta}-\frac{1}{q_1}\right)-1} \|u(t,\cdot)\|_{L^{q_1}} \right).$$

If $n > 2\sigma$, we may have that $1 - \frac{n}{\sigma}\left(\frac{1}{\bar r+\delta} - \frac{1}{q}\right) < -1$ for $\bar q \le q \le q_1$, so we can use $L^{\bar r+\delta} - L^q$ estimates only for $t \in [0, t/2]$, whereas for $t \in [t/2, t]$ we shall use $L^{r_\epsilon^\sharp} - L^q$ estimates, with $r_\epsilon^\sharp = \frac{2n}{n+2\sigma} + \varepsilon \in (1, 2)$, with $\varepsilon > 0$ sufficiently small such that $pr_\epsilon^\sharp \le q_1$. From now on we write just $r^\sharp$ instead $r_\epsilon^\sharp$. For $t \ge 2$ we split the integral as

$$\begin{aligned}\|Nu(t,\cdot)\|_{L^q} &\lesssim \int_0^{t/2} (t-s)^{1-\frac{n}{\sigma}\left(\frac{1}{\bar r+\delta}-\frac{1}{q}\right)} \||u(s,\cdot)|^p\|_{L^{\bar r+\delta}} ds \\ &\quad + \int_{t/2}^{t} (t-s)^{1-\frac{n}{\sigma}\left(\frac{1}{r^\sharp}-\frac{1}{q}\right)} \||u(s,\cdot)|^p\|_{L^{r^\sharp}} ds.\end{aligned}$$

Using that $\bar r < r^\sharp$ and $\bar q < p\bar r$, we conclude that $\bar q < p(\bar r+\delta) \le pr^\sharp \le q_1$ for $\delta > 0$ sufficiently small. Hence, it follows by interpolation

$$\|u(s,\cdot)\|^p_{L^{pr^\sharp}} \lesssim (1+t)^{p\left(1-\frac{n}{\sigma}\left(\frac{1}{\bar r+\delta}-\frac{1}{pr^\sharp}\right)\right)} \|u\|^p_{X(T)}.$$

and

$$\|u(s,\cdot)\|^p_{L^{p(\bar r+\delta)}} \lesssim (1+s)^{-\beta} \|u\|^p_{X(T)}.$$

where

$$\beta := p\left(\frac{n}{\sigma(\bar r+\delta)}\left(1-\frac{1}{p}\right) - 1\right).$$

Duo to $p > \frac{n+\bar r\sigma}{n-\bar r\sigma}$ and $\frac{n+\bar r\sigma}{n-\bar r\sigma}$ be an increasing function on $\bar r$, there exists a $\bar\delta > 0$ such that $p > \frac{n+(\bar r+\delta)\sigma}{n-(\bar r+\delta)\sigma} > \frac{n+\bar r\sigma}{n-\bar r\sigma}$ for all $0 < \delta < \bar\delta$. This implies that $\beta > 1$ and

$$\begin{aligned}\int_0^{t/2} (t-s)^{1-\frac{n}{\sigma}\left(\frac{1}{\bar r+\delta}-\frac{1}{q}\right)} \||u(s,\cdot)|^p\|_{L^{\bar r+\delta}} ds &\lesssim t^{1-\frac{n}{\sigma}\left(\frac{1}{\bar r+\delta}-\frac{1}{q}\right)} \int_0^{t/2} (1+s)^{-\beta} ds \|u\|^p_{X(T)} \\ &\lesssim t^{1-\frac{n}{\sigma}\left(\frac{1}{\bar r+\delta}-\frac{1}{q}\right)} \|u\|^p_{X(T)}.\end{aligned}$$

Now, thanks to $1-\frac{n}{\sigma}\left(\frac{1}{r^\sharp}-\frac{1}{q}\right)>-1$ we have

$$\int_{t/2}^t (t-s)^{1-\frac{n}{\sigma}\left(\frac{1}{r^\sharp}-\frac{1}{q}\right)}\||u(s,\cdot)|^p\|_{L^{r^\sharp}}ds \lesssim (1+t)^{p\left(1-\frac{n}{\sigma}\left(\frac{1}{\bar{r}+\delta}-\frac{1}{pr^\sharp}\right)\right)}\int_{t/2}^t (t-s)^{1-\frac{n}{\sigma}\left(\frac{1}{r^\sharp}-\frac{1}{q}\right)}ds\|u\|^p_{X(T)}$$

$$\lesssim (1+t)^{p\left(1-\frac{n}{\sigma}\left(\frac{1}{\bar{r}+\delta}-\frac{1}{pr^\sharp}\right)\right)+2-\frac{n}{\sigma}\left(\frac{1}{r^\sharp}-\frac{1}{q}\right)}\|u\|^p_{X(T)},$$

Using again that $p>\frac{n+\bar{r}\sigma}{n-\bar{r}\sigma}$ we conclude that

$$p\left(1-\frac{n}{\sigma}\left(\frac{1}{\bar{r}+\delta}-\frac{1}{pr^\sharp}\right)\right)+2-\frac{n}{\sigma}\left(\frac{1}{r^\sharp}-\frac{1}{q}\right)\leq 1-\frac{n}{\sigma}\left(\frac{1}{\bar{r}+\delta}-\frac{1}{q}\right)$$

and therefore

$$\|Nu(t,\cdot)\|_{L^q}\lesssim (1+t)^{1-\frac{n}{\sigma}\left(\frac{1}{\bar{r}+\delta}-\frac{1}{q}\right)}\|u\|^p_{X(T)}.$$

For $t\in[0,2]$, it is sufficient to use $L^{r^\sharp}-L^q$ estimate, namely,

$$\begin{aligned}\int_0^t \|K_1(t-s,\cdot)*|u(s,\cdot)|^p\|_{L^q}ds &\lesssim \int_0^t (t-s)^{1-\frac{n}{\sigma}\left(\frac{1}{r^\sharp}-\frac{1}{q}\right)}\||u(s,\cdot)|^p\|_{L^{r^\sharp}}ds\\ &\lesssim \int_0^t (t-s)^{1-\frac{n}{\sigma}\left(\frac{1}{r^\sharp}-\frac{1}{q}\right)}(1+s)^{-\beta}ds\|u\|^p_{X(T)}\\ &\lesssim \int_0^t (t-s)^{1-\frac{n}{\sigma}\left(\frac{1}{r^\sharp}-\frac{1}{q}\right)}ds\|u\|^p_{X(T)}\\ &\lesssim t^{2-\frac{n}{\sigma}\left(\frac{1}{r^\sharp}-\frac{1}{q}\right)}\|u\|^p_{X_\delta(T)}\lesssim \|u\|^p_{X(T)}.\end{aligned}$$

This concludes the proof.

Acknowledgements The first author has been has been partially supported by São Paulo Research Foundation (Fundação de Amparo à Pesquisa do Estado de São Paulo, FAPESP), grant 2017/19497-3. The second author was supported by the Coordenação de Aperfeiçoamento de Pessoal de Nível Superior – Brasil (CAPES) – Finance Code 001

Appendix

In the Appendix we list results of Harmonic Analysis.

The first ingredient is the celebrated Mikhlin-Hörmander multiplier theorem:

Theorem A.1 *Let* $1 < p < \infty$ *and* $k = \max\{[n(1/p - 1/2)] + 1, [n/2] + 1\}$. *Suppose that* $m \in \mathscr{C}^k(\mathbb{R}^n \setminus \{0\})$ *and*

$$\left|\partial_\xi^\beta m(\xi)\right| \leq C\,|\xi|^{-|\beta|}, \quad |\beta| \leq k.$$

Then $m \in M_p \doteq M_p^p$.

The next result is about translation invariant operators in $L^p = L^p(\mathbb{R}^n)$ spaces (see [6]).

Theorem A.2 *Let* f *be a measurable function. Moreover, we suppose the following relation with suitable positive constants* C *and* $b \in (1, \infty)$*:*

$$meas\,\{\xi \in \mathbb{R}^n : |f(\xi)| \geq l\} \leq Cl^{-b}.$$

Then $f \in M_p^q$ *if* $1 < p \leq 2 \leq q < \infty$ *and* $\frac{1}{p} - \frac{1}{q} = \frac{1}{b}$.

In [10] one can find the following result, that is useful tool to derive $L^q - L^q$ estimates.

Proposition A.1 (Berstein's inequality) *Let* $N > \frac{n}{2}$. *If* $f, D^N f \in L^2$, *then there exists a constant* $C > 0$ *such that*

$$\|f\|_{M_1} \leq C\|f\|_{L^2}^{1-\frac{n}{2N}}\|D^N f\|_{L^2}^{\frac{n}{2N}}.$$

In [8] one can find the following result, well known as Littman's lemma, that is a very useful tool to derive $L^r - L^q$ estimates on the dual line.

Theorem A.3 *Let us consider for* $\tau \geq \tau_0$, τ_0 *is a large positive number, the oscillating integral*

$$F^{-1}_{\eta\to x}\big(e^{-i\tau p(\eta)}v(\eta)\big).$$

The amplitude function $v = v(\eta)$ *is supposed to belong to* $C_0^\infty(\mathbb{R}^n)$ *with support in* $\{\eta \in \mathbb{R}^n : |\eta| \in [\frac{1}{2}, 2]\}$. *The function* $p = p(\eta)$ *is* C^∞ *in a neighborhood of the support of* v. *Moreover, the rank of the Hessian* $H_p(\eta)$ *is supposed to satisfy the assumption rank* $H_p(\eta) \geq k$ *on the support of* v. *Then the following* $L^\infty - L^\infty$ *estimate holds:*

$$\|F^{-1}_{\eta\to x}\big(e^{-i\tau p(\eta)}v(\eta)\big)\|_{L^\infty(\mathbb{R}^n_x)} \leq C(1+\tau)^{-\frac{k}{2}}\sum_{|\alpha|\leq L}\|D^\alpha_\eta v(\eta)\|_{L^\infty(\mathbb{R}^n_\eta)},$$

where L *is a suitable entire number.*

The next result about singular Fourier multipliers is due to Miyachi (see Theorem 4.1 in [7]):

Theorem A.4 *Let us consider Fourier multiplier*

$$m_{a,b}(\xi) = \frac{(1-\chi(\xi))e^{i|\xi|^a}}{|\xi|^b}, \qquad \xi \in \mathbb{R}^n, \qquad a>0, a\neq 1, \qquad b\in\mathbb{R},$$

where χ *is as in Notation 2. If* $1<p\leq q$*, then* $m\in M_p^q$ *if, and only if,* $\frac{1}{p}+\frac{1}{q}\leq 1$ *and* $\frac{1-a}{p}-\frac{1}{q}\leq\frac{b}{n}-\frac{a}{2}$ *or* $\frac{1}{p}+\frac{1}{q}\geq 1$ *and* $\frac{1}{p}-\frac{1-a}{q}\leq\frac{b}{n}+\frac{a}{2}$*. Moreover,* $m\in M_1^q$ *if, and only if,* $1-\frac{1-a}{q}<\frac{b}{n}+\frac{a}{2}$*.*

References

1. S. Cui, A. Guo, Solvability of the Cauchy problem of nonlinear beam equations in Besov spaces. Nonlinear Anal. **65**, 802–824 (2006)
2. M. D'Abbicco, M.R. Ebert, A new phenomenon in the critical exponent for structurally damped semi-linear evolution equations. Nonlinear Anal. **149**, 1–40 (2017)
3. M. D'Abbicco, S. Lucente, The beam equation with nonlinear memory. Z. Angew. Math. Phys. **67**, 60 (2016). http://dx.doi.org/10.1007/s00033-016-0655-x
4. L. D'Ambrosio, S. Lucente, Nonlinear Liouville theorems for Grushin and Tricomi operators. J. Differ. Equ. **123**, 511–541 (2003)
5. M.R. Ebert, M. Reissig, Methods for partial differential equations, *Qualitative Properties of Solutions, Phase Space Analysis, Semilinear Models* (Birkhäuser/Springer, Cham, 2018). https://doi.org/10.1007/978-3-319-66456-9
6. L. Hörmander, Estimates for translation invariant operators in L^p spaces. Acta Math. **104**, 93–140 (1960)
7. A. Miyachi, On some singular Fourier multipliers. J. Fac. Sci. Univ. Tokyo Sect. IA Math. **28**, 267–315 (1981)
8. H. Pecher, L^p-Abschätzungen und klassische Lösungen für nichtlineare Wellengleichungen. I. Math. Z. **150**, 159–183 (1976)
9. J. Peral, L^p estimates for the Wave Equation. J. Funct. Anal. **36**, 114–145 (1980)
10. S. Sjöstrand, On the Riesz means of the solutions of the Schrödinger equation. Annali della Scuola Normale Superiore di Pisa-Classe di Scienze **24**, 331–348 (1970)
11. W.A. Strauss, Nonlinear scattering theory at low energy. J. Funct. Anal. **41**, 110–133 (1981)
12. R. Strichartz, Convolutions with kernels having singularities on a sphere. Trans. Am. Math. Soc. **148**, 461–471 (1970)
13. S. Wang, Well-posedness and Ill-posedness of the Nonlinear Beam Equation. Thesis (Ph.D.), The University of New Mexico, p. 63 (2013). ISBN:978-1303-51754-9

Blow-Up or Global Existence for the Fractional Ginzburg-Landau Equation in Multi-dimensional Case

Luigi Forcella, Kazumasa Fujiwara, Vladimir Georgiev, and Tohru Ozawa

Abstract The aim of this work is to give a complete picture concerning the asymptotic behaviour of the solutions to fractional Ginzburg-Landau equation. In previous works, we have shown global well-posedness for the past interval in the case where spatial dimension is less than or equal to 3. Moreover, we have also shown blow-up of solutions for the future interval in one dimensional case. In this work, we summarise the asymptotic behaviour in the case where spatial dimension is less than or equal to 3 by proving blow-up of solutions for a future time interval in multidimensional case. The result is obtained via ODE argument by exploiting a new weighted commutator estimate.

L. Forcella
Bâtiment des Mathématiques, École Polytechnique Fédérale de Lausanne, Lausanne, Switzerland
e-mail: luigi.forcella@epfl.ch

K. Fujiwara (✉)
Centro di Ricerca Matematica Ennio De Giorgi, Scuola Normale Superiore, Pisa, Italy
e-mail: kazumasa.fujiwara@sns.it

V. Georgiev
Department of Mathematics, University of Pisa, Pisa, Italy

Faculty of Science and Engineering, Waseda University, Shinjuku-ku, Tokyo, Japan

IMI–BAS, Sofia, Bulgaria
e-mail: georgiev@dm.unipi.it

T. Ozawa
Department of Applied Physics, Waseda University, Shinjuku-ku, Tokyo, Japan
e-mail: txozawa@waseda.jp

M. D'Abbicco et al. (eds.), *New Tools for Nonlinear PDEs and Application*,
Trends in Mathematics, https://doi.org/10.1007/978-3-030-10937-0_6

1 Introduction

In this paper, we consider the following complex Ginzburg – Landau (CGL) equation in a future time interval

$$\begin{cases} i\partial_t u + Du = i|u|^{p-1}u, & t \in [0,T), \quad T > 0, \quad x \in \mathbb{R}^n, \\ u(0,x) = u_0(x), & x \in \mathbb{R}^n, \end{cases} \tag{1}$$

where u is a complex valued unknown function, $p > 1$, and $D = (-\Delta)^{1/2}$. The choice of D is closely connected with the recent attempts to develop fractional quantum mechanical approach (see [23]).

We shall observe some new interesting phenomena. On one hand, if we take a future time interval as in (1), then we shall obtain a blow-up result. If, instead, we take past time interval $(-T, 0]$, $T > 0$ in the place of the future time interval, then global small data existence for (1) can be proved and therefore we have a similarity to a diffusion type process.

Before giving the main results on the local and global well-posedness for (1), we introduce some notations. For a Banach space X and $1 \leq p \leq \infty$ let $L^p(\mathbb{R}^n; X)$ be a X-valued Lebesgue space of p-th power. We abbreviate $L^p(\mathbb{R}^n; \mathbb{C})$ as $L^p(\mathbb{R}^n)$. For $f, g \in L^2(\mathbb{R}^n)$, we define the inner product as

$$\langle f, g \rangle_{L^2(\mathbb{R}^n)} = \int_{\mathbb{R}^n} f(x)\overline{g}(x)dx.$$

For $s \in \mathbb{R}$, let $H^s(\mathbb{R}^n)$ be the usual inhomogeneous Sobolev space defined as $H^s(\mathbb{R}^n) = (1-\Delta)^{-s/2}L^2(\mathbb{R}^n)$. Let $\dot{H}^s(\mathbb{R}^n)$ be the usual homogeneous Sobolev space defined as $\dot{H}^s(\mathbb{R}^n) = (-\Delta)^{-s/2}L^2(\mathbb{R}^n)$. $H^s_{rad}(\mathbb{R}^n)$ is the restriction to radial functions of $H^s(\mathbb{R}^n)$. Lip refers to space of Lipschitz functions on euclidean space. For $f, g : A \subseteq \mathbb{R}^n \to [0,\infty)$, $f \lesssim g$ means that there exists $C > 0$ such that for any $a \in A$ $f(a) \leq Cg(a)$. Given two Banach spaces X, Y, $Y \hookrightarrow X$ means that $Y \subset X$ with continuous embedding. Moreover, we say that a Cauchy problem is locally well-posed forward in time in X, if for any X-valued initial data, there exists $T > 0$ and a Banach space $Y \hookrightarrow C([0,T]; X)$ such that there exists a unique solution to the Cauchy problem in Y and $\|u_n - u\|_Y \to 0$ as $\|u_{0,n} - u_0\|_X \to 0$, where u_n and u are solutions for the Cauchy problem for initial data u_0 and $u_{0,n}$, respectively (the last property goes under the name of *continuous dependence on the initial data*). We also say that a Cauchy problem is globally well-posed forward in time in X if the Cauchy problem is locally well-posed for any $T > 0$. Moreover, we also say that a Cauchy problem is globally well-posed in X with sufficiently small data, if we have the property above for sufficiently small data with respect to the X-norm.

Let us notice that Eq. (1) is invariant under the scale transformation

$$u_\lambda(t,x) = \lambda^{1/(p-1)}u(\lambda t, \lambda x)$$

with $\lambda > 0$. Then

$$\|u_{0,\lambda}\|_{\dot{H}^s(\mathbb{R}^n)} = \lambda^{1/(p-1)+s-n/2}\|u_0\|_{\dot{H}^s(\mathbb{R}^n)}$$

and with

$$s = s_{n,p} := n/2 - 1/(p-1) < n/2,$$

$\dot{H}^s$ norm of initial data is also invariant, for this $s_{n,p}$ is called scale critical exponent. We also call $p_{n,s} = 1 + 2/(n-2s)$ the $H^s(\mathbb{R}^n)$ scaling critical power. For any s, in the scaling subcritical case where $p < p_{n,s}$ or $s > s_{n,p}$, (1) is expected to have local solution for any $H^s(\mathbb{R}^n)$ initial data on the analogy of scaling invariant Schrödinger equation. For instance, we refer the reader to [4–6, 16, 17]. However, with power type nonlinearity without gauge invariance, semirelativistic equations could be not locally well-posed even in scaling subcritical case, see [10].

Here we recall local well-posedness results. It is worth mentioning that Borgna and Rial [2] showed that in one dimensional case, CGL equation with cubic nonlinearity is locally well-posed in $H^s(\mathbb{R})$ with $s > 1/2$. They constructed local solutions by a contraction argument based on the unitarity of the propagator and the Sobolev embedding $H^s(\mathbb{R}) \hookrightarrow L^\infty(\mathbb{R})$. Similarly, local solutions may be constructed in the case where uniform control of solutions holds, namely, in $H^s(\mathbb{R}^n)$ with $s > n/2$. On the other hand, for fixed p, $s_{n,p} < n/2$; therefore, the local well-posedness of (1) is expected in wider Sobolev spaces. Indeed, we have the following results that can be established using the approach in [12]:

Proposition 1 ([12]) *Let $n = 2$. For $p > 1$ and $3/4 < s < p < p_{2,s}$, the Cauchy problem* (1) *is locally well-posed in $H^s(\mathbb{R}^2)$.*

Proposition 2 ([12]) *Let $n \geq 3$ and u_0 be radial. For $1 < p < p_{n,1} = 1 + \frac{2}{n-2}$, the Cauchy problem* (1) *is locally well-posed in $H^1_{\text{rad}}(\mathbb{R}^n)$.*

Proposition 3 ([12]) *Let $n = 3$ and u_0 be radial. For $p = p_{3,1} = 3$, the Cauchy problem* (1) *is locally well-posed in $H^1_{\text{rad}}(\mathbb{R}^3)$ with sufficiently small $H^1_{\text{rad}}(\mathbb{R}^3)$ data.*

Remark 1 In Proposition 3, since the local existence result is based on a priori estimate of type

$$\|u\|_{X^1_{\text{rad}}(0,T)} \leq C_0 + C_1\|u\|^4_{X^1_{\text{rad}}(0,T)}$$

with C_1 which is independent of T, we restrict well-posedness to the small initial data.

We recall that in three dimensional case, $p = p_{3,1} = 3$ is a critical value in view of the result in [18]. However, the result in [18] treats non-gauge invariant nonlinearities having constant sign, for which the test function method works. The question of the existence of local and global solutions for $n \geq 3$ and $p \geq 1+2/(n-2)$ seems, at the best of our knowledge, still open.

Proposition 1 may be justified by a Strichartz estimate introduced by Nakamura and Ozawa in [26] or Ginibre and Velo [14]. We remark that they introduced the estimate to study Klein-Gordon equation and it was sufficient to consider Klein-Gordon equation in scaling subcritical case (see Lemma 1 below). On the other hand, for (1), local solutions cannot be constructed based on their Strichartz estimates in general subcritical case. Therefore, in order to consider the well-posedness in $H^1(\mathbb{R}^n)$ for $n \geq 3$, we put radial assumption and apply another Strichartz estimate introduced in [1] by the third author, Bellazzini and Visciglia. For details, see Sect. 2.

Next, we review the known blow-up result. In [11], the authors studied the blow-up of solutions to (1) in one dimensional case, by an ordinary differential equation (ODE) argument. In order to review their argument, we define a function space $hL^2(\mathbb{R}^n)$ by

$$hL^2(\mathbb{R}^n) = \{f : \text{mesurable and } \|\frac{1}{h} f\|_{L^2(\mathbb{R}^n)} < \infty\},$$

where h is a mesurable function. In their argument, an ordinary differential inequality (ODI) for the $hL^2(\mathbb{R})$ norm of solutions with some h are shown. In particular, we have the following:

Proposition 4 *Let h be a Lipschitz function satisfying $1/h \in L^\infty(\mathbb{R}) \cap L^2(\mathbb{R})$ and*

$$\left\| \frac{1}{h(\cdot)} \int_{\mathbb{R}} \langle \cdot - y \rangle^{-2} h(y) f(y) dy \right\|_{L^2(\mathbb{R})} \leq C \|f\|_{L^2(\mathbb{R})}. \tag{2}$$

Let $u_0 \in L^2(\mathbb{R})$ satisfy

$$\|\frac{1}{h} u_0\|_{L^2(\mathbb{R})} \geq C_1^{\frac{1}{p-1}} \|\frac{1}{h}\|_{L^2(\mathbb{R})}, \tag{3}$$

where $C_1 = \|1/h \cdot [D, h]\|_{L^2(\mathbb{R}) \to L^2(\mathbb{R})}$. If there is a solution $u \in C([0, T); hL^2(\mathbb{R}))$ to (1), then

$$\|\frac{1}{h} u(t)\|_{L^2(\mathbb{R})} \geq e^{-C_1 t/2} \left(\|\frac{1}{h} u_0\|_{L^2(\mathbb{R})}^{-p+1} + C_1^{-1} \|\frac{1}{h}\|_{L^2(\mathbb{R})}^{-p+1} \left\{ e^{-C_1 (p-1) t/2} - 1 \right\} \right)^{-\frac{1}{p-1}}. \tag{4}$$

Therefore, the lifespan is estimated by

$$T \leq -\frac{2}{p-1} C_1^{-1} \log \left(1 - C_1 \|\frac{1}{h}\|_{L^2(\mathbb{R})}^{p-1} \|\frac{1}{h} u_0\|_{L^2(\mathbb{R})}^{-p+1} \right).$$

Moreover, by scaling argument, the following statement is shown.

Corollary 1 ([11, Corollary 1]) *If* $p < 3$*, then any solutions to* (1) *with non trivial* $L^2(\mathbb{R})$ *initial data cannot stay in* $L^2(\mathbb{R})$ *globally.*

Remark 2 In the Corollary above, $p < 3$ stands for the condition in one dimensional case of the Fujita exponent generally defined in $\mathbb{R}^n$ by $p_F := 1 + 2/n$ (see also Corollary 2). Then the assumption of Corollary 1 is rewritten by $p < p_F$. Under this assumption, by scaling h, (3) holds for any non trivial $L^2(\mathbb{R})$ initial data u_0.

Remark 3 Condition (2) was required to guarantee the commutator estimate:

$$\|[D,h]f\|_{L^2(\mathbb{R})} \le C\|f\|_{L^2(\mathbb{R})}, \qquad \forall f \in L^2(\mathbb{R}). \tag{5}$$

We remark that Lenzmann and Schikorra [24, Theorem 6.1] showed that (5) holds for any Lipschitz function h, therefore, the assumption (2) can be omitted.

The commutator estimate (5) implies blow-up for solutions to (1) in the following manner. Let $v(t,x) = u(t,x)/h(x)$, where u is a solution to (1). Then, a straight computation shows that v satisfies

$$\begin{aligned} i\partial_t v + Dv + \frac{1}{h}[D,h]v &= i\frac{1}{h}\partial_t u + \frac{1}{h}Du \\ &= i\frac{1}{h}|u|^{p-1}u \\ &= i|h|^{p-1}|v|^{p-1}v. \end{aligned} \tag{6}$$

Therefore,

$$\begin{aligned} \frac{d}{dt}\|v(t)\|^2_{L^2(\mathbb{R})} &= 2\mathrm{Re}\langle v(t), \partial_t v(t)\rangle_{L^2(\mathbb{R})} \\ &= -2\mathrm{Im}\langle v(t), i\partial_t v(t)\rangle_{L^2(\mathbb{R})} \\ &= -2\mathrm{Im}\langle v(t), -Dv(t) - \frac{1}{h}[D,h]v(t) + i|h|^{p-1}|v(t)|^{p-1}v(t)\rangle_{L^2(\mathbb{R})} \\ &= 2\||h|^{(p-1)/(p+1)}v(t)\|^{p+1}_{L^{p+1}(\mathbb{R})} + 2\mathrm{Im}\langle v(t), \frac{1}{h}[D,h]v(t)\rangle_{L^2(\mathbb{R})}. \end{aligned} \tag{7}$$

By the Hölder inequality,

$$\|v(t)\|_{L^2(\mathbb{R})} \le \|\frac{1}{h}\|^{(p-1)/(p+1)}_{L^2(\mathbb{R})}\||h|^{(p-1)/(p+1)}v(t)\|_{L^{p+1}(\mathbb{R})},$$

which together with (7) implies

$$\frac{d}{dt}\|v(t)\|^2_{L^2(\mathbb{R})} \geq \|\frac{1}{h}\|^{-p+1}_{L^2(\mathbb{R})}\|v(t)\|^{p+1}_{L^2(\mathbb{R})} - \|\frac{1}{h}[D,h]\|_{L^2(\mathbb{R})\to L^2(\mathbb{R})}\|v(t)\|^2_{L^2(\mathbb{R})}. \tag{8}$$

Estimate (8) and Lemma 7 in Sect. 3 imply that if (3) holds and

$$\|\frac{1}{h}\cdot[D,h]\|_{L^2(\mathbb{R})\to L^2(\mathbb{R})} < \infty, \tag{9}$$

then $\|v(t)\|_{L^2(\mathbb{R})} = \|u(t)/h\|_{L^2(\mathbb{R})}$ blows up at a finite time. Therefore, if there exists $1/h \in L^2(\mathbb{R})$ satisfying (9), then the argument above works and blow-up of solutions to (1) is shown. In [11], (9) was shown by the boundedness assumption of $1/h$ and (5). We remark that (5) holds in more general situation; for example, in multidimensional case. We also remark that in [9] a generalization of (5) taking the form

$$\|[(-\mathscr{A})^{1/2},h]\|_{L^2(\mathbb{R}^n)\to L^2(\mathbb{R}^n)} \leq C\|h\|_{\dot{B}^1_{\infty,1}}$$

is shown, where $\dot{B}^1_{\infty,1}$ is the standard homogeneous Besov space and

$$\mathscr{A} := -\nabla\cdot A\nabla + V.$$

Here A is a smooth positive-definite $n\times n$ matrix and the real-valued potential V satisfies some weak integrability conditions. On the other hand, $h \in \mathrm{Lip}$ is a natural condition for (5). However, there exists some Lipschitz function h satisfying $1/h \in L^2(\mathbb{R}^n)$ only when $n = 1$. This means, we cannot consider blow-up phenomena in multi dimensional case based on (5).

In this paper, we show (9) with polynomial weights which are not Lipschitz in general. In particular, we show the following estimate:

Proposition 5 *Let $n \geq 1$ and $n/2 < q < n/2+1$. Then $\langle\cdot\rangle^{-q}[D,\langle\cdot\rangle^q]$ is bounded operator on $L^2(\mathbb{R}^n)$, where $\langle\cdot\rangle = (1+|x|^2)^{1/2}$.*

Remark 4 Obviously, if $n \geq 1$ and $n/2 < q < n/2+1$, then $\langle\cdot\rangle^{-q} \in L^2(\mathbb{R}^n)$. Moreover, only when $n = 1$, q can be 1.

Then, we have the following blow-up statement:

Proposition 6 *Let $n \geq 1$ and $n/2 < q < n/2+1$. Let $u_0 \in \langle\cdot\rangle^q L^2(\mathbb{R}^n)$ satisfy*

$$\|\langle x\rangle^{-q}u_0\|_{L^2(\mathbb{R}^n)} \geq C_2^{\frac{1}{p-1}}\|\langle x\rangle^{-q}\|_{L^2(\mathbb{R}^n)}, \tag{10}$$

where

$$C_2 = \|\langle\cdot\rangle^{-q}[D,\langle\cdot\rangle^q]\|_{L^2(\mathbb{R}^n)\to L^2(\mathbb{R}^n)}.$$

If there is a solution $u \in C([0,T); \langle\cdot\rangle^q L^2(\mathbb{R}^n))$ *to* (1)*, then*

$$\begin{aligned} &\|\langle\cdot\rangle^{-q} u(t)\|_{L^2(\mathbb{R}^n)} \\ &\ge e^{-C_2 t/2}\left(\|\langle\cdot\rangle^{-q} u_0\|_{L^2(\mathbb{R}^n)}^{-p+1} + C_2^{-1}\|\langle\cdot\rangle^{-q}\|_{L^2(\mathbb{R}^n)}^{-p+1}\left\{e^{-(p-1)C_2 t/2} - 1\right\}\right)^{-\frac{1}{p-1}}. \end{aligned} \tag{11}$$

Therefore, the lifespan is estimated by

$$T \le -\frac{2}{p-1} C_2^{-1} \log\left(1 - C_2 \|\langle\cdot\rangle^{-q}\|_{L^2(\mathbb{R}^n)}^{p-1} \|\langle\cdot\rangle^{-q} u_0\|_{L^2(\mathbb{R}^n)}^{-p+1}\right). \tag{12}$$

Corollary 2 *Let* $n \ge 1$*. If* $p < p_F := 1 + 2/n$*, then any solutions to* (1) *with non trivial* $L^2(\mathbb{R}^n)$ *initial data cannot exist globally.*

Remark 5 As Remark 2, under the condition, $p < p_F$, by scaling h, (10) holds for any non trivial $L^2(\mathbb{R}^n)$ data.

In [11], so as to show (5), higher frequency part of D is handled by the Coifman-Meyer estimate and lower frequency part is estimated by (2). We remark that (5) is regarded as a Kato-Ponce inequality. For related subjects, we refer the reader to [15, 19, 20, 25] and we remark that Fourier multiplier argument plays a critical role in these references. On the other hand, it seems not easy to obtain (9) based on a Fourier multiplier argument because of the weight function. Therefore, we show Proposition 5 by using the following representation of the commutator:

$$([D, \langle\cdot\rangle^q] f)(x) = C \cdot \text{P.V.} \int_{\mathbb{R}^n} \frac{(\langle x\rangle^q - \langle x+y\rangle^q) f(x+y)}{|y|^{n+1}} dy, \tag{13}$$

where $P.V.$ stands for Principal Value (for detail, we refer the reader to [8]). Combining (13) and the Calderón-Zygmund theory, we show (9) with non-Lipschitz weight functions.

Our next step is to study the global existence result for negative times of the following Cauchy problem:

$$\begin{cases} i\partial_t u + Du = i|u|^{p-1}u, & t \in (-T, 0], \quad T > 0, \quad x \in \mathbb{R}^n, \\ u(0,x) = u_0(x), & x \in \mathbb{R}^n. \end{cases} \tag{14}$$

Making the change of variables $t \to -t$, we reduce this problem to the future time interval for the Cauchy problem

$$\begin{cases} i\partial_t u - Du = -i|u|^{p-1}u, & t \in [0, T), \quad T > 0, \quad x \in \mathbb{R}^n, \\ u(0,x) = u_0(x), & x \in \mathbb{R}^n. \end{cases} \tag{15}$$

At least formally, (15) may be rewritten in the following integral equation:

$$u(t) = U(-t)u_0 - \int_0^t U(-t+t')|u(t')|^{p-1}u(t')dt', \tag{16}$$

where $U(t) = e^{itD}$.

Then, Propositions 1, 2, and 3 are valid for (15). Moreover, for (16), we can obtain the following a priori estimates that we include for completeness but detailed proofs can be found in [12].

Proposition 7 ([12]) *Let $n \in \mathbb{N}$ and $p > 1$. Let $u_0 \in L^2(\mathbb{R}^n)$ and $T > 0$. Let $u \in L^\infty(0, T; L^2(\mathbb{R}^n)) \cap L^p(0, T; L^{2p}(\mathbb{R}^n))$ be a solution to the integral equation* (16) *for the initial data u_0. Then, for any t_1, t_2 with $0 < t_1 < t_2 < T$,*

$$\|u(t_2)\|^2_{L^2(\mathbb{R}^n)} + 2\|u\|^{p+1}_{L^{p+1}(t_1,t_2;L^{p+1}(\mathbb{R}^n))} = \|u(t_1)\|^2_{L^2(\mathbb{R}^n)}.$$

Proposition 8 ([12]) *Let $n \in \mathbb{N}$ and $p > 1$. Let $u_0 \in H^1(\mathbb{R}^n)$ and $T > 0$. Let $u \in L^\infty(0, T; H^1(\mathbb{R}^n)) \cap L^{p-1}(0, T; L^\infty(\mathbb{R}^n))$ be a solution to the integral equation* (16) *for the initial data u_0. Then, for any t_1, t_2 with $0 \le t_1 < t_2 \le T$,*

$$\begin{aligned} &\|\nabla u(t_2)\|^2_{L^2(\mathbb{R}^n)} + 2\||u|^{\frac{p-1}{2}}\nabla u\|^2_{L^2(t_1,t_2;L^2(\mathbb{R}^n))} + \frac{p-1}{2}\||u|^{\frac{p-3}{2}}\nabla |u|^2\|^2_{L^2(t_1,t_2;L^2(\mathbb{R}^n))} \\ &= \|\nabla u(t_1)\|^2_{L^2(\mathbb{R}^n)}. \end{aligned} \tag{17}$$

Proposition 9 ([12]) *Let $n = 1, 2$, $p > 1$, $n/2 < s < \min\{2, p\}$, and $T > 0$. Let $u_0 \in H^s(\mathbb{R}^n)$ and $u \in L^\infty(0, T; H^s(\mathbb{R}^n)) \cap L^2(0, T; L^\infty(\mathbb{R}^n))$ be a solution to* (16) *for the initial data u_0. Then for any t_1, t_2 with $0 < t_1 < t_2 < T$,*

$$\|u(t_2)\|^2_{\dot{H}^s(\mathbb{R}^n)} \le \|u(t_1)\|^2_{\dot{H}^s(\mathbb{R}^n)} + C\int_{t_1}^{t_2} \|u(t)\|^{p-1}_{L^\infty(\mathbb{R}^n)}\|u(t)\|^2_{\dot{H}^s(\mathbb{R}^n)}dt.$$

Proposition 10 ([12]) *Let $1 \le n \le 3$, $u_0 \in H^2(\mathbb{R}^n)$ and $T > 0$. Let $u \in C((0, T); H^2(\mathbb{R}^n) \cap L^\infty(\mathbb{R}^n))$ be a solution to the integral equation* (16) *for the initial data u_0. Then, for any t_1, t_2 with $0 < t_1 < t_2 < T$,*

$$\begin{aligned} &\|u(t_2)\|^2_{\dot{H}^2(\mathbb{R}^n)} + 2\sum_{j,k=1}^{n}\int_{t_1}^{t_2} \|u(t)\partial_j\partial_k u(t)\|^2_{L^2(\mathbb{R}^n)}dt \\ &\le \|u(t_1)\|^2_{\dot{H}^2(\mathbb{R}^n)} + 2n^2(n+1)\int_{t_1}^{t_2} \|u(t)\|^{4-n}_{\dot{H}^1(\mathbb{R}^n)}\|u(t)\|^n_{\dot{H}^2(\mathbb{R}^n)}dt. \end{aligned} \tag{18}$$

Therefore, for (14) we have the following:

Proposition 11 *Under the conditions of Propositions* 1, 2, *and* 3, (14) *is globally well-posed.*

This paper is composed as follows. In Sect. 2, we show local well-posedness of (1) by means of Strichartz estimates of [1, 14, 26]. In Sect. 3, blow-up for (1) is shown with a weighted commutator estimate. In Sect. 4, a priori estimates for (14) are shown by a direct approach leading to the global well-posedness results.

2 Local Well-Posedness of (1)

This section is devoted to the proof of the local well-posedness for the Cauchy problem of (1), where $u_0(x) = u(0, x)$ is considered as initial datum. The proof is essentially the same as [12] but for the reader's convenience, we give a proof for Propositions 1, 2, and 3. Here we consider the corresponding integral equation:

$$u(t) = \Phi(u)(t) = U(t)u_0 + \int_0^t U(t-t')|u(t')|^{p-1}u(t')dt'. \tag{19}$$

where $U(t) = e^{itD}$.

2.1 Two Dimensional Case

In two dimensional case, the local well-posedness may be obtained by the following Strichartz estimates:

Lemma 1 ([26, Lemma 2.1], [14, Remark 3.2]) *Let* (q_1, r_1) *and* (q_2, r_2) *satisfy*

$$\frac{1}{r_j} = \frac{1}{2} - \frac{2}{q_j}, \quad 2 \le r_j \le \infty, \quad 4 \le q_j \le \infty$$

for $j = 1, 2$. *Then for* $s \in \mathbb{R}$,

$$\|U(t)\phi\|_{L^{q_1}(0,T;B_{r_1}^{s-\frac{3}{q_1}}(\mathbb{R}^2))} \lesssim \|\phi\|_{H^s(\mathbb{R}^2)},$$

$$\left\|\int_0^t U(t-t')h(t')dt'\right\|_{L^{q_1}(0,T;B_{r_1}^{s-\frac{3}{q_1}}(\mathbb{R}^2))} \lesssim \|h\|_{L^{q_2'}(0,T;B_{r_2'}^{s+\frac{3}{q_2}}(\mathbb{R}^2))},$$

where $B_p^s(\mathbb{R}^2) = B_{p,2}^s(\mathbb{R}^2)$ *is the usual inhomogeneous Besov space.*

Lemma 2 ([12, Lemma 3.2]) *Let $r > 2$, and $T > 0$. If*

$$s > \frac{3}{4} + \frac{1}{2r},$$

then $B_r^{s-\frac{3}{2}(\frac{1}{2}-\frac{1}{r})}(\mathbb{R}^2) \hookrightarrow L^\infty(\mathbb{R}^2)$.

We can now proceed with the proof of Proposition 1.

Proof (Proof of Proposition 1*)* At first we fix $3/4 < s < p < p_{2,s}$. Let (q_1, r_1) satisfy the conditions of Lemma 1, Lemma 2 and $q_1 > p - 1$. We remark that such a pair exists under the assumption $s < p < p_{2,s}$. Let $X^s(0, T) = L^\infty(0, T; H^s(\mathbb{R}^2)) \cap L^{q_1}(0, T; B_{r_1}^{s-3/q_1}(\mathbb{R}^2))$. Then, for a fixed T,

$$\begin{aligned} \|\Phi(u)\|_{X^s(0,T)} &\leq \|u_0\|_{H^s(\mathbb{R}^2)} + C\||u|^{p-1}u\|_{L^1(0,T;H^s(\mathbb{R}^2))} \\ &\leq \|u_0\|_{H^s(\mathbb{R}^2)} + CT^{1-(p-1)/q_1}\|u\|_{X^s(0,T)}^p, \end{aligned} \tag{20}$$

and

$$\begin{aligned} &\|\Phi(u) - \Phi(v)\|_{X^s(0,T)} \\ &\leq C\||u|^{p-1}u - |v|^{p-1}v\|_{L^1(0,T;H^s(\mathbb{R}^2))} \\ &\leq CT^{1-(p-1)/q_1}(\|u\|_{X^s(0,T)} + \|v\|_{X^s(0,T)})^{p-1}\|u - v\|_{X^s(0,T)} \\ &+ CT^{1-(p-1)/q_1}(\|u\|_{X^s(0,T)} + \|v\|_{X^s(0,T)})^{\max(1,p-1)}\|u - v\|_{X^s(0,T)}^{\min\{1,p-1\}}. \end{aligned}$$

This means that if T is sufficiently small, then Φ is a map from

$$B_{X^s(0,T)}(2\|u_0\|_{H^s(\mathbb{R}^2)}) := \left\{ f \in X^s(0, T) \mid \|f\|_{X^s(0,T)} \leq 2\|u_0\|_{H^s(\mathbb{R}^2)} \right\}.$$

into itself. Moreover, if $p \geq 2$, Φ is a contraction map in $X^s(0, T)$. If $p < 2$, Φ may not be a contraction map on $X^s(0, T)$ for any $T > 0$. On the other hand, it is not difficult to see that

$$\begin{aligned} &\|\Phi(u) - \Phi(v)\|_{L^\infty(0,T;L^2(\mathbb{R}^2))} \\ &\lesssim T^{1-(p-1)/q_1}(\|u\|_{X^s(0,T)} + \|v\|_{X^s(0,T)})^{p-1}\|u - v\|_{L^\infty(0,T;L^2(\mathbb{R}^2))}. \end{aligned} \tag{21}$$

Therefore (20) and (21) imply that if $u_1 \in B_{X^s(0,T)}(2\|u_0\|_{H^s(\mathbb{R}^2)})$ and $u_k = \Phi(u_{k-1})$ for $k \geq 2$, then there exists $u^* = \lim_{k\to\infty} u_k$ in $L^\infty(0, T; L^2(\mathbb{R}^2))$. Since $\Phi(u_k) \to \Phi(u^*)$ in $L^\infty(0, T; L^2(\mathbb{R}^2))$ as $k \to \infty$, u^* is a solution of (19). Moreover, since $X^s(0, T) \hookrightarrow L^\infty(0, T; H^s(\mathbb{R}^2))$, u^* is also in $L^\infty(0, T; H^s(\mathbb{R}^2))$, which and (20) imply

$$u^* \in L^{q_1}(0, T; B_{r_1}^{s-\frac{3}{q_1}}(\mathbb{R}^2)).$$

If $s > 1$, by the Gagliardo-Nirenberg inequality, for some $0 < \theta < 1$,

$$\|u - v\|_{L^\infty(0,T;L^\infty(\mathbb{R}^2))} \lesssim \|u - v\|^{\theta}_{L^\infty(0,T;L^2(\mathbb{R}^2))} \|u - v\|^{1-\theta}_{L^\infty(0,T;H^s(\mathbb{R}^2))}$$

and therefore the solution map depends continuously on the initial data in $H^s(\mathbb{R}^2)$. In the case where $s \le 1$, by (21), the solution map depends continuously on the initial data in $L^2(\mathbb{R}^2)$. We define $s_3, s_4 > 0$ so that they satisfy the following:

$$\max\left\{\frac{3}{4} + \frac{1}{2r_1}, s_4 - \frac{3}{4}(p-1)\right\} < s_3 < s_4 < \min\left\{s, s_3 + \frac{3}{4}\right\},$$

$$r_3 = \frac{3}{2}\left(s_3 - s_4 + \frac{3}{4}\right)^{-1},$$

and $q_3 = \frac{3}{s_4 - s_3}$, where (q_3, r_3) satisfy the condition of Lemma 1. Let u and v be solutions of (1) for initial data u_0 and v_0, respectively. Then by Lemma 1,

$$\begin{aligned} &\|u - v\|_{L^{q_1}(0,T;B^{s_3 - \frac{3}{q_1}}_{r_1}(\mathbb{R}^2))} \\ &\le \|u_0 - v_0\|_{H^{s_3}(\mathbb{R}^2)} + C\||u|^{p-1}u - |v|^{p-1}v\|_{L^{q_3'}(0,T;B^{s_4}_{r_3'}(\mathbb{R}^2))}. \end{aligned} \tag{22}$$

For $z_j \in \mathbb{C}$ with $j = 1, 2, 3, 4$, with $w_1 = z_2 - z_1$ and $w_2 = z_4 - z_3$,

$$\begin{aligned} &|z_4|^{p-1}z_4 - |z_3|^{p-1}z_3 - |z_2|^{p-1}z_2 + |z_1|^{p-1}z_1 \\ &= \frac{p+1}{2}\int_0^1 |z_3 + \theta w_2|^{p-1} d\theta w_2 - \frac{p+1}{2}\int_0^1 |z_1 + \theta w_1|^{p-1} d\theta w_1 \\ &+ \frac{p-1}{2}\int_0^1 |z_3 + \theta w_2|^{p-3}(z_3 + \theta w_2)^2 d\theta \overline{w_2} \\ &- \frac{p-1}{2}\int_0^1 |z_1 + \theta w_1|^{p-3}(z_1 + \theta w_1)^2 d\theta \overline{w_1}. \end{aligned}$$

Then a direct computation implies that

$$\begin{aligned} &\left||z_4|^{p-1}z_4 - |z_3|^{p-1}z_3 - |z_2|^{p-1}z_2 + |z_1|^{p-1}z_1\right| \\ &\lesssim (|z_3|^{p-1} + |z_4|^{p-1})|w_2 - w_1| \\ &+ \frac{p+1}{p}|w_1||z_3 - z_1|^{p-1} + \frac{1}{p}|w_1||z_4 - z_2|^{p-1} + (|z_3|^{p-1} + |z_4|^{p-1})|w_2 - w_1| \\ &+ |w_1||z_3 - z_1|^{p-1} + |w_1||z_4 - z_2|^{p-1}. \end{aligned}$$

Therefore,

$$
\begin{aligned}
&\Big\| |u(t,\cdot+h)|^{p-1}u(t,\cdot+h) - |v(t,\cdot+h)|^{p-1}v(t,\cdot+h) \\
&\quad - |u(t)|^{p-1}u(t) + |v(t)|^{p-1}v(t)\Big\|_{L^{r_3'}(\mathbb{R}^2)} \\
&= \Big\| |u(t,\cdot+h)|^{p-1}u(t,\cdot+h) - |u(t)|^{p-1}u(t) \\
&\quad - |v(t,\cdot+h)|^{p-1}v(t,\cdot+h) + |v(t)|^{p-1}v(t)\Big\|_{L^{r_3'}(\mathbb{R}^2)} \\
&\leq 4\|u(t)\|^{p-1}_{L^{\frac{2r_3(p-1)}{r_3-2}}(\mathbb{R}^2)} \|u(t,\cdot+h) - v(t,\cdot+h) - u(t) + v(t)\|_{L^2(\mathbb{R}^2)} \\
&\quad + \frac{2(p+2)}{p}\|v(t,\cdot+h) - v(t)\|_{L^2(\mathbb{R}^2)}\|u(t) - v(t)\|^{p-1}_{L^{\frac{2r_3(p-1)}{r_3-2}}(\mathbb{R}^2)},
\end{aligned}
$$

and this means

$$
\begin{aligned}
&\||u|^{p-1}u - |v|^{p-1}v\|_{L^{q_3'}(0,T;B^{s_4}_{r_3'}(\mathbb{R}^2))} \\
&\lesssim \Big\|\|u\|^{p-1}_{L^{\frac{2r_3(p-1)}{r_3-2}}(\mathbb{R}^2)}\|u-v\|_{H^{s_4}(\mathbb{R}^2)} + \|v\|_{H^{s_4}(\mathbb{R}^2)}\|u-v\|^{p-1}_{L^{\frac{2r_3(p-1)}{r_3-2}}(\mathbb{R}^2)}\Big\|_{L^{q_3'}(0,T)} \\
&\leq \Big\|\|u\|^{p-1-\frac{r_3-2}{r_3}}_{L^\infty(\mathbb{R}^2)}\|u\|^{\frac{r_3-2}{r_3}}_{L^2(\mathbb{R}^2)}\|u-v\|_{H^{s_4}(\mathbb{R}^2)}\Big\|_{L^{q_3'}(0,T)} \\
&\quad + \Big\|\|v\|_{H^{s_4}(\mathbb{R}^2)}\|u-v\|^{p-1-\frac{r_3-2}{r_3}}_{L^\infty(\mathbb{R}^2)}\|u-v\|^{\frac{r_3-2}{r_3}}_{L^2(\mathbb{R}^2)}\Big\|_{L^{q_3'}(0,T)} \\
&\leq \|u\|^{p-1-\frac{r_3-2}{r_3}}_{L^{q_3'(p-1-\frac{r_3-2}{r_3})}(0,T;L^\infty(\mathbb{R}^2))}\|u\|^{\frac{r_3-2}{r_3}}_{L^\infty(0,T;L^2(\mathbb{R}^2))}\|u-v\|_{L^\infty(0,T;H^{s_4}(\mathbb{R}^2))} \\
&\quad + \|v\|_{L^\infty(0,T;H^{s_4}(\mathbb{R}^2))}\|u-v\|^{p-1-\frac{r_3-2}{r_3}}_{L^{q_3'(p-1-\frac{r_3-2}{r_3})}(0,T;L^\infty(\mathbb{R}^2))}\|u-v\|^{\frac{r_3-2}{r_3}}_{L^\infty(0,T;L^2(\mathbb{R}^2))} \\
&\leq \|u\|^{p-1-\frac{r_3-2}{r_3}}_{L^{q_1}(0,T;L^\infty(\mathbb{R}^2))}\|u\|^{\frac{r_3-2}{r_3}}_{L^\infty(0,T;L^2(\mathbb{R}^2))}\|u-v\|_{L^\infty(0,T;H^{s_4}(\mathbb{R}^2))} \\
&\quad + \|v\|_{L^\infty(0,T;H^{s_4}(\mathbb{R}^2))}\|u-v\|^{p-1-\frac{r_3-2}{r_3}}_{L^{q_1}(0,T;L^\infty(\mathbb{R}^2))}\|u-v\|^{\frac{r_3-2}{r_3}}_{L^\infty(0,T;L^2(\mathbb{R}^2))},
\end{aligned}
$$

where $q_1, q_3 > 4 > q_3' > q_3'\left(p-1-\frac{r_3-2}{r_3}\right)$. This and (22) imply that $u \to v$ in $L^{q_1}(0,T;B^{s_3-\frac{3}{q_1}}_{r_1}(\mathbb{R}^2))$ as $u_0 \to v_0$ in $H^s(\mathbb{R}^2)$ because $u \to v$ in $(L^\infty(0,T;L^2(\mathbb{R}^2)))$ and u, v are uniformly bounded in $(L^\infty(0,T;H^s(\mathbb{R}^2)))$ as

$u_0 \to v_0$ in $H^s(\mathbb{R}^2)$. Moreover,

$$\begin{aligned}
&\|u - v\|_{L^\infty(0,T;H^s(\mathbb{R}^2))} \\
&\lesssim \|u_0 - v_0\|_{H^s(\mathbb{R}^2)} + (\|u_0\|_{H^s(\mathbb{R}^2)} + \|v_0\|_{H^s(\mathbb{R}^2)})\|u - v\|^{p-1}_{L^{p-1}(0,T;L^\infty(\mathbb{R}^2))} \\
&\lesssim \|u_0 - v_0\|_{H^s(\mathbb{R}^2)} + (\|u_0\|_{H^s(\mathbb{R}^2)} + \|v_0\|_{H^s(\mathbb{R}^2)})\|u - v\|^{p-1}_{L^{q_1}(0,T;B^{s_3-\frac{3}{q_1}}_{r_1}(\mathbb{R}^2))}.
\end{aligned}$$

Therefore, the solution map is also continuously dependent in $L^\infty(0, T; H^s(\mathbb{R}^2))$.

2.2 *The Case $n \geq 3$: Local H^1 Existence Result*

In the case where $n \geq 3$, the Strichartz estimate Lemma 1 doesn't seem sufficient to obtain a uniform control of solutions in the $H^1(\mathbb{R}^3)$ setting. So here, we consider radial data and use the following Strauss lemma.

Lemma 3 ([28, Theorems 1,2], [7, Proposition 1]) *Let $n \geq 2$ and let $1/2 < s < n/2$. Then for a radial function f*

$$\||\cdot|^{\frac{n}{2}-s} f\|_{L^\infty(\mathbb{R}^n)} \lesssim \|f\|_{\dot{H}^s_{\mathrm{rad}}(\mathbb{R}^n)}.$$

Since solutions are not uniformly controlled at the origin by the Strauss lemma above, we apply the following weighted Strichartz estimate:

Lemma 4 ([1, Propositions 2.2 and 2.3]) *Let $n \in \mathbb{N}$. Let $\delta > 0$ and $[x]_\delta = |x|^{1-\delta} + |x|^{1+\delta}$. The for any $q_1 \in [2, \infty]$ and $q_2 \in (2, \infty]$,*

$$\|[x]_\delta^{-1/q_1} U(t) f\|_{L^{q_1}(\mathbb{R};L^2(\mathbb{R}^n))} \lesssim \|f\|_{L^2(\mathbb{R}^n)},$$

$$\left\|[x]_\delta^{-1/q_1} \int_0^t U(t-t')F(t')dt'\right\|_{L^{q_1}(0,T;L^2(\mathbb{R}^n))} \lesssim \|[x]_\delta^{1/q_2} F\|_{L^{q_2'}(0,T;L^2(\mathbb{R}^n))}.$$

We can now prove Proposition 2.

Proof (Proof of Proposition 2) By using the uniform $H^1(\mathbb{R}^n)$ control obtained in (17), we reduce the proof to the local well-posedness in $H^1(\mathbb{R}^n)$. Let $\delta > 0$, $1/2 < s < 1$, and $2 < q_1, q_2 < \infty$ satisfy

$$-(p-1)\left(\frac{n}{2} - s\right) + \frac{1-\delta}{q_2} = -\frac{1-\delta}{q_1}. \tag{23}$$

We remark that there exist δ, q_1, q_2, s if $1 < p < 1 + 2/(n-2)$ since,

$$(p-1)\left(\frac{n}{2} - s\right) < 1 \Longrightarrow p < 1 + \frac{2}{n-2s} < 1 + \frac{2}{n-2}.$$

We define the norm $Y^1(T)$ as

$$\begin{aligned}\|u\|_{Y^1(T)} &= \|u\|_{L^\infty(0,T;H^1_{\mathrm{rad}}(\mathbb{R}^n))} \\ &\quad + \left\|[x]_\delta^{-1/q_1} u\right\|_{L^{q_1}(0,T;L^2_{\mathrm{rad}}(\mathbb{R}^n))} + \left\|[x]_\delta^{-1/q_1} \nabla u\right\|_{L^{q_1}(0,T;L^2_{\mathrm{rad}}(\mathbb{R}^n))}.\end{aligned}$$

Let $\psi \in \mathscr{S}(\mathbb{R}^n; [0, 1])$ be radial and satisfy

$$\psi(x) = \begin{cases} 1 & \text{if} \quad |x| \leq 1, \\ 0 & \text{if} \quad |x| \geq 2. \end{cases} \tag{24}$$

Then by Lemmas 3 and 4 and (23),

$$\begin{aligned}&\|\Phi(u)\|_{Y^1(T)} \\ &\lesssim \|u_0\|_{H^1_{\mathrm{rad}}(\mathbb{R}^n)} + \left\|\int_0^t U(t-t')\left(\psi|u(t')|^{p-1}u(t')\right) dt'\right\|_{Y^1(T)} \\ &+ \left\|\int_0^t U(t-t')\left((1-\psi)|u(t')|^{p-1}u(t')\right) dt'\right\|_{Y^1(T)} \\ &\lesssim \|u_0\|_{H^1_{\mathrm{rad}}(\mathbb{R}^n)} \\ &+ \||x|^{-(p-1)(\frac{n}{2}-s)+\frac{1-\delta}{q_2}} ||x|^{\frac{n}{2}-s}u|^{p-1}u\|_{L^{q_2'}(0,T;L^2_{\mathrm{rad}}(|x|\leq 2))} \\ &+ \||x|^{-(p-1)(\frac{n}{2}-s)+\frac{1-\delta}{q_2}} ||x|^{\frac{n}{2}-s}u|^{p-1}\nabla u\|_{L^{q_2'}(0,T;L^2_{\mathrm{rad}}(|x|\leq 2))} \\ &+ \||u|^{p-1}u\|_{L^1(0,T;L^2_{\mathrm{rad}}(|x|>1))} + \|\nabla(|u|^{p-1}u)\|_{L^1(0,T;L^2_{\mathrm{rad}}(|x|>1))} \\ &\lesssim \|u_0\|_{H^1_{\mathrm{rad}}(\mathbb{R}^n)} + T^{1-\frac{1}{q_1}-\frac{1}{q_2}} \|u\|^p_{Y^1(T)}\end{aligned}$$

and therefore for some T and R, Φ is a map from $B_{Y^1(T)}(R)$ into itself. Moreover,

$$\begin{aligned}&\|\Phi(u) - \Phi(v)\|_{Y^1(T)} \\ &\lesssim \||x|^{-\frac{1-\delta}{q_1}}(||x|^{\frac{n}{2}-s}u|^{p-1} - ||x|^{\frac{n}{2}-s}v|^{p-1})(|\nabla u| + |u|)\|_{L^{q_2'}(0,T;L^2_{\mathrm{rad}}(|x|\leq 2))} \\ &+ \||x|^{-\frac{1-\delta}{q_1}} ||x|^{\frac{n}{2}-s}v|^{p-1}(|\nabla(u-v)| + |u-v|)\|_{L^{q_2'}(0,T;L^2_{\mathrm{rad}}(|x|\leq 2))}\end{aligned}$$

$$+ \|(||x|^{\frac{n}{2}-s}u|^{p-1} - ||x|^{\frac{n}{2}-s}v|^{p-1})|x|^{-\frac{1+\delta}{q_1}}(|\nabla u| + |u|)\|_{L^1(0,T;L^2_{\mathrm{rad}}(|x|>1))}$$

$$+ \|||x|^{\frac{n}{2}-s}v|^{p-1}|x|^{-\frac{1+\delta}{q_1}}(|\nabla(u-v)| + |u-v|)\|_{L^1(0,T;L^2_{\mathrm{rad}}(|x|>1))}. \tag{25}$$

Then for $p \geq 2$, Φ is a contraction map on $B_{Y^1(T)}(R)$. Similarly, for $1 < p < 2$, we define the auxiliary norm $Y^0(T)$ as

$$\|u\|_{Y^0(T)} := \|u\|_{L^\infty(0,T;L^2_{\mathrm{rad}}(\mathbb{R}^n))} + \|[x]_\delta^{-1/q_1}u\|_{L^{q_1}(0,T;L^2_{\mathrm{rad}}(\mathbb{R}^n))}.$$

Then for $1 < p < 2$,

$$\begin{aligned}
&\|(\Phi(u) - \Phi(v))\|_{Y^0(T)}\\
&\lesssim \left\| [x]_\delta^{-1/q_1}\left(\left||x|^{\frac{n}{2}-s}v\right| + \left||x|^{\frac{n}{2}-s}v\right|\right)^{p-1}|u-v|\right\|_{L^{q_2'}(0,T;L^2_{\mathrm{rad}}(|x|\leq 2))}\\
&+ \left\|\left(\left||x|^{\frac{n}{2}-s}v\right| + \left||x|^{\frac{n}{2}-s}v\right|\right)^{p-1}|u-v|\right\|_{L^1(0,T;L^2_{\mathrm{rad}}(|x|>1))}\\
&\lesssim T^{1-\frac{1}{q_1}-\frac{1}{q_2}}(\|u\|_{Y^1(T)} + \|v\|_{Y^1(T)})^{p-1}\|u-v\|_{Y^0(T)}.
\end{aligned}$$

Therefore Φ is a contraction map on $Y^0(T)$ for some T and R, which implies that (1) posses a unique solution in $Y^1(T)$. Moreover, by Lemma 3 and (25), with some $0 < \theta < 1$, for solutions u and v of (4) for initial data u_0 and v_0, respectively,

$$\begin{aligned}
&\|u-v\|_{Y^1(T)}\\
&\lesssim \|u_0 - v_0\|_{H^1_{\mathrm{rad}}(\mathbb{R}^n)} + T^{1-\frac{1}{q_1}-\frac{1}{q_2}}(\|u\|_{Y^1(T)} + \|v\|_{Y^1(T)})^{p-1}\|u-v\|_{Y^1(T)}\\
&+ T^{1-\frac{1}{q_1}-\frac{1}{q_2}}(\|u\|_{Y^1(T)} + \|v\|_{Y^1(T)})\left\||x|^{\frac{n}{2}-s}(u-v)\right\|^{p-1}_{L^\infty(0,T;L^\infty_{\mathrm{rad}}(\mathbb{R}^n))}\\
&\lesssim \|u_0 - v_0\|_{H^1_{\mathrm{rad}}(\mathbb{R}^n)} + T^{1-\frac{1}{q_1}-\frac{1}{q_2}}(\|u\|_{Y^1(T)} + \|v\|_{Y^1(T)})^{p-1}\|u-v\|_{Y^1(T)}\\
&+ T(\|u\|_{Y^1(T)} + \|v\|_{Y^1(T)})\,\|u-v\|^{p-1}_{Y^1(T)}
\end{aligned}$$

and therefore $\|u-v\|_{Y^1(T)} \to 0$ as $\|u_0 - v_0\|_{H^1_{\mathrm{rad}}(\mathbb{R}^n)} \to 0$.

2.3 Three Dimensional Case, Small H^1 Data Solutions for $p = 3$

In the three dimensional scaling critical case, the weighted Strichartz estimate Lemma 4 doesn't seem sufficient to control solutions uniformly. So here, we transform (1) into the corresponding wave equation.

The Cauchy problem (1) with initial data $u(0) = u_0$ is rewritten as the following:

$$\begin{aligned}\Box u &= i(-i\partial_t + D)|u|^{p-1}u \\ &= i\frac{p+1}{2}|u|^{p-1}(Du - i|u|^{p-1}u) \\ &\quad - i\frac{p-1}{2}|u|^{p-3}u^2\overline{(Du - i|u|^{p-1}u)} + iD(|u|^{p-1}u) \\ &= i\left(D(|u|^{p-1}u) + \frac{p+1}{2}|u|^{p-1}Du - \frac{p-1}{2}|u|^{p-3}u^2 D\overline{u}\right) + p|u|^{2p-2}u \\ &=: F_p(u).\end{aligned}$$

Then the corresponding integral equation is the following:

$$\begin{aligned}u(t) = \cos(tD)u_0 &+ \frac{\sin(tD)}{D}(iDu_0 + |u_0|^{p-1}u_0) \\ &+ \int_0^t \frac{\sin((t-t')D)}{D}F_p(u)(t')dt'.\end{aligned} \tag{26}$$

For any radially symmetric function f, we define $\widetilde{f}$ as $\widetilde{f}(|x|) = f(x)$. Then for any radial data, (26) is rewritten as

$$\widetilde{u}(t) = \partial_t J[u_0](t) + J[iDu_0 + |u_0|^{p-1}u_0](t) + \int_0^t J[F_p(u)(t')](t-t')dt' \tag{27}$$

where

$$J[f](t,r) = \frac{1}{2r}\int_{|r-t|}^{r+t} \lambda\widetilde{f}(\lambda)d\lambda.$$

This transformation is justified as follows:

Lemma 5 ([12, Lemma 3.5]) *Let $1 < p \le 3$ and $u_0 \in H^1_{\mathrm{rad}}(\mathbb{R}^3)$ and $u \in C(0,T; H^1_{\mathrm{rad}}(\mathbb{R}^3))$ be the solution of (16). Then u is also the solution of (27).*

To obtain the uniform control, we use the estimates below regarding J. For any $f : [0,\infty) \to \mathbb{C}$, we define $A[f] : \mathbb{R} \to \mathbb{C}$ as $A[f](\lambda) = f(|\lambda|)$. See also [21].

Lemma 6 ([12, Lemma 3.6]) *Let* $f : [0,\infty) \to \mathbb{C}$. *Then*

$$\left\| \frac{1}{2\cdot} \int_{|\cdot - t|}^{\cdot + t} f(\lambda) d\lambda \right\|_{L^\infty(0,\infty)} \leq M[A[f]](t),$$

where M *is the Hardy-Littlewood-Maximal operator defined by*

$$M[h](x) = \sup_{r>0} \frac{1}{2r} \int_{|x-y|<r} |h(y)| dy$$

for $h : \mathbb{R} \to \mathbb{C}$.

Corollary 3 ([12, Corollary 3.7]) *Let* $f : \mathbb{R}^3 \to \mathbb{C}$ *be radial. Then*

$$\|J[f]\|_{L^2(0,T;L^\infty(\mathbb{R}^3))} \leq C \|f\|_{L^2_{\mathrm{rad}}(\mathbb{R}^3)}.$$

Corollary 4 ([12, Corollary 3.8]) *Let* $h : [0,\infty) \times \mathbb{R}^3 \to \mathbb{C}$ *be radial. Then*

$$\left\| \int_0^t J[h(t')](t-t') dt' \right\|_{L^2(0,T;L^\infty(0,\infty))} \leq C \|h\|_{L^1(0,T;L^2_{\mathrm{rad}}(\mathbb{R}^3))}.$$

Corollary 5 (Hardy, [12, Corollary 3.9]) *Let* $f \in C^1([0,\infty);\mathbb{C})$. *Then*

$$\left\| \frac{d}{dt} \left(\frac{1}{2r} \int_{|r-t|}^{r+t} \lambda f(\lambda) d\lambda \right) \right\|_{L^2(0,\infty;L^\infty(0,\infty))} \leq C \|rf'\|_{L^2(0,\infty)}.$$

Proof Let g be even extension of f.

$$\begin{aligned} \frac{d}{dt} \left(\frac{1}{2r} \int_{|r-t|}^{r+t} \lambda f(\lambda) d\lambda \right) &= \frac{(r+t)f(r+t) - (t-r)f(|r-t|)}{2r} \\ &= \frac{(r+t)f(r+t) - (t-r)g(t-r)}{2r} \\ &= \frac{1}{2r} \int_{-r}^{r} \{g(t+\tau) + (t+\tau)g'(t+\tau)\} d\tau. \end{aligned}$$

Then

$$\left\| \frac{d}{dt} \left(\frac{1}{2r} \int_{|r-t|}^{r+t} \lambda f(\lambda) d\lambda \right) \right\|_{L^2(0,\infty;L^\infty(0,\infty))} \leq \|M[g]\|_{L^2(\mathbb{R})} + \|M[\cdot g']\|_{L^2(\mathbb{R})}. \tag{28}$$

Therefore, (28) and the following Hardy estimate([22, (0.2)]) imply Corollary 5:

$$\|g\|_{L^2(\mathbb{R})} \lesssim \| \cdot g'\|_{L^2(\mathbb{R})}.$$

We can now give the proof of Proposition 3.

Proof (Proof of Proposition 3) Let

$$X^1_{\text{rad}}(0,T) = L^\infty(0,T;\, H^1_{\text{rad}}(\mathbb{R}^3)) \cap L^2(0,T;\, L^\infty_{\text{rad}}(\mathbb{R}^3)).$$

For $0 < T < 1$ and $p = 3$, By Corollaries 3, 4, 5, and the Hölder and Gagliardo-Nirenberg inequalities imply that, if initial data u_0 sufficiently small, then Φ maps $B_{X^1_{\text{rad}}(0,T)}(R)$ into itself with some T and R. Since

$$\begin{aligned}
&|F_3(u) - F_3(v)| \\
&= \left| i \left(D(|u|^2 u) - 2|u|^2 Du - u^2 D\overline{u} \right) + 3|u|^4 u \right. \\
&\left. - i \left(D(|v|^2 v) - 2|v|^2 Dv - v^2 D\overline{v} \right) - 3|v|^4 v \right| \\
&\lesssim |D(|u|^2 u - |v|^2 v)| + |u|^2 |D(u-v)| \\
&+ \left(\left| |u|^2 - |v|^2 \right| + \left| u^2 - v^2 \right| \right) |Dv| + \left| |u|^4 u - |v|^4 v \right|,
\end{aligned}$$

we have

$$\begin{aligned}
&\|F_3(u) - F_3(v)\|_{L^1(0,T;L^2_{\text{rad}}(\mathbb{R}^3))} \\
&\lesssim (\|u\|_{X^1_{\text{rad}}(0,T)} + \|v\|_{X^1_{\text{rad}}(0,T)})^2 \|u - v\|_{X^1_{\text{rad}}(0,T)} \\
&+ (\|u\|_{X^1_{\text{rad}}(0,T)} + \|v\|_{X^1_{\text{rad}}(0,T)})^4 \|u - v\|_{X^1_{\text{rad}}(0,T)}.
\end{aligned}$$

This means Φ is a contraction map on $B_{X^1_{\text{rad}}(0,T)}(R)$ for sufficiently small u_0.

3 Blow-Up for (1)

At first, we recall the following ODE argument:

Lemma 7 ([11, Lemma 2.1]) *Let $C_1, C_2 > 0$ and $q > 1$. If $f \in C^1([0,T);\mathbb{R})$ satisfies $f(0) > 0$ and*

$$f' + C_1 f = C_2 f^q \quad \textit{on } [0,T) \textit{ for some } T > 0,$$

then

$$f(t) = e^{-C_1 t} \left(f(0)^{-(q-1)} + C_1^{-1} C_2 e^{-C_1(q-1)t} - C_1^{-1} C_2 \right)^{-\frac{1}{q-1}}.$$

Moreover, if $f(0) > C_1^{\frac{1}{q-1}} C_2^{-\frac{1}{q-1}}$, then $T < -\frac{1}{C_1(q-1)} \log(1 - C_1 C_2^{-1} f(0)^{-q+1})$.

Next, we recall Calderón-Zygmund argument. We call K, a mesurable function on $\mathbb{R}^n$, Calderón-Zygmund (CZ) kernel if K satisfies

$$|K(x)| \leq |x|^{-n}, \quad |\nabla K(x)| \leq |x|^{-n+1}, \quad \int_{\varepsilon<|x|<R} K(x) = 0, \qquad 0 < \forall \varepsilon < \forall R.$$

Then CZ kernel is known to give a $L^p(\mathbb{R}^n)$ bounded operator as follows:

Lemma 8 ([3, Theorem 1]) *Let K be a CZ kernel. Then for $1 < p < \infty$, there exists a positive constant C such that*

$$\left\| \text{P.V.} \int_{\mathbb{R}^n} K(x-y) f(y) dy \right\|_{L^p(\mathbb{R}^n)} \leq C \|f\|_{L^p(\mathbb{R}^n)}$$

for any $f \in L^p(\mathbb{R}^n)$.

Now we are in position to show Proposition 5.

Proof Thanks to Lemma 7, it is enough to show

$$\left\| \langle \cdot \rangle^{-q} [D, \langle \cdot \rangle^q] \right\|_{L^2(\mathbb{R}^n) \to L^2(\mathbb{R}^n)} < \infty.$$

At first, We divide the operator into the following two pieces:

$$\langle x \rangle^{-q} [(-\Delta)^{1/2}, \langle x \rangle^q] = CT_1 + CT_2,$$

where ψ is a cut-off function defined by (24).

$$T_1(f)(x) = \langle x \rangle^{-q} \int_{\mathbb{R}^n} \frac{(1-\psi(y))(\langle x \rangle^q - \langle x+y \rangle^q)}{|y|^{n+1}} f(x+y) dy,$$

$$T_2(f)(x) = \langle x \rangle^{-q} \, \text{P.V.} \int_{\mathbb{R}^n} \frac{\psi(y)(\langle x \rangle^q - \langle x+y \rangle^q)}{|y|^{n+1}} f(x+y) dy.$$

In order to estimate T_1 by dividing into two pieces:

$$T_1 = T_3 + T_4,$$

where

$$T_3(f)(x) = \langle x \rangle^{-q} \int_{|x|\leq|y|} \frac{(1-\psi(y))(\langle x \rangle^q - \langle x+y \rangle^q)}{|y|^{n+1}} f(x+y) dy,$$

$$T_4(f)(x) = \langle x \rangle^{-q} \int_{|x|\geq|y|} \frac{(1-\psi(y))(\langle x \rangle^q - \langle x+y \rangle^q)}{|y|^{n+1}} f(x+y) dy.$$

By the Hölder and Young inequalities,

$$
\begin{aligned}
&\|T_3(f)\|_{L^2(\mathbb{R}^n)} \\
&\le (1+2^q)\left\|\langle x\rangle^{-q}\int_{|x|\le|y|}\frac{\langle y\rangle^q(1-\psi(y))}{|y|^{n+1}}f(x+y)dy\right\|_{L^2(\mathbb{R}^n)} \\
&\le (1+2^q)\|\langle\cdot\rangle^{-q}\|_{L^2(\mathbb{R}^n)}\left\|\int_{\mathbb{R}^n}\frac{\langle y\rangle^q(1-\psi(y))}{|y|^{n+1}}f(x+y)dy\right\|_{L^\infty(\mathbb{R}^n)} \\
&\le (1+2^q)\|\langle\cdot\rangle^{-q}\|_{L^2(\mathbb{R}^n)}\|\langle\cdot\rangle^q|\cdot|^{-n-1}(1-\psi)\|_{L^2(\mathbb{R}^n)}\|f\|_{L^2(\mathbb{R}^n)}.
\end{aligned}
$$

Similarly by the Young inequality,

$$
\begin{aligned}
\|T_4(f)\|_{L^2(\mathbb{R}^n)} &\le (1+2^q)\left\|\int_{\mathbb{R}^n}\frac{1-\psi(y)}{|y|^{n+1}}|f(x+y)|dy\right\|_{L^2(\mathbb{R}^n)} \\
&\le (1+2^q)\||\cdot|^{-n-1}(1-\psi)\|_{L^1(\mathbb{R}^n)}\|f\|_{L^2(\mathbb{R}^n)}.
\end{aligned}
$$

Next, in order to estimate T_2, we recall that

$$
\begin{aligned}
\langle x+y\rangle^q &= \langle x\rangle^q + \frac{q}{2}\langle x\rangle^{q-2}(|x+y|^2-|x|^2) + R_1(x,y), \\
&= \langle x\rangle^q + q\langle x\rangle^{q-2}x\cdot y + R_2(x,y),
\end{aligned} \tag{29}
$$

where $R_2(x,y) = R_1(x,y) + q\langle x\rangle^{q-2}|y|^2/2$ and

$$
R_1(x,y) = \frac{q(q-2)}{2^2}\int_{|x|^2}^{|x+y|^2}(1+\rho)^{q/2-2}(|x+y|^2-\rho)d\rho.
$$

By combining (13) and (29), we have

$$
T_2 = -qT_5 - T_6,
$$

where

$$
\begin{aligned}
T_5(f)(x) &= \frac{x}{\langle x\rangle^2}\cdot \text{P.V.}\int_{\mathbb{R}^n}\frac{y\psi(y)}{|y|^{n+1}}f(x+y)dy, \\
T_6(f)(x) &= \frac{1}{\langle x\rangle^q}\text{P.V.}\int_{\mathbb{R}^n}\frac{R_2(x,y)\psi(y)}{|y|^{n+1}}f(x+y)dy.
\end{aligned}
$$

It is easy to see that $K(y) = y|y|^{-n-1}\psi(y)$ is a CZ kernel. Therefore

$$
\|T_5(f)\|_{L^2(\mathbb{R}^n)} \le \left\|\text{P.V.}\int_{\mathbb{R}^n}\frac{y\psi(y)}{|y|^{n+1}}f(\cdot+y)dy\right\|_{L^2(\mathbb{R}^n)} \le C\|f\|_{L^2(\mathbb{R}^n)}.
$$

Moreover, since

$$\begin{aligned}|y|^{-n-1}|R_1(x,y)| &\le (\langle x\rangle^{q-2}+\langle x+y\rangle^{q-2})(|x+y|^2-|x|^2)^2|y|^{-n-1}\\ &\le (\langle x\rangle^{q-2}+\langle x+y\rangle^{q-2})(|x+y|+|x|)^2|y|^{-n+1},\end{aligned}$$

by the Young inequality,

$$\|T_6(f)\|_{L^2(\mathbb{R}^n)} \le C\left\|\int_{\mathbb{R}^n}\frac{\psi(y)}{|y|^{n-1}}f(x+y)dy\right\|_{L^2(\mathbb{R}^n)} \le \||\cdot|^{-n+1}\psi\|_{L^1(\mathbb{R}^n)}\|f\|_{L^2(\mathbb{R}^n)}.$$

4 A Priori Estimates

This last section is devoted to the proofs of Propositions 7, 8, 9, 10, and 11. The proofs are essentially the same in [12], but we report them here for sake of completeness.

Proof (Proof of Proposition 7*)* The proposition follows from a standard argument, so we omit the proof.

Proof (Proof of Proposition 8*)* The proposition follows from a standard argument, so we omit the proof.

Proof (Proof of Proposition 9*)* Here we give a direct proof based on the integral equation by using the method in [27].

$$\begin{aligned}&\|u(t_2)\|^2_{\dot{H}^s(\mathbb{R}^n)}\\ &= \|u(t_1)\|^2_{\dot{H}^s(\mathbb{R}^n)} - 2\mathrm{Re}\int_{t_1}^{t_2}\langle D^s(|u(t)|^{p-1}u(t)), D^s u(t)\rangle_{L^2(\mathbb{R}^n)}dt\\ &\le \|u(t_1)\|^2_{\dot{H}^s(\mathbb{R}^n)} + 2\int_{t_1}^{t_2}\|D^s(|u(t)|^{p-1}u(t))\|_{L^2(\mathbb{R}^n)}\|u(t)\|_{\dot{H}^s(\mathbb{R}^n)}dt\\ &\le \|u(t_1)\|^2_{\dot{H}^s(\mathbb{R}^n)} + C\int_{t_1}^{t_2}\|u(t)\|^{p-1}_{L^\infty(\mathbb{R}^n)}\|u(t)\|^2_{\dot{H}^s(\mathbb{R}^n)}dt,\end{aligned}$$

where we used the nonlinear estimate

$$\||f|^{p-1}f\|_{\dot{H}^s(\mathbb{R}^n)} \lesssim \|f\|^{p-1}_{L^\infty(\mathbb{R}^n)}\|f\|_{\dot{H}^s(\mathbb{R}^n)}$$

(see [13, Lemma 3.4]).

Proof (Proof of Proposition 10) Since $|u|^2u \in C((0,T); H^2(\mathbb{R}^n))$, the following calculation is justified by the Plancherel identity:

$$
\begin{aligned}
&\|u(t_2)\|^2_{\dot{H}^2(\mathbb{R}^n)}\\
&= \|u(t_1)\|^2_{\dot{H}^2(\mathbb{R}^n)} - 2\mathrm{Re}\int_{t_1}^{t_2} \Big\langle \Delta|u(t)|^2u(t), \Delta u(t)\Big\rangle_{L^2(\mathbb{R}^n)} dt\\
&= \|u(t_1)\|^2_{\dot{H}^2(\mathbb{R}^n)} - 2\mathrm{Re}\sum_{j,k=1}^{n}\int_{t_1}^{t_2} \Big\langle |u(t)|^2\partial_j\partial_k u(t), \partial_j\partial_k u(t)\Big\rangle_{L^2(\mathbb{R}^n)} dt\\
&- 4\mathrm{Re}\sum_{j,k=1}^{n}\int_{t_1}^{t_2} \Big\langle \partial_k u(t)\partial_j|u(t)|^2, \partial_j\partial_k u(t)\Big\rangle_{L^2(\mathbb{R}^n)} dt\\
&- 2\mathrm{Re}\sum_{j,k=1}^{n}\int_{t_1}^{t_2} \Big\langle \partial_j\partial_k|u(t)|^2, \overline{u(t)}\partial_j\partial_k u(t)\Big\rangle_{L^2(\mathbb{R}^n)} dt\\
&= \|u(t_1)\|^2_{\dot{H}^2(\mathbb{R}^n)} - 2\sum_{j,k=1}^{n}\int_{t_1}^{t_2} \|u(t)\partial_j\partial_k u(t)\|^2_{L^2(\mathbb{R}^n)}dt\\
&+ 2\sum_{j,k=1}^{n}\int_{t_1}^{t_2} \Big\langle \partial_j^2|u(t)|^2, |\partial_k u(t)|^2\Big\rangle_{L^2(\mathbb{R}^n)} dt\\
&- \sum_{j,k=1}^{n}\int_{t_1}^{t_2} \Big\langle \partial_j\partial_k|u(t)|^2, \partial_j\partial_k|u(t)|^2 - 2\mathrm{Re}(\overline{\partial_j u(t)}\partial_k u(t))\Big\rangle_{L^2(\mathbb{R}^n)} dt.
\end{aligned}
$$

By the Hölder, Young, and Sobolev inequalities,

$$
\begin{aligned}
&\|u(t_2)\|^2_{\dot{H}^2(\mathbb{R}^n)}\\
&\leq \|u(t_1)\|^2_{\dot{H}^2(\mathbb{R}^n)} - 2\sum_{j,k=1}^{n}\int_{t_1}^{t_2} \|u(t)\partial_j\partial_k u(t)\|^2_{L^2(\mathbb{R}^n)}dt\\
&+ 2n^2\sum_{k=1}^{n}\int_{t_1}^{t_2} \|\partial_k u(t)\|^4_{L^4(\mathbb{R}^n)}dt + 2\sum_{j,k=1}^{n}\int_{t_1}^{t_2} \|\partial_j u(t)\|^2_{L^4(\mathbb{R}^n)}\|\partial_k u(t)\|^2_{L^4(\mathbb{R}^n)}dt\\
&\leq \|u(t_1)\|^2_{\dot{H}^2(\mathbb{R}^n)} - 2\sum_{j,k=1}^{n}\int_{t_1}^{t_2} \|u(t)\partial_j\partial_k u(t)\|^2_{L^2(\mathbb{R}^n)}dt\\
&+ 2n^2(n+1)\int_{t_1}^{t_2} \|u(t)\|^{4-n}_{\dot{H}^1(\mathbb{R}^n)}\|u(t)\|^n_{\dot{H}^2(\mathbb{R}^n)}dt.
\end{aligned}
$$

We can now conclude the paper by showing Proposition 11.

Proof (Proof of Proposition 11*)* When $s = 1$ and when $s = 2$ and $p = 3$, a priori estimates shows the global well-posedness by the blow-up alternative argument. Here we consider the case where $p = 3$ and $1 < s < 2$. Let $[a]$ be the floor function of a. Let $T_1 = \min\{1, T_0\}$. By using the H^1 a priori estimate, for any $t > 0$,

$$\begin{aligned}\|u\|_{L^4(0,t;L^\infty(\mathbb{R}^2))} &\leq \sum_{k=0}^{[t/T_1]+1} \|u\|_{L^4(kT_1,(k+1)T_1;L^\infty(\mathbb{R}^2))} \\ &\leq \sum_{k=0}^{[t/T_1]+1} \|u\|_{X^1(kT_1,(k+1)T_1)} \\ &\leq 2T_1^{-1}(1+t)\|u_0\|_{H^1(\mathbb{R}^2)}.\end{aligned}$$

Then by using Proposition 10,

$$\begin{aligned}\|u(t)\|^2_{\dot{H}^s(\mathbb{R}^2)} &\lesssim \|u_0\|^2_{H^s(\mathbb{R}^2)} + \int_0^t \|u(t')\|^2_{L^\infty(\mathbb{R}^n)} \|u(t')\|^2_{\dot{H}^s(\mathbb{R}^2)} dt \\ &\lesssim \|u_0\|^2_{H^s(\mathbb{R}^2)} + \|u(t')\|^2_{L^4(0,t;L^\infty(\mathbb{R}^2))} \|u\|^2_{L^4(0,t;\dot{H}^s(\mathbb{R}^2))} \\ &\lesssim \|u_0\|^2_{H^s(\mathbb{R}^2)} + \|u_0\|^2_{H^1(\mathbb{R}^2)}(1+t)^2\|u\|^2_{L^4(0,t;\dot{H}^s(\mathbb{R}^2))}.\end{aligned}$$

This shows

$$\|u(t)\|^4_{\dot{H}^s(\mathbb{R}^2)} \lesssim \|u_0\|^4_{H^s(\mathbb{R}^2)} + \|u_0\|^4_{H^1(\mathbb{R}^2)}(1+t)^4\|u\|^4_{L^4(0,t;\dot{H}^s(\mathbb{R}^2))}.$$

Therefore Gronwall inequality imply the global well-posedness in $H^s(\mathbb{R}^2)$.

Acknowledgements V. Georgiev was supported in part by INDAM, GNAMPA – Gruppo Nazionale per l'Analisi Matematica, la Probabilità e le loro Applicazioni, by Institute of Mathematics and Informatics, Bulgarian Academy of Sciences and Top Global University Project, Waseda University and the Project PRA 2018 – 49 of University of Pisa. The authors are grateful to the referees for their helpful comments.

References

1. J. Bellazzini, V. Georgiev, N. Visciglia, Long time dynamics for semirelativistic NLS and half wave in arbitrary dimension. Math. Annalen. **371**(1–2), 707–740 (2018)
2. J.P. Borgna, D.F. Rial, Existence of ground states for a one-dimensional relativistic Schödinger equation. J. Math. Phys. **53**, 062301 (2012)
3. A.P. Calderón, A. Zygmund, On the existence of certain singular integrals. Acta Math. **88**, 85–139 (1952)
4. T. Cazenave, *Semilinear Schrödinger Equations*. Courant Lecture Notes in Mathematics, vol. 10 (American Mathematical Society/New York University/Courant Institute of Mathematical Sciences, New York; American Mathematical Society, Providence, 2003)

5. T. Cazenave, F.B. Weissler, Some remarks on the nonlinear Schrödinger equation in the critical case, in *Nonlinear Semigroups, Partial Differential Equations and Attractors*, ed. by T.L. Gill, W.W. Zachary (Washington, DC, 1987). Lecture Notes in Mathematics, vol. 1394 (Springer, Berlin, 1989), pp. 18–29
6. T. Cazenave, F.B. Weissler, The Cauchy problem for the critical nonlinear Schrödinger equation in H^s. Nonlinear analysis. Theory, methods & applications. Int. Multidiscip. J. Ser. A Theory Methods **14**, 807–836 (1990)
7. Y. Cho, T. Ozawa, Sobolev inequalities with symmetry. Commun. Contemp. Math. **11**, 355–365 (2009)
8. E. Di Nezza, G. Palatucci, E. Valdinoci, Hitchhiker's guide to the fractional Sobolev spaces. Bull. Sci. Math. **136**, 521–573 (2012)
9. L. Forcella, K. Fujiwara, V. Georgiev, T. Ozawa, Local well-posedness and blow-up for the half Ginzburg-Landau-Kuramoto equation with rough coefficients and potential. Discrete Contin. Dyn. Syst. **39**, 2661–2678 (2019). arXiv:1804.02524
10. K. Fujiwara, Remark on local solvability of the Cauchy problem for semirelativistic equations. J. Math. Anal. Appl. **432**, 744–748 (2015)
11. K. Fujiwara, V. Georgiev, T. Ozawa, Blow-up for self-interacting fractional Ginzburg-Landau equation. Dyn. Partial Differ. Equ. **15**, 175–182 (2018)
12. K. Fujiwara, V. Georgiev, T. Ozawa, On global well-posedness for nonlinear semirelativistic equations in some scaling subcritical and critical cases. arXiv: 1611.09674 (2016)
13. J. Ginibre, T. Ozawa, G. Velo, On the existence of the wave operators for a class of nonlinear Schrödinger equations. Ann. Inst. H. Poincaré Phys. Théor. **60**, 211–239 (1994)
14. J. Ginibre, G. Velo, Generalized Strichartz inequalities for the wave equation. J. Funct. Anal. **133**, 50–68 (1995)
15. L. Grafakos, S. Oh, The Kato-Ponce Inequality. Comm. Partial Differ. Equ. **39**, 1128–1157 (2014)
16. M. Ikeda, T. Inui, Some non-existence results for the semilinear Schrödinger equation without gauge invariance. J. Math. Anal. Appl. **425**, 758–773 (2015)
17. M. Ikeda, Y. Wakasugi, Small-data blow-up of L^2-solution for the nonlinear Schrödinger equation without gauge invariance. Differ. Integral Equ. **26**, 11–12 (2013)
18. T. Inui, Some nonexistence results for a semirelativistic Schrödinger equation with nongauge power type nonlinearity. Proc. Am. Math. Soc. **144**, 2901–2909 (2016)
19. T. Kato, G. Ponce, Commutator estimates and the Euler and Navier-Stokes equations. Comm. Pure Appl. Math. **41**, 891–907 (1988)
20. C. Kenig, G. Ponce, L. Vega, The Cauchy problem for the Korteweg-de Vries equation in Sobolev spaces of negative indices. Duke Math. J. **71**, 1–21 (1993)
21. S. Klainerman, M. Machedon, Space-time estimates for null forms and the local existence theorem. Comm. Pure Appl. Math. **46**, 1221–1268 (1993)
22. A. Kufner, B. Opic, *Hardy-Type Inequalities*. Pitman Research Notes in Mathematics Series (Longman Scientific & Technical, Harlow, 1990)
23. N. Laskin, Fractional quantum mechanics and Lévy path integrals. Phys. Lett. A. **268**, 298–305 (2000)
24. E. Lenzmann, A. Schikorra, Sharp commutator estimates via harmonic extensions. arXiv: 1609.08547
25. D. Li, On Kato-Ponce and fractional Leibniz. Rev. Mat. Iberoamericana. (in press). arXiv:1609.01780v2. It appeared on arXiv in 2016 and revised in 2018
26. M. Nakamura, T. Ozawa, The Cauchy problem for nonlinear Klein-Gordon equations in the Sobolev spaces. Publ. Res. Inst. Math. Sci. **37**, 255–293 (2001)
27. T. Ozawa, Remarks on proofs of conservation laws for nonlinear Schrödinger equations. Calc. Var. Partial Differ. Equ. **25**, 403–408 (2006)
28. W. Sickel, L. Skrzypczak, Radial subspaces of Besov and Lizorkin-Triebel classes: extended Strauss lemma and compactness of embeddings. J. Fourier Anal. Appl. **6**, 639–662 (2000)

Semilinear Damped Klein-Gordon Models with Time-Dependent Coefficients

Giovanni Girardi

Abstract We consider the following Cauchy problem for a wave equation with time-dependent damping term $b(t)u_t$ and mass term $m(t)^2u$, and a power nonlinearity $|u|^p$:

$$\begin{cases} u_{tt} - \Delta u + b(t)u_t + m^2(t)u = |u|^p, & t \geq 0,\ x \in \mathbb{R}^n, \\ u(0,x) = f(x), \quad u_t(0,x) = g(x). \end{cases}$$

We discuss how the interplay between an effective time-dependent damping term and a time-dependent mass term influences the decay rate of the solution to the corresponding linear Cauchy problem, in the case in which the damping term is dominated by the mass term, i.e. $\liminf_{t\to\infty} (m(t)/b(t)) > 1/4$.

Then we use the obtained estimates of solutions to linear Cauchy problems to prove that a unique global in-time energy solution to the Cauchy problem with power nonlinearity $|u|^p$ at the right-hand side of the equation exists for any $p > 1$, assuming small data in the energy space $(f, g) \in H^1 \times L^2$.

1 Introduction

In this paper, we look for global (in time) small data energy solutions to the Cauchy problem

$$\begin{cases} u_{tt} - \Delta u + b(t)u_t + m^2(t)u = h(u), & t \geq 0,\ x \in \mathbb{R}^n, \\ u(0,x) = f(x), \quad u_t(0,x) = g(x), \end{cases} \tag{1}$$

G. Girardi (✉)
Department of Mathematics, University of Bari, Bari, Italy
e-mail: giovanni.girardi@uniba.it

M. D'Abbicco et al. (eds.), *New Tools for Nonlinear PDEs and Application*, Trends in Mathematics, https://doi.org/10.1007/978-3-030-10937-0_7

where $b(t)u_t$ and $m(t)^2u$, with $b(t)$, $m(t) > 0$, respectively represent a damping and a mass term, and the power nonlinearity $h(u)$ verifies

$$h(0) = 0, \quad |h(u) - h(v)| \lesssim |u - v| \, (|u| + |v|)^{p-1} \tag{2}$$

for a given $p > 1$, for instance, $h(u) = |u|^p$. In particular, we consider the case in which the damping term is dominated by the mass term, i.e. $\liminf_{t\to\infty} (m(t)/b(t)) > 1/4$.

In order to do that, we derive suitable estimates for solutions to the corresponding linear Cauchy problem

$$\begin{cases} u_{tt} - \Delta u + b(t)u_t + m^2(t)u = 0, & t \geq 0, \ x \in \mathbb{R}^n, \\ u(0, x) = f(x), \quad u_t(0, x) = g(x), \end{cases} \tag{3}$$

and we apply a contraction argument to construct the solution to (1).

In [1, 6], the model without mass

$$\begin{cases} u_{tt} - \Delta u + b(t)u_t = |u|^p, & t \geq 0, \ x \in \mathbb{R}^n, \\ u(0, x) = f(x), \quad u_t(0, x) = g(x), \end{cases} \tag{4}$$

has been considered, and it has been proved that the critical exponent for global (in time) small data energy solutions to (4) remains the same as for the Cauchy problem with $b = 1$ (see [11, 14, 16, 19, 22, 26]), that is $1 + 2/n$. Here the assumption of *effectiveness* of the damping term was essential to derive suitable estimates for solutions to the corresponding linear Cauchy problem

$$\begin{cases} u_{tt} - \Delta u + b(t)u_t = 0, & t \geq 0, \ x \in \mathbb{R}^n, \\ u(0, x) = f(x), \quad u_t(0, x) = g(x). \end{cases} \tag{5}$$

In particular, global existence holds for $p > 1+2/n$ if initial data are assumed to be small in exponentially weighted energy spaces. In the subcritical and critical range, $1 < p \leq 1 + 2/n$, no global in time small data Sobolev solutions exist, under a suitable sign assumption [4] for the data. If smallness of the data is assumed only in the standard energy space $H^1 \times L^2$ and in L^1, then the same result holds in space dimension $n = 1, 2$. If also the additional L^1 smallness is dropped, then the critical exponent becomes $1 + 4/n$.

In this paper, by effectiveness of the damping term we mean that for a suitable large class of damping coefficients $b(t)$, the estimates obtained for (5) are the same obtained for the solution to the corresponding Cauchy problem for the heat equation

$$\begin{cases} b(t)v_t - \Delta v = 0, & t \geq 0, \ x \in \mathbb{R}^n, \\ v(0, x) = \varphi(x), \end{cases} \tag{6}$$

for suitable initial data $\varphi = \varphi(f, g, b)$ (see [25]). In the case of polynomial shape $b(t) = \mu(1+t)^k$, the damping is effective if $k \in (-1, 1]$ (see [18, 20, 24] for the corresponding global existence result), and partially effective if $k = -1$, according to which μ and which estimate are considered.

In fact, in this latter case $b(t) = \mu(1+t)^{-1}$, the critical exponent of global small data solutions to (4) remains $1 + 2/n$ if the positive coefficient μ is sufficiently large [2, 23], whereas it seems to increase to $\max\{p_S(n+\mu), 1 + 2/n\}$, as μ becomes smaller with respect to the space dimension, as conjectured in [5, 7] (see also [12, 17]), where p_S is the Strauss exponent for the semilinear undamped wave equation [10, 15, 21]. The overdamping case $b(t) = \mu(1+t)^k$, with $k > 1$ has been studied in [13]; in that case the authors prove the global (in time) existence of small data energy solutions for all $p > 1$.

In [8] the same Cauchy problem (1) is considered, in the case in which the damping term is effective and it dominates the mass term, i.e. $m(t) = o(b(t))$ as $t \to \infty$. Here, we show that under a simple condition on the interaction between $b(t)$ and $m(t)$, and assuming only small initial data in the energy space, one may find a scale of critical exponents, which continuously move from $1 + 4/n$ to 1, as the mass becomes more influent, with respect to the damping term. In particular, in that case for any small initial data $(f, g) \in H^1 \times L^2$ we obtain the following estimate for the solution to the non linear Cauchy problem (1):

$$\|u(t, \cdot)\|_{L^2} \leq C\, \gamma(t)\, \|(f, g)\|_{H^1 \times L^2},$$

where we define

$$\gamma(t) = \exp\left(- \int_0^t \frac{m^2(\tau)}{b(\tau)}\, d\tau \right). \tag{7}$$

Thus, the decreasing function $\gamma = \gamma(t)$ in (7) represents the influence on the estimates of the mass term with respect to the damping term.

On the other hand, it is well-known that global small data solutions to the Cauchy problem for the damped Klein-Gordon equation

$$\begin{cases} u_{tt} - \Delta u + u_t + u = h(u), & t \geq 0,\ x \in \mathbb{R}^n, \\ u(0, x) = f(x), \quad u_t(0, x) = g(x), \end{cases} \tag{8}$$

exist for any $p > 1$ (see [3] for more results), that is, the interplay of a damping term and a mass term may lead to an improvement in the critical exponent of (1). This interaction has been recently studied in the scale invariant case, $b(t) = \mu_1(1+t)^{-1}$ and $m(t) = \mu_2(1+t)^{-1}$ in [9].

The main purpose of this paper is to complete the study of Cauchy problems associated to the wave equation with damping term and mass term, by considering the case in which the influence of the mass term dominates the influence of the damping term, i.e. $\liminf_{t\to\infty} (m(t)/b(t)) > 1/4$.

Under this hypothesis, assuming the effectiveness of the damping term $b(t)u_t$, we obtain an exponential decay rate for the solution to the linear Cauchy problem (3) and so we are able to prove global existence (in time) of small data solutions to the Cauchy problem (1) for any $p > 1$. Our result generalizes the already known results for the Cauchy problem (8) with constant coefficients $b(t) = m(t) \equiv 1$.

The scheme of the paper is the following:

- in Sect. 2 we present the main results for the Cauchy problems (1) and (3);
- in Sect. 3 we derive estimates for solutions of the associated linear Cauchy problem (Theorem 1);
- in Sect. 4 we prove our result for the global in time existence of small data solutions (Theorems 2).

2 Main Results

We assume that the following assumption is satisfied for $b = b(t)$.

Hypothesis 2.1 *We assume that $b \in \mathscr{C}^1$, with $b(t) > 0$, is monotone and it holds:*

$$|b'(t)| = o(b(t)^2), \qquad as\ t \to \infty. \tag{9}$$

As a consequence of (9), we derive

$$tb(t) \to \infty, \qquad \text{as } t \to \infty. \tag{10}$$

Hypothesis 2.1 means that the damping term $b(t)u_t$ is *effective* according to the definition given in [25]. In particular, also the overdamping case can be considered, that is, the case in which $1/b(t)$ is integrable.

The fundamental assumption is that the influence of the mass term dominates the influence of the damping term in the equation, so that the presence of the damping has a minor influence on the profile of the solution, which is mainly determined by the appearance of the mass term on the wave equation.

Hypothesis 2.2 *We assume that $m \in \mathscr{C}^1$, with $m(t) > 0$, is monotone and it has controlled oscillations:*

$$|m'(t)| \leq C\,\frac{m(t)}{1+t}, \tag{11}$$

and that

$$\liminf_{t\to\infty} \frac{m(t)}{b(t)} > \frac{1}{4}. \tag{12}$$

Example 1 An example of monotonous coefficient function with oscillations allowed by (11) is given by:

$$m(t) = (1+t)^k\Big(1 + \frac{\sin t}{1+t}\Big),$$

for $k > 0$.

Remark 1 The hypothesis of monotonicity can be avoided assuming that the coefficient $m = m(t)$ behaves like a monotonic shape function (see [25]).

Example 2 An example of nonmonotonous coefficient function allowed by (11) is given by

$$m(t) = \Big(2 + \frac{\cos(\log(e+t))}{1+t}\Big)(1+t)^k$$

for $k > -1$.

Remark 2 Assumption (12) is only of technical nature in order to simplify the proof, but we expect to obtain an exponential decay rate for the solution to the Cauchy problem (1), also with a weaker assumption. In fact, in the case $m(t) = a \cdot b(t)$ with $a > 0$, it is easy to prove that the solution to our Cauchy problem (1) has an exponential decay rate for each $a > 0$.

Furthermore, in [8], we proved that the solution to our Cauchy problem (1) satisfies the following estimate

$$\|u(t,\cdot)\|_{L^2} \leq C \exp\left(-\int_0^t \frac{m^2(\tau)}{b(\tau)}\,d\tau\right) \|(f,g)\|_{H^1\times L^2},$$

in the case in which the mass term is dominated by the damping term; thus, we expect that the assumption $\liminf_{t\to\infty} m/b > \delta$ for some $\delta > 0$ is sufficient to obtain an exponential decay rate, in the case in which the damping term is dominated by the mass term.

Assuming initial data in the energy space $H^1(\mathbb{R}^n)\times L^2(\mathbb{R}^n)$, we derive the following decay estimates for the solution to the linear Cauchy problem (3).

Theorem 1 *Let $n \geq 1$. Let us assume that $b = b(t)$ satisfies Hypothesis 2.1 and $m = m(t)$ satisfies Hypothesis 2.2. Then, there exists $\delta > 0$ such that the solution $u = u(t, x)$ to the Cauchy problem (3) satisfies the following estimates:*

$$\|u(t, \cdot)\|_{L^2(\mathbb{R}^n)} \leq C \exp\Big(- \delta \int_0^t b(\tau)d\tau\Big)\|(f, g)\|_{H^1 \times L^2}, \tag{13}$$

$$\|\nabla u(t, \cdot)\|_{L^2(\mathbb{R}^n)} \leq C \exp\Big(- \delta \int_0^t b(\tau)d\tau\Big)\|(f, g)\|_{H^1 \times L^2}, \tag{14}$$

$$\|\partial_t u(t, \cdot)\|_{L^2(\mathbb{R}^n)} \leq C \exp\Big(- \delta \int_0^t b(\tau)d\tau\Big)\|(f, g)\|_{H^1 \times L^2}, \tag{15}$$

where the constant $C > 0$ does not depend on the data.

Under our assumptions on $b = b(t)$ and $m = m(t)$ we are able to prove the global (in time) existence of small data solutions for any $p > 1$. Moreover we obtain the required estimates for the solutions to the semilinear Cauchy problem. Thus, we conclude that $\{|u|^p,\ p > 1\}$ is not the correct scale to observe blow-up if we consider an effective dissipation and a dominating effective mass.

By considering small initial data in the energy space $H^1(\mathbb{R}^n) \times L^2(\mathbb{R}^n)$, we may state the following result.

Theorem 2 *Let $n \geq 1$ and assume that $b = b(t)$ satisfies Hypothesis 2.1 and $m = m(t)$ satisfies Hypothesis 2.2. Then, for any $p > 1$ and $p \leq 1 + 2/(n-2)$ if $n \geq 3$, there exists ϵ_0 such that for any initial data $(f, g) \in H^1 \times L^2$ with*

$$\|(f, g)\|_{H^1 \times L^2} \leq \epsilon_0,$$

there exists a unique energy solution $u = u(t, x)$ of Cauchy problem (1) in $C([0, \infty), H^1) \cap C^1([0, \infty), L^2)$. Moreover, there exists $\delta > 0$ such that this solution satisfies the following estimates:

$$\|u(t, \cdot)\|_{L^2(\mathbb{R}^n)} \leq C \exp\Big(- \delta \int_0^t b(\tau)d\tau\Big)\|(f, g)\|_{H^1 \times L^2}, \tag{16}$$

$$\|\nabla u(t, \cdot)\|_{L^2(\mathbb{R}^n)} \leq C \exp\Big(- \delta \int_0^t b(\tau)d\tau\Big)\|(f, g)\|_{H^1 \times L^2}, \tag{17}$$

$$\|\partial_t u(t, \cdot)\|_{L^2(\mathbb{R}^n)} \leq C \exp\Big(- \delta \int_0^t b(\tau)d\tau\Big)\|(f, g)\|_{H^1 \times L^2}, \tag{18}$$

where C does not depend on the initial data.

The estimates (16), (17), and (18) are consistent with the estimates (13), (14), and (15) for solutions to the linear Cauchy problem (3).

Remark 3 For the same Cauchy problem (1) without the mass term, the global existence for small data solution for each $p > 1$ has been proved in [13], in the overdamping case. Thus, in this case the presence of the mass term gives no influence to the critical exponent, but it produces an exponential decay rate for the solution to the non linear Cauchy problem (1).

Instead, in the case in which $1/b(t)$ is not integrable we proved (see [8]) that the critical exponent moves from $1 + 4/n$ to 1 as the mass becomes more influential. Thus, we already expected to have the global existence of small data solutions for each $p > 1$, in the case in which the mass term dominates the damping term.

2.1 Examples

The easiest class of coefficients $b(t)$ and $m(t)$ which can be considered are of polynomial type.

Example 3 Let

$$b(t) = \mu(1+t)^k, \qquad m(t) = \nu(1+t)^\ell,$$

for some $\mu, \nu > 0$ and $k, \ell \in \mathbb{R}$.

Then, Hypothesis 2.1 holds if, and only if, $k \in (-1, +\infty)$. On the other hand, Hypothesis 2.2 holds if, and only if, $\ell > k$ or $\ell = k$ and $\nu > (1/4)\mu$.

Example 4 Let

$$b(t) = \mu(1+t)^k (\log(e+t))^a, \qquad m(t) = \nu(1+t)^\ell (\log(e+t))^b,$$

for some $\mu, \ell > 0$, and $k, \ell, a, b \in \mathbb{R}$.

Then, Hypothesis 2.1 holds if, and only if, either $k \in (-1, +\infty)$, or $k = -1$ and $a > 0$. On the other hand, Hypothesis 2.2 holds if, and only if, either $\ell > k$, or $\ell = k$ and $b \geq a$ with $\nu > (1/4)\mu$ if $b = a$.

3 Decay Estimates for Solutions to Linear Cauchy Problems

In order to prove Theorem 2, we plan to apply Duhamel's principle. However, due to the presence of time-dependent coefficients, the equation in (3) is not invariant by time translations. Having this in mind, we derive decay estimates for the solution to a family of parameter-dependent Cauchy problems

$$\begin{cases} u_{tt} - \Delta u + b(t)u_t + m^2(t)u = 0, & t \geq s, \\ u(s,x) = f(s,x), \quad u_t(s,x) = g(s,x), \end{cases} \tag{19}$$

where $s \geq 0$, obtaining decay rates which depend on both t and s. At $s = 0$, we will obtain the estimates for solutions of the linear Cauchy problem, stated in Theorem 1. On the other hand, for any $s > 0$ we may assume $f(s, \cdot) = 0$, and we prove the following result.

Lemma 1 *Let $b = b(t)$ satisfy Hypothesis 2.1 and $m = m(t)$ satisfy Hypothesis 2.2. Let $f(s, \cdot) = 0$ and $g(s, \cdot) \in L^2(\mathbb{R}^n)$. Then, there exists $\delta > 0$ such that the solution $u = u(t, x)$ to the Cauchy problem (19) satisfies the following estimates for $t \geq s \geq 0$:*

$$\|u(t, \cdot)\|_{L^2} \leq C \frac{1}{m(s)} \exp\Big(- \delta \int_s^t b(\tau) d\tau \Big) \|g(s, \cdot)\|_{L^2}, \tag{20}$$

$$\|\nabla u(t, \cdot)\|_{L^2} \leq C \exp\Big(- \delta \int_s^t b(\tau) d\tau \Big) \|g(s, \cdot)\|_{L^2}, \tag{21}$$

$$\|\partial_t u(t, \cdot)\|_{L^2(\mathbb{R}^n)} \leq C \exp\Big(- \delta \int_s^t b(\tau) d\tau \Big) \|g(s, \cdot)\|_{L^2}, \tag{22}$$

where C does not depend on s and on the data.

Following the approach in [25], we transform our Cauchy problem (3) with dissipation and mass terms in a Cauchy problem with time dependent mass. By applying the Fourier transform, we obtain

$$\hat{u}_{tt} + |\xi|^2 \hat{u} + b(t) \hat{u}_t + m^2(t) \hat{u} = 0.$$

Let

$$\hat{v}(t, \xi) := \lambda(t) \hat{u}(t, \xi), \qquad \lambda(t) := \exp\Big(\frac{1}{2} \int_0^t b(\tau) d\tau \Big).$$

The function $\hat{v}$ solves

$$\hat{v}_{tt} + M(t, \xi) \hat{v} = 0, \tag{23}$$

where

$$M(t, \xi) = |\xi|^2 - \frac{1}{4} b^2(t) - \frac{1}{2} b'(t) + m^2(t). \tag{24}$$

We notice that, in particular,

$$\hat{v}(0, \xi) = \hat{f}(\xi), \quad \hat{v}_t(0, \xi) = \frac{b(0)}{2} \hat{f}(\xi) + \hat{g}(\xi).$$

Let us introduce the symbol

$$\langle\xi\rangle_{m(t)} = \sqrt{|\xi|^2 + m(t)^2}.$$

We introduce the micro-energy $V = (\langle\xi\rangle_{m(t)}\hat{v}, D_t\hat{v})^T$. Since $D_t = \frac{1}{i}\partial_t$ we obtain that V satisfies the system of first order

$$D_t V = A(t,\xi)V = \left[\begin{pmatrix} 0 & \langle\xi\rangle_{m(t)} \\ \langle\xi\rangle_{m(t)} & 0 \end{pmatrix} + \begin{pmatrix} \frac{D_t\langle\xi\rangle_{m(t)}}{\langle\xi\rangle_{m(t)}} & 0 \\ -\frac{b(t)^2+2b'(t)}{4\langle\xi\rangle_{m(t)}} & 0 \end{pmatrix}\right] V. \tag{25}$$

We prove the following result.

Lemma 2 *The fundamental solution $\mathscr{E} = \mathscr{E}(t,s,\xi)$ to the system* (25) *satisfies the following estimates for each $t \geq s \geq 0$ and $\xi \in \mathbb{R}^n$:*

$$\|\mathscr{E}(t,s,\xi)\| \lesssim \left(\frac{\langle\xi\rangle_{m(t)}}{\langle\xi\rangle_{m(s)}}\right)^{\pm\frac{1}{2}} \exp\Big(\int_s^t cb(\tau)d\tau\Big),$$

where $c < \frac{1}{2}$, and $\pm$ stays for $+$ or, respectively, for $-$ if $m = m(t)$ is increasing or, respectively, decreasing.

Proof We introduce

$$P = \begin{pmatrix} 1 & -1 \\ 1 & 1 \end{pmatrix}, \quad P^{-1} = \frac{1}{2}\begin{pmatrix} 1 & 1 \\ -1 & 1 \end{pmatrix},$$

and we set $V^{(0)} := P^{-1}V$. Thus, we arrive at the system

$$D_t V^{(0)} = [D(t,\xi) + R(t,\xi)]V^{(0)},$$

where

$$D(t,\xi) = \begin{pmatrix} \langle\xi\rangle_{m(t)} & 0 \\ 0 & -\langle\xi\rangle_{m(t)} \end{pmatrix};$$

$$R(t,\xi) = \frac{1}{2}\frac{D_t\langle\xi\rangle_{m(t)}}{\langle\xi\rangle_{m(t)}}\begin{pmatrix} 1 & -1 \\ -1 & 1 \end{pmatrix} - \frac{b^2(t) + 2b'(t)}{8\langle\xi\rangle_{m(t)}}\begin{pmatrix} 1 & -1 \\ 1 & -1 \end{pmatrix}.$$

For a sufficiently large $t_0 > 0$, due to Hypothesis (9) and (12), there exists $c < \frac{1}{2}$ such that for each $\tau > t_0$ it holds:

$$\left|\frac{b^2(\tau) + 2b'(\tau)}{8m(\tau)}\right| < cb(\tau).$$

Therefore, using $\langle\xi\rangle_{m(t)} > m(t)$ we get

$$\int_s^t \frac{b^2(\tau)}{8\langle\xi\rangle_{m(\tau)}}d\tau + \int_s^t \frac{|b'(\tau)|}{4\langle\xi\rangle_{m(\tau)}}d\tau < \int_s^t cb(\tau)d\tau.$$

This concludes the proof for $s \geq t_0$. For $t \leq t_0$, it is clear that $\mathscr{E} = \mathscr{E}(t, s, \xi)$ is bounded, so that the proof of the lemma follows by combining the two cases.

3.1 Proof of Theorem **1** *and Lemma* **1**

To prove Theorem 1 and Lemma 1, we first notice that the solution to (19) verifies the pointwise estimates

$$|\hat{u}(t,\xi)| \lesssim \frac{\lambda(s)}{\lambda(t)} \frac{\|\mathscr{E}(t,s,\xi)\|}{\langle\xi\rangle_{m(t)}} (\langle\xi\rangle_{m(s)}|\hat{f}(s,\xi)| + |\hat{g}(s,\xi)|),$$
$$|\hat{u}_t(t,\xi)| \lesssim \frac{\lambda(s)}{\lambda(t)} \|\mathscr{E}(t,s,\xi)\| \, (\langle\xi\rangle_{m(s)}|\hat{f}(s,\xi)| + |\hat{g}(s,\xi)|).$$

In the second estimate, we used $\hat{u}_t = (\lambda(s)/\lambda(t))(\hat{v}_t - (b(t)/2)\,\hat{v})$, and the property $b(t) < \langle\xi\rangle_{m(t)}$ for $t > t_0$ sufficiently large.
It is straight-forward to prove the following lemma:

Lemma 3 *Let $k \in \mathbb{R}$. Then, for each $\epsilon > 0$, for all $t \geq s \geq 0$ it holds*

$$\left(\frac{m(s)}{m(t)}\right)^k \lesssim \left(\frac{\lambda(t)}{\lambda(s)}\right)^\epsilon. \tag{26}$$

Proof *Notice that (26) is trivially satisfied if $k\,m'(t) > 0$.*
Let us consider the case $k\,m'(t) < 0$. It holds

$$\left(\frac{m(s)}{m(t)}\right)^k = \exp\Big(-k\int_s^t \frac{m'(\tau)}{m(\tau)}d\tau\Big).$$

Thus, it is sufficient to prove that there exists $t_0 > 0$ such that for each $\tau > t_0$ it holds

$$-k\frac{m'(\tau)}{m(\tau)} \leq \frac{\epsilon}{2}b(\tau).$$

But this is true for t_0 sufficiently large, due to hypothesis (11) *and the property* $tb(t) \to \infty$.

By Lemma 2, using (26), we get the following pointwise estimates for the solution

$$|\hat{u}(t,\xi)| \lesssim \left(\frac{\lambda(s)}{\lambda(t)}\right)^{\delta} \Big(|\hat{f}(s,\xi)| + \frac{1}{\langle\xi\rangle_{m(s)}}|\hat{g}(s,\xi)|\Big),$$

$$|\hat{u}_t(t,\xi)| \lesssim \left(\frac{\lambda(s)}{\lambda(t)}\right)^{\delta} (\langle\xi\rangle_{m(s)}|\hat{f}(s,\xi)| + |\hat{g}(s,\xi)|),$$

for some $\delta > 0$.

The proofs of Theorem 1 and Lemma 1 follow immediately. In particular, we notice $\langle\xi\rangle_{m(s)} > m(s)$ and, respectively $|\xi| < \langle\xi\rangle_{m(s)}$, to prove (20) and, respectively (21).

4 Proof of Theorem 2

A function $u = u(t,x)$ solves the Cauchy problem (1) in a suitable space $X = X(T)$ of Sobolev solutions if and only if

$$u(t,x) = u^{\text{lin}}(t,x) + (Nu)(t,x),$$

where u^{lin} is the solution to the Cauchy problem (3), and

$$Nu(t,x) = \int_0^t \Phi(t,s,\cdot) *_{(x)} h(u(s,\cdot))(x)ds,$$

in X, where by $\Phi(t,s,\cdot)*_{(x)}h(u(s,\cdot))(x)$ is the solution to the Cauchy problem (19) with $f = 0$ and $g(s,\cdot) = h(u(s,\cdot))$.

To prove Theorem 2, we will rely on a standard contraction argument in the solution space $\{\mathscr{C}^1([0,T),H^1) \times \mathscr{C}([0,T),L^2)\}_{T>0}$, equipped with a suitable norm, defined accordingly to the decay estimates for the solutions to the corresponding linear Cauchy problems with vanishing right-hand side obtained in Theorem 1.

For any $T > 0$, we define the Banach spaces

$$X_0(T) = C([0,T],H^1), \qquad X(T) = X_0(T) \cap C^1([0,T],L^2).$$

We will fix a suitable norm on $X(T)$ such that

$$\|u^{\text{lin}}\|_{X(T)} \leq C\|(f,g)\|_{H^1\times L^2}. \tag{27}$$

Then, we will prove that

$$\|Nu\|_{X(T)} \leq C\|u\|^p_{X_0(T)}, \tag{28}$$

$$\|Nu - Nv\|_{X(T)} \leq C\|u - v\|_{X(T)}\big(\|u\|^{p-1}_{X_0(T)} + \|v\|^{p-1}_{X_0(T)}\big), \tag{29}$$

with a constant $C > 0$, independent of T. From condition (28) it follows that N maps $X_0(T)$ into $X(T)$.

Due to the inequalities (28) and (29), we can apply the Banach's fixed point theorem to prove that there exists a uniquely determined solution to Cauchy problem (1), in $X(T)$, provided that $\|(f, g)\|_{H^1\times L^2}$ in (27) is sufficiently small. Since the constants in (27), (28), and (29) do not depend on T, the solution is global (in time).

To prove Theorem 2, we equip $X_0(T)$ and $X(T)$ with the norms

$$\|u\|_{X_0(T)} := \sup_{0\leq\tau\leq T} \lambda(\tau)^{2\delta}[\|u(\tau,\cdot)\|_{L^2} + \|\nabla u(\tau,\cdot)\|_{L^2}],$$

$$\|u\|_{X(T)} := \sup_{0\leq\tau\leq T} \lambda(\tau)^{2\delta}[\|u(\tau,\cdot)\|_{L^2} + \|\nabla u(\tau,\cdot)\|_{L^2} + \|u_\tau(\tau,\cdot)\|_{L^2}],$$

where δ is given by Theorem 1.

By Theorem 1, we conclude (27). As a consequence of Gagliardo-Nirenberg inequality, any function $u \in X_0(T)$ verifies the inequality

$$\|u(\tau,\cdot)\|_{L^q} \leq C\,\lambda(\tau)^{-2\delta}\,\|u\|_{X_0(T)}, \tag{30}$$

for any $\tau \in [0, T]$, for any $q \in [2, \infty)$ if $n = 1, 2$ and for any $q \in [2, 2n/(n-2)]$ if $n \geq 3$.

Let $j, \ell = 0, 1$ with $j + \ell \leq 1$. Then

$$\Big\|\nabla^j \partial_t^\ell Nu(t,\cdot)\Big\|_{L^2} \leq \int_0^t \Big\|\nabla^j \partial_t^\ell \Phi(t,s,\cdot) *_{(x)} h(u(s,\cdot))\Big\|_{L^2} ds.$$

By Lemma 1, we get

$$\Big\|\nabla^j \partial_t^\ell Nu(t,\cdot)\Big\|_{L^2} \leq C \int_0^t \frac{1}{m(s)^{1-j-\ell}} \left(\frac{\lambda(s)}{\lambda(t)}\right)^{2\delta} \|h(u(s,\cdot))\|_{L^2} ds. \tag{31}$$

Using $|h(u)| \lesssim |u|^p$ and (30), noticing that $2p \leq 2n/(n-2)$ if $n \geq 3$, we may estimate

$$\|h(u(s,\cdot))\|_{L^2} \lesssim \|u(s,\cdot)\|^p_{L^{2p}} \lesssim \lambda(s)^{-2\delta p}\|u\|^p_{X_0(T)}. \tag{32}$$

Thus, using condition (12), we can estimate the right-hand side of (31) as follows:

$$\begin{aligned}
&\|u\|^p_{X_0(t)} \frac{1}{\lambda(t)^{2\delta}} \int_0^t \frac{\lambda(s)^{2\delta(1-p)}}{b(s)^{1-j-\ell}} ds \\
&\lesssim \|u\|^p_{X_0(t)} \frac{1}{\lambda(t)^{2\delta}} \int_0^t b(s) \exp\Big(-c \int_0^s b(\tau) d\tau \Big) ds,
\end{aligned}$$

for some constant $c > 0$. We obtain the integrability by using the change of variables $r = \int_0^s b(\tau) d\tau$.
Let $c > 0$. The last inequality follows by integrating the following estimate:

$$-\delta(p-1)b(s) - (2-j-\ell)\frac{b'(s)}{b(s)} \leq -cb(s).$$

This is equivalent to

$$(c - \delta(p-1))b(s)^2 \leq (2-j-\ell)b'(s).$$

This inequality becomes trivial, for c sufficiently small, if $b = b(t)$ is an increasing function. If $b = b(t)$ is a decreasing function, then this inequality follows, for c small enough, by taking account of the condition $|b'(t)| \sim o(b(t)^2)$. This concludes the proof of the estimate (28).
We proceed similarly to prove (29). In particular, we replace (32) by

$$\begin{aligned}
\|h(u(s,\cdot)) - h(v(s,\cdot))\|_{L^2} &\lesssim \Big\| |u(s,\cdot) - v(s,\cdot)| \left(|u(s,\cdot)|^{p-1} + |v(s,\cdot)|^{p-1} \right) \Big\|_{L^2} \\
&\lesssim \|u(s,\cdot) - v(s,\cdot)\|_{L^{2p}} \Big\| |u(s,\cdot)|^{p-1} + |v(s,\cdot)|^{p-1} \Big\|_{L^{2p'}} \\
&\lesssim \lambda(s)^{-2\delta p} \|u - v\|_{X_0(T)} \left(\|u\|^{p-1}_{X_0(T)} + \|v\|^{p-1}_{X_0(T)} \right).
\end{aligned}$$

This concludes the proof.

References

1. M. D'Abbicco, Small data solutions for semilinear wave equations with effective damping. Discret. Contin. Dyn. Syst. 183–191, Supplement (2013)
2. M. D'Abbicco, The threshold of effective damping for semilinear wave equations. Math. Methods Appl. Sci. **38**(6), 1032–1045 (2015). http://dx.doi.org/10.1002/mma.3126
3. M. D'Abbicco, Asymptotics for damped evolution operators with mass-like terms. Complex Anal. Dyn. Syst. VI Contemp. Math. **653**, 93–116 (2015). http://dx.doi.org/10.1090/conm/653/13181
4. M. D'Abbicco, S. Lucente, A modified test function method for damped wave equations. Adv. Nonlinear Stud. **13**, 867–892 (2013)

5. M. D'Abbicco, S. Lucente, NLWE with a special scale-invariant damping in odd space dimension. Discret. Contin. Dyn. Syst. AIMS Proc. 312–319 (2015). http://dx.doi.org/10.3934/proc.2015.0312
6. M. D'Abbicco, S. Lucente, M. Reissig, Semilinear wave equations with effective damping. Chinese Ann. Math. **34B**(3), 345–380 (2013). http://dx.doi.org/10.1007/s11401-013-0773-0
7. M. D'Abbicco, S. Lucente, M. Reissig, A shift in the critical exponent for semilinear wave equations with a not effective damping. J. Differ. Equ. **259**, 5040–5073 (2015). http://dx.doi.org/10.1016/j.jde.2015.06.018
8. M. D'Abbicco, G. Girardi, M. Reissig, A scale of critical exponents for semilinear waves with time-dependent damping and mass terms. Nonlinear Anal. **179**, 15–40 (2019). https://doi.org/10.1016/j.na.2018.08.006
9. W.N. do Nascimento, A. Palmieri, M. Reissig, Semi-linear wave models with power non-linearity and scale invariant time-dependent mass and dissipation. Math. Nachr. **290**, 1779–1805 (2017)
10. V. Georgiev, H. Lindblad, C.D. Sogge, Weighted Strichartz estimates and global existence for semilinear wave equations. Am. J. Math. **119**, 1291–1319 (1997)
11. R. Ikehata, M. Ohta, Critical exponents for semilinear dissipative wave equations in $\mathbb{R}^N$. J. Math. Anal. Appl. **269**, 87–97 (2002)
12. M. Ikeda, M. Sobajima, Life-span of solutions to semilinear wave equation with time-dependent critical damping for specially localized initial data. M. Math. Ann. (2018). https://doi.org/10.1007/s00208-018-1664-1
13. M. Ikeda, Y. Wakasugi, A remark on the global existence for the semi-linear damped wave equation in the overdamping case. arXiv:1708.08044 (2017)
14. R. Ikehata, Y. Mayaoka, T. Nakatake, Decay estimates of solutions for dissipative wave equations in $\mathbb{R}^N$ with lower power nonlinearities. J. Math. Soc. Jpn. **56**(2), 365–373 (2004)
15. F. John, Blow-up of solutions of nonlinear wave equations in three space dimensions. Manuscripta Math. **28**, 235–268 (1979)
16. T.T. Li, Y. Zhou, Breakdown of solutions to $\Box u + u_t = |u|^{1+\alpha}$. Discret. Contin. Dyn. Syst. **1**, 503–520 (1995)
17. N.A. Lai, H. Takamura, K. Wakasa, Blow-up for semilinear wave equations with the scale invariant damping and super-Fujita exponent. J. Differ. Equ. **263**, 5377–5394 (2017)
18. J. Lin, K. Nishihara, J. Zhai, Critical exponent for the semilinear wave equation with time-dependent damping. Discret. Contin. Dyn. Syst. **32**(12), 4307–4320 (2012). http://dx.doi.org/10.3934/dcds.2012.32.4307
19. A. Matsumura, On the asymptotic behavior of solutions of semi-linear wave equations. Publ. RIMS. **12**, 169–189 (1976)
20. K. Nishihara, Asymptotic behavior of solutions to the semilinear wave equation with time-dependent damping. Tokyo J. Math. **34**, 327–343 (2011)
21. W. Strauss, Nonlinear scattering theory at low energy. J. Funct. Anal. **41**, 110–133 (1981)
22. G. Todorova, B. Yordanov, Critical exponent for a nonlinear wave equation with damping. J. Differ. Equ. **174**, 464–489 (2001)
23. Y. Wakasugi, Critical exponent for the semilinear wave equation with scale invariant damping, in *Fourier Analysis* ed. by M. Ruzhansky, V. Turunen (Trends in Mathematics, Springer, Basel, 2014), pp. 375–390
24. Y. Wakasugi, Scaling variables and asymptotic profiles for the semilinear damped wave equation with variable coefficients. J. Math. Anal. Appl. **447**, 452–487 (2017)
25. J. Wirth, Wave equations with time-dependent dissipation II. Effective dissipation. J. Differ. Equ. **232**, 74–103 (2007)
26. Q.S. Zhang, A blow-up result for a nonlinear wave equation with damping: the critical case. C. R. Acad. Sci. Paris Sér. I Math. **333**, 109–114 (2001)

Wave-Like Blow-Up for Semilinear Wave Equations with Scattering Damping and Negative Mass Term

Ning-An Lai, Nico Michele Schiavone, and Hiroyuki Takamura

Abstract In this paper we establish blow-up results and lifespan estimates for semilinear wave equations with scattering damping and negative mass term for subcritical power, which are the same as that of the corresponding problem without mass term, and also the same as that of the corresponding problem without both damping and mass term. For this purpose, we have to use the comparison argument twice, due to the damping and mass term, in additional to a key multiplier. Finally, we get the desired results by an iteration argument.

1 Introduction

In this paper, we consider the Cauchy problem for semilinear wave equations with scattering damping and negative mass term

$$\begin{cases} u_{tt} - \Delta u + \dfrac{\mu_1}{(1+t)^\beta} u_t - \dfrac{\mu_2}{(1+t)^{\alpha+1}} u = |u|^p, & \text{in } \mathbf{R}^n \times [0, T), \\ u(x,0) = \varepsilon f(x), \quad u_t(x,0) = \varepsilon g(x), & x \in \mathbf{R}^n, \end{cases} \tag{1}$$

where $\mu_1, \mu_2 > 0, \alpha > 1, \beta > 1, n \in \mathbf{N}$ and $\varepsilon > 0$ is a "small" parameter.

N.-A. Lai
Department of Mathematics, Institute of Nonlinear Analysis, Lishui University, Lishui City, China
e-mail: ninganlai@lsu.edu.cn

N. M. Schiavone
Department of Mathematics, Università di Roma "La Sapienza", Rome, Italy
e-mail: schiavone@mat.uniroma1.it

H. Takamura (✉)
Mathematical Institute, Tohoku University, Aoba, Japan
e-mail: hiroyuki.takamura.a1@tohoku.ac.jp

M. D'Abbicco et al. (eds.), *New Tools for Nonlinear PDEs and Application*, Trends in Mathematics, https://doi.org/10.1007/978-3-030-10937-0_8

We call the term $\mu_1 u_t/(1+t)^\beta$ ($\beta > 1$) scattering damping, due to the reason that the solution of the following Cauchy problem

$$\begin{cases} u^0_{tt} - \Delta u^0 + \dfrac{\mu}{(1+t)^\beta} u^0_t = 0, & \text{in } \mathbf{R}^n \times [0,\infty), \\ u^0(x,0) = u_1(x), \quad u^0_t(x,0) = u_2(x), & x \in \mathbf{R}^n, \end{cases} \tag{2}$$

scatters to that of the free wave equation when $\beta > 1$ and $t \to \infty$. In fact, according to the works of Wirth [22–24], we may classify the damping for different values of β into four cases, as shown in the next table.

Range of β	Classification
$\beta \in (-\infty, -1)$	Overdamping
$\beta \in [-1, 1)$	Effective
$\beta = 1$	Scaling invariant if $\mu \in (0, 1) \Rightarrow$ non-effective
$\beta \in (1, \infty)$	Scattering

If we come to the nonlinear problem with power nonlinearity, thus

$$\begin{cases} u_{tt} - \Delta u + \dfrac{\mu}{(1+t)^\beta} u_t = |u|^p, & \text{in } \mathbf{R}^n \times [0,\infty), \\ u(x,0) = u_1(x), \quad u_t(x,0) = u_2(x), & x \in \mathbf{R}^n, \end{cases} \tag{3}$$

we want to determine the long time behaviour of the solution according to the different value of p, n and even μ. Ikeda and Wakasugi [9] proved global existence for (3) for all $p > 1$ when $\beta < -1$. For $\beta \in [-1, 1)$, due to the work [3, 5–7, 11, 13, 21, 26], we know that problem (3) admits a critical power $p_F(n) := 1 + 2/n$ (Fujita power), which means that for $p \in (1, p_F(n)]$ the solution will blow up in a finite time, while for $p \in (p_F(n), \infty)$ we have global existence. Obviously, in this case the critical is exactly the same as that of the Cauchy problem of semilinear heat equation

$$u_t - \Delta u = |u|^p,$$

and so we call it admits "heat-like" behaviour.

For the case $\beta = 1$ in (3), we say that the damping is scale invariant, due to the reason that the equation in the corresponding linear problem (2) is invariant under the following scaling transformation

$$\widetilde{u^0}(x,t) := u^0(\sigma x, \sigma(1+t) - 1), \ \sigma > 0.$$

It is a bit sophisticated for the scale invariant nonlinear problem (3), since the size of the positive constant μ will also have an effect on the long time behaviour of the solution. Generally speaking, according to the known results [1, 2, 4, 8, 12, 18–20], it is believed that if μ is large enough, then the critical power is related to the Fujita power, while if μ is relatively small, then the critical power is related to the Strauss power, i.e. $p_S(n)$, which is denoted to be the positive root of the following quadratic equation

$$\gamma(p,n) := 2 + (n+1)p - (n-1)p^2 = 0,$$

and which is also the critical power of the small data Cauchy problem of the semilinear wave equation

$$u_{tt} - \Delta u = |u|^p.$$

It means that for relatively small μ we have "wave-like" behaviour. Unfortunately, we are not clear of the exact threshold determined by the value μ between the "heat-like" and "wave-like" phenomenon till now.

For the scattering case ($\beta > 1$), one expects that problem (3) admits the long time behaviour as that of the corresponding problem without damping. In [10], Lai and Takamura obtained the blow-up results for

$$1 < p < \begin{cases} p_S(n) & \text{for } n \geq 2, \\ \infty & \text{for } n = 1 \end{cases}$$

and the upper bound of the lifespan estimate

$$T \leq C\varepsilon^{-2p(p-1)/\gamma(p,n)}.$$

What is more, when $n = 1, 2$ and

$$\int_{\mathbf{R}^n} g(x)dx \neq 0,$$

they established an improved upper bound of the lifespan for $1 < p < 2, n = 2$ and $p > 1, n = 1$. However, it remains to determine the exact critical power for (3) with $\beta > 1$.

Recently, the small data Cauchy problem for semilinear wave equation with scale-invariant damping and mass and power non-linearity, i.e.,

$$\begin{cases} u_{tt} - \Delta u + \dfrac{\mu_1}{1+t}u_t + \dfrac{\mu_2^2}{(1+t)^2}u = |u|^p, & \text{in } \mathbf{R}^n \times [0,\infty), \\ u(x,0) = u_1(x), \quad u_t(x,0) = u_2(x), & x \in \mathbf{R}^n, \end{cases} \tag{4}$$

attracts more and more attention. Denote

$$\delta := (\mu_1 - 1)^2 - 4\mu_2^2. \tag{5}$$

Then in [14] and [15] a blow-up result was established for

$$1 < p \le p_F\left(n + \frac{\mu_1 - 1 - \sqrt{\delta}}{2}\right)$$

assuming $\delta \ge 0$, by using two different approaches. Furthermore, in [14] they improved the result for $\delta = 1$ to

$$1 < p \le \max\left\{p_S(n+\mu_1),\ p_F\left(n + \frac{\mu_1}{2} - 1\right)\right\}.$$

Recently, Palmieri and Reissig [16] generalized the blow-up result for $n \ge 1$ and $\delta \in (0, 1]$ to the following power:

$$\begin{cases} p < p_{\mu_1,\mu_2}(n) := \max\left\{p_S(n+\mu_1),\ p_F\left(n + \frac{\mu_1}{2} - \frac{\sqrt{\delta}}{2}\right)\right\}, \\ p = p_{\mu_1,\mu_2}(n) = p_F\left(n + \frac{\mu_1}{2} - \frac{\sqrt{\delta}}{2}\right), \\ p = p_{\mu_1,\mu_2}(n) = p_S(n+\mu_1), \quad \text{for } n = 2. \end{cases}$$

We note that a transform by $v := (1+t)^{\mu_1/2}u$ changes the equation in (4) into

$$v_{tt} - \Delta v + \frac{1-\delta}{4(1+t)^2}v = \frac{|v|^p}{(1+t)^{\mu_1(p-1)/2}},$$

so that the assumption of $\delta \in (0, 1]$ implies the non-negativeness of the mass term in this equation.

In this paper, we are going to study the small data Cauchy problem of semilinear wave equations with power nonlinearity, scattering damping and mass term with negative sign, thus, problem (1). Blow-up results and lifespan estimates will be established for $1 < p < p_S(n)$, which are the same as that in the work [10]. We could say that we experience a double phenomenon of scattering, due to the damping term and the mass term. For the proof, we will borrow the idea from [10], by introducing a key multiplier to absorb the damping term and establishing an iteration frame. However, we have to deal with the mass term. Due to the negative sign, we use a comparison argument to eliminate the effect from the mass term. Although the calculations in this work hold for any mass exponent $\alpha \in \mathbf{R}$, we suppose that it satisfies $\alpha > 1$ because otherwise we have shorter lifespan estimates due to the effect of the negative mass term. This analysis will appear in our forthcoming paper.

2 Main Result

Before the statement of our main results, we first denote the energy and weak solutions of problem (1).

Definition 1 We say that u is an energy solution of (1) over $[0, T)$ if

$$u \in C([0,T), H^1(\mathbf{R}^n)) \cap C^1([0,T), L^2(\mathbf{R}^n)) \cap C((0,T), L^p_{loc}(\mathbf{R}^n)) \tag{6}$$

satisfies $u(x,0) = \varepsilon f(x)$ in $H^1(\mathbf{R}^n)$ and $u_t(x,0) = \varepsilon g(x)$ in $L^2(\mathbf{R}^n)$, and

$$\begin{aligned}
&\int_{\mathbf{R}^n} u_t(x,t)\phi(x,t)dx - \int_{\mathbf{R}^n} \varepsilon g(x)\phi(x,0)dx \\
&+ \int_0^t ds \int_{\mathbf{R}^n} \{-u_t(x,s)\phi_t(x,s) + \nabla u(x,s)\cdot\nabla\phi(x,s)\}\,dx \\
&+ \int_0^t ds \int_{\mathbf{R}^n} \frac{\mu_1}{(1+s)^\beta} u_t(x,s)\phi(x,s)dx - \int_0^t ds \int_{\mathbf{R}^n} \frac{\mu_2}{(1+s)^{\alpha+1}} u(x,s)\phi(x,s)dx \\
&= \int_0^t ds \int_{\mathbf{R}^n} |u(x,s)|^p \phi(x,s)dx
\end{aligned} \tag{7}$$

with any test function $\phi \in C_0^\infty(\mathbf{R}^n \times [0,T))$ and for any $t \in [0,T)$.

Employing the integration by parts in the above equality and letting $t \to T$, we got the definition of the weak solution of (1), that is

$$\begin{aligned}
&\int_{\mathbf{R}^n\times[0,T)} u(x,s)\Bigg\{\phi_{tt}(x,s) - \Delta\phi(x,s) - \frac{\partial}{\partial s}\left(\frac{\mu_1}{(1+s)^\beta}\phi(x,s)\right) \\
&\quad - \frac{\mu_2}{(1+s)^{\alpha+1}}\phi(x,s)\Bigg\}dxds \\
&= \int_{\mathbf{R}^n} \mu_1\varepsilon f(x)\phi(x,0)dx - \int_{\mathbf{R}^n} \varepsilon f(x)\phi_t(x,0)dx + \int_{\mathbf{R}^n} \varepsilon g(x)\phi(x,0)dx \\
&\quad + \int_{\mathbf{R}^n\times[0,T)} |u(x,s)|^p\phi(x,s)dxds.
\end{aligned} \tag{8}$$

Definition 2 As in the introduction, set

$$\gamma(p,n) := 2 + (n+1)p - (n-1)p^2$$

and, for $n \geq 2$, define $p_S(n)$ the positive root of the quadratic equation $\gamma(p,n) = 0$, the so-called Strauss exponent, that is

$$p_S(n) = \frac{n+1+\sqrt{n^2+10n-7}}{2(n-1)}.$$

Note that if $n = 1$, then $\gamma(p, 1) = 2 + 2p$ and we can set $p_S(1) := +\infty$.

Now we announce our main results.

Theorem 1 *Let $n = 1$ and $p > 1$, or $n \geq 2$ and $1 < p < p_S(n)$. Assume that both $f \in H^1(\mathbf{R}^n)$ and $g \in L^2(\mathbf{R}^n)$ are non-negative, and at least one of them does not vanish identically. Suppose that u is an energy solution of* (1) *on* $[0, T)$ *that satisfies*

$$\operatorname{supp} u \subset \{(x, t) \in \mathbf{R}^n \times [0, \infty) \colon |x| \leq t + R\} \tag{9}$$

with some $R \geq 1$. Then, there exists a constant $\varepsilon_0 = \varepsilon_0(f, g, n, p, \mu_1, \beta, R) > 0$ which is independent of μ_2, such that T has to satisfy

$$T \leq C\varepsilon^{-2p(p-1)/\gamma(p,n)} \tag{10}$$

for $0 < \varepsilon \leq \varepsilon_0$, where C is a positive constant independent of ε.

In low dimensions ($n = 1, 2$), with some additional hypothesis, we may have improvements on the lifespan estimates as follows.

Theorem 2 *Let $n = 2$ and $1 < p < 2$. Assume that both $f \in H^1(\mathbf{R}^2)$ and $g \in L^2(\mathbf{R}^2)$ are non-negative and that g does not vanish identically. Then the lifespan estimate* (10) *is replaced by*

$$T \leq C\varepsilon^{-(p-1)/(3-p)}. \tag{11}$$

Theorem 3 *Let $n = 1$ and $p > 1$. Assume that both $f \in H^1(\mathbf{R}^1)$ and $g \in L^2(\mathbf{R}^1)$ are non-negative and that g does not vanish identically. Then the lifespan estimate* (10) *is replaced by*

$$T \leq C\varepsilon^{-(p-1)/2}. \tag{12}$$

Theorem 4 *Let $n = p = 2$. Suppose that $\alpha \leq \beta$ and*

$$\mu_2 \geq \begin{cases} \frac{\beta\mu_1}{2} & \text{if } \alpha = \beta, \\ \frac{\beta\mu_1}{2} \frac{\beta-1}{2\beta-\alpha-1} \left(4\frac{\mu_1^2}{\mu_2}\frac{\beta-\alpha}{\beta-1}\right)^{\frac{\beta-\alpha}{2\beta-\alpha-1}} & \text{if } \alpha < \beta. \end{cases} \tag{13}$$

Assume that $f \equiv 0$ and $g \in C^2(\mathbf{R}^2)$ is non-negative and does not vanish identically. Suppose also that u is a classical solution of (1) *on* $[0, T)$ *with the support property* (9)*. Then, T satisfies*

$$T \leq Ca(\varepsilon) \tag{14}$$

where $a = a(\varepsilon)$ is a number satisfying

$$a^2\varepsilon^2 \log(1 + a) = 1. \tag{15}$$

Remark 1 In Theorem 1, we require that at least one of the initial data does not vanish identically, which is weaker than that in the corresponding result (Theorem 2.1) in [10].

Remark 2 Observe that:

- Equation (11) is stronger than (10) by the fact that $1 < p < 2$ is equivalent to

$$\frac{p-1}{3-p} < \frac{2p(p-1)}{\gamma(p,2)};$$

- Equation (12) is stronger than (10) by the fact that $p > 1$ is equivalent to

$$\frac{p-1}{2} < \frac{2p(p-1)}{\gamma(p,1)};$$

- Equation (14) is stronger than (10) by the fact that when $n = p = 2$

$$a(\varepsilon) < \varepsilon^{-1} = \varepsilon^{-2\cdot 2(2-1)/\gamma(2,2)}$$

  for sufficiently small ε.

3 Lower Bound for Derivative of the Functional

Following the idea in [10], we introduce the multiplier

$$m(t) := \exp\left(\mu_1 \frac{(1+t)^{1-\beta}}{1-\beta}\right). \tag{16}$$

Clearly

$$1 \geq m(t) \geq m(0) > 0 \quad \text{for } t \geq 0. \tag{17}$$

Moreover, let us define the functional

$$F_0(t) := \int_{\mathbf{R}^n} u(x,t)dx,$$

then

$$F_0(0) = \varepsilon \int_{\mathbf{R}^n} f(x)dx, \quad F_0'(0) = \varepsilon \int_{\mathbf{R}^n} g(x)dx$$

are non-negative due to the hypothesis of positiveness on the initial data. Our final target is to establish a lower bound for $F_0(t)$.

Let us start finding the lower bound of the derivative of the functional, i.e., $F_0'(t)$. Due to (9), choosing the test function $\phi = \phi(x,s)$ in (7) to satisfy $\phi \equiv 1$ in $\{(x,s) \in \mathbf{R}^n \times [0,t] : |x| \le s+R\}$, we get

$$\int_{\mathbf{R}^n} u_t(x,t)dx - \int_{\mathbf{R}^n} u_t(x,0)dx + \int_0^t ds \int_{\mathbf{R}^n} \frac{\mu_1}{(1+s)^\beta} u_t(x,s)dx$$
$$= \int_0^t \int_{\mathbf{R}^n} \frac{\mu_2}{(1+s)^{\alpha+1}} u(x,s)dx + \int_0^t ds \int_{\mathbf{R}^n} |u(x,s)|^p dx,$$

which yields by taking derivative with respect to t

$$F_0''(t) + \frac{\mu_1}{(1+t)^\beta} F_0'(t) = \frac{\mu_2}{(1+t)^{\alpha+1}} F_0(t) + \int_{\mathbf{R}^n} |u(x,t)|^p dx. \tag{18}$$

Here we note that (18) can be established by regularity assumption on the solution. Multiplying both sides of (18) with $m(t)$ yields

$$\{m(t)F_0'(t)\}' = m(t)\frac{\mu_2}{(1+t)^{\alpha+1}} F_0(t) + m(t)\int_{\mathbf{R}^n} |u(x,t)|^p dx. \tag{19}$$

Integrating the above equality over $[0,t]$ we get

$$\begin{aligned} F_0'(t) = {} & \frac{m(0)}{m(t)} F_0'(0) + \frac{1}{m(t)} \int_0^t m(s) \frac{\mu_2}{(1+s)^{\alpha+1}} F_0(s)ds \\ & + \frac{1}{m(t)} \int_0^t m(s)ds \int_{\mathbf{R}^n} |u(x,s)|^p dx. \end{aligned} \tag{20}$$

To get the lower bound for F_0', we need the positiveness of F_0, and this can be obtained by a comparison argument. However, since we assume that at least one of the initial data does not vanish identically, we have to consider the following two cases.

Case 1: $f \ge 0 (\not\equiv 0)$, $g \ge 0$. This means that $F_0(0) > 0$, $F_0'(0) \ge 0$. By the continuity of F_0, it is positive at least for small time. Suppose that t_0 is the smallest zero point of F_0, such that $F_0 > 0$ in $[0, t_0)$. Then, integrating (20) over this interval we have

$$\begin{aligned} 0 = F_0(t_0) = {} & F_0(0) + m(0)F_0'(0) \int_0^{t_0} \frac{ds}{m(s)} \\ & + \int_0^{t_0} \frac{ds}{m(s)} \int_0^s m(r) \frac{\mu_2}{(1+r)^{\alpha+1}} F_0(r)dr \\ & + \int_0^{t_0} \frac{ds}{m(s)} \int_0^s m(r)dr \int_{\mathbf{R}^n} |u(x,r)|^p dx > 0, \end{aligned}$$

which leads to a contradiction, and hence $F(t)$ is positive all the time.

Case 2: $f \geq 0, g \geq 0 (\not\equiv 0)$. This imply that $F_0(0) \geq 0$, $F_0'(0) > 0$. We apply the same argument as in the first case to F_0'. Suppose that t_0 is the smallest zero point of F_0', such that F_0' is positive on the interval $[0, t_0)$. Therefore F_0 is strictly monotone increasing on the same interval, and hence positive due to $F_0(0) \geq 0$. Letting $t = t_0$ in (20), we again come to a contradiction. Therefore F_0' is always strictly positive, and hence $F_0(t) > 0$ holds for all $t > 0$.

Coming back to (20), using the positivity of F_0, the boundedness of $m(t)$ and that $F_0'(0) \geq 0$, we obtain the lower bound for F_0' as

$$F_0'(t) \geq m(0) \int_0^t \int_{\mathbf{R}^n} |u(x,s)|^p dxds \quad \text{for } t \geq 0. \tag{21}$$

4 Lower Bound for the Weighted Functional

Set

$$F_1(t) := \int_{\mathbf{R}^n} u(x,t)\psi_1(x,t)dx,$$

where ψ_1 is the test function introduced by Yordanov and Zhang [25]

$$\psi_1(x,t) := e^{-t}\phi_1(x), \quad \phi_1(x) := \begin{cases} \displaystyle\int_{S^{n-1}} e^{x\cdot\omega} dS_\omega & \text{for } n \geq 2, \\ e^x + e^{-x} & \text{for } n = 1. \end{cases}$$

Lemma 1 (Inequality (2.5) of Yordanov and Zhang [25])

$$\int_{|x| \leq t+R} [\psi_1(x,t)]^{p/(p-1)} dx \leq C(1+t)^{(n-1)\{1-p/(2(p-1))\}}, \tag{22}$$

where $C_1 = C_1(n,p,R) > 0$.

Next we aim to establish the lower bound for F_1. From the definition of energy solution (7), we have that

$$\begin{aligned} &\frac{d}{dt}\int_{\mathbf{R}^n} u_t(x,t)\phi(x,t)dx + \int_{\mathbf{R}^n} \{-u_t(x,t)\phi_t(x,t) - u(x,t)\Delta\phi(x,t)\} dx \\ &+ \int_{\mathbf{R}^n} \frac{\mu_1}{(1+t)^\beta} u_t(x,t)\phi(x,t)dx - \int_{\mathbf{R}^n} \frac{\mu_2}{(1+t)^{\alpha+1}} u(x,t)\phi(x,t)dx \\ &= \int_{\mathbf{R}^n} |u(x,t)|^p \phi(x,t)dx. \end{aligned}$$

Multiplying both sides of the above equality with $m(t)$ yields

$$
\begin{aligned}
&\frac{d}{dt}\left\{m(t)\int_{\mathbf{R}^n} u_t(x,t)\phi(x,t)dx\right\}\\
&+m(t)\int_{\mathbf{R}^n}\{-u_t(x,t)\phi_t(x,t)-u(x,t)\Delta\phi(x,t)\}\,dx\\
&=m(t)\int_{\mathbf{R}^n}\frac{\mu_2}{(1+t)^{\alpha+1}}u(x,t)\phi(x,t)dx+m(t)\int_{\mathbf{R}^n}|u(x,t)|^p\phi(x,t)dx,
\end{aligned}
$$

integrating which over $[0,t]$ yields

$$
\begin{aligned}
&m(t)\int_{\mathbf{R}^n} u_t(x,t)\phi(x,t)dx-m(0)\varepsilon\int_{\mathbf{R}^n} g(x)\phi(x,0)dx\\
&-\int_0^t ds\int_{\mathbf{R}^n} m(s)u_t(x,s)\phi_t(x,s)dx-\int_0^t ds\int_{\mathbf{R}^n} m(s)u(x,s)\Delta\phi(x,s)dx\\
&=\int_0^t ds\int_{\mathbf{R}^n} m(s)\frac{\mu_2}{(1+s)^{\alpha+1}}u(x,s)\phi(x,s)dx\\
&+\int_0^t ds\int_{\mathbf{R}^n} m(s)|u(x,s)|^p\phi(x,s)dx.
\end{aligned}
$$

Integrating by parts the first term in the second line of the above equality, we have

$$
\begin{aligned}
&m(t)\int_{\mathbf{R}^n} u_t(x,t)\phi(x,t)dx-m(0)\varepsilon\int_{\mathbf{R}^n} g(x)\phi(x,0)dx\\
&-m(t)\int_{\mathbf{R}^n} u(x,t)\phi_t(x,t)dx+m(0)\varepsilon\int_{\mathbf{R}^n} f(x)\phi_t(x,0)dx\\
&+\int_0^t ds\int_{\mathbf{R}^n} m(s)\frac{\mu_1}{(1+s)^{\beta}}u(x,s)\phi_t(x,s)dx\\
&+\int_0^t ds\int_{\mathbf{R}^n} m(s)u(x,s)\phi_{tt}(x,s)dx-\int_0^t ds\int_{\mathbf{R}^n} m(s)u(x,s)\Delta\phi(x,s)dx\\
&=\int_0^t ds\int_{\mathbf{R}^n} m(s)\frac{\mu_2}{(1+s)^{\alpha+1}}u(x,s)\phi(x,s)dx\\
&+\int_0^t ds\int_{\mathbf{R}^n} m(s)|u(x,s)|^p\phi(x,s)dx.
\end{aligned}
\tag{23}
$$

Setting

$$
\phi(x,t)=\psi_1(x,t)=e^{-t}\phi_1(x)\quad\text{on }\operatorname{supp}u,
$$

then we have

$$\phi_t = -\phi, \quad \phi_{tt} = \Delta\phi \quad \text{on supp}\, u.$$

Hence we obtain from (23)

$$\begin{aligned} m(t)\{F_1'(t) + 2F_1(t)\} &= m(0)\varepsilon \int_{\mathbf{R}^n} \{f(x) + g(x)\}\phi_1(x)dx \\ &\quad + \int_0^t m(s)\left\{\frac{\mu_1}{(1+s)^\beta} + \frac{\mu_2}{(1+s)^{\alpha+1}}\right\} F_1(s)ds \\ &\quad + \int_0^t ds \int_{\mathbf{R}^n} m(s)|u(x,s)|^p dx, \end{aligned}$$

which implies

$$\begin{aligned} F_1'(t) + 2F_1(t) &\geq \frac{m(0)}{m(t)} C_{f,g}\varepsilon + \frac{1}{m(t)} \int_0^t m(s)\left\{\frac{\mu_1}{(1+s)^\beta} + \frac{\mu_2}{(1+s)^{\alpha+1}}\right\} F_1(s)ds \\ &\geq m(0)C_{f,g}\varepsilon + \int_0^t m(s)\left\{\frac{\mu_1}{(1+s)^\beta} + \frac{\mu_2}{(1+s)^{\alpha+1}}\right\} F_1(s)ds, \end{aligned} \tag{24}$$

where

$$C_{f,g} := \int_{\mathbf{R}^n} \{f(x) + g(x)\}\phi_1(x)dx > 0.$$

Integrating the above inequality over $[0, t]$ after a multiplication with e^{2t}, we get

$$\begin{aligned} e^{2t}F_1(t) &\geq F_1(0) + m(0)C_{f,g}\varepsilon \int_0^t e^{2s}ds \\ &\quad + \int_0^t e^{2s}ds \int_0^s m(r)\left\{\frac{\mu_1}{(1+r)^\beta} + \frac{\mu_2}{(1+r)^{\alpha+1}}\right\} F_1(r)dr. \end{aligned} \tag{25}$$

Applying a comparison argument, we have that $F_1(t) > 0$ for $t > 0$. Again, we should consider two cases due to the hypothesis on the data.

Case 1: $f \geq 0 (\not\equiv 0)$, $g \geq 0$. In this case $F_1(0) = C_{f,0}\varepsilon > 0$. The continuity of F_1 yields that $F_1(t) > 0$ for small $t > 0$. If there is the nearest zero point t_0 to $t = 0$ of F_1, then (25) gives a contradiction at t_0.

Case 2: $f \geq 0$, $g \geq 0 (\not\equiv 0)$. If $f \not\equiv 0$, we are in the previous case. If $f \equiv 0$, then $F_1(0) = 0$, $F_1'(0) = C_{0,g}\varepsilon > 0$. By the continuity of F_1', we have that F_1' is strictly positive for small t, hence there exists some $t_1 > 0$ such that $F_1' > 0$ over $[0, t_1]$. Then F_1 is strictly monotone increasing on this interval, and then strictly positive on $(0, t_1]$. Now, suppose by contradiction that $t_2 (> t_1)$ is the smallest zero point of F_1, and so $F_1 > 0$ on $(0, t_2)$. Then we claim that $F_1'(t_2) \leq 0$. If not, by continuity, F_1'

is strictly positive in a small interval $(t_3, t_2]$ for some time t_3 satisfying $0 < t_3 < t_2$. This implies that F_1 is strictly monotone increasing on $(t_3, t_2]$ and then negative due to the fact that $F_1(t_2) = 0$, a contradiction. We then verify the claim $(F_1'(t_2) \le 0)$. Letting $t = t_2$ in the inequality (24), noting the fact that $F_1(t_2) = 0$, $F_1'(t_2) \le 0$ and $F_1 \ge 0$ on $[0, t_2]$, we come to a contradiction. And we show that $F_1 > 0$ for $t > 0$ also in this case.

Therefore, coming back to (25), we may ignore the last term, and then we have

$$e^{2t}F_1(t) \ge F_1(0) + m(0)C_{f,g}\varepsilon \int_0^t e^{2s}ds \ge \frac{1}{2}m(0)C_{f,g}\varepsilon(e^{2t} - 1),$$

from which, finally, we get the lower bound of $F_1(t)$ in the form

$$F_1(t) > \frac{1-e^{-2}}{2}m(0)C_{f,g}\varepsilon \quad \text{for } t \ge 1. \tag{26}$$

Remark 3 Note that we have to cut off the time because f can vanish and so $F_1(0)$ can be equal to 0, due to our assumption on the data. If f is not identically equal to zero, then the lower bound of F_1, i.e. (26), holds for all $t \ge 0$.

5 Lower Bound for the Functional

By Hölder inequality and using the compact support of the solution (9), we have

$$\int_{\mathbf{R}^n} |u(x,t)|^p dx \ge C_2(1+t)^{-n(p-1)}|F_0(t)|^p \quad \text{for } t \ge 0, \tag{27}$$

where $C_2 = C_2(n, p, R) > 0$. Plugging this inequality into (21) and then integrating it over $[0, t]$, we have

$$F_0(t) \ge C_3 \int_0^t ds \int_0^s (1+r)^{-n(p-1)} F_0(r)^p dr \quad \text{for } t \ge 0, \tag{28}$$

where $C_3 := C_2 m(0) > 0$.

Moreover, by Hölder inequality, Lemma 1 and estimate (26), we get

$$\begin{aligned}\int_{\mathbf{R}^n} |u(x,t)|^p dx &\ge \left(\int_{\mathbf{R}^n} |\psi_1(x,t)|^{p/(p-1)}dx\right)^{1-p} |F_1(t)|^p \\ &\ge C_1^{1-p}\left(\frac{1-e^{-2}}{2}m(0)C_{f,g}\right)^p \varepsilon^p(1+t)^{(n-1)(1-p/2)} \quad \text{for } t \ge 1.\end{aligned}$$

Plugging this inequality into (21) we have

$$F_0'(t) \geq C_4\varepsilon^p \int_1^t (1+s)^{(n-1)(1-p/2)} ds \quad \text{for } t \geq 1, \tag{29}$$

where

$$C_4 := m(0)C_1^{1-p}\left(\frac{1-e^{-2}}{2} m(0)C_{f,g}\right)^p > 0.$$

Integrating (29) over $[1, t]$, we obtain

$$\begin{aligned} F_0(t) &\geq C_4\varepsilon^p \int_1^t ds \int_1^s (1+r)^{(n-1)(1-p/2)} dr \\ &\geq C_4\varepsilon^p (1+t)^{-(n-1)p/2} \int_1^t ds \int_1^s (r-1)^{n-1} dr \\ &= \frac{C_4}{n(n+1)}\varepsilon^p (1+t)^{-(n-1)p/2}(t-1)^{n+1} \quad \text{for } t \geq 1. \end{aligned} \tag{30}$$

6 Iteration Argument

Now we come to the iteration argument to get the upper bound of the lifespan estimates. First we make the ansatz that $F_0(t)$ satisfies

$$F_0(t) \geq D_j(1+t)^{-a_j}(t-1)^{b_j} \quad \text{for } t \geq 1, \quad j = 1, 2, 3, \ldots \tag{31}$$

with positive constants D_j, a_j, b_j, which will be determined later. Due to (30), note that (31) is true when $j = 1$ with

$$D_1 = \frac{C_4}{n(n+1)}\varepsilon^p, \quad a_1 = (n-1)\frac{p}{2}, \quad b_1 = n+1. \tag{32}$$

Plugging (31) into (28), we have

$$\begin{aligned} F_0(t) &\geq C_3 D_j^p \int_1^t ds \int_1^s (1+r)^{-n(p-1)-pa_j}(r-1)^{pb_j} dr \\ &\geq C_3 D_j^p (1+t)^{-n(p-1)-pa_j} \int_1^t ds \int_1^s (r-1)^{pb_j} dr \\ &\geq \frac{C_3 D_j^p}{(pb_j+2)^2}(1+t)^{-n(p-1)-pa_j}(t-1)^{pb_j+2} \quad \text{for } t \geq 1. \end{aligned}$$

So we can define the sequences $\{D_j\}_{j\in\mathbf{N}}$, $\{a_j\}_{j\in\mathbf{N}}$, $\{b_j\}_{j\in\mathbf{N}}$ by

$$D_{j+1} \geq \frac{C_3 D_j^p}{(pb_j+2)^2}, \quad a_{j+1} = pa_j + n(p-1), \quad b_{j+1} = pb_j + 2 \tag{33}$$

to establish

$$F_0(t) \geq D_{j+1}(1+t)^{-a_{j+1}}(t-1)^{b_{j+1}} \quad \text{for } t \geq 1.$$

It follows from (32) and (33) that for $j = 1, 2, 3, \ldots$

$$a_j = p^{j-1}\left((n-1)\frac{p}{2} + n\right) - n, \quad b_j = p^{j-1}\left(n+1+\frac{2}{p-1}\right) - \frac{2}{p-1}.$$

Employing the inequality

$$b_{j+1} = pb_j + 2 \leq p^j\left(n+1+\frac{2}{p-1}\right)$$

in (33), we have

$$D_{j+1} \geq C_5 \frac{D_j^p}{p^{2j}}, \tag{34}$$

where

$$C_5 := \frac{C_3}{\left(n+1+\frac{2}{p-1}\right)^2} > 0.$$

From (34) it holds that

$$\begin{aligned}
\log D_j &\geq p\log D_{j-1} - 2(j-1)\log p + \log C_5 \\
&\geq p^2 \log D_{j-2} - 2\big(p(j-2) + (j-1)\big)\log p + (p+1)\log C_5 \\
&\geq \cdots \\
&\geq p^{j-1}\log D_1 - \sum_{k=1}^{j-1} 2p^{k-1}(j-k)\log p + \sum_{k=1}^{j-1} p^{k-1}\log C_5 \\
&= p^{j-1}\left(\log D_1 - \sum_{k=1}^{j-1}\frac{2k\log p - \log C_5}{p^k}\right),
\end{aligned}$$

which yields that

$$D_j \geq \exp\left\{p^{j-1}\left(\log D_1 - S_p(j)\right)\right\},$$

where

$$S_p(j) := \sum_{k=1}^{j-1} \frac{2k\log p - \log C_5}{p^k}.$$

We know that $\sum_{k=0}^{\infty} x^k = 1/(1-x)$ and $\sum_{k=1}^{\infty} kx^k = x/(1-x)^2$ when $|x| < 1$. Then

$$S_p(\infty) := \lim_{j\to\infty} S_p(j) = \log\{C_5^{p/(1-p)} p^{2p/(1-p)^2}\}.$$

Moreover $S_p(j)$ is a sequence definitively increasing with j. Hence we obtain that

$$D_j \geq \exp\left\{p^{j-1}\left(\log D_1 - S_p(\infty)\right)\right\}, \quad \text{for } j \text{ sufficiently large.}$$

Turning back to (31), we have

$$F_0(t) \geq (1+t)^n (t-1)^{-2/(p-1)} \exp\left(p^{j-1} J(t)\right) \quad \text{for } t \geq 1, \tag{35}$$

where

$$J(t) = -\left((n-1)\frac{p}{2} + n\right)\log(1+t) + \left(n+1+\frac{2}{p-1}\right)\log(t-1) \\ + \log D_1 - S_p(\infty).$$

For $t \geq 2$, by the definition of $J(t)$, we have

$$\begin{aligned} J(t) \geq &-\left((n-1)\frac{p}{2} + n\right)\log(2t) + \left(n+1+\frac{2}{p-1}\right)\log\left(\frac{t}{2}\right) \\ &+ \log D_1 - S_p(\infty) \\ = &\frac{\gamma(p,n)}{2(p-1)}\log t + \log D_1 - \left((n-1)\frac{p}{2} + 2n + 1 + \frac{2}{p-1}\right)\log 2 - S_p(\infty) \\ = &\log\left(t^{\gamma(p,n)/\{2(p-1)\}} D_1\right) - C_6, \end{aligned}$$

where

$$C_6 := \left((n-1)\frac{p}{2} + 2n + 1 + \frac{2}{p-1}\right)\log 2 + S_p(\infty).$$

Thus, if

$$t > C_7 \varepsilon^{-2p(p-1)/\gamma(p,n)}$$

with

$$C_7 := \left(\frac{n(n+1)e^{C_6+1}}{C_4}\right)^{2(p-1)/\gamma(p,n)} > 0,$$

then we get $J(t) > 1$, and this in turn gives that $F_0(t) \to \infty$ by letting $j \to \infty$ in (35). Since we assume that $t \geq 2$ in the above iteration argument, we require

$$0 < \varepsilon \leq \varepsilon_0 := \left(\frac{C_7}{2}\right)^{\frac{\gamma(p,n)}{2p(p-1)}}.$$

Therefore we get the desired upper bound,

$$T \leq C_7 \varepsilon^{-2p(p-1)/\gamma(p,n)}$$

for $0 < \varepsilon \leq \varepsilon_0$, and hence we finish the proof of Theorem 1.

7 Proof for Theorems 2 and 3

To prove the theorems in low dimensions, we proceed similarly as for Theorem 1, but we change the first step of the iteration argument to get the desired improvement.

From (20), using (17) and noting that F_0 is positive, we have

$$F_0'(t) \geq \frac{m(0)}{m(t)} F_0'(0) \geq C_8 \varepsilon,$$

where

$$C_8 := m(0) \int_{\mathbf{R}^n} g(x)dx > 0$$

due to the assumption on g. The above inequality implies that

$$F_0(t) \geq C_8 \varepsilon t \quad \text{for } t \geq 0. \tag{36}$$

By (27) and (36), we have

$$\int_{\mathbf{R}^2} |u(x,t)|^p dx \geq C_9 \varepsilon^p (1+t)^{-n(p-1)} t^p, \tag{37}$$

with $C_9 := C_2 C_8^p > 0$. Plugging (37) into (21) and integrating it over $[0, t]$ we come to

$$\begin{aligned} F_0(t) &\geq m(0)C_9\varepsilon^p \int_0^t ds \int_0^s (1+r)^{-n(p-1)} r^p dr \\ &\geq m(0)C_9\varepsilon^p (1+t)^{-n(p-1)} \int_0^t ds \int_0^s r^p dr \\ &= C_{10}\varepsilon^p (1+t)^{-n(p-1)} t^{p+2} \quad \text{for } t \geq 0 \end{aligned} \tag{38}$$

with

$$C_{10} := \frac{m(0)C_9}{(p+1)(p+2)} > 0.$$

Remark 4 Note that the inequality (38) improves the lower bound of (30) for $n = 2$ and $1 < p < 2$, and for $n = 1$ and $p > 1$. Hence we may establish the improved lifespan estimate as stated in Theorems 2 and 3.

In a similar way as in the last section, we define our iteration sequences, $\{\widetilde{D}_j\}, \{\widetilde{a}_j\}, \{\widetilde{b}_j\}$, such that

$$F_0(t) \geq \widetilde{D}_j (1+t)^{-\widetilde{a}_j} t^{\widetilde{b}_j} \quad \text{for } t \geq 0 \text{ and } j = 1, 2, 3, \dots \tag{39}$$

with positive constants, $\widetilde{D}_j, \widetilde{a}_j, \widetilde{b}_j$, and

$$\widetilde{D}_1 = C_{10}\varepsilon^p, \quad \widetilde{a}_1 = n(p-1), \quad \widetilde{b}_1 = p+2.$$

Combining (28) and (39), we have

$$\begin{aligned} F_0(t) &\geq C_3 \widetilde{D}_j^p \int_0^t ds \int_0^s (1+r)^{-n(p-1)-p\widetilde{a}_j} r^{p\widetilde{b}_j} dr \\ &\geq \frac{C_3 \widetilde{D}_j^p}{(p\widetilde{b}_j + 2)^2} (1+t)^{-n(p-1)-p\widetilde{a}_j} t^{p\widetilde{b}_j+2} \quad \text{for } t \geq 0. \end{aligned}$$

So the sequences satisfy

$$\begin{aligned} \widetilde{a}_{j+1} &= p\widetilde{a}_j + n(p-1), \\ \widetilde{b}_{j+1} &= p\widetilde{b}_j + 2, \\ \widetilde{D}_{j+1} &\geq \frac{C_3 \widetilde{D}_j^p}{(p\widetilde{b}_j + 2)^2}, \end{aligned}$$

which means that

$$\begin{aligned}\widetilde{a}_j &= np^j - n,\\ \widetilde{b}_j &= \frac{p+1}{p-1}p^j - \frac{2}{p-1},\\ \widetilde{D}_{j+1} &\geq C_{11}\frac{\widetilde{D}_j^p}{p^{2j}},\end{aligned}$$

where $C_{11} := C_3(p-1)^2/[p(p+1)]^2$, from which we get

$$\log \widetilde{D}_j \geq p^{j-1}\left(\log \widetilde{D}_1 - \sum_{k=1}^{j-1}\frac{2k\log p - \log C_{11}}{p^k}\right).$$

Then proceeding as above we have

$$\begin{aligned}F_0(t) &\geq \widetilde{D}_j(1+t)^{n-np^j}t^{p^j(p+1)/(p-1)-2/(p-1)}\\ &\geq (1+t)^n t^{-2/(p-1)}\exp\left(p^{j-1}\widetilde{J}(t)\right),\end{aligned}$$

where

$$\widetilde{J}(t) := -np\log(1+t) + \left(p\,\frac{p+1}{p-1}\right)\log t + \log \widetilde{D}_1 - \widetilde{S}_p(\infty)$$

and

$$\widetilde{S}_p(\infty) = \log\{C_{11}^{p/(1-p)}p^{2p/(1-p)^2}\}.$$

Estimating $\widetilde{J}(t)$ for $t \geq 1$ we get

$$\begin{aligned}\widetilde{J}(t) &\geq -np\log(2t) + \left(p\,\frac{p+1}{p-1}\right)\log t + \log \widetilde{D}_1 - \widetilde{S}_p(\infty)\\ &= \frac{\gamma(p,n)-2}{p-1}\log t + \log \widetilde{D}_1 - \widetilde{S}_p(\infty) - np\log 2,\end{aligned}$$

and then we obtain that

$$\widetilde{J}(t) \geq \log\left(t^{(\gamma(p,n)-2)/(p-1)}\widetilde{D}_1\right) - C_{12} \quad \text{for } t \geq 1,$$

where $C_{12} := \widetilde{S}_p(\infty) + np\log 2$. In particular,

$$\gamma(p,n) - 2 = \begin{cases} p(3-p) & \text{if } n = 2, \\ 2p & \text{if } n = 1. \end{cases}$$

By the definition of $\widetilde{D}_1$, proceeding in the same way as that in the previous section, we get the lifespan estimate in Theorem 2 when $n = 2$, and the lifespan estimate in Theorem 3 when $n = 1$.

8 Proof for Theorem 4

Let us come back to our initial equation (1), with $n = p = 2$. In this case we introduce another multiplier

$$\lambda(t) := \exp\left(\frac{\mu_1}{2}\frac{(1+t)^{1-\beta}}{1-\beta}\right), \tag{40}$$

which yields

$$\lambda'(t) = \frac{\mu_1}{2(1+t)^\beta}\lambda(t)$$

and

$$\lambda''(t) = \left(\frac{\mu_1^2}{4(1+t)^{2\beta}} - \frac{\beta\mu_1}{2(1+t)^{\beta+1}}\right)\lambda(t).$$

Introducing a new unknown function by

$$w(x,t) := \lambda(t)u(x,t),$$

then it is easy to get

$$w_t = \frac{\mu_1}{2(1+t)^\beta}\lambda u + \lambda u_t$$

and

$$w_{tt} = \frac{\mu_1^2}{4(1+t)^{2\beta}}\lambda u - \frac{\beta\mu_1}{2(1+t)^{\beta+1}}\lambda u + \frac{\mu_1}{(1+t)^\beta}\lambda u_t + \lambda u_{tt}.$$

With this in hand the equation (1) can be transformed to

$$\begin{cases} w_{tt} - \Delta w = Qw + \lambda^{-1}|w|^2 \\ w(x,0) = 0, \quad w_t(x,0) = \lambda(0)\varepsilon g(x) \end{cases} \tag{41}$$

where

$$Q = Q(t) := \frac{\mu_1^2}{4(1+t)^{2\beta}} - \frac{\beta\mu_1}{2(1+t)^{\beta+1}} + \frac{\mu_2}{(1+t)^{\alpha+1}}.$$

A key property of the function Q is its positivity. Indeed, we can write this function as $Q = \widetilde{Q}/(1+t)^{\beta+1}$, where

$$\widetilde{Q} = \widetilde{Q}(t) := \frac{\mu_1^2}{4(1+t)^{\beta-1}} - \frac{\beta\mu_1}{2} + \frac{\mu_2}{(1+t)^{\alpha-\beta}},$$

and so it is enough to check the positivity of $\widetilde{Q}$. If $\alpha = \beta$, then $\widetilde{Q}$ is strictly decreasing to $\mu_2 - \beta\mu_1/2$, that is positive by our assumption. If $\alpha < \beta$, than we can easily find the minimum t_0 of $\widetilde{Q}$, that is

$$t_0 = -1 + \left(\frac{\mu_1^2(\beta-1)}{4\mu_2(\beta-\alpha)}\right)^{\frac{1}{2\beta-\alpha-1}},$$

and verify that the condition in (13) is equivalent to $\widetilde{Q}(t_0) \geq 0$.

Remark 5 Observe that:

- when $\alpha < \beta$, the condition (13) can be replaced by the more strong but easier condition

$$\mu_2 \geq \frac{\mu_1^2}{4}\frac{\beta-1}{\beta-\alpha},$$

that is equivalent to ask that $t_0 \leq 0$, so that $\widetilde{Q}$ is increasing and positive for $t > 0$;
- when $\alpha > \beta$, $\widetilde{Q}$ is strictly decreasing to $-\beta\mu_1/2 < 0$, and then we have no chance to achieve the positivity of this function for all the time.

Remark 6 We can rewrite the function Q also as

$$Q(t) = \frac{1}{4(1+t)^2}\left[\left(\frac{\mu_1}{(1+t)^{\beta-1}} - \beta\right)^2 + \frac{4\mu_2}{(1+t)^{\alpha-1}} - \beta^2\right],$$

which implies some connection with the definition (5) of δ in the scale invariant case ($\beta = 1$) with positive mass and $\alpha = 1$.

Now, it is well-known that our integral equation is of the form

$$\begin{aligned} w(x,t) = & \frac{\lambda(0)\varepsilon}{2\pi} \int_{|x-y|\le t} \frac{g(y)}{\sqrt{t^2-|x-y|^2}} dy \\ & + \frac{1}{2\pi} \int_0^t d\tau \int_{|x-y|\le t-\tau} \frac{Q(\tau)w(y,\tau) + \lambda^{-1}(\tau)|w(y,\tau)|^2}{\sqrt{(t-\tau)^2-|x-y|^2}} dy. \end{aligned} \tag{42}$$

Before we can move forward, we need the positivity of the solution.

Lemma 2 *Under the assumption of Theorem* 4, *the solution* w *of* (41) *is positive.*

Proof Let $\widetilde{w} = \widetilde{w}(x,t)$ be the classical solution of the Cauchy problem

$$\begin{cases} \widetilde{w}_{tt} - \Delta \widetilde{w} = Q|\widetilde{w}| + \lambda^{-1}|\widetilde{w}|^2, & \text{in } \mathbf{R}^n \times [0,\infty), \\ \widetilde{w}(x,0) = 0, \quad \widetilde{w}_t(x,0) = \lambda(0)\varepsilon g(x), & x \in \mathbf{R}^n. \end{cases}$$

It is clear from the analogous of (42) for $\widetilde{w}$ that this function is positive, and then satisfies the system (41). But u is the unique solution of (1), and so $w = \lambda u$ is the unique solution of (41). Then $w \equiv \widetilde{w} \ge 0$.

By Lemma 2, we can neglect the second term on the right-hand side of (42). Using the relation $|y| \le R$, $|x| \le t + R$ due to the support property in the first term on the right-hand side, from which the inequalities

$$\begin{aligned} t - |x-y| \le t - ||x| - |y|| \le t - |x| + R \quad \text{for } |x| \ge R, \\ t + |x-y| \le t + |x| + R \le 2(t+R), \end{aligned}$$

we obtain that

$$w(x,t) \ge \frac{\lambda(0)\varepsilon}{2\sqrt{2}\pi\sqrt{t+R}\sqrt{t-|x|+R}} \int_{|x-y|\le t} g(y)dy \quad \text{for } |x| \ge R.$$

If we assume $|x| + R \le t$, which implies $|x-y| \le t$ for $|y| \le R$, we get

$$\int_{|x-y|\le t} g(y)dy = \|g\|_{L^1(\mathbf{R}^2)},$$

and then we obtain

$$w(x,t) \ge \frac{\lambda(0)\,\|g\|_{L^1(\mathbf{R}^2)}}{2\sqrt{2}\pi\sqrt{t+R}\sqrt{t-|x|+R}} \varepsilon \quad \text{for } R \le |x| \le t - R. \tag{43}$$

Defining the functional

$$W(t) := \int_{\mathbf{R}^2} w(x,t)dx,$$

we reach to

$$W''(t) = Q(t)W(t) + \lambda^{-1}(t) \int_{\mathbf{R}^2} |w(x,t)|^2 dx.$$

Noting that W is also positive by Lemma 2 (or by the fact that $W = \lambda F$), then we have

$$W''(t) \geq \lambda^{-1}(t) \int_{\mathbf{R}^2} |w(x,t)|^2 dx \geq \int_{R \leq |x| \leq t-R} |w(x,t)|^2 dx \quad \text{for } t \geq 2R,$$

where we used the fact that $\lambda^{-1}(t) > 1$. Plugging (43) into the right-hand side of the above inequality, we have

$$W''(t) \geq \frac{\lambda(0)^2 \, \|g\|^2_{L^1(\mathbf{R}^2)}}{8\pi^2(t+R)} \varepsilon^2 \int_{R \leq |x| \leq t-R} \frac{1}{t-|x|+R} dx,$$

which yields

$$W''(t) \geq \frac{\lambda(0)^2 \, \|g\|^2_{L^1(\mathbf{R}^2)}}{4\pi(t+R)} \varepsilon^2 \int_R^{t-R} \frac{r}{t-r+R} dr \quad \text{for } t \geq 2R.$$

Then, the rest of the demonstration is exactly the same as that of Theorem 4.1 in [17], and we omit the details here.

Remark 7 We want to emphasize that the results stated in our four Theorems are still true if we have no damping term, that is if $\mu_1 = 0$. In fact, a key point in our proofs was to introduce multipliers to absorb this term. If $\mu_1 = 0$, then $m \equiv \lambda \equiv 1$ and the demonstrations proceed analogously. In this case we do not need any additional condition on μ_2 in Theorem 4, but it is enough to ask $\mu_2 > 0$.

Acknowledgements The first author is partially supported by Zhejiang Province Science Foundation (LY18A010008), NSFC (11501273, 11726612), Chinese Postdoctoral Science Foundation (2017M620128, 2018T110332), CSC(201708330548), the Scientific Research Foundation of the First-Class Discipline of Zhejiang Province (B)(201601). The second author is partially supported by the Global Thesis study award in 2016–2017, University of Bari. And he is also grateful to Future University Hakodate for hearty hospitality during his stay there, 12/01/2018–04/04/2018. This work was prepared when the second author was enrolled as MSc student at University of Bari. The third author is partially supported by the Grant-in-Aid for Scientific Research (C) (No.15K04964) and (B)(No.18H01132), Japan Society for the Promotion of Science, and Special Research Expenses in FY2017, General Topics (No.B21), Future University Hakodate. This work started when the third author was working in Future University Hakodate.

References

1. M. D'Abbicco, The threshold of effective damping for semilinear wave equations. Math. Methods Appl. Sci. **38**, 1032–1045 (2015)
2. M. D'Abbicco, S. Lucente, A modified test function method for damped wave equations. Adv. Nonlinear Stud. **13**, 867–892 (2013)
3. M. D'Abbicco, S. Lucente, M. Reissig, Semi-linear wave equations with effective damping. Chin. Ann. Math. Ser. B **34**, 345–380 (2013)
4. M. D'Abbicco, S. Lucente, M. Reissig, A shift in the Strauss exponent for semilinear wave equations with a not effective damping. J. Differ. Equ. **259**, 5040–5073 (2015)
5. K. Fujiwara, M. Ikeda, Y. Wakasugi, Estimates of lifespan and blow-up rate for the wave equation with a time-dependent damping and a power-type nonlinearity. Funkcial. Ekvac. (2016). arXiv:1609.01035
6. M. Ikeda, T. Inui, The sharp estimate of the lifespan for the semilinear wave equation with time-dependent damping. arXiv:1707.03950 (2017)
7. M. Ikeda, T. Ogawa, Lifespan of solutions to the damped wave equation with a critical nonlinearity. J. Differ. Equ. **261**, 1880–1903 (2016)
8. M. Ikeda, M. Sobajima, Life-span of solutions to semilinear wave equation with time-dependent critical damping for specially localized initial data. Math. Ann. (2018). https://doi.org/10.1007/s00208-018-1664-1
9. M. Ikeda, Y. Wakasugi, Global well-posedness for the semilinear damped wave equation with time dependent damping in the overdamping case. arXiv:1708.08044 (2017)
10. N.-A. Lai, H. Takamura, Blow-up for semilinear damped wave equations with subcritical exponent in the scattering case. Nonlinear Anal. **168**, 222–237 (2018)
11. J. Lin, K. Nishihara, J. Zhai, Critical exponent for the semilinear wave equation with time-dependent damping. Discrete Contin. Dyn. Syst. Ser. A **32**, 4307–4320 (2012)
12. N.-A. Lai, H. Takamura, K. Wakasa, Blow-up for semilinear wave equations with the scale invariant damping and super-Fujita exponent. J. Differ. Equ. **263(9)**, 5377–5394 (2017)
13. K. Nishihara, Asymptotic behaviour of solutions to the semilinear wave equation with time-dependent damping. Tokyo J. Math. **34**, 327–343 (2011)
14. W. Nunes do Nascimento, A. Palmieri, M. Reissig, Semi-linear wave models with power non-linearity and scale-invariant time-dependent mass and dissipation. Math. Nachr. (2016). https://doi.org/10.1002/mana.201600069
15. A. Palmieri, Global existence of solutions for semi-linear wave equation with scale-invariant damping and mass in exponentially weighted spaces. J. Math. Anal. Appl. **461**(2), 1215–1240 (2018)
16. A. Palmieri, M. Reissig, Fujita versus Strauss-a never ending story (2017). arXiv:1710.09123v1
17. H. Takamura, Improved Kato's lemma on ordinary differential inequality and its application to semilinear wave equations. Nonlinear Anal. Theory Methods Appl. **125**, 227–240 (2015)
18. Z. Tu, J. Lin, A note on the blowup of scale invariant damping wave equation with sub-Strauss exponent (2017). arXiv:1709.00866
19. Z. Tu, J. Lin, Life-span of semilinear wave equations with scale-invariant damping: critical Strauss exponent case (2017). arXiv:1711.00223
20. Y. Wakasugi, Critical exponent for the semilinear wave equation with scale invariant damping, in *Fourier Analysis*. Trends in Mathematics (Birkhäuser/Springer, Cham, 2014), pp. 375–390
21. Y. Wakasugi, Scaling variables and asymptotic profiles for the semilinear damped wave equation with variable coefficients. J. Math. Anal. Appl. **447**, 452–487 (2017)
22. J. Wirth, Solution representations for a wave equation with weak dissipation. Math. Methods Appl. Sci. **27**, 101–124 (2004)
23. J. Wirth, Wave equations with time-dependent dissipation. I. Non-effective dissipation. J. Differ. Equ. **222**, 487–514 (2006)

24. J. Wirth, Wave equations with time-dependent dissipation. II. Effective dissipation. J. Differ. Equ. **232**, 74–103 (2007)
25. B. Yordanov, Q.S. Zhang, Finite time blow up for critical wave equations in high dimensions. J. Funct. Anal. **231**, 361–374 (2006)
26. Q.S. Zhang, A blow-up result for a nonlinear wave equation with damping: the critical case. C. R. Math. Acad. Sci. Paris, Sér. I **333**, 109–114 (2001)

4D Semilinear Weakly Hyperbolic Wave Equations

Sandra Lucente

Abstract In this paper we exploit the 4Dimensional weakly hyperbolic equation

$$u_{tt} - a(t)\Delta u = -b(t)|u|^{p-1}u\,.$$

We establish a global existence of radial solutions in a subcritical range of p. This range depends on the zero of $a(t)$ and $b(t)$. In particular we deal with $a(t) = |t - t_0|^{\lambda_1}$ and $b(t) = |t - t_0|^{\lambda_2}$ with $\lambda_1, \lambda_2 \geq 0$. In the case $\lambda_1 = 2$ the radial assumption can be omitted.

1 Introduction

In the last decades many efforts have been done for describing existence and properties of the solutions of wave-type equations with time-variable coefficients. In particular the decay properties of linear wave-type equations with time-variable coefficients and the existence for small data solution of semilinear corresponding equations have been analyzed in many aspects (see [6] and the contained references). On the contrary few results concern the semilinear case with large data. In this paper we add another step in this direction.

We consider the global existence result for the following Cauchy problem

$$\begin{cases} u_{tt} - |t_0 - t|^{\lambda_1}\Delta u = -|t_0 - t|^{\lambda_2}|u|^{p-1}u, & (t,x) \in \mathbb{R}_+ \times \mathbb{R}^4 \\ u(0,x) = u_0(x), & x \in \mathbb{R}^4 \\ \partial_t u(0,x) = u_1(x), & x \in \mathbb{R}^4 \end{cases} \tag{1}$$

S. Lucente (✉)
Department of Mathematics, University of Bari, Bari, Italy
e-mail: sandra.lucente@uniba.it

M. D'Abbicco et al. (eds.), *New Tools for Nonlinear PDEs and Application*,
Trends in Mathematics, https://doi.org/10.1007/978-3-030-10937-0_9

with $t_0 > 0$, $\lambda_1, \lambda_2 \geq 0$, $p > 1$. According to the heuristic argument contained in [15], the critical exponent of this equation with large data is attended to be

$$p_c(\lambda_1, \lambda_2, n) = 1 + \frac{4(\frac{\lambda_2}{2} + 1)}{n(\frac{\lambda_1}{2} + 1) - 2}. \tag{2}$$

For $n = 4$, this means

$$p_c(\lambda_1, \lambda_2, 4) = 1 + \frac{\lambda_2 + 2}{\lambda_1 + 1}.$$

In particular for $\lambda_1 = 0$ one has

$$p_c(0, \lambda_2, 4) = 3 + \lambda_2.$$

In turn for $\lambda_2 = 0$ this gives the classical exponent

$$p_c(0, 0, 4) = 3.$$

The case $\lambda_1 = \lambda_2 = 0$ has a long history: starting from 1961 with $p < p_c(0, 0, 3) = 5$ studied by Jörgens in [11], this result was extended in higher dimensions twenty years later by Brenner and von Whal in [1]. While reading this paper, we see that energy estimates is enough to treat the case $n = 3$, while for $n \geq 4$ the $L^p - L^q$ estimates with $p \neq q$ seems necessary.

The critical case $p = p_c(0, 0, 3) = 5$ was treated in small data context by Rauch [18]. Struwe in [22] and [23] considered radial data, and finally Grillakis [9, 10] removed such assumption. The critical case in higher dimensions can be found in [12] and [21]. Again to consider the high dimensions, it is necessary to use the $L^p - L^q$ estimates. An exception is given by the result of Kim Lee [14] where the case $n = 4$ is related only by energy method.

For the supercritical 3D case, some sufficient conditions in term of the Sobolev norm of the solution have been given by Kenig and Merle in [13].

Coming back to (1), with $\lambda_2 = 0$, we recall that the weakly hyperbolic semilinear results started from the paper [4] where Jorgens's theorem is generalized. The critical three-dimensional case $p = p_c(\lambda_1, \lambda_2, 3)$ has been studied in [7] and [15] with a smallness assumption on the initial data. In some sense, in these papers, Rauch's result has been extended. In [16] such smallness assumption is removed and the radial case is considered. Concerning other dimensions we can quote [5] for the case $n = 1, 2$, though the in 2D case the exponent do not reach the conjectured critical case. In Galstian's paper [8], one can find some results in one dimension.

To the best of our knowledge, in the weakly hyperbolic setting and higher dimension few results are known. Some non-existence results for $u_{tt} - |t|^\lambda \Delta u = f(u)$, requires wrong sign nonlinear term (see [3] and [2]). Concerning global existence, one needs Strichartz estimates for Grushin type operator like $\partial_{tt} - t^\ell \Delta$. These are known only for $\ell \in \mathbb{N}$ and $t \to \infty$ (see Reissig [19] and Yagdjian [24]).

On the contrary, while studying (1), we take not-integer λ_1 and $t \to t_0$. In [17] the toy model $\lambda_1 = 2$, $\lambda_2 = 4$ and $p = 2$ is considered. In the present paper, we enlarge the ranges of these parameters still remaining in the subcritical case but we pay a radial assumption. In a forthcoming paper we will consider the more delicate critical small data case.

Theorem 1 *Let $(u_0, u_1) \in \mathscr{C}^\infty(\mathbb{R}^4) \times C^\infty(\mathbb{R}^4)$ radial compactly supported function. Let $T > 0$. There exists a unique radial solution $u \in C^2([0, T] \times \mathbb{R}^4)$ for the Cauchy Problem* (1) *provided*

$$2 \le p < 1 + \frac{\lambda_2 + 2}{\lambda_1 + 1} \le 3 \tag{3}$$

provided

$$\lambda_1 - 1 < \lambda_2 \le 2\lambda_1 . \tag{4}$$

In order to emphasize where the radial assumption comes into play, we will not consider radial solution until Sect. 3.4. Then in Sect. 4 we will explain how the radial assumption can be omitted if $\lambda_1 = 2$.

1.1 Notation

We write $A \lesssim B$ if there exists a positive constant C such that $A(y) \le CB(y)$ for all y in the intersection of the domains of A and B.
Let $\langle \xi \rangle := \sqrt{1 + |\xi|^2}$. For fractional Sobolev spaces, we use the norm

$$\|f\|_{H^s(\mathbb{R}^4)} = \|\langle \xi \rangle^s \hat{u}\|_{L^2(\mathbb{R}^4)} = \|\langle D \rangle^s \hat{u}\|_{L^2(\mathbb{R}^4)} .$$

2 Preliminary Results

In [5], one can find a direct proof of the local existence and uniqueness for

$$u_{tt} - a(t)\Delta u = f(t, x, u)$$

with $a(t) \ge 0$ a continuous piecewise C^2 function with zero of finite order. By Remark 2.2 of [5] we see that this equation obeys to the finite speed of propagation property, though we cannot have strong Huyghens principle. In particular the solution is compactly supported in space variable.

Let $T > 0$. If $T \in [0, t_0[$, and $t \in [0, T]$, then we have

$$a(t) := (t_0 - t)^{\lambda_1} > 0\,, \quad b(t) := (t_0 - t)^{\lambda_2} > 0.$$

In particular the equation in (1) is strictly hyperbolic in $[0, T]$ and classical theory applies. From semilinear point of view we need a critical or subcritical strictly hyperbolic assumption, this means $p \le 3$. Since in (4) we put

$$\lambda_2 \le 2\lambda_1\,,$$

this condition is satisfied. According to [1], we see that there exists a unique solution $u \in C^2([0, T] \times \mathbb{R}^4)$. Suppose we can prolong this solution up to $T = t_0$, then the strictly hyperbolic argument leads to the solution in any subinterval $[t_0, T_1]$ with $T_1 > 0$.

In order to prove Theorem 1, it remains to consider the case $T = t_0$. In particular it is enough to prove that for any $x \in \mathbb{R}^4$ we have

$$\lim_{t \to t_0^-} |u(t, x)| < +\infty\,. \tag{5}$$

2.1 Energy Estimates

It worthy to mention that in $[0, t_0]$, the functions $a(t)$ and $b(t)$ decrease. Let us denote the energy density of the solution to

$$u_{tt}(t, x) - (t_0 - t)^{\lambda_1} \Delta u(t, x) = -(t_0 - t)^{\lambda_2} |u(t, x)|^{p-1} u(t, x) \tag{6}$$

by means of

$$e(u)(t, x) = \frac{1}{2}|u_t(t, x)|^2 + (t_0 - t)^{\lambda_1} \frac{|\nabla u(t, x)|^2}{2} + (t_0 - t)^{\lambda_2} \frac{|u(t, x)|^{p+1}}{p+1}. \tag{7}$$

Multiplying by u_t the Eq. (6), one gets

$$\partial_t e(u) - (t_0 - t)^{\lambda_1} \operatorname{div}(u_t \nabla u) = -\lambda_1 (t_0 - t)^{\lambda_1 - 1} \frac{|\nabla u|^2}{2} - \lambda_2 (t_0 - t)^{\lambda_2 - 1} \frac{|u|^{p+1}}{p+1} \tag{8}$$

with negative right side. For the energy we write

$$E(u)(t) = \int_{\mathbb{R}^4} e(u)(t, x) dx\,.$$

We denote the initial energy by

$$E_0 := E(u)(0) = \int_{\mathbb{R}^4} e(u)(0,x)dx\,.$$

After integration by parts from (8), we deduce $E(u)(t) - E(u)(0) \leq 0$ that is

$$E(u)(t) \leq E_0 \quad \forall t \in [0, t_0]. \tag{9}$$

Combining this information with the Sobolev embedding theorem

$$H^s(\mathbb{R}^4) \hookrightarrow L^\infty(\mathbb{R}^4)\,, \quad s > 2\,,$$

the aim (5) reduces to find a continuous function $C(t)$ defined on $[0, t_0]$ such that

$$||u(t,\cdot)||_{H^s_x(\mathbb{R}^4)} \leq C(t) \quad \text{with } s > 2. \tag{10}$$

Let us introduce the s-energy:

$$E_s(t) = \frac{1}{2}||u_t(t,\cdot)||^2_{H^s_x(\mathbb{R}^4)} + \frac{1}{2}(t_0 - t)^{\lambda_1}||\nabla u(t,\cdot)||^2_{H^s_x(\mathbb{R}^4)} + \frac{1}{2}||u(t,\cdot)||^2_{H^s_x(\mathbb{R}^4)}.$$

Deriving in time, commuting Δ with $\langle D\rangle^s$, the formal operator calculus gives

$$\begin{aligned} E_s'(t) &= -\frac{1}{2}\lambda_1(t_0 - t)^{\lambda_1 - 1}||\nabla u(t,\cdot)||^2_{H^s_x(\mathbb{R}^4)} \\ &\quad + \int_{\mathbb{R}^4} \langle D\rangle^s u_t \left(\langle D\rangle^s u - (t_0 - t)^{\lambda_2} \langle D\rangle^s (|u|^{p-1}u) \right) dx \end{aligned}$$

and finally

$$E_s'(t) \leq ||\partial_t u(t,\cdot)||_{H^s_x(\mathbb{R}^4)} \left(||u(t,\cdot)||_{H^s_x(\mathbb{R}^4)} + (t_0 - t)^{\lambda_2} |||u(t,\cdot)|^{p-1} u(t,\cdot)||_{H^s_x(\mathbb{R}^4)} \right).$$

By using a standard approximation argument, this inequality holds for the solution of (1). Let us recall the Moser type inequality

$$|||f|^{p-1} f||_{H^s(\mathbb{R}^n)} \leq C\, ||f||_{H^s(\mathbb{R}^n)} ||f||^{p-1}_{L^\infty(\mathbb{R}^n)}.$$

which holds for $s > 0$ if

$$p \text{ is integer} \quad \text{or} \quad s < p \quad \text{or} \quad s < p + 1/2 \text{ if } s \text{ is integer}\,.$$

This can be seen in [20]. This gives the restriction

$$p \geq 2\,.$$

Suppose that we are able to find that, for all $t \in [0, t_0[$ and $x \in \mathbb{R}^4$, it holds

$$|u(t,x)| \lesssim (t_0 - t)^{-\beta} \quad \text{for} \quad \beta < \frac{\lambda_2 + 1}{p-1} \quad \text{and } p \geq 2\,, \tag{11}$$

then we get

$$E_s'(t) \lesssim E_s(t)(1 + (t_0 - t)^{\lambda_2}(t_0 - t)^{-\beta(p-1)})\,.$$

By the Gronwall's lemma, we conclude

$$||u(t,\cdot)||^2_{H^s_x(\mathbb{R}^4)} \leq E_s(t) \lesssim E_s(0)e^{t-t_0}e^{-\frac{(t_0-t)^{\lambda_2-\beta(p-1)+1}}{\lambda_2-\beta(p-1)+1}}\,.$$

This relation implies (10) and hence (5). Our aim is now the pointwise estimate (11).

2.2 *The Liouville Transformation*

We associate to $a(t) = (t_0 - t)^{\lambda_1}$ the function ϕ which satisfies

$$\begin{cases} \phi'(S) = a(\phi(S))^{-1/2} & S \in [0, T_0), \\ \phi(0) = 0, \end{cases}$$

with

$$T_0 = \int_0^{t_0} a(s)^{1/2}\,ds = \frac{t_0^{\Lambda}}{\Lambda} \quad \text{with} \quad \Lambda = \frac{\lambda_1 + 2}{2}\,.$$

Hence

$$\phi(T) = t_0 - (\Lambda(T_0 - T))^{\frac{1}{\Lambda}}\,, \tag{12}$$

or equivalently

$$(t_0 - \phi(T)) = (\Lambda(T_0 - T))^{\frac{1}{\Lambda}}\,. \tag{13}$$

In particular

$$\phi'(T) = (\Lambda(T_0 - T))^{\frac{1-\Lambda}{\Lambda}} = (\Lambda(T_0 - T))^{-\frac{\lambda_1}{\lambda_1+2}} \tag{14}$$

and

$$a(\phi(T)) = (\Lambda(T_0 - T))^{-2\frac{1-\Lambda}{\Lambda}} = (\Lambda(T_0 - T))^{\frac{2\lambda_1}{\lambda_1+2}} . \tag{15}$$

Following [15], we can check that if u solves (1) in $[0, t_0)$, then the function

$$\begin{aligned} w(T, x) &= (\phi'(T))^{-1/2} u(\phi(T), x) \\ &= a(\phi(T))^{1/4} u(\phi(T), x) \\ &= (\Lambda(T_0 - T))^{\frac{\lambda_1}{2(\lambda_1+2)}} u(\phi(T), x), \end{aligned} \tag{16}$$

defined in $[0, T_0)$, solves the equation

$$w_{TT} - \Delta w = -\left(\frac{\Lambda^2 - 1}{4\Lambda^2}(T_0 - T)^{-2} w + (\Lambda(T_0 - T))^{-\frac{\alpha(\lambda_1,\lambda_2,p)}{\Lambda}} |w|^{p-1} w \right), \tag{17}$$

where

$$\alpha(\lambda_1, \lambda_2, p) = \frac{3+p}{4}\lambda_1 - \lambda_2 .$$

Concerning the initial data, we have

$$w(0, x) = t_0^{\frac{\lambda_1}{4}} u_0(x) , \tag{18}$$

$$\partial_t w(0, x) = -\frac{\lambda_1}{4} t_0^{-\frac{\lambda_1}{4}-1} u_0(x) + t_0^{-\frac{\lambda_1}{4}} u_1(x). \tag{19}$$

Moreover, starting from (11), our aim is to prove that for any $(T, x) \in [0, T_0[\times \mathbb{R}^4$, it holds

$$|w(T, x)| \lesssim (\phi'(T))^{-1/2} (t_0 - \phi(T))^{-\beta} .$$

Recalling (13) and (14), given $p \geq 2$, we reduce the proof of Theorem 1 to find

$$\beta < \frac{\lambda_2 + 1}{p - 1} , \tag{20}$$

such that

$$|w(T, x)| \lesssim (T_0 - T)^{-\frac{4\beta - \lambda_1}{2(\lambda_1+2)}} . \tag{21}$$

2.3 Representation Formula

In order to have a representation formula of the solution (17) with initial data (18), (19), we use the notation of [14]. Let us fix $\overline{x} \in \mathbb{R}^4$ and put

$$y = x - \overline{x}.$$

Defined

$$[w] = [w](T, y) = w(T - |y|, y + \overline{x}),$$

one has

$$\begin{aligned}
\nabla[w] &= [\nabla w] - [\partial_T w]\frac{y}{|y|},\\
\partial_T[w] &= [\partial_T w],\\
\nabla[\partial_T w] &= [\nabla \partial_T w] - [\partial_{TT}^2 w]\frac{y}{|y|}.
\end{aligned}$$

We can deduce that

$$\begin{aligned}
\Delta[w] &= [\Delta w] - 2[\nabla \partial_T w]\cdot\frac{y}{|y|} + [\partial_{TT}^2 w] - \frac{3}{|y|}[\partial_T w]\\
&= [\Delta w] - 2\nabla[\partial_T w]\cdot\frac{y}{|y|} - [\partial_{TT}^2 w] - \frac{3}{|y|}[\partial_T w].
\end{aligned}$$

It follows

$$\Delta[w] + 2\nabla[\partial_T w]\cdot\frac{y}{|y|} + \frac{3}{|y|}[\partial_T w] = [\Delta w] - [\partial_{TT}^2 w].$$

As conclusion

$$\nabla\cdot\left\{\frac{1}{|y|^2}[\nabla w] + \frac{y}{|y|^3}[\partial_T w] + 2\frac{y}{|y|^4}[w]\right\} = \frac{[\Delta w]}{|y|^2} - \frac{[\partial_{TT}^2 w]}{|y|^2} - \frac{1}{|y|^3}[\partial_T w].$$

Integrating on a D domain, by using (17), we have

$$\begin{aligned}
&\int_D \nabla\cdot\left\{\frac{1}{|y|^2}[\nabla w] + \frac{y}{|y|^3}[\partial_T w] + 2\frac{y}{|y|^4}[w]\right\}dy\\
&= \int_D -\frac{1}{|y|^3}[\partial_T w] + \frac{1}{|y|^2}\frac{\Lambda^2 - 1}{4\Lambda^2}[(T_0 - T)^{-2}w]\,dy\\
&+ \int_D \frac{1}{|y|^2}\left[(\Lambda(T_0 - T))^{-\frac{\alpha(\lambda_1,\lambda_2,p)}{\Lambda}}|w|^{p-1}w\right]dy.
\end{aligned}$$

Let us fix $\overline{z} = (\overline{T}, \overline{x}) \in [0, T_0[\times\mathbb{R}^4$ and $0 < \varepsilon < \overline{T}$. In what follows we consider

$$D = \{\varepsilon \le |x - \overline{x}| \le \overline{T}\}.$$

Being $y = x - \overline{x}$, rewriting the integral identity in $T = \overline{T}$, by divergence theorem we have

$$\begin{aligned}
&\int_{|y|=\overline{T}} \frac{1}{|y|^2}\left\{\frac{y}{|y|}\cdot\nabla w(0, y+\overline{x}) + \partial_t w(0, y+\overline{x}) + \frac{2}{|y|}w(0, y+\overline{x})\right\}d\sigma_y \\
&= \int_{|y|=\epsilon} \frac{1}{|y|^2}\left\{\frac{y}{|y|}\cdot\nabla w(\overline{T}-\epsilon, y+\overline{x}) + \partial_T w(\overline{T}-\epsilon, y+\overline{x}) + \frac{2}{|y|}w(\overline{T}-\epsilon, y+\overline{x})\right\}d\sigma_y \\
&+ \int_{\epsilon\le|y|\le\overline{T}} \frac{1}{|y|^2}\left\{-\frac{1}{|y|}\partial_T w(\overline{T}-|y|, y+\bar{x})\right\}dy \\
&+ \int_{\epsilon\le|y|\le\overline{T}} \frac{1}{|y|^2}\left\{\frac{\Lambda^2-1}{4\Lambda^2}(T_0-\overline{T}+|y|)^{-2}w(\overline{T}-|y|, y+\overline{x})\right\}dy \\
&+ \int_{\epsilon\le|y|\le\overline{T}} \frac{1}{|y|^2}\left(\Lambda(T_0-\overline{T}+|y|)\right)^{-\frac{\alpha(\lambda_1,\lambda_2,p)}{\Lambda}} |w(\overline{T}-|y|, x)|^{p-1}w(\overline{T}-|y|, y+\overline{x})dy
\end{aligned}$$

For $\epsilon \to 0$ we obtain

$$\begin{aligned}
&\int_{|y|=\epsilon} \frac{1}{|y|^2}\left\{\frac{y}{|y|}\cdot\nabla w + \partial_T w + \frac{2}{|y|}w(\right\}(\overline{T}-\epsilon, y+\overline{x})d\sigma_y \\
&= \int_{|z|=1} \frac{\epsilon^3}{\epsilon^2}\left\{z\cdot\nabla w + \partial_t w + \frac{2}{\epsilon}w\right\}(\overline{T}-\epsilon, \overline{x}+\epsilon z)d\sigma_z \to 2\omega_4 w(\overline{T}, \overline{x})
\end{aligned}$$

Here $\omega_4 = 2\pi^2$ is the measure of the unit sphere in $\mathbb{R}^4$. We can conclude

$$w(\overline{T}, \overline{x}) = w_D(\overline{x}) + w_L(\overline{T}, \overline{x}) + w_M(\overline{T}, \overline{x}) + w_N(\overline{T}, \overline{x}),$$

where

$$\begin{aligned}
w_D(\overline{T}, \overline{x}) &= \frac{1}{4\pi^2}\int_{|y|=\overline{T}} t_0^{\frac{\lambda_1}{4}}\frac{y}{|y|^3}\cdot\nabla u_0(y+\overline{x}) + \frac{2}{|y|^3}t_0^{\frac{\lambda_1}{4}}u_0(y+\overline{x})d\sigma_y \\
&+ \frac{1}{4\pi^2}\int_{|y|=\overline{T}} -\frac{\lambda_1}{4}t_0^{-\frac{\lambda_1}{4}-1}\frac{u_0(y+\overline{x})}{|y|^2} + t_0^{-\frac{\lambda_1}{4}}u_1(y+\overline{x})d\sigma_y;
\end{aligned}$$

the linear part is given by

$$\begin{aligned}
w_L(\overline{T}, \overline{x}) &= \frac{1}{4\pi^2}\int_{|y|\le\overline{T}} \frac{1}{|y|^3}\partial_T w(\overline{T}-|y|, y+\overline{x})dy \\
&= \frac{1}{4\pi^2\sqrt{2}}\int_0^{\overline{T}}\int_{|x-\overline{x}|=\overline{T}-R} (\overline{T}-R)^{-3}\partial_R w(R, x)d\sigma_x dR;
\end{aligned}$$

the mass term with time-singular coefficient is

$$w_M(\overline{T},\overline{x}) = -\frac{\Lambda^2-1}{16\Lambda^2\pi^2}\int_{|y|\le \overline{T}}\frac{1}{|y|^2}(T_0-\overline{T}+|y|)^{-2}w(\overline{T}-|y|,y+\overline{x})dy$$
$$= -\frac{\Lambda^2-1}{16\Lambda^2\pi^2\sqrt{2}}\int_0^{\overline{T}}\int_{|x-\overline{x}|=\overline{T}-R}(T_0-R)^{-2}(\overline{T}-R)^{-2}w(R,x)d\sigma_x dR\,;$$

the nonlinear part is

$$w_N(\overline{T},\overline{x}) =$$
$$= \frac{-1}{4\pi^2}\int_{|y|\le \overline{T}}\frac{1}{|y|^2}\left(\Lambda(T_0-\overline{T}+|y|)\right)^{-\frac{\alpha(\lambda_1,\lambda_2,p)}{\Lambda}}(|w|^{p-1}w)(\overline{T}-|y|,y+\overline{x})dy$$
$$= \frac{-1}{4\pi^2\sqrt{2}}\int_0^{\overline{T}}\int_{|x-\overline{x}|=\overline{T}-R}(\Lambda(T_0-R))^{-\frac{\alpha(\lambda_1,\lambda_2,p)}{\Lambda}}(\overline{T}-R)^{-2}(|w|^{p-1}w)(R,x)d\sigma_x dR\,.$$

3 Proof of Theorem 1

The main idea is to use Euler integral equation, see [4].

Lemma 1 *Let $\gamma > 0$ and $\delta > 1$. Fixed $r_0 > t > 0$, we considered the integral equation*

$$y(t) = \gamma + \delta(\delta-1)\int_0^t (t-s)(r_0-s)^{-2}y(s)ds\,.$$

For any and $t \in (0, r_0)$, its solution satisfies

$$y(t) \le C(\gamma,\delta,r_0)(r_0-t)^{1-\delta}\,.$$

The proof of this Lemma is not difficult: formally we derive the Euler ODE from the integral expression:

$$y''(t) = \delta(\delta-1)(r_0-t)^{-2}y(t)$$

with $y(0) = \gamma$ and $y'(0) = 0$ and $t \in (0, r_0)$.

In order to prove (21)–(20), we set

$$\mu(R) := \sup_{[0,R]\times\mathbb{R}^4}|w(S,x)|.$$

and prove that

$$\mu(\overline{T}) \le C_1(\overline{T}) + C_2\int_0^{\overline{T}}(\overline{T}-R)(T_0-R)^{-2}\mu(R)\,dR \tag{22}$$

for a positive function $C_1(\overline{T})$ bounded for $\bar{T} \to T_0$, and suitable $C_2 > 0$ independent of $\bar{T}$. Once we establish this, we have

$$\mu(\overline{T}) \lesssim (T_0 - \overline{T})^{1-\delta}, \quad \delta = \frac{1 + \sqrt{1 + 4C_2}}{2}.$$

In order to gain (21), we need one of the following conditions on $C_2 > 0$:

$$0 \leq \frac{\sqrt{1 + 4C_2} - 1}{2} \leq \frac{4\beta - \lambda_1}{2(\lambda_2 + 1)} \leq 1 \tag{23}$$

or

$$0 \leq \frac{\sqrt{1 + 4C_2} - 1}{2} \leq 1 \leq \frac{4\beta - \lambda_1}{2(\lambda_2 + 1)}. \tag{24}$$

or

$$1 \leq \frac{\sqrt{1 + 4C_2} - 1}{2} \leq \frac{4\beta - \lambda_1}{2(\lambda_2 + 1)}, \tag{25}$$

with $\beta < \frac{\lambda_2 + 1}{p-1}$.

3.1 The Estimate for the Initial Term

The estimate

$$w_D(\overline{x}) \leq C_D, \tag{26}$$

is trivial and gives the constant function on $[0, T_0]$ with value $C_D > 0$ independent on $\bar{T}$.

3.2 The Estimate for the Mass Term

Directly we have

$$|w_M(\overline{T}, \overline{x})| \leq \frac{\Lambda^2 - 1}{8\Lambda^2\sqrt{2}} \int_0^{\overline{T}} (T_0 - R)^{-2}(\overline{T} - R)\mu(R)dR, \tag{27}$$

with

$$C_M := \frac{\Lambda^2 - 1}{8\Lambda^2\sqrt{2}}.$$

This value will be important in the estimate of C_2.

3.3 Estimates for the Nonlinear Part

Let us start with Hölder inequality:

$$|w_N(\overline{T},\overline{x})| \leq \tag{28}$$
$$\leq \frac{1}{4\pi^2\sqrt{2}}\int_0^{\overline{T}}\int_{|x-\overline{x}|=\overline{T}-R}(\Lambda(T_0-R))^{-\frac{\alpha(\lambda_1,\lambda_2,p)}{\Lambda}}(\overline{T}-R)^{-2}|w(R,x)|^p d\sigma_x dR$$
$$\leq \frac{(2\pi^2)^{\frac{2}{p+1}}}{4\pi^2\sqrt{2}}\mu(\overline{T})\left(\int_0^{\overline{T}}(\overline{T}-R)^{3-(p+1)}(\Lambda(T_0-R))^{m_1}dR\right)^{\frac{2}{p+1}}\times \tag{29}$$
$$\times\left(\int_0^{\overline{T}}\int_{|x-\overline{x}|=\overline{T}-R}(\Lambda(T_0-R))^{m_2}|w(R,x)|^{p+1}d\sigma_x dR\right)^{\frac{p-1}{p+1}},$$

where

$$m_1\frac{2}{p+1}+m_2\frac{p-1}{p+1}=-\frac{\alpha(\lambda_1,\lambda_2,p)}{\Lambda}=\frac{4\lambda_2-\lambda_1(p+3)}{2(\lambda_1+2)}. \tag{30}$$

By energy estimates we will control the last factor.

Let us introduce some notations. Fixed $t_1, t_2, t \in [0,\overline{t}]$, we put

$$K_{t_2}^{t_1}(\overline{z})=\left\{z=(t,x)\in[t_1,t_2]\times\mathbb{R}^4 \mid |x-\overline{x}|\leq\phi^{-1}(\overline{t})-\phi^{-1}(t)\right\},$$
$$K_{t_1}(\overline{z})=K_{t_1}^{\overline{t}}(\overline{z}),\quad K(\overline{z})=K_0^{\overline{t}}(\overline{z})$$
$$M_{t_2}^{t_1}(\overline{z})=\left\{z=(t,x)\in[t_1,t_2]\times\mathbb{R}^4 \mid |x-\overline{x}|=\phi^{-1}(\overline{t})-\phi^{-1}(t)\right\},$$
$$M(\overline{z})=M_0^{\overline{t}}(\overline{z}),$$
$$D(t:\overline{z})=\{x\in\mathbb{R}^4 \mid z=(t,x)\in K(\overline{z})\}.$$

Due to (12), for $\overline{t}=t_0$, we have

$$K(t_0,\overline{x})=\left\{z=(t,x)\in[0,t_0]\times\mathbb{R}^4\,\middle|\,|x-\overline{x}|\leq\frac{(t_0-t)^\Lambda}{\Lambda}\right\},$$
$$M(t_0,\overline{x})=\left\{z=(t,x)\in[0,t_0]\times\mathbb{R}^4\,\middle|\,|x-\overline{x}|=\frac{(t_0-t)^\Lambda}{\Lambda}\right\}.$$

Recalling (7), we write the localized energy

$$E(u:D(t:\overline{z}))=\int_{D(t:\overline{z})}e(u)(t,x)dx.$$

The energy flux is given by

$$d_{\overline{z}}(u)(t,x) = \frac{1}{2}\left|\partial_t u(t,x) - (t_0-t)^{\frac{\lambda_1}{2}}\frac{x-\overline{x}}{|x-\overline{x}|}\cdot\nabla u(t,x)\right|^2 + (t_0-t)^{\lambda_2}\frac{|u(t,x)|^{p+1}}{p+1}\,.$$

Lemma 2 *Fixed* $\overline{z}=(\overline{t},\overline{x})\in[0,t_0]\times\mathbb{R}^4$, *let* $u\in C^2(K_0^{\overline{t}}(\overline{z}))$ *be the solution of* (1). *For any* $0\le t_1<t_2<\overline{t}\le t_0$, *it holds*

$$E(u:D(t_1:\overline{z})) = E(u:D(t_2:\overline{z})) + \int_{M_{t_1}^{t_2}(\overline{z})}\frac{d_{\overline{z}}(u)}{\sqrt{1+(t_0-t)^{-\lambda_1}}}do$$
$$+\int_{K_{t_1}^{t_2}(\overline{z})}\frac{\lambda_1}{2}(t_0-t)^{\lambda_1-1}|\nabla u|^2 + \lambda_2(t_0-t)^{\lambda_2-1}\frac{|u|^{p+1}}{p+1}dxdt\,. \tag{31}$$

Proof We integrate (7) on $K_{t_1}^{t_2}(\overline{z})$. By divergence theorem, we have

$$\int_{D(t_2:\overline{z})}e(u)dx - \int_{D(t_1:\overline{z})}e(u)dx + \int_{M_{t_1}^{t_2}(\overline{z})}e(u)n_t do = \int_{K_{t_1}^{t_2}(\overline{z})}\partial_t e(u)dz$$
$$= \int_{K_{t_1}^{t_2}(\overline{z})}(t_0-t)^{\lambda_1}(\nabla u_t\nabla u + \Delta u)dxdt - \int_{K_{t_1}^{t_2}(\overline{z})}\lambda_1(t_0-t)^{\lambda_1-1}\frac{|\nabla u|^2}{2}dxdt$$
$$-\int_{K_{t_1}^{t_2}(\overline{z})}\lambda_2(t_0-t)^{\lambda_2-1}\frac{|u|^{p+1}}{p+1}dxdt\,.$$

Hence

$$E(u:D(t_1:\overline{z})) = E(u:D(t_2:\overline{z})) + \int_{M_{t_1}^{t_2}(\overline{z})}e(u)n_t - (t_0-t)^{\lambda_1}u_t\nabla u\cdot n_x\, do$$
$$+\int_{K_{t_1}^{t_2}(\overline{z})}\frac{\lambda_1}{2}(t_0-t)^{\lambda_1-1}|\nabla u|^2 + \lambda_2(t_0-t)^{\lambda_2-1}\frac{|u|^{p+1}}{p+1}dxdt\,.$$

The outward normal to this cone is given by

$$n(t,x) = \frac{1}{\sqrt{1+(\phi'(\phi^{-1}(t)))^2}}\left(1,\phi'(\phi^{-1}(t))\frac{x-\overline{x}}{|x-\overline{x}|}\right)$$
$$= \frac{1}{\sqrt{1+(t_0-t)^{-\lambda_1}}}\left(1,(t_0-t)^{-\frac{\lambda_1}{2}}\frac{x-\overline{x}}{|x-\overline{x}|}\right)\,.$$

It follows that

$$\sqrt{1+(t_0-t)^{-\lambda_1}}\left(e(u)n_t-(t_0-t)^{\lambda_1}\partial_t u\nabla u\cdot n_x\right)$$
$$=\frac{1}{2}|\partial_t u|^2+(t_0-t)^{\lambda_1}\frac{|\nabla u|^2}{2}+(t_0-t)^{\lambda_2}\frac{|u|^{p+1}}{p+1}-(t_0-t)^{\frac{\lambda_1}{2}}\partial_t u\nabla u\cdot\frac{x-\overline{x}}{|x-\overline{x}|}$$
$$=d_{\overline{z}}(u)(t,x)\,.$$

We get the thesis. □

Being $\lambda_1, \lambda_2 \geq 0$, recalling that the equation is defocusing, from this lemma we deduce that $t \in [0, \overline{t}[\to E(u : D(t : \overline{z}))$ is a decreasing function. By using (9) we conclude

$$E(u : D(t : \overline{z})) \leq E_0 \quad \text{for any } t \in [0, \overline{t}], \overline{t} \in [0, t_0]\,.$$

Moreover

$$\int_{\phi^{-1}(t_1)}^{\phi^{-1}(t_2)}\int_{|x-\overline{x}|=\phi^{-1}(\overline{t})-R} d_{\overline{z}}(u)(\phi(R),x)d\sigma_x dR=\int_{M_{t_1}^{t_2}(\overline{z})}\frac{d_{\overline{z}}(u)}{\sqrt{1+(t_0-t)^{-\lambda_1}}}do\leq E_0\,.$$

In particular

$$\int_{\phi^{-1}(t_1)}^{\phi^{-1}(t_2)}\int_{|x-\overline{x}|=\phi^{-1}(t)-R}(t_0-\phi(R))^{\lambda_2}|u(\phi(R),x)|^{p+1}d\sigma_x dR\leq(p+1)E_0\,.$$

In terms of $T_1, T_2, \overline{T} \in [0, T_0]$ and $w(R, x)$ this means

$$\int_{T_1}^{T_2}\int_{|x-\overline{x}|=\overline{T}-R}(\Lambda(T_0-R))^{\frac{4\lambda_2-\lambda_1(p+1)}{2(\lambda_1+2)}}|w(R,x)|^{p+1}d\sigma_x dR\leq(p+1)E_0\,.$$

This suggest to take in (29)

$$m_2=\frac{4\lambda_2-\lambda_1(p+1)}{2(\lambda_1+2)}\,,\qquad m_1=\frac{2\lambda_2-\lambda_1(p+1)}{\lambda_1+2}$$

obtaining

$$|w_N(\overline{T},\overline{x})|\leq\frac{\mu(\overline{T})}{2\sqrt{2}}\left(\frac{p+1}{2\pi^2}E_0\right)^{\frac{p-1}{p+1}}\left(\int_0^{\overline{T}}(\overline{T}-R)^{2-p}(\Lambda(T_0-R))^{m_1}dR\right)^{\frac{2}{p+1}}\,.$$

The convergence of

$$\int_0^{\overline{T}}(\overline{T}-R)^{2-p}(\Lambda(T_0-R))^{m_1}dR$$

for $\bar{T} \to T_0$ require exactly

$$2 - p + m_1 > -1 \quad \Leftrightarrow \quad p < p_c.$$

In particular given $0 < \varepsilon < 1$ there exists $\bar{T}_\varepsilon > 0$ such that, splitting the integral domain in (28) as $[0, \bar{T}] = [0, \bar{T}_\varepsilon] \cup [\bar{T}_\varepsilon, \bar{T}]$, we can estimate

$$|w_N(\overline{T}, \overline{x})| \le C_\varepsilon(T_\varepsilon) + \varepsilon\mu(\overline{T}). \tag{32}$$

This means

$$C_\varepsilon(T_\varepsilon) = \frac{(\mu(T_\varepsilon))^p}{2\sqrt{2}} \int_0^{\overline{T}_\varepsilon} (\Lambda(T_0 - R))^{-\frac{\alpha(\lambda_1,\lambda_2,p)}{\Lambda}} (T_0 - R)dR\,,$$

and

$$\frac{1}{2\sqrt{2}}\left(\frac{p+1}{2\pi^2}E_0\right)^{\frac{p-1}{p+1}} \left(\int_{\bar{T}_\varepsilon}^{\overline{T}} (\overline{T} - R)^{2-p}\,(\Lambda(T_0 - R))^{m_1}\,dR\right)^{\frac{2}{p+1}} < \varepsilon\,.$$

In the last quantity no singularity appears; we will absorb this constant in $C_1(\overline{T})$.

3.4 The Estimate for the Linear Term

First we change variable setting, we put

$$\xi(R, z) = \bar{x} + (\bar{T} - R)z\,.$$

Let $\partial_{rad} = \frac{z}{|z|} \cdot \nabla_z$, since $|z| = 1$ it follows

$$(\partial_R w)(R, \xi(R, z)) = \partial_R(w(R, \xi(R, z))) - (\partial_{rad} w)(R, \xi(R, z))\,.$$

We have

$$\begin{aligned} w_L(\overline{T}, \overline{x}) &= \frac{1}{4\pi^2\sqrt{2}} \int_0^{\bar{T}} \int_{|z|=1} \frac{(\partial_R w)(R, \xi(R, z))}{(\bar{T} - R)^3} (\bar{T} - R)^3\, d\sigma_z\, dR \\ &= \frac{1}{4\pi^2\sqrt{2}} \int_0^{\bar{T}} \int_{|z|=1} \partial_R(w(R, \xi(R, z)))\, d\,\sigma_z\, dR \\ &\quad - \frac{1}{4\pi^2\sqrt{2}} \int_0^{\bar{T}} \int_{|z|=1} \partial_{rad} w(R, \xi(R, z)) d\sigma_z\, dR \;:= I + II. \end{aligned}$$

Changing order of integration, since $\xi(\bar{T}, z) = \bar{x}$, we get

$$|I| \leq \frac{1}{4\pi^2\sqrt{2}} \int_{|z|=1} w(\bar{T}, \bar{x}) d\sigma_z + \frac{1}{4\pi^2\sqrt{2}} \int_{|z|=1} w(0, \bar{T}z + \bar{x}) d\sigma_z \leq \frac{\mu(\bar{T})}{2\sqrt{2}} + \frac{\mu(0)}{2\sqrt{2}}.$$

In order to estimate II the radial assumption comes into play.

First we notice that, being u_0, u_1 radial initial data, since the equation is invariant with respect to the space rotations, the uniqueness result implies that the local solution is radial in the space variable.

Once we prove that the only possible blow-up point is into the origin, then (11) is trivially satisfied for any $x \neq 0$. We shall prove this in Lemma 3. Hence we deal only with II when $\bar{x} = 0$ and we can rewrite

$$\begin{aligned} II &= \frac{1}{4\pi^2\sqrt{2}} \int_0^{\bar{T}} \int_{|z|=1} \partial_{rad} w(R, (\bar{T} - R)z) d\sigma_z dR \\ &= \frac{1}{4\pi^2\sqrt{2}} \int_0^{\bar{T}} (\bar{T} - R)^{-1} \int_{|z|=1} \partial_{rad} (w(R, (\bar{T} - R)|z|)) d\sigma_z dR \end{aligned}$$

In particular we can use the estimate

$$\int_{|z|\leq 1} \partial_{rad} g(z)\, dz \leq n w_n \|g\|_{L^\infty(|z|\leq 1)},$$

which holds for any radial function g and $n \geq 1$. We arrive at

$$|II| \leq \frac{2}{\sqrt{2}} \int_0^{\bar{T}} (\bar{T} - R)^{-1} \mu(R) dR.$$

Fixed $\varepsilon > 0$, considered $T_\varepsilon > 0$ as in (29), can take $T_\varepsilon < T_2 < \bar{T}$ such that

$$|II| \leq \frac{2}{\sqrt{2}} \mu(T_2) \ln\left(1 + \frac{T_2}{\bar{T} - T_2}\right) + \frac{2}{\sqrt{2}} \int_{T_2}^{\bar{T}} (\bar{T} - R)(T_0 - R)^{-2} \mu(R) dR.$$

As a conclusion we have

$$\begin{aligned} |w_L(\bar{T}, 0)| \leq{}& \frac{1}{2\sqrt{2}} \mu(\bar{T}) + \\ &+ C_L(T_2, \bar{T}) + \frac{1}{2\sqrt{2}} \int_0^{\bar{T}} (\bar{T} - R)(T_0 - R)^{-2} \mu(R) dR, \end{aligned} \tag{33}$$

with

$$C_L(T_2, \bar{T}) = \frac{\mu(0)}{2\sqrt{2}} + \ln\left(1 + \frac{T_2}{\bar{T} - T_2}\right)\mu(T_2)\,.$$

It remains to prove the following lemma

Lemma 3 *If $\bar{x} \in \mathbb{R}^4$ such that*

$$\lim_{t \to t_0} |u(t, \bar{x})| = +\infty \tag{34}$$

then $\bar{x} = 0$.

Proof From (34) we deduce that for any $M > 1$ there exists $0 < t_M < t_0$ such that for any $s \in [t_M, t_0)$ it holds

$$|u(s, \bar{x})| \geq M\,.$$

Due to the continuity of the local solution we can find $R(s) > 0$ such that

$$|u(s, x)| \geq M \quad |x - \bar{x}| \leq R(s)\,.$$

Let

$$R_M = \min\left\{\frac{t_0^{\Lambda}}{2^{\Lambda}\Lambda}, \sup_{s \in [t_M, t_0)} R(s)\right\}.$$

It follows that

$$\int_{|x-\bar{x}| \leq \frac{(t_0-s)\Lambda}{\Lambda}} |u(s, x)|^{p+1} dx \geq \frac{2\pi^2 (t_0 - s)^{4\Lambda}}{\Lambda^4}$$

for any $s \geq \max\{t_M, t_0/2\}$.

Assume by absurd that $\bar{x} \neq 0$. Given $K \in \mathbb{N}$ we can choose $t_K \geq 0$ and K distinct elements $x_1, \ldots, x_K \in \mathbb{R}^4$ such that

$$\begin{aligned} &|x_h| = |\bar{x}| \quad h = 1, \ldots, K \\ &D(t : z_k) \cap D(t : z_h) = \emptyset \quad h \neq k, \quad z_h = (t_0, x_h), \quad t \in [t_K, t_0)\,. \end{aligned}$$

For any $h \in \mathbb{N}$ we can recursively associate $M(h) > 0$ such that $t_h \geq \max\{t_{h-1}, t_{M(h-1)}\}$ and

$$\frac{M(h)^{p+1}}{p+1} \frac{2\pi^2 (t_0 - t_h)^{4\Lambda + \lambda_2}}{\Lambda^4} \geq 1\,.$$

We have

$$K \le \sum_{h=1}^{K} (t_0 - t_h)^{\lambda_2} \int_{|x-\bar{x}| \le \frac{(t_0-t_h)\Lambda}{\Lambda}} \frac{|u(t_h, x)|^{p+1}}{p+1} dx$$

By using the decreasing property of the local energy, having in mind (9), we conclude

$$\begin{aligned} K &\le \sum_{h=1}^{K} (t_0 - t_1)^{\lambda_2} \int_{|x-x_h| \le \frac{(t_0-t_1)\Lambda}{\Lambda}} \frac{|u(t_1, x)|^{p+1}}{p+1} dx \\ &\le \int_{\bigcup_{h=1}^{K} D(t_1 : z_h)} e(u)(t_1) dx \le E(u)(t_1) \le E_0. \end{aligned}$$

For $K \to \infty$ we find the contradiction. □

3.5 Proof of Theorem 1, Conclusion

Summarizing (26), (27), (32), (33), for any fixed $\bar{T} < T_0$ and $0 < \varepsilon < 1$ there exist $T_2 > T_\varepsilon > 0$ such that

$$\begin{aligned} \left(1 - \varepsilon - \frac{1}{2\sqrt{2}}\right) \mu(\bar{T}) &\le C_D + C_L(T_2, \bar{T}) + C_\varepsilon(T_\varepsilon) + \\ &+ \left(\frac{\Lambda^2 - 1}{8\Lambda^2\sqrt{2}} + \frac{2}{\sqrt{2}}\right) \int_0^{\bar{T}} (T_0 - R)^{-2} (\bar{T} - R) \mu(R) dR\,. \end{aligned}$$

This means that (22) is satisfied with

$$\begin{aligned} C_1(\varepsilon, \bar{T}) &= 2\sqrt{2}(C_D + C_N(T_\varepsilon) + C_L(T_2, \bar{T}))/(2\sqrt{2} - 1 - \varepsilon)\,, \\ C_2(\varepsilon) &= \frac{\Lambda^2 - 1}{4\Lambda^2(2\sqrt{2} - 1 - \varepsilon)} + \frac{2}{\sqrt{2}}\,. \end{aligned}$$

We see that $C_1(\varepsilon, \bar{T})$ is continuous for $\bar{T}$ in a neighborhood of $\bar{T}_0$.

Let us fix $0 < \varepsilon < 2\sqrt{2} - 2$. We gain $2\sqrt{2} - 1 - \varepsilon \ge 1$ and $C_2(\varepsilon) \le 2 - \frac{2}{\sqrt{2}}$. Hence (25) is excluded. Distinguishing the case

$$\frac{4\beta - \lambda_1}{\lambda_2 + 1} \le 2 \quad \text{or} \quad \frac{4\beta - \lambda_1}{\lambda_2 + 1} \ge 2$$

we see that one between (23), (24) is satisfied. This conclude the proof.

4 Remark, for the Speciale Case $\lambda_1 = 2$

The radial assumption only comes in the estimate of II in Sect. 3.4. We see that

$$
\begin{aligned}
|II| &\leq \frac{1}{4\pi^2\sqrt{2}} \int_0^{\bar{T}} a(\phi(R))^{\frac{1}{4}} \int_{|z|=1} z \cdot \nabla u(\phi(R), \bar{x} + (\bar{T} - R)z) d\sigma_z dR \\
&\leq \frac{1}{4\pi^2\sqrt{2}} \left(\int_0^{\bar{T}} \int_{|z|=1} a(\phi(R))^{-\frac{1}{2}} d\sigma_z dR \right)^{1/2} \times \\
&\times \left(\int_0^{\bar{T}} \int_{|z|=1} a(\phi(R)) |\nabla u|^2 (\phi(R), \bar{x} + (\bar{T} - R)z) d\sigma_z dR \right)^{1/2} \\
&\leq \frac{1}{4\pi^2} E_0^{\frac{1}{2}} \left(2\pi^2 \int_0^{\bar{T}} (2(\bar{T} - R))^{-\frac{1}{2}} dR \right)^{\frac{1}{2}}.
\end{aligned}
$$

Hence

$$
|w_L(\bar{T}, \bar{x})| \leq \frac{\mu(\bar{T})}{2\sqrt{2}} + C_L \tag{35}
$$

with C_L depending on the initial data, and this leads directly to the conclusion of the proof of Theorem 1, as in Sect. 3.5.

For $\lambda_1 \neq 2$ we cannot apply such argument since only in this case the transformation $\xi(R, z)$ and the Liouville transformation correlate:

$$
a(\phi(T)) = (2(T_0 - T))
$$

hence

$$
\begin{aligned}
a(\phi(R))\partial_{rad} w(R, \bar{x} + (T_0 - R)z) = &= (2(T_0 - R)^{5/4} \partial_{rad} u(\phi(R), \bar{x} + (T_0 - R)z) \\
&= a(\phi(T))^{1/4} \partial_{rad}(u(\phi(R), \bar{x} + (T_0 - R)z)).
\end{aligned}
$$

In particular we prove the following extension of the result contained in [17].

Theorem 2 *Let* $(u_0, u_1) \in \mathscr{C}^\infty(\mathbb{R}^4) \times C^\infty(\mathbb{R}^4)$ *compactly supported function. Let* $T > 0$. *There exists a unique solution* $u \in C^2([0, T] \times \mathbb{R}^4)$ *for the Cauchy Problem* (1) *provided*

$$
\lambda_1 = 2, \quad 1 < \lambda_2 \leq 4, \quad 2 \leq p < \frac{\lambda_2 + 5}{3}. \tag{36}
$$

In addition, we believe that also in critical case, with this approach, radial assumption will be necessary. In particular for $\lambda_1 = \lambda_2 = 0$ and $p = 3$ the proof contained in [14] seems not complete in the estimate the linear term, that is where the radial assumption is here used.

Acknowledgements The Author is grateful to the organizers of Special interest group IGPDE and Sessions *Recent progress in evolution equations* and *Nonlinear PDE* in the 11th ISAAC Congress at Linneuniversitetet in Sweden. The Author is partially supported by INdAM – GNAMPA Project 2017 Equazioni di tipo dispersivo e proprietà asintotiche.

References

1. P. Brenner, W. von Wahl, Global classical solution of nonlinear wave equations. Math. Z. **176**, 87–121 (1981)
2. M. D'Abbicco, S. Lucente, A modified test function method for damped wave equations. Adv. Nonlinear Stud. **13**, 867–892 (2013)
3. L. D'Ambrosio, S. Lucente, Nonlinear Liouville theorems for Grushin and Tricomi operators. J. Differ. Equ. **123**, 511–541 (2003)
4. P. D'Ancona, A note on a theorem of Jörgens. Math. Z. **218**, 239–252 (1995)
5. P. D'Ancona, A. Di Giuseppe, Global existence with large data for a nonlinear weakly hyperbolic equation. Math. Nachr. **231**, 5–23 (2001)
6. M.R. Ebert, M. Reissig, *Methods for Partial Differential Equations* (Springer, Cham, 2018)
7. L. Fanelli, S. Lucente, The critical case for a semilinear weakly hyperbolic equation. Electron. J. Differ. Equ. **2004**(101), 1–13 (2004)
8. A. Galstian, Global existence for the one-dimensional second order semilinear hyperbolic equations. J. Math. Anal. Appl. **344**, 76–98 (2008)
9. M.G. Grillakis, Regularity and asymptotic behavior of the wave equation with a critical nonlinearity. Ann. Math. **132**, 485–509 (1990)
10. M.G. Grillakis, Regularity for wave equation with a critical nonlinearity. Commun. Pure Appl. Math. **95**, 749–774 (1992)
11. K. Jörgens, Das Anfangswertproblem im Großem fr̈ eine Klasse nichtlinearen Wellengleichungen. Math. Z. **77**, 295–308 (1961)
12. L. Kapitanski, The Cauchy problem for a semilinear wave equation, III. J. Sov. Math. **62**, 2619–2645 (1992)
13. C.E. Kenig, F. Merle, Nondispersive radial solutions to energy supercritical non-linear wave equations, with applications. Am. J. Math. **133**, 1029–1065 (2011)
14. J. Kim, C.H. Lee, The globally regular solutions of semilinear wave equation with a critical nonlinearity. J. Korean Math. Soc. **31**, 25–278 (1994)
15. S. Lucente, On a class of semilinear weakly hyperbolic equations. Annali dell'Università di Ferrara. **52**, 317–335 (2006)
16. S. Lucente, Large data solutions for critical semilinear weakly hyperbolic equations, in *Proceeding of the Conference Complex Analysis & Dinamical System VI*, vol. 653 (American Mathematical Society, 2015), pp. 251–276
17. S. Lucente, E. Marrone, The toy model of 4D semilinear weakly hyperbolic wave equations, in *Analysis, Probability, Applications, and Computation* (in press)
18. J. Rauch, The u^5-Klein-Gordon equation, in *Nonlinear Partial Differentia Equations and Their Applications: College de France Seminar, Vol. I*. Research Notes in Mathematics, vol. 53 (Pitman, Boston, 1981), pp. 335–364

19. M. Reissig, On $L^p - L^q$ estimates for solutions of a special weakly hyperbolic equation, in *Nonlinear Evolution Equations and Infinite-Dimensional Dynamical Systems*, ed. by L. Ta-Tsien (World Scientific, Singapore/River Edge, 1997), pp. 153–164
20. T. Runst, W. Sickel, *Sobolev Spaces of Fractional Order, Nemitskij Operators and Nonlinear Partial Differential Equations* (Walther de Gryter, Berlin, 1996)
21. J. Shatah, M. Struwe, Regularity result for nonlinear wave equation. Ann. Math. **138**, 503–518 (1993)
22. M. Struwe, Globally regular solutions to the u^5-Klein-Gordon equation. Ann. Sci. Norm. Super. Pisa Cl. Sci. IV Ser. **15**, 495–513 (1988)
23. M. Struwe, Semilinear wave equations. Bull. Am. Math. Soc. **26**, 53–84 (1992)
24. K. Yagdjian, A note on the fundamental solution for the Tricomi-type equation in the hyperbolic domain. J. Differ. Equ. **206**, 227–252 (2004)

Smoothing and Strichartz Estimates to Perturbed Magnetic Klein-Gordon Equations in Exterior Domain and Some Applications

Kiyoshi Mochizuki and Sojiro Murai

Abstract This paper is based on the talk of the first author at ISAAC Congress 2017 at Växjö, Sweden. We deal with the smoothing and Strichartz estimates to magnetic Klein-Gordon equations with time-dependent perturbations in exterior domain. Also, the smoothing estimates are applied to establish a scattering of solutions for small perturbations.

1 Introduction and Main Results

Let Ω be an exterior domain in $\mathbf{R}^n$ ($n \geq 2$) with smooth boundary $\partial\Omega$ which is star-shaped with respect to the origin 0 (the case $\Omega = \mathbf{R}^n$ is not excluded when $n \geq 3$). We consider in Ω the Klein-Gordon equation

$$\partial_t^2 w - \Delta_b w + m^2 w + \beta_0(x,t)\partial_t w + \beta(x,t) \cdot \nabla w + c(x,t)w = 0 \tag{1}$$

with Dirichlet boundary condition

$$w(x,t) = 0, \quad (x,t) \in \partial\Omega \times \mathbf{R}, \tag{2}$$

K. Mochizuki (✉)
Chuo University, Kasuga, Bunkyo-ku, Tokyo, Japan

Tokyo Metropolitan University (emeritus), Hachioji, Japan
e-mail: mochizuk@math.chuo-u.ac.jp

S. Murai
Tokyo Metropolitan College of Industrial Technology, Tokyo, Japan
e-mail: murai@metro-cit.ac.jp

M. D'Abbicco et al. (eds.), *New Tools for Nonlinear PDEs and Application*, Trends in Mathematics, https://doi.org/10.1007/978-3-030-10937-0_10

where $i = \sqrt{-1}$, $\partial_t = \partial/\partial t$, Δ_b is the magnetic Laplacian

$$\Delta_b = \nabla_b \cdot \nabla_b = \sum_{j=1}^{n} (\partial_j + ib_j(x))^2$$

with $\partial_j = \partial/\partial x_j$, m is a positive constant, $b_j(x)$ $(j = 1, \cdots, n)$ are real-valued smooth functions of $x \in \mathbf{R}^n$, and $\beta_0(x,t)$, $\beta(x,t) = (\beta_1(x,t), \cdots, \beta_n(x,t))$ and $c(x,t)$ are complex-valued smooth functions of $(x,t) \in \mathbf{R}^n \times \mathbf{R}$.

The functions $b(x) = (b_1(x), \cdots, b_n(x))$ represents a magnetic potential. Thus, the magnetic field is defined by its rotation $\nabla \times b(x) = \{\partial_j b_k(x) - \partial_k b_j(x)\}_{j<k}$. We require

$$|\nabla \times b(x)| \leq \epsilon_0 (1 + [r])^{-2}, \quad r = |x|. \tag{A1}$$

Here ϵ_0 is a small positive constant and

$$[r] = \begin{cases} r, & when\ n \geq 3 \\ r(1 + \log r/r_0), & when\ n = 2 \end{cases}$$

for a fixed $r_0 > 0$ satisfying $\partial\Omega \subset \{x; |x| > r_0\}$. For the coefficients of the perturbation term, we require the following:

$$|\beta_j(x,t)|, |\nabla\beta_j(x,t)|, |c(x,t)| \leq \epsilon_1 (1+[r])^{-2} + \eta(t) \ (j = 0, 1, \cdots, n), \tag{A2}$$

where ϵ_1 is a small positive constant and $\eta(t)$ is a positive L^1-function of $t \in \mathbf{R}$.

Equation (1) is rewritten in the system to the pair $\{w, w_t\}$ $(w_t = \partial_t w)$:

$$\partial_t \begin{pmatrix} w \\ w_t \end{pmatrix} = \begin{pmatrix} 0 & 1 \\ \Delta_b - m^2 & 0 \end{pmatrix} \begin{pmatrix} w \\ w_t \end{pmatrix} - \begin{pmatrix} 0 & 0 \\ \beta(x,t)\cdot\nabla + c(x,t) & \beta_0(x,t) \end{pmatrix} \begin{pmatrix} w \\ w_t \end{pmatrix}.$$

It is considered in the energy space $\mathscr{H} = H^1_{b,D} \times L^2$, where $L^2 = L^2(\Omega)$ is the usual L^2-space with inner-product and norm

$$(f, g) = \int_\Omega f(x)\overline{g(x)}dx, \quad \|f\| = \sqrt{(f, f)},$$

and $H^1_{b,D}$ is the completion of $C_0^\infty(\Omega)$ with norm

$$\|f\|^2_{H^1_{b,D}} = \int_\Omega \{|\nabla_b f(x)|^2 + |f(x)|^2\}dx$$

(in case $b \equiv 0$ we simply write H_D^1 for $H_{0,D}^1$). The norm of $\mathscr{H}$ is then defined by

$$\|\{f_1, f_2\}\|_{\mathscr{H}}^2 = \frac{1}{2}\{\|f_1\|_{H_{b,D}^1}^2 + \|f_2\|^2\}. \tag{3}$$

We define the operator M in $\mathscr{H}$ by

$$M = i\begin{pmatrix} 0 & 1 \\ \Delta_b - m^2 & 0 \end{pmatrix},$$

with domain

$$\mathscr{D}(M) = \left\{ f = \{f_1, f_2\} \in [H_{\mathrm{loc}}^2 \cap H_{b,D}^1] \times H_{b,D}^1; \Delta_b f_1 \in L^2 \right\}. \tag{4}$$

Then it forms a self–adjoint operator in $\mathscr{H}$, and (1), (2) is represented as

$$i\partial_t u = Mu + V(t)u \ \text{ in } \mathscr{H}, \tag{5}$$

where $u = \{w, w_t\}$ and

$$V(t) = -i\begin{pmatrix} 0 & 0 \\ \beta(x,t)\cdot\nabla + c(x,t) & \beta_0(x,t) \end{pmatrix}.$$

Moreover, by use of the unitary group of operators $\{e^{-itM}; t \in \mathbf{R}\}$ in $\mathscr{H}$, (5) with initial data $u(0) = f = \{f_1, f_2\} \in \mathscr{H}$ reduces to the integral equation

$$u(t) = e^{-itM}f - i\int_0^t e^{-i(t-\tau)M}V(\tau)u(\tau)d\tau. \tag{6}$$

Let X be the weighted energy space

$$X = \Bigg\{ f(x) = \{f_1(x), f_2(x)\}; \\ \|f\|_X^2 = \frac{1}{2}\int_\Omega (1+[r]^2)^{-1}\{|\nabla_b f_1|^2 + m^2|f_1|^2 + |f_2|^2\}dx < \infty \Bigg\}. \tag{7}$$

For an interval $I \subset \mathbf{R}$ and a Banach space W, we denote by $L_t^p(I; W)$ $(p \geq 1)$, the space of all W-valued functions $h(t)$ satisfying

$$\|h\|_{L_t^p(I;W)} = \left(\int_I \|h(t)\|_W^p dt\right)^{1/p} < \infty.$$

Similarly, $C(I; W)$ denotes the space of all W-valued continuous functions of $t \in I$. In case $I = \mathbf{R}$ we simply write $L_t^p W$ for $L_t^p(\mathbf{R}; W)$. Further, we denote by $\mathscr{B}(W)$ the space of bounded operators on W.

Our results for smoothing properties and scattering are summarized in the following three theorems.

Theorem 1 *Assume* (A1) *with small* ϵ_0. *Then there exists* $C_0 > 0$ *such that for each* $h(t) \in L^2(\mathbf{R}_\pm; X')$ *and* $f \in \mathscr{H}$, *we have*

$$\left\| \int_0^t e^{-i(t-\tau)M} h(\tau) d\tau \right\|_{L_t^2 X}^2 \leq C_0^2 \|h\|_{L_t^2 X'}^2, \tag{8}$$

$$\left\| \int_0^t e^{i\tau M} h(\tau) d\tau \right\|_{L_t^\infty \mathscr{H}}^2 \leq 2C_0 \|h\|_{L_t^2 X'}^2, \tag{9}$$

$$\|e^{-itM} f\|_{L_t^2 X}^2 \leq 2C_0 \|f\|_{\mathscr{H}}^2. \tag{10}$$

Theorem 2 *Assume* (A1) *and* (A2) *with small* ϵ_0 *and* ϵ_1. *Then for each* $f \in \mathscr{H}$ *there exists a unique solution* $u(t) \in C(\mathbf{R}; \mathscr{H})$ *to the integral equation* (6). *Let* $U(t,s)$, $s, t \in \mathbf{R}_\pm$, *be the evolution operator which maps* $u(s)$ *to* $u(t) = U(t,s)u(s)$. *Then there exists* $C_1 > 0$ *such that*

$$\|U(\cdot, s)g\|_{L_t^2 X}^2 \leq C_1 \|g\|_{\mathscr{H}}^2. \tag{11}$$

Theorem 3 *Under the same conditions as above, we have*

(i) $\{U(t,s)\}_{t,s\in\mathbf{R}}$ *is a family of uniformly bounded operators in* $\mathscr{H}$:

$$\sup_{t,s\in\mathbf{R}} \|U(t,s)\|_{\mathscr{B}(\mathscr{H})} = C_U < \infty.$$

(ii) *For every* $s \in \mathbf{R}_\pm$, *there exits the strong limit*

$$Z^\pm(s) = s - \lim_{t\to\pm\infty} e^{-i(-t+s)M} U(t,s).$$

(iii) *The operator* $Z^\pm = Z^\pm(0)$ *satisfies*

$$w - \lim_{s\to\pm\infty} Z^\pm U(0,s) e^{-isM} = I \ (weak\ limit).$$

(iv) *If* ϵ_1 *is chosen smaller also to satisfy* $\epsilon_1 \max\{1, m^{-1}\}\sqrt{2C_0C_1} < 1$, *then* $Z^\pm : \mathscr{H} \longrightarrow \mathscr{H}$ *is a bijection on* $\mathscr{H}$. *Thus, the scattering operator* $S = Z^+(Z^-)^{-1}$ *is well defined and also gives a bijection on* $\mathscr{H}$.

Next, we consider Strichartz estimates to the restricted problem with Δ_b replaced by the usual Laplacian Δ:

$$\partial_t^2 w - \Delta w + m^2 w + \beta_0(x,t)\partial_t w + \beta(x,t)\cdot\nabla w + c(x,t)w = 0 \text{ in } \Omega \tag{12}$$

with boundary condition (2), requiring the following conditions on the complex coefficients:

$$|\beta_j(x,t)|, |\nabla\beta_j(x,t)|, |c(x,t)| \le \epsilon_2(1+r)^{-2(1+\delta)} \ (j=0,1,\cdots,n), \tag{A3}$$

where $\epsilon_2 > 0$ is a small constants and $0 < \delta < 1$.

We mean by Δ_D the Laplacian acting in Ω with Dirichlet boundary conditions. So, in the following Δ is restricted to the Laplacian acting in the whole $\mathbf{R}^n$. Moreover, for the sake of simplicity we write

$$H = \sqrt{-\Delta_D + m^2}, \quad H_0 = \sqrt{-\Delta + m^2}.$$

Note that the solution $w(t)$ of (12), (2) with initial data $f = \{f_1, f_2\} \in \mathcal{H}$ verifies the integral equation

$$w(t) = \cos(tH)f_1 + H^{-1}\sin(tH)f_2 + \int_0^t H^{-1}\sin((t-\tau)H)[Vw]_2(\tau)d\tau, \tag{13}$$

where

$$[Vw]_2(x,t) = \beta_0(x,t)\partial_t w + \beta(x,t)\cdot\nabla w + c(x,t)w. \tag{14}$$

The Strichartz estimate for $w(t)$ is then given by the following:

Theorem 4 *Let (p,q) be Schrödinger admissible, i.e.,*

$$\frac{2}{p} + \frac{n}{q} = \frac{n}{2}, \ 2 < p \le \infty, \ 2 \le q \le \frac{2n}{n-2}, \ q \ne \infty,$$

and let $\sigma = n\left(\frac{1}{2} - \frac{1}{q}\right)$. Then there exists $C_2 > 0$ such that

$$\|e^{-itH}g\|_{L_t^p L^q} \le C_2\|g\|_{H^\sigma}. \tag{15}$$

Thus, under $(A3)$ we have

$$\|Hw\|_{L_t^p L^q} + \|w_t\|_{L_t^2 L^q} \le C_3\{\|f_1\|_{H_D^{\sigma+1}} + \|f_2\|_{H_D^\sigma}\}. \tag{16}$$

Remark 1 In case $\Omega = \mathbf{R}^n$, $n \geq 2$, the following Strichartz estimate is proved in D'Ancona-Fanelli [3] (see also Marshall- Straus-Wainger [6]):

$$\|e^{-itH_0} g\|_{L_t^p L^q} \leq C_2 \|g\|_{H^\gamma}, \tag{17}$$

where $\gamma = \dfrac{n+2}{2}\left(\dfrac{1}{2} - \dfrac{1}{q}\right)$. The estimate similar to (16) with σ replaced by γ is proved in Mochizuki [7] when $\Omega = \mathbf{R}^n$ with $n \geq 3$. So, the loss σ seems unnecessarily large when $n \geq 3$. It should be improved in a natural way.

Estimate (15) will be proved in §5. Then number σ appears there when (17) is modified to the case of exterior domain.

The proof of the first 3 theorems are set up based on Mochizuki-Murai [10]. The existence of the additional term $b(x,t) \cdot \nabla w$ in (1) causes no serious difficulties (see [7]).

2 Proof of Theorem 1

Theorem 1 represents smoothing properties of the Klein-Gordon evolution e^{-itM}. As is well known (e.g., Reed-Simon [12]), this theorem is derived from the uniform resolvent estimates for the operator M:

Proposition 1 *For $\kappa \in \mathbf{C}\backslash\mathbf{R}$ put $\mathscr{R}(\kappa) = (M - \kappa)^{-1}$. If ϵ_0 in $(A1)$ is chosen small enough, then there exists $C_0 > 0$ such that*

$$\sup_{\kappa \in \mathbf{C}\backslash\mathbf{R}} \|\mathscr{R}(\kappa) f\|_X \leq C_0 \|f\|_{X'},$$

for each $f \in X'$, where X' is the dual space of X with respect to $\mathscr{H}$.

For the magnetic Laplacian $L = -\Delta_b$ in L^2 with domain

$$\mathscr{D}(L) = \{u(x) \in L^2 \cap H^2_{\rm loc}(\overline{\Omega});\ -\Delta_b u \in L^2 \text{ and } u|_{\partial\Omega} = 0\}$$

and $\kappa \in \mathbf{C}_+$, put $R_m(\kappa^2) = (L + m^2 - \kappa^2)^{-1}$. Then we have

$$\mathscr{R}(\kappa) = \begin{pmatrix} \kappa & i \\ i(\Delta_b - m^2) & \kappa \end{pmatrix} R_m(\kappa^2),$$

and it follows that

$$|(\mathscr{R}(\kappa) f, g)_{\mathscr{H}}| = \frac{1}{2}\Big[|(\nabla_b\{\kappa R_m(\kappa^2) f_1 + i R_m(\kappa^2) f_2\}, \nabla_b g_1)$$
$$+(m^2\{\kappa R_m(\kappa^2) f_1 + i R_m(\kappa^2) f_2\}, g_1)$$

$$+(\{i(\Delta_b - m^2)R_m(\kappa^2)f_1 + \kappa R_m(\kappa^2)f_2\}, g_2)|\Big] \tag{18}$$

$$\leq \frac{1}{2}\Big[\{|\kappa|\|\xi\nabla_b(R_m(\kappa^2)f_1)\| + \|\xi\nabla_b R_m(\kappa^2)f_2\|\}\|\xi^{-1}\nabla_b g_1\|$$

$$+m^2\{|\kappa|\|\xi R_m(\kappa^2)f_1\| + \|\xi R_m(\kappa^2)f_2\|\}\|\xi^{-1}g_1\|$$

$$+\{\|\xi\Delta_b(R_m(\kappa^2)f_1)\| + m^2\|\xi R_m(\kappa^2)f_1\| + |\kappa|\|\xi R_m(\kappa^2)f_2\|\}\|\xi^{-1}g_2\|\Big].$$

Thus, for the proof of Proposition 1, necessary estimates of $R_m(\kappa^2)$ are summarized in the following lemma:

Lemma 1 *We put* $\xi(r) = (1 + [r])^{-1}$. *Then there exists* $C > 0$ *such that*

$$(1 + |\kappa|)\|\xi R_m(\kappa^2)f\| + \|\xi\nabla_b(R_m(\kappa^2)f)\| \leq C\|\xi^{-1}f\|, \tag{19}$$

$$\|\xi\Delta_b(R_m(\kappa^2)f)\| \leq C\{\|\xi^{-1}\nabla_b f\| + \|\xi^{-1}f\|\}, \tag{20}$$

$$|\kappa|\|\xi\nabla(R_m(\kappa^2)f)\| \leq C\{\|\xi^{-1}\nabla_b f\| + \|\xi^{-1}f\|\} \tag{21}$$

for each $\kappa \in \mathbf{C}_+$ *and* f *satisfying* $\xi^{-1}f, \xi^{-1}\nabla_b f \in L^2$.

Proof The estimate (19) is already proved (see Mochizuki [8, 9] for $n \geq 3$ and Mochizuki-Nakazawa [11] for $n = 2$).

To show (20) we start from the equation

$$\xi\Delta_b(R_m(\kappa^2)g) = \Delta_b(\xi R_m(\kappa^2)g) - 2\nabla_b \cdot \{(\nabla\xi)R_m(\kappa^2)g\} + (\Delta\xi)R_m(\kappa^2)g.$$

Put $\mathbf{f} = (\nabla\xi)R_m(\kappa^2)g$. Then since $\mathbf{f}|_{\partial\Omega} = \mathbf{0}$, we have

$$|(\nabla_b \cdot \mathbf{f}, h)| = |(\mathbf{f}, -\nabla_b h)| \leq \|\mathbf{f}\|\|h\|_{\dot{H}_b^1},$$

where $\dot{H}_b^1$ is the completion of $C_0^\infty(\Omega)$ with respect to the norm $\|\nabla_b h\|$. Let $\dot{H}_b^{-1}$ denote the dual space of $\dot{H}_b^1$. Then we have $\|\nabla_b \cdot \mathbf{f}\|_{\dot{H}_b^{-1}} \leq \|\mathbf{f}\|$, and hence

$$\|\xi\Delta_b(R_m(\kappa^2)g)\|_{\dot{H}_b^{-1}}$$

$$\leq \|\Delta_b(\xi R_m(\kappa^2)g)\|_{\dot{H}_b^{-1}} + 2\|(\nabla\xi)R_m(\kappa^2)g\| + \|(\Delta\xi)R_m(\kappa^2)g\|_{\dot{H}_b^{-1}}$$

Here, since

$$\Delta_b(\xi R_m(\kappa^2)g) = \nabla_b \cdot \{(\nabla\xi)R_m(\kappa^2)g + \xi\nabla_b(R_m(\kappa^2)g)\}$$

and $|\nabla\xi| \le C|\xi|$, it follows from (19) that

$$\|\Delta_b(\xi R_m(\kappa^2)g)\|_{\dot{H}_b^{-1}} \le \|(\nabla\xi)R_m(\kappa^2)g\| + \|\xi\nabla_b(R_m(\kappa^2)g)\| \le C\|\xi^{-1}g\|.$$

On the other hand, noting $|\Delta\xi| \le C[r]^{-1}\xi$, we can use the Hardy inequality

$$\int \frac{[n-2]^2|f|^2}{4[r]^2}dxt \le \int |\tilde{x}\cdot\nabla_b f|^2 dx \tag{22}$$

(see [8, 10]), where $\tilde{x} = x/|x|$, and $[n-2] = n-2$ when $n \ge 3$, and $= 1$ when $n = 2$, to obtain

$$\|(\Delta\xi)R_m(\kappa^2)g\|_{\dot{H}_b^{-1}} \le C\|\xi R_m(\kappa^2)g\| \le C\|\xi^{-1}g\|.$$

In fact

$$\begin{aligned}&|((\Delta\xi)R_m(\kappa^2)g, f)|\\ &\le |([r](\Delta\xi)R_m(\kappa^2)g, [r]^{-1}f)| \le C\|\xi R_m(\kappa^2)g\|\|f\|_{\dot{H}_b^1}.\end{aligned}$$

These lead us to the inequality

$$\|\xi\Delta_b(R_m(\kappa^2)g)\|_{\dot{H}_b^{-1}} \le C\|\xi^{-1}g\|.$$

The estimate (20) then follows since we have

$$\begin{aligned}|(\Delta_b(R_m(\kappa^2)f, g))| = |(R_m(\kappa^2)\Delta_b f, g)| &= |(\xi^{-1}f, \xi\Delta_b(R_m(\overline{\kappa}^2)g))|\\ &\le C\|\xi^{-1}f\|_{\dot{H}_b^1}\|\xi^{-1}g\|.\end{aligned}$$

Next note that

$$\kappa^2 R_m(\kappa^2)f = -f - (\Delta_b - m^2)R_m(\kappa^2)f.$$

Then the use of (19) and (20) shows

$$\|\xi(\Delta_b - m^2)R_m(\kappa^2)f\| \le C\{\|\xi^{-1}\nabla_b f\| + \|\xi^{-1}f\|\}.$$

Since $\|\xi f\| \le \|\xi^{-1}f\|$, this proves

$$|\kappa|^2\|\xi R_m(\kappa^2)f\| \le C\{\|\xi^{-1}\nabla_b f\| + \|\xi^{-1}f\|\}. \tag{23}$$

By use of (23), (19) and (20) we have

$$
\begin{aligned}
&|\kappa|^2\|\xi\nabla_b(R_m(\kappa^2)f)\|^2 \\
&\quad= -|\kappa|^2(\{\xi\Delta_b(R_m(\kappa^2)f) + 2\nabla\xi\cdot\nabla_b(R_m(\kappa^2)f)\}, \xi R_m(\kappa^2)f) \\
&\quad\le \{\|\xi\Delta_b(R_m(\kappa^2)f)\| + 2\|\nabla\xi\cdot\nabla_b(R_m(\kappa^2)f)\|\}|\kappa|^2\|\xi R_m(\kappa^2)f\| \\
&\quad\le C\{\|\xi^{-1}\nabla_b f\| + \|\xi^{-1}f\|\}\{\|\xi^{-1}\nabla_b f\| + \|\xi^{-1}f\|\}.
\end{aligned}
$$

which proves (21).

3 Proof of Theorem 2

First note that the perturbation $V(t)$ satisfies the following

Lemma 2 *Under* $(A2)$ *we have*

$$
|(V(t)u, v)_{\mathscr{H}}| \le \tilde{\eta}(t)\|u\|_{\mathscr{H}}\|v\|_{\mathscr{H}} + \tilde{\epsilon}_1\|u\|_X\|v\|_X,
$$

where $\tilde{\eta}(t) = \max\{1, m^{-1}\}\eta(t)$, $\tilde{\epsilon}_1 = \max\{1, m^{-1}\}\epsilon_1$.

In fact

$$
\begin{aligned}
|(V(t)u, v)_{\mathscr{H}}| &= \frac{1}{2}\left|\int_\Omega \{b(x,t)\cdot\nabla u_1 + c(x,t)u_1 + b_0(x,t)u_2\}\overline{v_2}dx\right| \\
&\le \frac{1}{2}\int_\Omega (\eta(t) + \epsilon_1(1+[r])^{-2})\{|\nabla u_1| + |u_1| + |u_2|\}|v_2|dx \\
&\le \max\{1, m^{-1}\}\{\eta(t)\|u\|_{\mathscr{H}}\|v\|_{\mathscr{H}} + \epsilon_1\|u\|_X\|v\|_X\}.
\end{aligned}
$$

With this lemma we consider the integral equation (6) when $t > 0$.

For $0 \le s \le T \le \infty$ let $I_s = [s, T]$. We do not exclude $T = \infty$ and write $\mathbf{R}_s = [s, \infty)$. Let $Y(I_s)$ be the space of functions $v(t) \in BC(I_s; \mathscr{H}) \cap L^2_t(I_s; X)$ (BC means the space of bounded continuous functions) such that

$$
\|v\|_{Y(I_s)} = \sup_{t\in I_s}\|v(t)\|_{\mathscr{H}} + \|v\|_{L^2(I_s;X)} < \infty. \tag{24}
$$

We put

$$
\Phi_s v(t) = \int_s^t e^{-i(t-s)M}V(s)v(s)ds, \quad v(t) \in Y(I_s).
$$

Then by use of Theorem 1 and Lemma 2 we can prove the

Lemma 3 *$\Phi_s \in \mathscr{B}(Y(I_s))$ and we have*

$$\sup_{t\in I_s} \|\Phi_s v(t)\|_{\mathscr{H}} \leq \|\tilde{\eta}\|_{L^1(I_s)} \sup_{t\in I_s} \|v(t)\|_{\mathscr{H}} + \tilde{\epsilon}_1\sqrt{2C_0}\|v\|_{L^2(I_s;X)}, \tag{25}$$

$$\|\Phi_s v\|_{L^2(I_s;X)} \leq 2\sqrt{2C_0}\|\tilde{\eta}\|_{L^1(I_s)} \sup_{t\in I_s} \|v(t)\|_{\mathscr{H}} + 3\tilde{\epsilon}_1 C_0\|v\|_{L^2(I_s;X)}. \tag{26}$$

Now, since $\tilde{\eta}(t) \in L^1(\mathbf{R}_+)$, we can choose $0 < \delta \leq 1$ and $\sigma > 0$ to satisfy

$$(1 + 2\sqrt{2C_0})\|\tilde{\eta}\|_{L^1(I_s)} < 1 \tag{27}$$

if $|I_s| = |T - s| \leq \delta$ or $I_s = \mathbf{R}_s$ with $s \geq \sigma$. So, if ϵ_1 is chosen small enough to satisfy $\tilde{\epsilon}_1(2\sqrt{2C_0} + 3C_0) < 1$, then it follows from (24), (25) and (26) that

$$\begin{aligned}\|\Phi_s v\|_{Y(I_s)} &\leq \max\{(1 + 2\sqrt{2C_0})\|\tilde{\eta}\|_{L^1(I_s)}, \tilde{\epsilon}_1(2\sqrt{2C_0} + 3C_0)\}\|v\|_{Y(I_s)} \\ &< \|v\|_{Y(I_s)}.\end{aligned} \tag{28}$$

Note also that $u_0(t) = e^{-i(t-s)M}f$ satisfies

$$\|u_0\|_{Y(I_s)} \leq (1 + \sqrt{2C_0})\|f\|_{\mathscr{H}}.$$

Then the successive approximation method is applied to obtain

Lemma 4 *For each fixed I_s satisfying* (27)*, the integral equation*

$$u(t) = e^{-i(t-s)M}f - i\int_s^t e^{-i(t-\tau)M}V(\tau)u(\tau)d\tau \tag{29}$$

has a solution $u(t) \in Y(I_s)$ and it satisfies

$$\|u\|_{Y(I_s)} = \sup_{t\in I_s} \|u(t)\|_{\mathscr{H}} + \|u\|_{L^2(I_s;X)} \leq C_{\delta,\sigma}\|f\|_{\mathscr{H}} \tag{30}$$

for some $C_{\delta,\sigma} > 0$ independent of f.

Proof (Proof of Theorem 2*)* For δ and σ given in (27) we choose integer N to satisfy $N\delta \geq \sigma$, and divide $\mathbf{R}_+$ into $N + 1$ subintervals

$$I_{s_j} = [s_j, s_{j+1}]\ (j = 0, 1, \cdots, N-1),\ \text{ and } I_{s_N} = \mathbf{R}_{s_N},$$

where $s_j = j\delta\ (j = 0, 1, \cdots, N)$.

Then by Lemma 4 the solution of (29) with $f = u(s_j)$ is constructed in each interval I_{s_j}, and by putting together, a global solution of (6) is obtained. Moreover, the above argument and (30) imply assertion (11) to hold with $C_1 = (N+1)C_{\delta,\sigma}^N$.

The uniqueness of solutions in $C(\mathbf{R}; \mathscr{H})$ follows from the inequality

$$\|\Phi_0 v(t)\|_{\mathscr{H}} \leq \left| \int_0^t \|V(\tau)v(\tau)\|_{\mathscr{H}} d\tau \right| \leq \left| \int_0^t \{\tilde{\eta}(\tau) + \tilde{\epsilon}_1\} \|v(\tau)\|_{\mathscr{H}} d\tau \right|.$$

If $v(t)$ satisfies (6) with $f = 0$, then this implies

$$\frac{d}{dt}\left[e^{-\int_0^t \{\tilde{\eta}(\tau)+\tilde{\epsilon}_1\} d\tau} \int_0^t \{\tilde{\eta}(\tau) + \tilde{\epsilon}_1\} \|v(\tau)\|_{\mathscr{H}} d\tau \right] \leq 0.$$

Integrating both sides, we conclude that $\|v(t)\|_{\mathscr{H}} = 0$ in $\mathbf{R}_+$.

4 Proof of Theorem 3

We put $u(t,s) = U(t,s)f$, $u_0(t-s) = e^{-i(t-s)M} f_0$. Then we have from (6)

$$(u(t,s), u_0(t-s))_{\mathscr{H}} = (f, f_0)_{\mathscr{H}} - i \int_s^t (V(\tau)u(\tau,s), u_0(\tau - s))_{\mathscr{H}} d\tau.$$

In the right side we apply the inequality of Lemma 3. It then follows from (10) and (11) that for any $\sigma, t \in \mathbf{R}_+$,

$$\begin{aligned} &|(u(t,s), u_0(t-s))_{\mathscr{H}} - (u(\sigma,s), u_0(\sigma - s))_{\mathscr{H}}| \\ &\leq \left| \int_\sigma^t \tilde{\eta}(\tau) \|u(\tau,s)\|_{\mathscr{H}} \|u_0(\tau - s)\|_{\mathscr{H}} d\tau \right| \\ &+ \tilde{\epsilon}_1 \left| \int_\sigma^t \|u(\tau,s)\|_X^2 d\tau \right|^{1/2} \left| \int_\sigma^t \|u_0(\tau - s)\|_X^2 d\tau \right|^{1/2}. \end{aligned} \tag{31}$$

All the assertions of the theorem are verified from this inequality.

Proof (Proof of Theorem 3*)*

(i) We put $\sigma = s$ in (31). Then by (10) and (11)

$$\begin{aligned} |(u(t,s), u_0(t-s))_{\mathscr{H}} - (f, f_0)_{\mathscr{H}}| &\leq \left| \int_s^t \tilde{\eta}(\tau) \|u(\tau,s)\|_{\mathscr{H}} \|u_0(\tau - s)\|_{\mathscr{H}} d\tau \right| \\ &+ \tilde{\epsilon}_1 \sqrt{2C_0C_1} \|f\|_{\mathscr{H}} \|f_0\|_{\mathscr{H}}. \end{aligned}$$

Since $e^{-i(t-s)\Lambda}$ is unitary, it follows that

$$\|u(t,s)\|_{\mathcal{H}} \le (1+\tilde{\epsilon}_1\sqrt{2C_0C_1})\|f\|_{\mathcal{H}} + \int_s^t \tilde{\eta}(\tau)\|u(\tau,s)\|_{\mathcal{H}} d\tau.$$

The requirement $\eta(t) \in L^1(\mathbf{R})$ and the Gronwall inequality show the assertion with

$$C_U = (1+\tilde{\epsilon}_1\sqrt{2C_0C_1})e^{\|\tilde{\eta}\|_{L^1}}.$$

(ii) Noting (i), we have from (31), (10) and (11)

$$|(u(t,s),u_0(t-s))_{\mathcal{H}} - (u(\sigma,s),u_0(\sigma-s))_{\mathcal{H}}| \le \left\{C_U\|f\|_{\mathcal{H}}\left|\int_\sigma^t \tilde{\eta}(\tau)d\tau\right| + \right.$$

$$\left. + \tilde{\epsilon}_1\left|\int_\sigma^t \|u(\tau,s)\|_X^2 d\tau\right|^{1/2}\sqrt{2C_0}\right\}\|f_0\|_{\mathcal{H}}.$$

Here, for fixed any $s \in \mathbf{R}_\pm$, $\left\{\cdots\right\} \to 0$ as σ, $t \to \pm\infty$. Thus, $e^{-i(s-t)M}U(t,s)$ converges strongly in $\mathcal{H}$ as $t \to \pm\infty$.

(iii) Let $\sigma = s$ and $t \to \pm\infty$ in (31). Then noting (i) and (11), we have

$$|(Z^\pm(s)f,f_0)_{\mathcal{H}} - (f,f_0)_{\mathcal{H}}| \le \|f\|_{\mathcal{H}}\left\{C_U\left|\int_s^{\pm\infty}\tilde{\eta}(\tau)d\tau\right|\|f_0\|_{\mathcal{H}} + \right.$$

$$\left. + \tilde{\epsilon}_1\sqrt{C_1}\left|\int_s^{\pm\infty}\|u_0(\tau-s)\|_X^2 d\tau\right|^{1/2}\right\}. \tag{32}$$

Choose here $f = e^{-isM}g$ and $f_0 = e^{-isM}g_0$. Then

$$|(\{e^{isM}Z^\pm(s)e^{-isM} - I\}g,g_0)_{\mathcal{H}}| \le \|e^{-isM}g\|_{\mathcal{H}}\left\{C_U\left|\int_s^{\pm\infty}\tilde{\eta}(\tau)d\tau\right| \times \right.$$

$$\left. \times \|e^{-isM}g_0\|_{\mathcal{H}} + \tilde{\epsilon}_1\sqrt{2C_0}\left|\int_s^{\infty}\|e^{-i\tau M}g_0\|_X^2 d\tau\right|^{1/2}\right\}.$$

g and g_0 being arbitrary, this implies that as $s \to \pm\infty$,

$$Z^\pm U(0,s)e^{-isM} = e^{isM}Z^\pm(s)e^{-isM} \to I \text{ weakly in } \mathcal{H}.$$

(iv) Note that (32) and (10) imply

$$|(\{Z^{\pm}(s)-I\}f, f_0)_{\mathscr{H}}| \le \left\{\left|\int_s^{\pm\infty} \tilde{\eta}(\tau)d\tau\right| C_U + \tilde{\epsilon}_1\sqrt{2C_0C_1}\right\}\|f\|_{\mathscr{H}}\|f_0\|_{\mathscr{H}}.$$

Since $\tilde{\epsilon}_1\sqrt{2C_1C_0} < 1$, we can choose $\pm s > 0$ sufficiently large to satisfy

$$\left|\int_s^{\pm\infty} \tilde{\eta}(\tau)d\tau\right| C_U + \tilde{\epsilon}_1\sqrt{2C_0C_1} < 1.$$

Thus, $\|Z_{\pm}(s) - I\|_{\mathscr{B}(\mathscr{H})} < 1$ and $Z^{\pm}(s)$ gives a bijection on $\mathscr{H}$. The same property of $Z^{\pm}$ then easily follows.

5 Proof of Theorem 4

In order to show the Strichartz estimate (15), the smoothing properties (10) for the evolution e^{-itM} and (11) for the solution to (12) are not sufficient. The Strichartz estimate (17) and the smoothing property for the free evolution $e^{-itM_0}f$, where

$$M_0 = i\begin{pmatrix} 0 & 1 \\ H_0^2 & 0 \end{pmatrix},$$

are also indispensable. This last property is used in the following form:

Proposition 2 *Put $\mu = \mu(r) = (1+r)^{-1-\delta}$. Then there exists $C > 0$ independent of $f \in L^2(\mathbf{R}^2)$*

$$\|\mu(r)e^{-itH_0}f\|_{L_t^2L^2} \le C\|f\|. \tag{33}$$

This proposition is included in (10) if $n \ge 3$. So, let us derive necessary results when $n = 2$.

Lemma 5 *Let $R_0(\kappa^2) = (-\Delta - \kappa^2)^{-1}$, $\mathrm{Im}\kappa > 0$, in $L^2(\mathbf{R}^2)$. Then there exists $C > 0$ independent of κ and $f \in C_0^\infty(\mathbf{R}^2)$ such that*

$$\|\mu^{1/2}\kappa R_0(\kappa^2)f\| + \|\mu^{1/2}\nabla R_0(\kappa^2)f\| \le C\|\mu^{-1/2}f\|, \tag{34}$$

$$\sup_{\kappa\in\mathbf{C}_+} \|\mu\{R_0(\kappa^2) - R_0(\bar{\kappa}^2)\}f\| \le C\|\mu^{-1}f\|. \tag{35}$$

Proof The first estimate (34) is proved in Barcelo-Ruiz-Vega [1]. The second estimate is easy if we note that $R_0(\kappa^2)$ forms an integral operator

$$R_0(\kappa^2)f = \int_{\mathbf{R}^2} \frac{i}{4}H_0^{(1)}(\kappa|x-y|)f(y)dy,$$

where $H_0^{(1)}(\kappa|x|)$ is the Hankel function of the first kind verifying the asymptotic expansion

$$\lim_{\kappa\to 0}\left\{H_0^{(1)}(\kappa|x|)+\frac{1}{2\pi}\log\kappa\right\}=\frac{1}{2\pi}\log\frac{1}{|x|}.$$

Lemma 6 *Let $\mathscr{R}_0(\kappa)=(M_0-\kappa)^{-1}$. Let $A:\mathscr{H}\to L^2$, $A^*:L^2\to\mathscr{H}$ be defined by*

$$Af=\mu(r)H_0f_1 \; for\; f=\{f_1,f_2\},\quad A^*g=\{2H_0^{-1}\mu(r)g,0\}\; for\; g\in L^2.$$

Then there exists $C>0$ independent of $\kappa\in\mathbf{C}_+$ and $g\in L^2$ such that

$$\|A\{\mathscr{R}_0(\kappa)-\mathscr{R}_0(\bar\kappa)\}A^*g\|\le C\|g\|, \tag{36}$$

and hence we have

$$\|Ae^{-itM_0}f\|_{L_t^2L^2}\le C\|f\|_{\mathscr{H}}. \tag{37}$$

Proof Note that

$$A\{\mathscr{R}_0(\kappa)-\mathscr{R}_0(\bar\kappa)\}A^*g=\mu(r)\{\kappa R_0(\kappa^2-m^2)-\bar\kappa R_0(\bar\kappa^2-m^2)\}\mu(r)g. \tag{38}$$

We put $J=\kappa R_0(\kappa^2-m^2)-\bar\kappa R_0(\bar\kappa^2-m^2)$. Then

$$J=\bar\kappa\{R_0(\kappa^2-m^2)-R_0(\bar\kappa^2-m^2)\}+\frac{2i\mathrm{Im}\kappa}{\sqrt{\kappa^2-m^2}}\sqrt{\kappa^2-m^2}R_0(\kappa^2-m^2),\quad |\kappa|\le 2m,$$

$$=\frac{\kappa}{\sqrt{\kappa^2-m^2}}\sqrt{\kappa^2-m^2}R_0(\kappa^2-m^2)-\frac{\bar\kappa}{\sqrt{\bar\kappa^2-m^2}}\sqrt{\bar\kappa^2-m^2}R_0(\kappa^2-m^2),\quad |\kappa|>2m,$$

where $\left|\frac{2i\mathrm{Im}\kappa}{\sqrt{\kappa^2-m^2}}\right|$ is bounded in $|\kappa|\le 2m$ and $\frac{\kappa}{\sqrt{\kappa^2-m^2}}$, $\frac{\bar\kappa}{\sqrt{\bar\kappa^2-m^2}}$ are bounded in $|\kappa|>2m$. In account of these estimates, (34) and (35) are applied to (38) to obtain (36).

The lemma is complete since (37) is equivalent to (36) (see Kato [4] or Kato-Yajima [5]).

Proof (Proof of Proposition 2) The first component of $e^{-itM_0}f$ is

$$w(t)=\cos(tH_0)f_1+H_0^{-1}\sin(tH_0)f_2.$$

So, if we choose $f = \{H_0^{-1}g, 0\}$ and $f = \{0, g\}$ for $g \in L^2$ in (37), then

$$\left| \int_0^{\pm\infty} \|\mu(r)\cos(tH_0)g\|^2 dt \right| \le C\|g\|^2$$

and

$$\left| \int_0^{\pm\infty} \|\mu(r)\sin(tH_0)g\|^2 dt \right| \le C\|g\|^2.$$

Summing up these inequalities, we obtain (33).

The smoothing property for $e^{-itH}f$ is used in the following form:

Proposition 3 *Let $W = L^2$ or $= H_D^1$. Then we have*

$$\|(1+[r])^{-1}e^{-itH}f\|_{L_t^2 W} \le C_W\|f\|_W. \tag{39}$$

Proof We return to (10):

$$\|e^{-itM}f\|_{L_t^2 X} \le C\|f\|_{\mathscr{H}}.$$

Choose here $f = \{g, 0\}$ and $f = \{0, Hg\}$. Then

$$\int_{-\infty}^{\infty}\int_{\Omega}(1+[r])^{-2}\{|\nabla\cos(tH)g|^2 + m^2|\cos(tH)g|^2 + |H\sin(tH)g|^2\}dxdt$$

$$\le C\int_{\Omega}\{|\nabla g|^2 + m^2|g|^2\}dx,$$

$$\int_{-\infty}^{\infty}\int_{\Omega}(1+[r])^{-2}\{|\nabla\sin(tH)g|^2 + m^2|\sin(tH)|^2 + |H\cos(tH)g|^2\}dxdt$$

$$\le C\int_{\Omega}|Hg|^2 dx.$$

Combining these to inequalities, we obtain

$$\int_{-\infty}^{\infty}\int_{\Omega}(1+[r])^{-2}\{|\nabla e^{-itH}g|^2 + m^2|e^{-itH}g|^2 + |He^{-itH}g|^2\}dxdt \le C\|g\|_{H_D^1}^2,$$

which proves (39) with $W = H_D^1$.

The case $W = L^2$ is similarly proved if we choose $f = \{H^{-1}g, 0\}$ and $f = \{0, g\}$ in (10).

The following lemma is known as Christ-Kiselev theorem ([2]).

Lemma 7 *Let X, Y be Banach spaces and Let* $Th(t) = \int_{-\infty}^{\infty} K(t,s)h(s)ds$ *be a bounded operator from* $L_t^\alpha X$ *to* $L_t^\beta Y$. *If* $\alpha < \beta$, *then* $\tilde{T}h(t) = \int_0^t K(t,s)h(s)ds$ *is also a bounded operator, and we have* $\|\tilde{T}\| \le C(\alpha,\beta)\|T\|$.

Proof (Proof of Theorem 4) To show (15) let $\chi = \chi(x)$, $x \in \mathbf{R}^n$, be a C^∞-function whose support is restricted in a neighborhood of $\mathbf{R}^n \backslash \Omega$, and we decompose $e^{-itH}f$ into two parts:

$$e^{-itH}f = \chi e^{-itH}f + (1-\chi)e^{-itH}f \equiv v_1(t) + v_2(t).$$

As for $v_1(t)$ we have from (39) with $W = H_D^1$ that

$$\|v_1\|_{L_t^2 H_D^1} \le \|(1+[r])^{-1}e^{-itH}f\|_{L_t^2 H_D^1} \le C\|f\|_{H_D^1}.$$

Interpolating this and the energy inequality $\|v_1\|_{L_t^\infty L2} \le C\|f\|_{L^2}$, we conclude

$$\|v_1\|_{L^p H_D^{2/p}} \le C\|f\|_{H_D^{2/p}}.$$

Choosing $\dfrac{1}{q} = \dfrac{1}{2} - \dfrac{2}{np}$, we can apply Sobolev embedding to obtain

$$\|v_1(t)\|_{L^p L^q} \le C\|f\|_{H_D^{2/p}}. \tag{40}$$

Our $\sigma = 2/p$ is a result of this inequality.

Next consider the function $v_2(x)$. Note that it satisfies the initial-value problem in $\mathbf{R}^n$

$$\partial_t^2 v_2 + H_0^2 v_2 = g(x,t),$$

$$v_2(0) = (1-\chi)f, \ \ \partial_t v_2(0) = -(1-\chi)iHf,$$

where

$$g(x,t) = (\Delta\chi)e^{-itH}f - 2\nabla\cdot((\nabla\chi)e^{-itH}f).$$

The Duhamel principle then asserts that

$$v_2(t) = \cos(tH_0)(1-\chi)f - iH_0^{-1}\sin(tH_0)(1-\chi)Hf + \int_0^t H_0^{-1}\sin\{(t-\tau)H_0\}g(\tau)d\tau.$$

The free Strichartz estimate (17) shows

$$\|e^{-itH_0}(1-\chi)f\|_{L_t^p L^q} + \|H_0^{-1}e^{-itH_0}(1-\chi)Hf\|_{L_t^p L^q} \le C\|f\|_{H_D^\gamma}. \tag{41}$$

As for the third term, note the following inequalities

$$\left\| \int_0^\infty H_0^{-1} e^{-i(t-\tau)H_0} g(\tau) d\tau \right\|_{L_t^p H^{q,-\gamma}} \le C \left\| H_0^{-1} \int_0^\infty e^{i\tau H_0} (\Delta\chi) e^{-i\tau H} f d\tau \right\|$$

$$+ C \left\| H_0^{-1} \nabla \cdot \int_0^\infty e^{i\tau H_0} 2(\nabla\chi) \cdot \nabla\} e^{-i\tau H} f d\tau \right\|$$

$$\le C \|(1+r)^{1+\delta} \{|\Delta\chi| + 2|\nabla\chi|\} e^{-i\tau H} f\|_{L_t^2 L^2} \le C\|f\|.$$

To show the second inequality we have used the boundedness in L^2 of H_0^{-1} and $H_0^{-1}\nabla$, and then the dual inequality of (33). The last inequality is a result of (39) with $W = L^2$.

Now, Lemma 7 with $\alpha = 2$, $\beta = p$ and $X = Y = H_q^{-\gamma}$ implies the following

$$\left\| \int_0^t H_0^{-1} e^{-i(t-\tau)H_0} g(\tau) d\tau \right\|_{L_t^p L^q} \le C(2, p)\|f\| \le C\|f\|_{H_D^\gamma}. \tag{42}$$

Since $\sigma \ge \gamma$, summarizing (40), (41) and (42), we conclude the assertion (15).

To enter into the proof of (16), we return to Eq. (13). The function $Hw(t)$ verifies

$$Hw(t) = \cos(tH)Hf_1 - i\sin(tH)f_2 - \int_0^t \sin((t-\tau)H)[Vw]_2(\tau)d\tau.$$

Hence, the inequality (15) assures the estimate

$$\|\cos(tH)Hf_1\|_{L_t^p L^q} + \|\sin(tH)f_2\|_{L_t^p L^q} \le C\{\|Hf_1\|_{H_D^\sigma} + \|f_2\|_{H_D^\sigma}\}.$$

To establish the estimate for the third term, we use (17) and also the dual estimate of (39) with $W = L^2$. Then

$$\left\| \int_0^\infty e^{-i(t-\tau)H} h(\tau) d\tau \right\|_{L^p H^{q,-\sigma}} \le \left\| \int_0^\infty e^{i\tau H} h(\tau) d\tau \right\| \le C\|\mu(r)^{-1} h\|_{L_t^2 L^2}.$$

Choose $h = [Vw]_2(t)$ and remember (15). Then noting the smoothing estimate (11) for $U(t,0)f$, we have

$$\|\mu(r)^{-1}[Vw]_2\|_{L_t^2 L^2} \le C\|U(t,0)f\|_{L_t^2 X} \le C\|f\|_{\mathscr{H}}.$$

Thus, Lemma 7 with $\alpha = 2$ and $\beta = p$ is applied to conclude the desired inequality also to the third term.

References

1. J.A. Barcelo, A. Ruiz, L. Vega, J.A. Barcelo, A. Ruiz, L. Vega, Weighted estimates for the Helmholtz equation and some applications. J. Funct. Anal. **150**, 356–382 (1997)
2. M. Christ, A. Kiselev, Maximal functions associated to filtrations. J. Funct. Anal. **179**, 409–425 (2001)
3. P. D'Ancona, L. Fanelli, Strichartz and smoothing estimates for dispersive equations with magnetic potentials. Commun. Partial Differ. Equ. **33**, 1082–1112 (2008)
4. T. Kato, Wave operators and similarity for some nonselfadjoint operators. Math. Ann. **162**, 258–279 (1966)
5. T. Kato, K. Yajima, Some examples of smooth operators and the associated smoothing effect. Rev. Math. Phys. **1**, 481–496 (1989)
6. B. Marshall, W.A. Strauss, S. Wainger, $L^p - L^q$ estimates for the Klein-Gordon equation. J. Math. Pures Appl. **59**, 417–440 (1980)
7. K. Mochizuki, Resolvent estimates and scattering problems for Schrödinger, Klein-Gordon and wave equations, in *Progress in Partial Differential Equations*. Proceedings in Mathematics and Statistics, ed. by M. Reissig, M. Ruzhansky, vol. 44 (Springer, Heidelberg, 2013), pp. 201–221
8. K. Mochizuki, Uniform resolvent estimates for magnetic Schödinger operators and smoothing effects for related evolution equations. Publ. Res. Inst. Math. Sci. **46**, 741–754 (2010)
9. K. Mochizuki, Resolvent estimates for magnetic Schrödinger operators and their applications to related evolution equations. Rend. Inst. Math. Univ. Trieste **42**(Suppl.), 143–164 (2010)
10. K. Mochizuki, S. Murai, Smoothing properties and scattering for the magnetic Schrödinger and Klein-Gordon equations in an exterior domain with time-dependent perturbations. Publ. Res. Inst. Math. Sci. **53**, 371–393 (2017)
11. K. Mochizuki, H. Nakazawa, Uniform resolvent estimates for magnetic Schrödinger operators in 2D exterior domain and their applications to related ecolution equations. Publ. Res. Inst. Math. Sci. **51**, 319–336 (2015)
12. M. Reed, B. Simon, *Methods of Modern Mathematical Physics*, vol. 3 (Academic, San Diego, 1979)

The Cauchy Problem for Dissipative Wave Equations with Weighted Nonlinear Terms

Makoto Nakamura and Hidemitsu Wadade

Abstract The Cauchy problem for dissipative wave equations with weighted nonlinear terms is considered. The nonlinear terms are power type with a singularity at the origin of Coulomb type. The local and global solutions are shown in the energy class by the use of the Caffarelli-Kohn-Nirenberg inequality. The exponential type nonlinear terms are also considered in the critical two-spatial dimensions.

1 Introduction

In this paper, we consider local and global energy solutions for the Cauchy problem of dissipative wave equations with nonlinear terms which have a space-singularity at the origin. Let $n \ge 1$, $0 \le s \le 1$, $s < n/2$. Let Q be any fixed function which satisfies

$$C_*|\xi|^2 \le Q(\xi) \le C^*|\xi|^2 \tag{1}$$

for any $\xi \in \mathbb{R}^n$ for some constants $C_* > 0$, $C^* > 0$, and put $\Delta_Q := -F^{-1}Q(\xi)F$ for the Fourier and inverse Fourier transform F and F^{-1}. We note Δ_Q is a generalization of the Laplacian $\Delta := \sum_{j=1}^n \partial^2/\partial x_j^2$ and $\Delta_Q = \Delta$ when $Q(\xi) = |\xi|^2$. Our Cauchy problem is given by

$$\begin{cases} (\partial_t^2 - \Delta_Q + \partial_t)u(t,x) + \dfrac{f(u(t,x))}{|x|^s} = 0 \quad \text{for } (t,x) \in [0,T) \times \mathbb{R}^n \\ u(0,\cdot) = u_0(\cdot) \in H^1(\mathbb{R}^n), \quad \partial_t u(0,\cdot) = u_1(\cdot) \in L^2(\mathbb{R}^n), \end{cases} \tag{2}$$

M. Nakamura (✉)
Faculty of Science, Yamagata University, Yamagata, Japan
e-mail: nakamura@sci.kj.yamagata-u.ac.jp

H. Wadade
Faculty of Mechanical Engineering, Institute of Science and Engineering, Kanazawa University, Kanazawa, Japan
e-mail: wadade@se.kanazawa-u.ac.jp

M. D'Abbicco et al. (eds.), *New Tools for Nonlinear PDEs and Application*, Trends in Mathematics, https://doi.org/10.1007/978-3-030-10937-0_11

where u_0, u_1 and f are real-valued functions. To denote power type nonlinear terms of order p, we define the following set $N(p)$. We note that $f(u) = \lambda|u|^{p-1}u$ and $f(u) = \lambda|u|^p$ for $\lambda \in \mathbb{R}$ satisfy $f \in N(p)$.

Definition 1.1 Let $p \geq 1$. We denote by $N(p)$ the set of function f from $\mathbb{R}$ to $\mathbb{R}$ which satisfies $f(0) = 0$ and

$$|f(u) - f(v)| \leq C \max_{w=u,v} |w|^{p-1}|u - v| \tag{3}$$

for any u and $v \in \mathbb{R}$, where $C > 0$ is a constant independent of u and v.

The nonlinear term in (2) has a singularity at the origin when $s > 0$. The case $n = 3$, $s = 1$ and $f(u) = -u$ is known as the Coulomb potential. The case $n \geq 1$, $s = 2$ and $f(u) = \lambda u$ for $\lambda \in \mathbb{R}$ is known as the square-inverse potential (see [3] for elliptic equations, [2] for heat equation, [4] and [30] for Schödinger equations, [13] for Klein-Gordon equations). The nonlinear terms of the form $|u|^{p-1}u/|x|^s$ are also considered for elliptic equations and p-Laplace equations (see [14, 24] and the references therein). Our aim in this paper is to analyze how the singularity affects on the Cauchy problem and give a unified way to control it in the framework of energy solutions for dissipative wave equations. When there is no singularity ($s = 0$), the power type nonlinear terms $f(u)$ could be handled by the standard Sobolev embeddings $H^1(\mathbb{R}^n) \hookrightarrow L^q(\mathbb{R}^n)$ for $\max\{0, 1/2-1/n\} \leq 1/q \leq 1/2$ with $(n, q) \neq (2, \infty)$. When there is the singularity ($s > 0$), we use the Caffarelli-Kohn-Nirenberg inequality Lemma 3.1, below, and we show the following result. Here, $C_b(I, X) = C(I, X) \cap L^\infty(I, X)$ and $C_b^1(I, X) := \{u \in C^1(I, X) \cap L^\infty(I, X) : \partial_t u \in C_b(I, X)\}$ for any interval $I \subset \mathbb{R}$ and normed space X.

Theorem 1.2 *Let $n \geq 1$, $0 \leq s \leq 1$, $s < n/2$. Let p satisfy*

$$1 \leq p \begin{cases} < \infty & \text{if } n = 1, 2 \\ \leq 1 + \frac{2(1-s)}{n-2} & \text{if } n \geq 3. \end{cases} \tag{4}$$

Let $f \in N(p)$. Then we have the following results.

(1) *For any u_0 and u_1, there exists $T = T(\|u_0\|_{H^1(\mathbb{R}^n)} + \|u_1\|_{L^2(\mathbb{R}^n)}) > 0$ and a unique solution $u \in C_b([0, T), H^1(\mathbb{R}^n)) \cap C_b^1([0, T), L^2(\mathbb{R}^n))$ of* (2).
(2) *If $n = 1, 2$, $\|u_0\|_{H^1(\mathbb{R}^n)} + \|u_1\|_{L^2(\mathbb{R}^n)}$ is sufficiently small, and $1+2(2-s)/n \leq p$, then the solution u of* (1) *is a global solution, namely, we are able to take $T = \infty$.*

Remark 1.3 The scaling argument shows that the critical index for the upper bound for p is given by $p(\mu) = 1 + 2(2 - s)/(n - 2\mu)$ for the Cauchy problems $(\partial_t - \Delta)u(t, x) + |u(t, x)|^p/|x|^s = 0$ with $u(0, \cdot) \in \dot{H}^\mu(\mathbb{R}^n)$, and $(\partial_t^2 - \Delta)u(t, x) + |u(t, x)|^p/|x|^s = 0$ with $u(0, \cdot) \in \dot{H}^\mu(\mathbb{R}^n)$, $\partial_t u(0, \cdot) \in \dot{H}^{\mu-1}(\mathbb{R}^n)$, where $\dot{H}^\mu(\mathbb{R}^n)$ denotes the homogeneous Sobolev space of order $\mu \in \mathbb{R}$. In this sense, the upper bound for (2) is expected to be $p(1)$. However, $p(1)$ is not achieved in (4), while the lower bound $1 + 2(2 - s)/n$ in (2) equals to $p(0)$ and it is the critical index for the $L^2(\mathbb{R}^n)$ theory.

Remark 1.4 We have considered the single nonlinear term f in Theorem 1.2. W are also able to consider the sum of nonlinear terms by the analogous proof of the theorem. Let us consider only the case of double nonlinear terms. Namely, let $n \geq 1$, $0 \leq s_j \leq 1$, $s_j < n/2$ for $j = 1, 2$, and let p_1 and p_2 satisfy

$$1 \leq p_j \begin{cases} < \infty & \text{if } n = 1, 2 \\ \leq 1 + \frac{2(1-s_j)}{n-2} & \text{if } n \geq 3 \end{cases} \tag{5}$$

for $j = 1, 2$. Let $f_1 \in N(p_1)$ and $f_2 \in N(p_2)$. Then the result (1) in Theorem 1.2 holds with (2) replaced by

$$\begin{cases} (\partial_t^2 - \Delta_Q + \partial_t)u(t,x) + \dfrac{f_1(u(t,x))}{|x|^{s_1}} + \dfrac{f_2(u(t,x))}{|x|^{s_2}} = 0 \quad \text{for } (t,x) \in [0,\infty) \times \mathbb{R}^n \\ u(0,\cdot) = u_0(\cdot) \in H^1(\mathbb{R}^n), \quad \partial_t u(0,\cdot) = u_1(\cdot) \in L^2(\mathbb{R}^n). \end{cases} \tag{6}$$

The result (2) in Theorem 1.2 also holds with $1 + 2(2 - s)/n \leq p$ replaced by $1+2(2-s_j)/n \leq p_j$ for $j = 1, 2$. Analogous remark is also valid for the following theorems.

Theorem 1.2 shows that any growth order p of polynomial type is subcritical for $n = 1, 2$. The next theorem shows that any growth order of C^1 type is also subcritical when $n = 1$. This result follows from the embedding $H^1(\mathbb{R}) \hookrightarrow L^\infty(\mathbb{R})$, by which we are able to treat any growth order of the nonlinear term at infinity $|u| \to \infty$.

Theorem 1.5 *Let $g \in C^1(\mathbb{R})$ be a real-valued function. When $n = 1$, Theorem 1.2 is also true even if we replace $f(u)$ with $f(u)g(u)$.*

When $n = 2$, the embedding $H^1(\mathbb{R}^2) \hookrightarrow L^\infty(\mathbb{R}^2)$ does not hold and it is critical embedding. The next theorem shows that the exponential growth order is also subcritical. We use the weighted Gagliardo-Nirenberg interpolation inequality Lemma 3.5, below, to prove it. When $\Delta_Q = \Delta$, the exponential nonlinear terms without singularities ($s = 0$) have been considered for Schrödinger equations in [10, 39], wave equations in [25, 40], Klein-Gordon equations in [27, 41], heat equations in [26], complex Ginzburg-Landau equations and dissipative wave equations in [38], damped Klein-Gordon equations in [1]. We put

$$D := F^{-1}\sqrt{Q(\xi)}F, \tag{7}$$

which is a generalization of $|\nabla|$.

Theorem 1.6 *Let $n = 2$, $0 \leq s < 1$. Let $\alpha > 0$, $\lambda \in \mathbb{R}$. Let $f(u) = \lambda u(e^{\alpha u^2} - 1)$. Then we have the following results.*

(1) *If* $\|\nabla u_0\|_{L^2(\mathbb{R}^2)} < \{2\pi(1-s)/\alpha\}^{1/2}$, *and* $T > 0$ *is sufficiently small, then* (2) *has a unique local solution* $u \in C_b([0,T), H^1(\mathbb{R}^2)) \cap C_b^1([0,T), L^2(\mathbb{R}^2))$.
(2) *If* $\|u_0\|_{H^1(\mathbb{R}^2)} + \|u_1\|_{L^2(\mathbb{R}^2)}$ *is sufficiently small, then the solution* u *of (1) is a global solution, namely, we are able to take* $T = \infty$.

Remark 1.7 The index 2 of $e^{\alpha u^2}$ seems to be optimal in the class $H^1(\mathbb{R}^2)$ in view of the Trudinger inequality (see [47]). When $e^{\alpha u^2}$ is replaced by $e^{\alpha|u|^{2-\varepsilon}}$ for $0 < \varepsilon \le 2$, the size restriction for data in (1) of Theorem 1.6 could be removed.

Finally, we consider the global solutions for large data when our equation satisfies the energy conservation law, which becomes dissipative by $\lambda \ge 0$.

Theorem 1.8 *Let* $n \ge 1$, $0 \le s \le 1$, $s < n/2$. *Let* $\lambda \ge 0$. *Let* $f(u)$ *be given by the following* (1) *or* (2).

(1) *We put* $f(u) = \lambda|u|^{p-1}u$, *where* p *satisfies* (4).
(2) *Let* $n = 2$, $\alpha > 0$. *We put* $f(u) = \lambda u(e^{\alpha u^2} - 1)$. *We assume*

$$\int_{\mathbb{R}^2} \left(|Du_0|^2 + u_1^2 + \frac{\lambda(e^{\alpha u_0^2} - 1 - \alpha u_0^2)}{\alpha|x|^s} \right) dx \le \frac{2\pi(1-s)C_*}{\alpha}, \tag{8}$$

where C_* *is the constant in* (1).

Then for any u_0 *and* u_1, (2) *has a unique global solution* $u \in C_b([0,\infty), H^1(\mathbb{R}^n)) \cap C_b^1([0,\infty), L^2(\mathbb{R}^n))$.

The global existence or blowing up of solutions for the Cauchy problem (2) have been considered by many authors for power type nonlinear terms with $s = 0$ and $\Delta_Q = \Delta$. See for example [19, 20, 22, 28, 31, 32, 36, 38, 42–44, 48, 50]. For the detailed review of the history, we refer to [45] by Nishihara. The analysis for nonlinear terms with singularities ($s > 0$) are considered for elliptic equations extensively where the singularities appear in a domain or on its boundary (see [6–8, 12, 16–18, 21, 23, 34, 35, 49]). We consider the nonstationary problem and we give a unified way to construct energy solutions for $s \ge 0$ based on the classical energy estimates, the Caffarelli-Kohn-Nirenberg inequality and the Gagliardo-Nirenberg inequality. For the latter inequality, we refer to [29] and [37] for recent results. We refer to [11, 15] for nonlinear Schrödinger equations and nonlinear Klein-Gordon equations. We remark that the dissipative wave equation has both aspects of the heat equation and the wave equation, and the linear and nonlinear estimates in this paper are applicable to the perturbation of equations and nonlinear terms. The arguments in this paper would provide a unified method to treat nonlinear terms with singular weights for parabolic or hyperbolic equations. We refer to the book [9] by Cherrier and Milani for a unified approach for the Cauchy problems of quasilinear parabolic and hyperbolic equations with variable coefficients based on energy estimates, and we consider the case of singular nonlinear terms in this paper. We put $\nabla := (\partial_1, \cdots, \partial_n)$, $\nabla_{t,x} := (\partial_t, \nabla)$, and $D_{t,x} := (\partial_t, D)$, where D is defined by (7). The notation $a \lesssim b$ denotes the inequality $a \leq Cb$ for a positive constant C which is not essential for our arguments.

This paper is organized as follows. In Sects. 2 and 3, we prepare linear estimates and nonlinear estimates, respectively. In Sect. 4, we prepare a priori estimates on the energy, which are used in Sect. 5. In Sect. 5, we prove the theorems.

2 Estimates for Linear Terms

We prepare the following two fundamental results. Let $0 < T \leq \infty$. We consider the Cauchy problem

$$\begin{cases} (\partial_t^2 - \Delta_Q + \partial_t)u(t,x) + h(t,x) = 0 \quad \text{for } (t,x) \in [0,T) \times \mathbb{R}^n \\ u(0,\cdot) = u_0(\cdot), \quad \partial_t u(0,\cdot) = u_1(\cdot). \end{cases} \tag{9}$$

First, we consider the energy estimates. The solution of (9) satisfies the following energy estimates.

$$\begin{aligned} &\|u\|^2_{L^\infty((0,T),H^1(\mathbb{R}^n))} + \|\partial_t u\|^2_{L^\infty((0,T),L^2(\mathbb{R}^n))} + \|\nabla_{t,x} u\|^2_{L^2((0,T)\times\mathbb{R}^n)} \\ &\quad \leq C\|u_0\|^2_{H^1(\mathbb{R}^n)} + C\|u_1\|^2_{L^2(\mathbb{R}^n)} + C\int_0^T \int_{\mathbb{R}^n} (|u| + |\partial_t u|)|h| dxdt, \end{aligned} \tag{10}$$

where the constant $C > 0$ is independent of u, T, u_0, u_1 and h. Indeed, by the multiplication of $\partial_t u$ to $(\partial_t^2 - \Delta_Q + \partial_t)u = h$, the integration by t and x variables, and the divergence theorem, we have

$$\begin{aligned} &\|Du(t,\cdot)\|^2_{L^2(\mathbb{R}^n)} + \|\partial_t u(t,\cdot)\|^2_{L^2(\mathbb{R}^n)} + 2\|\partial_t u\|^2_{L^2((0,t)\times\mathbb{R}^n)} \\ &\quad = \|Du(0,\cdot)\|^2_{L^2(\mathbb{R}^n)} + \|\partial_t u(0,\cdot)\|^2_{L^2(\mathbb{R}^n)} - 2\int_0^t \int_{\mathbb{R}^n} \partial_t u\, h\, dxds. \end{aligned} \tag{11}$$

Similarly, the multiplication of u to the equation yields

$$\begin{aligned} &\|u(t,\cdot)\|^2_{L^2(\mathbb{R}^n)} + 2\|Du\|^2_{L^2((0,t)\times\mathbb{R}^n)} = \|u(0,\cdot)\|^2_{L^2(\mathbb{R}^n)} + 2\|\partial_t u\|^2_{L^2((0,t)\times\mathbb{R}^n)} \\ &\quad - 2\int_0^t \int_{\mathbb{R}^n} u\, h\, dxds + 2\int_{\mathbb{R}^n} u(0,\cdot)\partial_t u(0,\cdot) dx - 2\int_{\mathbb{R}^n} u(t,\cdot)\partial_t u(t,\cdot) dx, \end{aligned} \tag{12}$$

where the last two terms are bounded by

$$\|u(0,\cdot)\|_{L^2(\mathbb{R}^n)}^2+\|\partial_t u(0,\cdot)\|_{L^2(\mathbb{R}^n)}^2+2\|\partial_t u(t,\cdot)\|_{L^2(\mathbb{R}^n)}^2+\|u(t,\cdot)\|_{L^2(\mathbb{R}^n)}^2/2. \tag{13}$$

Combining the above estimates and $C_*|\xi|^2 \le Q(\xi) \le C^*|\xi|^2$, we obtain (10).

Second, we rewrite the Cauchy problem (9) as the integral equation. For $\xi \in \mathbb{R}^n$, we put $[\xi] := (Q(\xi)-1/4)^{1/2}$ if $Q(\xi) \ge 1/4$ and $[\xi] := i(1/4-Q(\xi))^{1/2}$ if $Q(\xi) < 1/4$. We put $\omega := F^{-1}[\xi]F$ and

$$K(t) := e^{-t/2}\,\frac{\sin t\omega}{\omega} = e^{-t/2}\,F^{-1}\frac{\sin t[\xi]}{[\xi]}F \tag{14}$$

for $t \ge 0$. We regard the differential equation (9) as the integral equation of the form

$$u(t,\cdot) = \partial_t K(t)u_0 + K(t)(u_0+u_1) - \int_0^t K(t-\tau)h(\tau,\cdot)d\tau. \tag{15}$$

Lemma 2.1 *Let $u_0 \in H^1(\mathbb{R}^n)$, $u_1 \in L^2(\mathbb{R}^n)$, $h \in L^1((0,T),L^2(\mathbb{R}^n))$. Then u given by* (15) *satisfies $u \in C([0,T],H^1(\mathbb{R}^n))$, $\partial_t u \in C([0,T],L^2(\mathbb{R}^n))$.*

Proof We put $\langle\xi\rangle := \sqrt{1+\xi^2}$. We have $K(\cdot)g \in C([0,\infty),H^1(\mathbb{R}^n))$ for $g \in L^2(\mathbb{R}^n)$ by the Lebesgue convergence theorem,

$$\begin{aligned}&\|K(t+\varepsilon)g-K(t)g\|_{H^1(\mathbb{R}^n)}\\ &= \left\|\langle\xi\rangle\Big(e^{-(t+\varepsilon)/2}\frac{\sin(t+\varepsilon)[\xi]}{[\xi]} - e^{-t/2}\frac{\sin t[\xi]}{[\xi]}\Big)Fg\right\|_{L^2(\mathbb{R}^n)}\end{aligned} \tag{16}$$

and $|\sin t[\xi]/[\xi]| \lesssim e^{t/2}\langle\xi\rangle^{-1}$ for $t \ge 0$ and $\varepsilon \in \mathbb{R}$. So that, we also have $\int_0^\cdot K(\cdot-\tau)h(\tau)d\tau \in C([0,T],H^1(\mathbb{R}^n))$ for $h \in L^1((0,T),L^2(\mathbb{R}^n))$. Similarly, we have

$$e^{-t/2}\cos t\omega g \in C([0,\infty),L^2(\mathbb{R}^n))$$

for $g \in L^2(\mathbb{R}^n)$ by $|\cos t[\xi]| \le e^{t/2}$, which shows $\partial_t K(\cdot)g \in C([0,\infty),L^2(\mathbb{R}^n))$ for $g \in L^2(\mathbb{R}^n)$. Therefore, we obtain $u \in C([0,T],H^1(\mathbb{R}^n))$. Since $\partial_t^2 K(\cdot)g \in C([0,\infty),L^2(\mathbb{R}^n))$ for $g \in H^1(\mathbb{R}^n)$ by $\partial_t^2 K(\cdot)g = \Delta_Q K(\cdot)g - \partial_t K(\cdot)g$, we also obtain $\partial_t u \in C([0,T],L^2(\mathbb{R}^n))$. □

Especially, the next result follows quickly from Lemma 2.1.

Corollary 2.2 *Let u be the solution of* (9) *with $h = 0$. Namely, u is the free solution for the dissipative wave equation. Then the following estimate holds.*

$$\lim_{t\to 0}\|\nabla u(t,\cdot)-\nabla u_0(\cdot)\|_{L^2(\mathbb{R}^n)} = 0 \tag{17}$$

3 Estimates for Nonlinear Terms

We collect several estimates for nonlinear terms.

Lemma 3.1 (Caffarelli, Kohn, Nirenberg [5]) *Let $n \ge 1$, $1 \le p, q, r < \infty$, $0 \le a \le 1$. Let real numbers σ, α, β, γ satisfy*

$$\frac{1}{p} + \frac{\alpha}{n} > 0, \quad \frac{1}{q} + \frac{\beta}{n} > 0, \quad \frac{1}{r} + \frac{\gamma}{n} > 0, \quad \gamma = a\sigma + (1-a)\beta. \tag{18}$$

Then the inequality

$$\| |x|^{\gamma} u \|_{L^r(\mathbb{R}^n)} \le C \| |x|^{\alpha} \nabla u \|^{a}_{L^p(\mathbb{R}^n)} \| |x|^{\beta} u \|^{1-a}_{L^q(\mathbb{R}^n)} \tag{19}$$

holds for any nonconstant function u if and only if

$$\begin{cases} \dfrac{1}{r} + \dfrac{\gamma}{n} = a\left(\dfrac{1}{p} + \dfrac{\alpha - 1}{n}\right) + (1-a)\left(\dfrac{1}{q} + \dfrac{\beta}{n}\right) \\ 0 \le \alpha - \sigma \text{ if } a > 0 \\ \alpha - \sigma \le 1 \text{ if } a > 0 \text{ with } \frac{1}{p} + \frac{\alpha-1}{n} = \frac{1}{r} + \frac{\gamma}{n}. \end{cases} \tag{20}$$

We also refer to [33] for the generalization of Caffarelli-Kohn-Nirenberg's inequality to the inequalities involving higher order derivatives. The following lemma follows from the above lemma when we put $p = q = 2$, $\alpha = \beta = 0$, and σ as $\gamma = a\sigma$.

Lemma 3.2 *Let $n \ge 1$, and let r, γ satisfy*

$$0 < \frac{1}{r} \le \frac{1}{2}, \quad \frac{1}{2} - \frac{1}{n} \le \frac{1}{r}, \quad n\left(\frac{1}{2} - \frac{1}{r}\right) - 1 \le \gamma \le 0, \quad -\frac{n}{r} < \gamma. \tag{21}$$

Then there exists a constant $C > 0$ such that the inequality

$$\| |x|^{\gamma} u \|_{L^r(\mathbb{R}^n)} \le C \| \nabla u \|^{a}_{L^2(\mathbb{R}^n)} \| u \|^{1-a}_{L^2(\mathbb{R}^n)} \tag{22}$$

holds for any nonconstant function u, where $a := n(1/2 - 1/r) - \gamma$ satisfies $0 \le a \le 1$.

By the use of the above lemma, we show the following estimates for nonlinear terms.

Lemma 3.3 *Let $n \ge 1$, $0 \le s \le 1$, $s < n/2$,*

$$1 \le p \begin{cases} < \infty & \text{if } n = 1, 2 \\ \le 1 + \frac{2(1-s)}{n-2} & \text{if } n \ge 3 \end{cases} \tag{23}$$

We put $a := n(p-1)/2p + s/p$. *Let* $f \in N(p)$. *Then there exists a constant* $C > 0$ *such that the following inequalities hold.*

(1)
$$\left\| \frac{f(u)}{|x|^s} \right\|_{L^1((0,T),L^2(\mathbb{R}^n))} \le C \|\nabla u\|^{ap}_{L^{ap}((0,T),L^2(\mathbb{R}^n))} \|u\|^{(1-a)p}_{L^\infty((0,T),L^2(\mathbb{R}^n))} \tag{24}$$

(2)
$$\begin{aligned} &\left\| \frac{f(u)-f(v)}{|x|^s} \right\|_{L^1((0,T),L^2(\mathbb{R}^n))} \\ &\quad \le C \max_{w=u,v} \|\nabla w\|^{a(p-1)}_{L^{ap}((0,T),L^2(\mathbb{R}^n))} \|w\|^{(1-a)(p-1)}_{L^\infty((0,T),L^2(\mathbb{R}^n))} \\ &\quad\quad \cdot \|\nabla(u-v)\|^a_{L^{ap}((0,T),L^2(\mathbb{R}^n))} \|u-v\|^{1-a}_{L^\infty((0,T),L^2(\mathbb{R}^n))} \end{aligned} \tag{25}$$

for any $T > 0$ *and any nonconstant functions* u *and* v.

In Lemma 3.3, we remark that $0 \le a \le 1$ holds. And $ap \ge 2$ holds if and only if $1 + 2(2-s)/n \le p$.

Proof (1) By Lemma 3.2, we have

$$\left\| \frac{f(u)}{|x|^s} \right\|_{L^2(\mathbb{R}^n)} \le C \left\| \frac{u}{|x|^{s/p}} \right\|^p_{L^{2p}(\mathbb{R}^n)} \le C \|\nabla u\|^{ap}_{L^2(\mathbb{R}^n)} \|u\|^{(1-a)p}_{L^2(\mathbb{R}^n)}. \tag{26}$$

Applying the Hölder inequality in time variable, we obtain the required result. The proof of (2) follows similarly by the use of (3). □

Corollary 3.4 *Let* $g \in C^1(\mathbb{R})$ *be a real-valued function. Under the same assumption of Lemma* 3.3, *the following estimates hold.*

(1)
$$\begin{aligned} &\left\| \frac{f(u)g(u)}{|x|^s} \right\|_{L^1((0,T),L^2(\mathbb{R}))} \\ &\quad \lesssim C \|\nabla u\|^{ap}_{L^{ap}((0,T),L^2(\mathbb{R}))} \|u\|^{(1-a)p}_{L^\infty((0,T),L^2(\mathbb{R}))} G(\|u\|_{L^\infty((0,T)\times\mathbb{R})}) \end{aligned} \tag{27}$$

(2)
$$\begin{aligned} &\left\| \frac{f(u)g(u)-f(v)g(v)}{|x|^s} \right\|_{L^1((0,T),L^2(\mathbb{R}))} \\ &\quad \le C \max_{w=u,v} \|\nabla w\|^{a(p-1)}_{L^{ap}((0,T),L^2(\mathbb{R}))} \|w\|^{(1-a)(p-1)}_{L^\infty((0,T),L^2(\mathbb{R}))} \\ &\quad \cdot \|\nabla(u-v)\|^a_{L^{ap}((0,T),L^2(\mathbb{R}))} \|u-v\|^{1-a}_{L^\infty((0,T),L^2(\mathbb{R}))} \\ &\quad\quad \cdot \max_{w=u,v} G(\|w\|_{L^\infty((0,T)\times\mathbb{R})}), \end{aligned} \tag{28}$$

where $G(\rho) := \max\{\max_{|z|\le\rho} |g(z)|,\ \rho \max_{|z|\le\rho} |g'(z)|\}$ *for* $\rho \ge 0$.

Proof The result (1) follows from the proof of Lemma 3.3 analogously by the use of $|g(u)| \le G(|u|)$. The result (2) follows from the bound

$$|f(u)g(u) - f(v)g(v)| \le C \max_{w=u,v} |w|^{p-1}|u-v| \max_{w=u,v} G(|w|). \tag{29}$$

□

Lemma 3.5 ([29, Corollary 1.6]) *Let* $0 \le t < 2$, $\beta > (4\pi e(2-t))^{-1/2}$. *Then there exists a constant* $r = r(t, \beta) \ge 2$ *such that the inequality*

$$\left\| \frac{u}{|x|^{t/q}} \right\|_{L^q(\mathbb{R}^2)} \le \beta q^{1/2} \|u\|_{L^2(\mathbb{R}^2)}^{(2-t)/q} \|\nabla u\|_{L^2(\mathbb{R}^2)}^{1-(2-t)/q} \tag{30}$$

holds for any $u \in H^1(\mathbb{R}^2)$ *and any* q *with* $r \le q < \infty$.

Lemma 3.6 *Let* $0 \le s < 1$, $T > 0$, $\beta > (8\pi e(1-s))^{-1/2}$. *Let* $\alpha > 0$, $\lambda \in \mathbb{R}$. *We put*

$$f(u) := \lambda u(e^{\alpha u^2} - 1). \tag{31}$$

Then there exists a constant $C > 0$ *such that the following estimates hold.*

$$\text{(1)} \quad \left\| \frac{f(u)}{|x|^s} \right\|_{L^1((0,T),L^2(\mathbb{R}^2))} \le C \sum_{j=1}^{\infty} a_u(j) \|\nabla u\|_{L^2((0,T)\times\mathbb{R}^2)}^2 \cdot \|\nabla u\|_{L^\infty((0,T),L^2(\mathbb{R}^2))}^s \|u\|_{L^\infty((0,T),L^2(\mathbb{R}^2))}^{1-s} \tag{32}$$

$$\text{(2)} \quad \left\| \frac{f(u)-f(v)}{|x|^s} \right\|_{L^1((0,T),L^2(\mathbb{R}^2))} \le C \max_{w=u,v} \sum_{j=1}^{\infty} j a_w(j) \|\nabla w\|_{L^2((0,T)\times\mathbb{R}^2)}^{1+(1-s)/p(j)} \cdot \|\nabla w\|_{L^\infty((0,T),L^2(\mathbb{R}^2))}^s \|w\|_{L^\infty((0,T),L^2(\mathbb{R}^2))}^{(1-s)2j/p(j)} \cdot \|\nabla(u-v)\|_{L^2((0,T)\times\mathbb{R}^2)}^{1-(1-s)/p(j)} \|u-v\|_{L^\infty((0,T),L^2(\mathbb{R}^2))}^{(1-s)/p(j)}, \tag{33}$$

where $p(j) := 2j+1$, $a_u(j) := \alpha^j \beta^{p(j)} (2p(j))^{p(j)/2} \|\nabla u\|_{L^\infty((0,T),L^2(\mathbb{R}^2))}^{p(j)-3} / j!$, *and the series* $\sum_{j=1}^{\infty} a_u(j)$ *and* $\sum_{j=1}^{\infty} j a_u(j)$ *are finite if*

$$\alpha \|\nabla u\|_{L^\infty((0,T),L^2(\mathbb{R}^2))}^2 < 2\pi(1-s). \tag{34}$$

Proof (1) We use the expansion $|f(u)| \lesssim \sum_{j=1}^{\infty} \frac{\alpha^j}{j!}|u|^{p(j)}$ to have

$$\left\| \frac{f(u)}{|x|^s} \right\|_{L^2(\mathbb{R}^2)} \lesssim \sum_{j=1}^{\infty} \frac{\alpha^j}{j!} \left\| \frac{u}{|x|^{s/p(j)}} \right\|_{L^{2p(j)}(\mathbb{R}^2)}^{p(j)}. \tag{35}$$

For any fixed β with $\beta > (8\pi e(1-s))^{-1/2}$, we have by Lemma 3.5 the estimate

$$\left\| \frac{u}{|x|^{s/p(j)}} \right\|_{L^{2p(j)}(\mathbb{R}^2)} \leq \beta(2p(j))^{1/2} \|\nabla u\|_{L^2(\mathbb{R}^2)}^{1-(1-s)/p(j)} \|u\|_{L^2(\mathbb{R}^2)}^{(1-s)/p(j)} \tag{36}$$

for sufficiently large j, where we are also able to obtain the similar bound for small j by Lemma 3.2. By the Hölder inequality in time variable, we obtain the required result. The proof of (2) follows similarly if we use the bound

$$|f(u) - f(v)| \lesssim \max_{w=u,v} \sum_{j=1}^{\infty} \frac{\alpha^j}{(j-1)!} |w|^{2j} |u - v|. \tag{37}$$

Finally, the condition (34) yields the convergence of the series since $\lim_{j\to\infty} a_u(j+1)/a_u(j) < 1$. □

4 A Priori Estimates

We prepare a priori estimates for global solutions for large data.

Lemma 4.1 *Let $T > 0$, $n \geq 1$. Let u be the solution of*

$$\begin{cases} (\partial_t^2 - \Delta_Q + \partial_t)u(t,x) + h(x, u(t,x)) = 0 \ \textit{for} \ (t,x) \in [0,T) \times \mathbb{R}^n \\ u(0,\cdot) = u_0(\cdot), \ \ \partial_t u(0,\cdot) = u_1(\cdot). \end{cases} \tag{38}$$

We assume that there exists a function H such that $h(x,u) = \partial H(x,u)/\partial u$. We put

$$E(u)(t) := \int_{\mathbb{R}^n} (\partial_t u(t,x))^2 + |Du(t,x)|^2 + 2H(x, u(t,x))dx. \tag{39}$$

Then we have the following results.

$$(1) \quad E(u)(t) + 2\|\partial_t u\|_{L^2((0,t)\times\mathbb{R}^n)}^2 = E(u)(0) \quad \textit{for} \ \ 0 \leq t < T. \tag{40}$$

(2) *Let $(u_0, u_1) \neq (0,0)$. We assume that H satisfies $H(x,u) \geq 0$ for any $(x,u) \in \mathbb{R}^{n+1}$, and $u = 0$ if $H(x,u) = 0$. Then*

$$\|Du_0\|_{L^2(\mathbb{R}^n)}^2 < E(u)(0), \quad \|Du\|_{L^\infty((0,T),L^2(\mathbb{R}^n))}^2 < E(u)(0). \tag{41}$$

Moreover, for any fixed t_0 with $0 \le t_0 < T$, if v is the free solution of

$$\begin{cases} (\partial_t^2 - \Delta_Q + \partial_t)v(t,x) = 0 \text{ for } (t,x) \in [t_0,\infty) \times \mathbb{R}^n \\ v(t_0,\cdot) = u(t_0,\cdot), \quad \partial_t v(t_0,\cdot) = \partial_t u(t_0,\cdot), \end{cases} \tag{42}$$

then

$$\|\partial_t v(t,\cdot)\|_{L^2(\mathbb{R}^n)}^2 + \|Dv(t,\cdot)\|_{L^2(\mathbb{R}^n)}^2 + 2\|\partial_t v\|_{L^2((t_0,t)\times\mathbb{R}^n)}^2 \le E(u)(0) - 2\|\partial_t u\|_{L^2((0,t_0)\times\mathbb{R}^n)}^2 \tag{43}$$

for any $t_0 \le t < \infty$, and the bound $\|Dv\|_{L^\infty((t_0,T),L^2(\mathbb{R}^n))}^2 < E(u)(0)$ holds.

Proof The result (1) follows easily from $h = \partial H/\partial u$, and the multiplication of $\partial_t u$ to the first equation in (38). We prove (2) in the following. $\|Du_0\|_{L^2(\mathbb{R}^n)}^2 < E(u)(0)$ follows from the definition of $E(u)(0)$, $(u_0,u_1) \ne (0,0)$, and $u_0 = 0$ if $h(x,u_0) = 0$. So that, if $\|Du\|_{L^\infty((0,T),L^2(\mathbb{R}^n))}^2 = E(u)(0)$, there exists a nondecreasing sequence $\{t_j\}_{j=1}^\infty$ such that $0 < t_j \le t_\infty := \lim_{j\to\infty} t_j \le T$, $E(u)(0) = \lim_{j\to\infty} \|Du(t_j,\cdot)\|_{L^2(\mathbb{R}^n)}^2$. This means

$$\lim_{j\to\infty} \int_{\mathbb{R}^n} \Big(\partial_t u(t_j,\cdot)^2 + 2H(x,u(t_j,x))\Big)\,dx = 0 \quad \text{and} \quad \|\partial_t u\|_{L^2((0,t_\infty)\times\mathbb{R}^n)} = 0 \tag{44}$$

by (1) and $H \ge 0$. The latter shows $u = u_0$ on $[0,t_\infty)$, and $u_1 = 0$, therefore the former shows $u_0 = 0$ since $H(x,u_0) = 0$ yields $u_0 = 0$, which is a contradiction to $(u_0,u_1) \ne (0,0)$. We have shown $\|Du\|_{L^\infty((0,T),L^2(\mathbb{R}^n))}^2 < E(u)(0)$. Next, we show (43). We use (1) for v to have

$$E(v)(t) + 2\|\partial_t v\|_{L^2((t_0,t)\times\mathbb{R}^n)}^2 = E(v)(t_0) \quad \text{for } t_0 \le t < \infty. \tag{45}$$

Since we have

$$E(v)(t_0) \le E(u)(t_0) = E(u)(0) - 2\|\partial_t u\|_{L^2((0,t_0)\times\mathbb{R}^n)}^2, \tag{46}$$

we obtain (43). We show the final part of the lemma. If $\|Dv\|_{L^\infty((t_0,T),L^2(\mathbb{R}^n))}^2 = E(u)(0)$, then there exists a nondecreasing sequence $\{t_j\}_{j=1}^\infty$ such that $t_0 < t_j \le t_\infty := \lim_{j\to\infty} t_j \le T$, $E(u)(0) = \lim_{j\to\infty} \|Dv(t_j,\cdot)\|_{L^2(\mathbb{R}^n)}^2$ since $\|Dv(t_0,\cdot)\|_{L^2(\mathbb{R}^n)}^2 = \|Du(t_0,\cdot)\|_{L^2(\mathbb{R}^n)}^2 < E(u)(0)$ by (41). This means

$$\|\partial_t u\|_{L^2((0,t_0)\times\mathbb{R}^n)} = \|\partial_t v\|_{L^2((t_0,t_\infty)\times\mathbb{R}^n)} = 0 \tag{47}$$

by (43). So that, $v = u_0$ on $[t_0, t_\infty)$ and it yields $\|Du_0\|^2_{L^2(\mathbb{R}^n)} = E(u)(0)$, which is a contradiction to (41). □

5 Proof of Theorems

We prove the theorems in the introduction.

5.1 *Proof of Theorem 1.2*

Let $0 < T \le \infty$, $0 < R < \infty$. We define the complete metric space $X(T, R)$ by $X(T, R) := \{u \in \mathcal{S}'(\mathbb{R}^{n+1}) \mid \|u\|_X \le R\}$, where

$$\|u\|_X := \max\{\|u\|_{L^\infty((0,T),H^1(\mathbb{R}^n))}, \|\partial_t u\|_{L^\infty((0,T),L^2(\mathbb{R}^n))}, \|D_{t,x}u\|_{L^2((0,T)\times\mathbb{R}^n)}\}. \tag{48}$$

We show a map Φ defined by

$$\Phi(u)(t) := \partial_t K(t)u_0 + K(t)(u_0 + u_1) - \int_0^t K(t-\tau)\frac{f(u(\tau, x))}{|x|^s}d\tau \tag{49}$$

is a contraction on $X(T, R)$ for some T and R. By (10), we have

$$\|\Phi(u)\|_X \lesssim \|u_0\|_{H^1} + \|u_1\|_{L^2} + \left\|\frac{f(u)}{|x|^s}\right\|_{L^1_t L^2_x}. \tag{50}$$

We estimate the last term by Lemma 3.3 as

$$\left\|\frac{f(u)}{|x|^s}\right\|_{L^1_t L^2_x} \lesssim \|\nabla u\|^{ap}_{L^{ap}_t L^2_x} \|u\|^{(1-a)p}_{L^\infty_t L^2_x}, \tag{51}$$

where $a := \{n(p-1) + 2s\}/2p$. Since $2 \le ap$ when $1 + 2(2-s)/n \le p$, we have

$$\|\Phi(u)\|_X \le C_0(\|u_0\|_{H^1} + \|u_1\|_{L^2}) + C\begin{cases} T\|u\|^p_X \\ \|u\|^p_X \text{ if } 1 + 2(2-s)/n \le p \end{cases} \tag{52}$$

for some constants $C_0 > 0$ and $C > 0$, which are independent of u, where we have used $\|\nabla u\|_{L^{ap}_t L^2_x} \le T^{1/ap}\|\nabla u\|_{L^\infty L^2}$ for the upper inequality. Similarly, we also have

$$\|\Phi(u) - \Phi(v)\|_X \le C \max_{w=u,v} \|w\|^{p-1}_X \begin{cases} T\|u-v\|_X \\ \|u-v\|_X \text{ if } 1 + 2(2-s)/n \le p. \end{cases} \tag{53}$$

Putting $R = 2C_0(\|u_0\|_{H^1} + \|u_1\|_{L^2})$, we conclude that Φ is a contraction mapping on $X(T, R)$ for sufficiently small $T > 0$ for any data, and moreover $T = \infty$ for sufficiently small data when $1 + 2(2-s)/n \le p$.

Next, we show the uniqueness of the solution. Let $v \in C_b([0, T), H^1(\mathbb{R}^n)) \cap C_b^1([0, T), L^2(\mathbb{R}^n))$ be another solution. We define $t_0 := \inf\{t \in [0, T) : u(t) \neq v(t)\}$ and show $t_0 < T$ yields a contradiction. Let $t_0 < T$. By the continuity of u and v, we have $(u - v)(t_0) = \partial_t(u - v)(t_0) = 0$. Similarly to (53), we have

$$\|u - v\|_{X((t_0,t_0+\varepsilon))} \le C\varepsilon \max_{w=u,v} \|w\|^{p-1}_{X((t_0,t_0+\varepsilon))} \|u - v\|_{X((t_0,t_0+\varepsilon))} \tag{54}$$

for sufficiently small $\varepsilon > 0$, where $\|\cdot\|_{X((t_0,t_0+\varepsilon))}$ is defined by (48) with $(0, T)$ replaced by $(t_0, t_0 + \varepsilon)$. Since $\|v\|_{X((t_0,t_0+\varepsilon))} < \infty$ by $L^\infty((t_0, t_0 + \varepsilon)) \hookrightarrow L^2((t_0, t_0 + \varepsilon))$, we obtain $\|u - v\|_{X((t_0,t_0+\varepsilon))} = 0$ for sufficiently small $\varepsilon > 0$, which shows $u = v$ on $[t_0, t_0 + \varepsilon)$ and a contradiction to the definition of t_0. □

5.2 *Proof of Theorem 1.5*

The proof of Theorem 1.5 follows in a similar way to that of Theorem 1.2. The exception is that we use Corollary 3.4 to derive

$$\left\| \frac{f(u)g(u)}{|x|^s} \right\|_{L_t^1 L_x^2} \lesssim G(C\|u\|_{L^\infty H^1}) \|\nabla u\|^{ap}_{L^{ap} L^2} \|u\|^{(1-a)p}_{L^\infty L^2}, \tag{55}$$

where we have used the embedding $H^1(\mathbb{R}) \hookrightarrow L^\infty(\mathbb{R})$.

5.3 *Proof of Theorem 1.6*

Let u_F be the free solution given by $u_F(t, \cdot) := \partial_t K(t)u_0 + K(t)(u_0 + u_1)$. For any function u_N, let $\Psi(u_N)$ be given by

$$\Psi(u_N)(t) := -\int_0^t K(t - \tau) \frac{f((u_F + u_N)(\tau, x))}{|x|^s} d\tau. \tag{56}$$

Then the fixed point $u_N = \Psi(u_N)$ satisfies $u = u_F + u_N$ for the solution u of (2). Here, u_N denotes the nonlinear part of u. We show Ψ is a contraction mapping on $X(T, R)$ in the following. By (10), we have

$$\|u_F\|_X \lesssim \|u_0\|_{H^1} + \|u_1\|_{L^2}, \tag{57}$$

and by Lemma 3.6, we also have

$$\|\Psi(u_N)\|_X \lesssim \left\| \frac{f(u_F+u_N)}{|x|^s} \right\|_{L_t^1 L_x^2}$$

$$\lesssim \sum_{j=1}^{\infty} a_{u_F+u_N}(j) \|\nabla(u_F+u_N)\|_{L_t^2 L_x^2}^2 \|\nabla(u_F+u_N)\|_{L_t^\infty L_x^2}^s \|u_F+u_N\|_{L_t^\infty L_x^2}^{1-s}$$

$$\lesssim \min\{1,T\} \sum_{j=1}^{\infty} a_{u_F+u_N}(j) \|u_F+u_N\|_X^3. \tag{58}$$

Similarly, we also have

$$\|\Psi(u_N)-\Psi(v_N)\|_X \lesssim \left\| \frac{f(u_F+u_N)-f(u_F+v_N)}{|x|^s} \right\|_{L_t^1 L_x^2}$$

$$\lesssim \max_{w=u_F+u_N, u_F+v_N} \sum_{j=1}^{\infty} j a_w(j) \|\nabla w\|_{L_t^2 L_x^2}^{1+(1-s)/p(j)} \|\nabla w\|_{L_t^\infty L_x^2}^s \|w\|_{L_t^\infty L_x^2}^{2j(1-s)/p(j)}$$

$$\cdot \|\nabla(u_N-v_N)\|_{L_t^2 L_x^2}^{1-(1-s)/p(j)} \|u_N-v_N\|_{L_t^\infty L_x^2}^{(1-s)/p(j)}$$

$$\lesssim \min\{1,T\} \max_{w=u_F+u_N, u_F+v_N} \sum_{j=1}^{\infty} j a_w(j) \|w\|_X^2 \|u_N-v_N\|_X. \tag{59}$$

Here, the series $\sum_{j=1}^{\infty} a_{u_F+u_N}(j)$ and $\sum_{j=1}^{\infty} j a_w(j)$ converge in finite provided

$$\max_{w=u_F+u_N, u_F+v_N} \|\nabla w\|_{L_t^\infty L_x^2}^2 < \frac{2\pi(1-s)}{\alpha}, \tag{60}$$

which is satisfied if

$$\|\nabla u_F\|_{L_t^\infty L_x^2}^2 < \frac{2\pi(1-s)}{\alpha} \tag{61}$$

and $R > 0$ is sufficiently small. By Corollary 2.2, $\|\nabla u_F(t,\cdot)\|_{L^2}$ converges to $\|\nabla u_0\|_{L^2}$. So that, (61) holds if $T > 0$ is sufficiently small. Therefore, we obtain (1) in the theorem. The result (2) follows since (61) holds for small data by the energy estimate (10).

Next, we show the uniqueness of the solution. The solution u obtained by the above argument is written as $u = u_F + u_N$, where u_N is the fixed point of Ψ and satisfies $\|u_N\|_X \le R$. Since R is sufficiently small and u_F satisfies (61), we have

$$\|\nabla u\|_{L^\infty((0,T),L^2(\mathbb{R}^n))}^2 < 2\pi(1-s)/\alpha. \tag{62}$$

Let $v \in C_b([0,T), H^1(\mathbb{R}^2)) \cap C_b^1([0,T), L^2(\mathbb{R}^2))$ be another solution. We define $t_0 := \inf\{t \in [0,T) : u(t) \neq v(t)\}$ and show $t_0 < T$ yields a contradiction. Let $t_0 < T$. By the continuity of u and v, we have $(u-v)(t_0) = \partial_t(u-v)(t_0) = 0$. Let u_{F,t_0} be the free solution given by $u_{F,t_0}(t) := \partial_t K(t-t_0)u(t_0) + K(t-t_0)(u(t_0)+\partial_t u(t_0))$. Similarly to (59), we have

$$\|u-v\|_{X((t_0,t_0+\varepsilon))} \leq C\min\{1,\varepsilon\} \max_{w=u,v} \sum_{j=1}^{\infty} j a_w(j) \cdot \|w\|^2_{X((t_0,t_0+\varepsilon))} \|u-v\|_{X((t_0,t_0+\varepsilon))} \tag{63}$$

for sufficiently small $\varepsilon > 0$. Since $\|v\|_{X((t_0,t_0+\varepsilon))}$ tends to $\|u\|_{X((t_0,t_0+\varepsilon))}$ as ε tends to 0, we have $\sum_{j=1}^{\infty} j a_v(j) < \infty$ by (62) for sufficiently small $\varepsilon > 0$. So that, we obtain $\|u-v\|_{X((t_0,t_0+\varepsilon))} = 0$ for sufficiently small $\varepsilon > 0$, which shows $u = v$ on $[t_0, t_0+\varepsilon)$ and a contradiction to the definition of t_0.

5.4 *Proof of Theorem 1.8*

When $\lambda = 0$, the solution is a free solution and exists globally in time. We assume $\lambda > 0$ in the following. We put

$$H(x,u) := \begin{cases} \lambda |u|^{p+1}/(p+1)|x|^s & \text{for } (1) \\ \lambda(e^{\alpha u^2} - 1 - \alpha u^2)/2\alpha |x|^s & \text{for } (2). \end{cases} \tag{64}$$

Then we have $\partial H/\partial u = f(u)/|x|^s$, $H \geq 0$, and $u = 0$ if $H(x,u) = 0$. We are able to use Lemma 4.1.

(1) We use (40) to have $\|D_{t,x}u\|^2_{L^\infty((0,t),L^2(\mathbb{R}^n))} \leq E(u)(0)$. Combining this and the trivial inequality

$$\|u(t,\cdot)\|_{L^2} \leq \|u(0,\cdot)\|_{L^2} + \int_0^t \|\partial_t u(s,\cdot)\|_{L^2} ds, \tag{65}$$

we have

$$\|u(t,\cdot)\|_{H^1} + \|\partial_t u(t,\cdot)\|_{L^2} = O(t) \tag{66}$$

as $t \to \infty$. So that, we obtain the global solutions by the continuation of the local solutions obtained by Theorem 1.2.

(2) We consider the case $(u_0,u_1) \neq (0,0)$ since the solution $u = 0$ is global in the case $(u_0,u_1) = (0,0)$. We note $E(u)(0) \leq 2\pi(1-s)C_*/\alpha$ by (8), and it yields $\|\nabla u_0\|^2_{L^2} < 2\pi(1-s)/\alpha$ by (41). By Theorem 1.6, we have a local in time solution u on $[0,T)$ for some $T > 0$. Let T^* be the supremum of such T. By (65)

and $\|\partial_t u\|^2_{L^\infty((0,t),L^2)} \le E(u)(0)$, we have

$$\|u(t,\cdot)\|_{L^2} \le \|u(0,\cdot)\|_{L^2} + t(E(u)(0))^{1/2} \tag{67}$$

for $0 \le t < T^*$. We assume $T^* < \infty$ and show a contradiction in the following. First, we claim

$$\|\partial_t u\|_{L^2((0,t)\times\mathbb{R}^n)} \ne 0 \quad \text{for } 0 < t < T^*. \tag{68}$$

Indeed, if this does not hold, then u is a stationary solution and it exists globally, a contradiction to $T^* < \infty$. Next, for any fixed sufficiently small $\varepsilon > 0$ such that $T^* - \varepsilon \ge T^*/2$, let v be the solution of (42) with $t_0 := T^* - \varepsilon$. By (43), we have

$$\begin{aligned}\|Dv\|^2_{L^\infty((T^*-\varepsilon,\infty),L^2)} &\le E(u)(0) - 2\|\partial_t u\|^2_{L^2((0,T^*-\varepsilon)\times\mathbb{R}^n)}\\ &\le E(u)(0) - 2\|\partial_t u\|^2_{L^2((0,T^*/2)\times\mathbb{R}^n)}.\end{aligned} \tag{69}$$

This estimate shows $\|Dv\|^2_{L^\infty((T^*-\varepsilon,\infty),L^2)}$ is strictly smaller than $E(u)(0)$ uniformly on ε by (68). And (67) shows that $\|u(t,\cdot)\|_{L^2}$ is also bounded from above uniformly on ε. So that, the argument in the proof of Theorem 1.6 guarantees that we are able to continue the solution u beyond T^* starting from $T^* - \varepsilon$, which is a contradiction to the definition of T^*. We note that the corresponding estimate (60) holds for v by (68) and (69), so that the existence time of u starting from t_0 is bounded from below uniformly on ε.

Next, we show the uniqueness of the solution. The uniqueness for (1) follows from that of Theorem 1.2. Let us consider the case for (2). Let u be the global solution obtained by the above argument. Let $v \in C_b([0,\infty), H^1(\mathbb{R}^2)) \cap C_b^1([0,\infty), L^2(\mathbb{R}^2))$ be another solution. We define $t_0 := \inf\{t \in [0,\infty) : u(t) \ne v(t)\}$ and show $t_0 < \infty$ yields a contradiction. Let $t_0 < \infty$. By the continuity of u and v, we have $(u-v)(t_0) = \partial_t(u-v)(t_0) = 0$. Since $E(u)(t_0) \le 2\pi(1-s)C_*/\alpha$, we have $\|Du\|^2_{L^\infty((0,\infty),L^2(\mathbb{R}^2))} < E(u)(0)$ by Lemma 4.1, which yields $\|\nabla v(t_0)\|^2_{L^2(\mathbb{R}^2)} < 2\pi(1-s)/\alpha$. So that, the uniqueness result of Theorem 1.6 shows $u = v$ on $[t_0, t_0+\varepsilon)$ for sufficiently small $\varepsilon > 0$, which is a contradiction to the definition of t_0.

Appendix

In this section, we give a rigorous proof of the energy estimates given by (10), which has been proved there by formal calculation. Since the equation in (2) has a singularity at $x = 0$, the standard density argument by the use of $C^\infty(\mathbb{R}^n)$ in spatial variables is not valid. We give its modification for the completeness of our argument. We refer to the paper [46] by Shatah and Struwe for the rigorous proof of the energy estimates to treat energy solutions of critical semilinear wave equations without the singularity ($s = 0$). We show the corresponding result to the case of singular nonlinear terms ($s > 0$). To start with, we prepare the following basic estimate.

Lemma A.1 *Let $n \geq 1$, $0 < T < \infty$. Let $u, v \in C([0, T], L^2(\mathbb{R}^n))$. Assume that the first derivatives $\partial_t u$ and $\partial_t v$ exist in $L^2(\mathbb{R}^n)$ for any $t \in [0, T]$. Then the equality*

$$\partial_t \langle u(t, \cdot), v(t, \cdot)\rangle = \langle \partial_t u(t, \cdot), v(t, \cdot)\rangle + \langle u(t, \cdot), \partial_t v(t, \cdot)\rangle \tag{70}$$

holds for any $t \in [0, T]$, where $\langle f, g\rangle := \int_{\mathbb{R}^n} f(x)\overline{g(x)}dx$ for $f, g \in L^2(\mathbb{R}^n)$.

Proof For $\varepsilon \in \mathbb{R}$, we put

$$\begin{aligned} I := (\langle u(t+\varepsilon, \cdot), v(t+\varepsilon, \cdot)\rangle - \langle u(t, \cdot), v(t, \cdot)\rangle) /\varepsilon \\ - \langle \partial_t u(t, \cdot), v(t, \cdot)\rangle - \langle u(t, \cdot), \partial_t v(t, \cdot)\rangle \end{aligned} \tag{71}$$

and bound it by

$$\begin{aligned} |I| \leq \|(u(t+\varepsilon, \cdot) - u(t, \cdot))/\varepsilon - \partial_t u(t, \cdot)\|_{L^2(\mathbb{R}^n)} \|v(t+\varepsilon, \cdot)\|_{L^2(\mathbb{R}^n)} \\ + \|u(t, \cdot)\|_{L^2(\mathbb{R}^n)} \|(v(t+\varepsilon, \cdot) - v(t, \cdot))/\varepsilon - \partial_t v(t, \cdot)\|_{L^2(\mathbb{R}^n)} \\ + \|\partial_t u(t, \cdot)\|_{L^2(\mathbb{R}^n)} \|v(t+\varepsilon, \cdot) - v(t, \cdot)\|_{L^2(\mathbb{R}^n)}. \end{aligned} \tag{72}$$

By the assumption, we have $\lim_{\varepsilon\to 0} I = 0$ and obtain the required result. □

We start from the following estimates for the strong solutions.

Lemma A.2 *Let $n \geq 1$, $0 < T < \infty$. Let $u \in C([0, T], H^2(\mathbb{R}^n)) \cap C^1([0, T], H^1(\mathbb{R}^n))$ and $h \in L^1((0, T), L^2(\mathbb{R}^n))$ be real-valued functions. Assume that the second derivative $\partial_t^2 u$ exists and satisfies*

$$(\partial_t^2 - \Delta_Q + \partial_t)u(t, \cdot) + h(t, \cdot) = 0 \tag{73}$$

in $L^2(\mathbb{R}^n)$ for any $t \in [0, T]$. Then the following inequalities hold for any $t \in [0, T]$.

$$\begin{aligned} (1)\ \|D_{t,x}u(t, \cdot)\|^2_{L^2(\mathbb{R}^n)} + 2\|\partial_t u\|^2_{L^2((0,t)\times\mathbb{R}^n)} \\ + 2\int_0^t \int_{\mathbb{R}^n} \partial_t u(s, x)h(s, x)dxds = \|D_{t,x}u(0, \cdot)\|^2_{L^2(\mathbb{R}^n)} \end{aligned} \tag{74}$$

$$\begin{aligned} (2)\ \|u(t, \cdot)\|^2_{L^2(\mathbb{R}^n)} + \|D_{t,x}u(t, \cdot)\|^2_{L^2(\mathbb{R}^n)} + 2\|D_{t,x}u\|^2_{L^2((0,t)\times\mathbb{R}^n)} \\ + 4\int_0^t \int_{\mathbb{R}^n} u(s, x)h(s, x)dxds + 10\int_0^t \int_{\mathbb{R}^n} \partial_t u(s, x)h(s, x)dxds \\ \leq 4\|u(0, \cdot)\|^2_{L^2(\mathbb{R}^n)} + 7\|D_{t,x}u(0, \cdot)\|^2_{L^2(\mathbb{R}^n)} \end{aligned} \tag{75}$$

Proof (1) Taking the L^2 inner products of both sides of (73) with $\partial_t u$, and using Lemma A.1 and $2\langle \partial_t u, \Delta_Q u\rangle = -\partial_t \|Du\|^2_{L^2(\mathbb{R}^n)}$, we obtain

$$\partial_t \left(\|D_{t,x}u\|^2_{L^2(\mathbb{R}^n)} \right) + 2\|\partial_t u\|^2_{L^2(\mathbb{R}^n)} + 2\langle \partial_t u, h\rangle = 0. \tag{76}$$

This leads to $\partial_t \|D_{t,x}u\|^2 \in L^1((0,T))$ by the assumption on u and h, and we obtain the required equality by the integration in time variable.

(2) Similarly, taking the L^2 inner products of both sides of (73) with u, we have

$$\partial_t(2\langle u, \partial_t u\rangle + \|u\|^2_{L^2(\mathbb{R}^n)}) - 2\|\partial_t u\|^2_{L^2(\mathbb{R}^n)} + 2\|Du\|^2_{L^2(\mathbb{R}^n)} + 2\langle u, h\rangle = 0. \tag{77}$$

This leads to $\partial_t(2\langle u, \partial_t u\rangle + \|u\|^2_{L^2(\mathbb{R}^n)}) \in L^1((0,t))$, and we obtain

$$\begin{aligned} &\|u(t,\cdot)\|^2_{L^2(\mathbb{R}^n)} + 2\|Du\|^2_{L^2((0,t)\times\mathbb{R}^n)} + 2\int_0^t \langle u(s,\cdot), h(s,\cdot)\rangle ds \\ &= \|u(0,\cdot)\|^2_{L^2(\mathbb{R}^n)} + 2\|\partial_t u\|^2_{L^2((0,t)\times\mathbb{R}^n)} + 2\langle u(0,\cdot), \partial_t u(0,\cdot)\rangle - 2\langle u(t,\cdot), \partial_t u(t,\cdot)\rangle. \end{aligned} \tag{78}$$

Since the last term is bounded by $\|u(t,\cdot)\|^2_{L^2(\mathbb{R}^n)}/2 + 2\|\partial_t u(t,\cdot)\|^2_{L^2(\mathbb{R}^n)}$, the required inequality follows from this equality and (1). □

Next, we consider the energy estimates for the energy solutions.

Lemma A.3 *Let $n \geq 1$, $0 < T < \infty$. Let $u \in C([0,T], H^1(\mathbb{R}^n)) \cap C^1([0,T], L^2(\mathbb{R}^n))$ and $h \in L^2((0,T)\times\mathbb{R}^n)$ be real-valued functions. Assume that the second derivative $\partial_t^2 u$ exists and satisfies* (73) *in $H^{-1}(\mathbb{R}^n)$ for any $t \in [0,T]$. Then the results in Lemma* A.2 *hold for any $t \in [0,T]$.*

Proof Let $\{\varphi_j\}_{j=1}^\infty$ be a mollifier on $\mathbb{R}^n$. Put $u_j := u * \varphi_j$ and $h_j := h * \varphi_j$ for $1 \leq j < \infty$. By the assumption on u and h, $\{u_j\}_{j=1}^\infty$ and $\{h_j\}_{j=1}^\infty$ satisfy $u_j \in C([0,T], H^2(\mathbb{R}^n))$, $\partial_t(u_j) = (\partial_t u) * \varphi_j \in C([0,T], H^1(\mathbb{R}^n))$ and $h_j \in L^1((0,T), L^2(\mathbb{R}^n))$. Moreover, $\partial_t^2(u_j) = (\partial_t^2 u) * \varphi_j$ exists in $L^2(\mathbb{R}^n)$ and satisfies $(\partial_t^2 - \Delta_Q + \partial_t)u_j + h_j = 0$ in $L^2(\mathbb{R}^n)$ on $[0,T]$. So that, we have the results in Lemma A.2 with u and h replaced by u_j and h_j. It is easy to check by the Lebesgue convergence theorem that $\|u_j(t,\cdot)\|_{L^2(\mathbb{R}^n)}$, $\|D_{t,x}u_j(t,\cdot)\|_{L^2(\mathbb{R}^n)}$ and $\|D_{t,x}u_j\|_{L^2((0,t)\times\mathbb{R}^n)}$ converge to $\|u(t,\cdot)\|_{L^2(\mathbb{R}^n)}$, $\|D_{t,x}u(t,\cdot)\|_{L^2(\mathbb{R}^n)}$ and $\|D_{t,x}u\|_{L^2((0,t)\times\mathbb{R}^n)}$ as $j \to \infty$. The term $\int_0^t \int_{\mathbb{R}^n} \partial_t u_j h_j dx ds$ converges to $\int_0^t \int_{\mathbb{R}^n} \partial_t u h dx ds$ by

$$\begin{aligned} &\left| \int_0^t \int_{\mathbb{R}^n} \partial_t u_j(s,x) h_j(s,x) - \partial_t u(s,x) h(s,x) dx ds \right| \\ &\leq \|\partial_t(u_j - u)\|_{L^2(S_t)} \|h_j\|_{L^2(S_t)} + \|\partial_t u\|_{L^\infty((0,t),L^2(\mathbb{R}^n))} \|h_j - h\|_{L^1((0,t),L^2(\mathbb{R}^n))} \end{aligned} \tag{79}$$

and $h \in L^2(S_t)$, where $S_t := (0,t) \times \mathbb{R}^n$. Similarly, the term $\int_0^t \int_{\mathbb{R}^n} u_j h_j dxds$ converges to $\int_0^t \int_{\mathbb{R}^n} uhdxds$. Therefore, we have obtained the required results. □

Next, we consider the potential of the inhomogeneous term.

Lemma A.4 *Let $n \geq 1$, $0 < T < \infty$. Let $H = H(x,u)$ be a function on $\mathbb{R}^n \times \mathbb{R}$ such that $H(x,\cdot) \in C^1(\mathbb{R},\mathbb{R})$ for almost everywhere $x \in \mathbb{R}^n$. Assume that there exists a nondecreasing and nonnegative function M on $[0,\infty)$ such that*

$$\left\| \frac{\partial H}{\partial u}(\cdot, u(\cdot)) \right\|_{L^2(\mathbb{R}^n)} \leq M(\|u\|_{H^1(\mathbb{R}^n)}) \tag{80}$$

$$\left\| \frac{\partial H}{\partial u}(\cdot, u(\cdot)) - \frac{\partial H}{\partial u}(\cdot, v(\cdot)) \right\|_{L^2(\mathbb{R}^n)} \leq M(\|u\|_{H^1(\mathbb{R}^n)} + \|v\|_{H^1(\mathbb{R}^n)}) \|u - v\|_{H^1(\mathbb{R}^n)} \tag{81}$$

for any $u, v \in H^1(\mathbb{R}^n)$. Then the estimate

$$\int_0^t \int_{\mathbb{R}^n} \partial_t u(s,x) \frac{\partial H}{\partial u}(x, u(s,x)) dxds = \int_{\mathbb{R}^n} H(x, u(t,x)) dx - \int_{\mathbb{R}^n} H(x, u(0,x)) dx \tag{82}$$

holds for any $t \in [0,T]$ and $u \in C([0,T], H^1(\mathbb{R}^n)) \cap C^1([0,T], L^2(\mathbb{R}^n))$.

Proof For $\varepsilon \in \mathbb{R}$, we put I as

$$I := \frac{1}{\varepsilon} \int_{\mathbb{R}^n} H(x, u(t+\varepsilon, x)) - H(x, u(t,x)) dx - \int_{\mathbb{R}^n} \frac{\partial H}{\partial u}(x, u(t,x)) \partial_t u(t,x) dx. \tag{83}$$

By the assumption on H, we are able to write

$$H(x, u(t+\varepsilon, x)) - H(x, u(t,x)) = \int_0^1 \frac{\partial H}{\partial u}(x, u_\theta) d\theta (u(t+\varepsilon, x) - u(t,x)), \tag{84}$$

where $u_\theta := (1-\theta)u(t,x) + \theta u(t+\varepsilon, x)$. Inserting this equality, we have

$$|I| \leq \left\| \int_0^1 \frac{\partial H}{\partial u}(\cdot, u_\theta) d\theta \right\|_{L^2(\mathbb{R}^n)} \cdot \left\| \frac{u(t+\varepsilon, \cdot) - u(t, \cdot)}{\varepsilon} - \partial_t u(t, \cdot) \right\|_{L^2(\mathbb{R}^n)} + \left\| \int_0^1 \frac{\partial H}{\partial u}(\cdot, u_\theta) - \frac{\partial H}{\partial u}(\cdot, u) d\theta \right\|_{L^2(\mathbb{R}^n)} \cdot \|\partial_t u(t, \cdot)\|_{L^2(\mathbb{R}^n)} \cdot \tag{85}$$

So that, by the assumption on u and H, we obtain $\lim_{\varepsilon\to 0}|I| = 0$, which shows

$$\partial_t \int_{\mathbb{R}^n} H(x, u(t,x))dx = \int_{\mathbb{R}^n} \frac{\partial H}{\partial u}(x, u(t,x))\partial_t u(t,x)dx. \tag{86}$$

Since the right hand side is integrable on $[0, T]$ by

$$\int_{\mathbb{R}^n} \left| \frac{\partial H}{\partial u}(x, u(t,x))\partial_t u(t,x) \right| dx \le M(\|u(t,\cdot)\|_{H^1(\mathbb{R}^n)})\|\partial_t u(t,\cdot)\|_{L^2(\mathbb{R}^n)}, \tag{87}$$

we obtain the required result by the fundamental theorem of calculus. □

By Lemmas A.3 and A.4, we quickly obtain the following result.

Corollary A.5 *Under the assumption of Lemmas* A.3 *and* A.4, *if* h *and* H *satisfy*

$$h(t,x) = \frac{\partial H}{\partial u}(x, u(t,x)), \tag{88}$$

then the estimate

$$\int_{\mathbb{R}^n} |D_{t,x}u(t,x)|^2 + 2H(x, u(t,x))dx + 2\|\partial_t u\|^2_{L^2((0,t)\times\mathbb{R}^n)}$$
$$\le \int_{\mathbb{R}^n} |D_{t,x}u(0,x)|^2 + 2H(x, u(0,x))dx \tag{89}$$

holds for any $t \in [0, T]$.

Now, we describe that our energy solutions satisfy the energy estimates. Let n, s, p and f satisfy the assumption in Theorem 1.2. For $T > 0$, put $X := L^\infty((0,T), H^1(\mathbb{R}^n))$. Let $u_0 \in H^1(\mathbb{R}^n)$ and $u_1 \in L^2(\mathbb{R}^n)$ be arbitrarily fixed. For any $u \in X$, we put

$$\Phi(u)(t) := \partial_t K(t)u_0 + K(t)(u_0 + u_1) - \int_0^t K(t-\tau)h(\cdot, u(\tau,\cdot))d\tau, \tag{90}$$

where $h(x, u(t,x)) := \frac{f(u(t,x))}{|x|^s}$ and $K(\cdot)$ is defined by (14). We note that K satisfies

$$\|\partial_t K(t)u_0\|_{H^1(\mathbb{R}^n)} \lesssim \|u_0\|_{H^1(\mathbb{R}^n)}, \quad \|K(t)(u_0+u_1)\|_{H^1(\mathbb{R}^n)} \lesssim \|u_0+u_1\|_{L^2(\mathbb{R}^n)}, \tag{91}$$

and h satisfies

$$\|h(\cdot, u(t, \cdot))\|_{L^2(\mathbb{R}^n)} \lesssim \|u(t, \cdot)\|_{H^1(\mathbb{R}^n)}^p \tag{92}$$

$$\|h(\cdot, u(t, \cdot)) - h(\cdot, v(t, \cdot))\|_{L^2(\mathbb{R}^n)} \lesssim \max_{w=u,v} \|w(t, \cdot)\|_{H^1(\mathbb{R}^n)}^{p-1} \|u(t, \cdot) - v(t, \cdot)\|_{H^1(\mathbb{R}^n)} \tag{93}$$

for any $u, v \in X$ and $t \in [0, T]$ from the proof of Lemma 3.3. So that, we obtain the fixed point u of Φ in X for sufficiently small $T = T(\|u_0\|_{H^1(\mathbb{R}^n)} + \|u_1\|_{L^2(\mathbb{R}^n)}) > 0$. Lemma 2.1 shows that $u \in C([0, T], H^1(\mathbb{R}^n)) \cap C^1([0, T], L^2(\mathbb{R}^n))$. And $\partial_t^2 u$ exists and satisfies

$$(\partial_t^2 - \Delta_Q + \partial_t)u(t, \cdot) + h(\cdot, u(t, \cdot)) = 0 \tag{94}$$

in $H^{-1}(\mathbb{R}^n)$ for any $t \in [0, T]$. So that, we have the required energy estimates (74) and (75) by Lemma A.3. Moreover when $f(u) = \lambda|u|^{p-1}u$ with $\lambda > 0$, we put H by the first line in (64). Then H satisfies (88) and the assumption of Lemma A.4 by (92) and (93). So that, we obtain (89) by Corollary A.5. We have described the energy estimates corresponding to Theorem 1.2 and (1) of Theorem 1.8. The energy estimates for the other theorems are similarly obtained.

Acknowledgements The author is thankful to the anonymous referee for several comments to revise the paper.

References

1. L. Aloui, S. Ibrahim, K. Nakanishi, Exponential energy decay for damped Klein-Gordon equation with nonlinearities of arbitrary growth. Commun. Partial Differ. Equ. **36**(5), 797–818 (2011)
2. P. Baras, J.A. Goldstein, The heat equation with a singular potential. Trans. Am. Math. Soc. **284**(1), 121–139 (1984)
3. H. Brezis, L. Dupaigne, A. Tesei, On a semilinear elliptic equation with inverse-square potential. Selecta Math. (N.S.) **11**(1), 1–7 (2005)
4. N. Burq, F. Planchon, J.G. Stalker, A.S. Tahvildar-Zadeh, Strichartz estimates for the wave and Schrödinger equations with the inverse-square potential. J. Funct. Anal. **203**(2), 519–549 (2003)
5. L. Caffarelli, R. Kohn, L. Nirenberg, First order interpolation inequalities with weights. Compositio Math. **53**(3), 259–275 (1984)
6. F. Catrina, Z.Q. Wang, On the Caffarelli-Kohn-Nirenberg inequalities. C. R. Acad. Sci. Paris Sér. I Math. **330**(6), 437–442 (2000)
7. F. Catrina, Z.Q. Wang, On the Caffarelli-Kohn-Nirenberg inequalities: sharp constants, existence (and nonexistence), and symmetry of extremal functions. Commun. Pure Appl. Math. **54**(2), 229–258 (2001)

8. J.L. Chern, C.S. Lin, Minimizers of Caffarelli-Kohn-Nirenberg inequalities on domains with the singularity on the boundary. Arch. Ration. Mech. Anal. **197**, 401–432 (2010)
9. P. Cherrier, A. Milani, *Linear and Quasi-Linear Evolution Equations in Hilbert Spaces*. Graduate Studies in Mathematics, vol. 135 (American Mathematical Society, Providence, 2012), xviii+377pp.
10. J. Colliander, S. Ibrahim, M. Majdoub, N. Masmoudi, Energy critical NLS in two space dimensions. J. Hyperbolic Differ. Equ. **6**(3), 549–575 (2009)
11. V. Combet, F. Genoud, Classification of minimal mass blow-up solutions for an L^2 critical inhomogeneous NLS. J. Evol. Equ. **16**(2), 483–500 (2016)
12. H. Egnell, Positive solutions of semilinear equations in cones. Trans. Am. Math. Soc. **330**(1), 191–201 (1992)
13. Z. Gan, Cross-constrained variational methods for the nonlinear Klein-Gordon equations with an inverse square potential. Commun. Pure Appl. Anal. **8**(5), 1541–1554 (2009)
14. J.P. García Azorero, I. Peral Alonso, Hardy inequalities and some critical elliptic and parabolic problems. J. Differ. Equ. **144**(2), 441–476 (1998)
15. V. Georgiev, S. Lucente, Focusing NLKG equation with singular potential. Commun. Pure Appl. Anal. **17**(4), 1387–1406 (2018)
16. N. Ghoussoub, X.S. Kang, Hardy-Sobolev critical elliptic equations with boundary singularities. Ann. Inst. H. Poincaré Anal. Non Linéaire **21**(6), 767–793 (2004)
17. N. Ghoussoub, F. Robert, Concentration estimates for Emden-Fowler equations with boundary singularities and critical growth. IMRP Int. Math. Res. Pap. **21867**, 1–85 (2006)
18. N. Ghoussoub, F. Robert, The effect of curvature on the best constant in the Hardy-Sobolev inequalities. Geom. Funct. Anal. **16**(6), 1201–1245 (2006)
19. N. Hayashi, E.I. Kaikina, P.I. Naumkin, Damped wave equation with super critical nonlinearities. Differ. Integr. Equ. **17**(5–6), 637–652 (2004)
20. N. Hayashi, E.I. Kaikina, P.I. Naumkin, Damped wave equation in the subcritical case. J. Differ. Equ. **207**(1), 161–194 (2004)
21. J. Hernández, F.J. Mancebo, J.M. Vega, Positive solutions for singular nonlinear elliptic equations. Proc. R. Soc. Edinb. Sect. A **137**(1), 41–62 (2007)
22. T. Hosono, T. Ogawa, Large time behavior and L^p-L^q estimate of solutions of 2-dimensional nonlinear damped wave equations, J. Differ. Equ. **203**, 82–118 (2004)
23. C.H. Hsia, C.S. Lin, H. Wadade, Revisiting an idea of Brézis and Nirenberg. J. Funct. Anal. **259**, 1816–1849 (2010)
24. T-S. Hsu, H-L. Lin, Multiple positive solutions for singular elliptic equations with weighted Hardy terms and critical Sobolev-Hardy exponents. Proc. R. Soc. Edinb. Sect. A **140**(3), 617–633 (2010)
25. S. Ibrahim, R. Jrad, Strichartz type estimates and the well-posedness of an energy critical 2D wave equation in a bounded domain. J. Differ. Equ. **250**(9), 3740–3771 (2011)
26. S. Ibrahim, R. Jrad, M. Majdoub, T. Saanouni, Local well posedness of a 2D semilinear heat equation. Bull. Belg. Math. Soc. Simon Stevin **21**(3), 535–551 (2014)
27. S. Ibrahim, M. Majdoub, N. Masmoudi, Global solutions for a semilinear, two-dimensional Klein-Gordon equation with exponential-type nonlinearity. Commun. Pure Appl. Math. **59**(11), 1639–1658 (2006)
28. R. Ikehata, Y. Miyaoka, T. Nakatake, Decay estimates of solutions for dissipative wave equations in $\mathbb{R}^N$ with lower power nonlinearities. J. Math. Soc. Jpn. **56**(2), 365–373 (2004)
29. M. Ishiwata, M. Nakamura, H. Wadade, On the sharp constant for the weighted Trudinger-Moser type inequality of the scaling invariant form. Ann. Inst. H. Poincaré Anal. Non Linéaire **31**(2), 297–314 (2014)
30. T. Kato, Schrödinger operators with singular potentials, in *Proceedings of the International Symposium on Partial Differential Equations and the Geometry of Normed Linear Spaces*, Jerusalem (1972). Israel J. Math. **13**, 135–148 (1972/1973)
31. S. Kawashima, M. Nakao, K. Ono, On the decay property of solutions to the Cauchy problem of the semilinear wave equation with a dissipative term. J. Math. Soc. Jpn. **47**(4), 617–653 (1995)

32. T.T. Li, Y. Zhou, Breakdown of solutions to $\Box u + u_t = |u|^{1+\alpha}$. Discret. Contin. Dyn. Syst **1**(4), 503–520 (1995)
33. C.S. Lin, Interpolation inequalities with weights. Commun. Partial Differ. Equ. **11**(14), 1515–1538 (1986)
34. C.S. Lin, H. Wadade, Minimizing problems for the Hardy-Sobolev type inequality with the singularity on the boundary. Tohoku Math. J. (2) **64**(1), 79–103 (2012)
35. C.S. Lin, Z.Q. Wang, Symmetry of extremal functions for the Caffarelli-Kohn-Nirenberg inequalities. Proc. Am. Math. Soc. **132**(6), 1685–1691 (2004)
36. A. Matsumura, On the asymptotic behavior of solutions of semi-linear wave equations. Publ. Res. Inst. Math. Sci. **12**(1), 169–189 (1976/1977)
37. S. Nagayasu, H. Wadade, Characterization of the critical Sobolev space on the optimal singularity at the origin. J. Funct. Anal. **258**(11), 3725–3757 (2010)
38. M. Nakamura, Small global solutions for nonlinear complex Ginzburg-Landau equations and nonlinear dissipative wave equations in Sobolev spaces. Rev. Math. Phys. **23**(8), 903–931 (2011)
39. M. Nakamura, T. Ozawa, Nonlinear Schrödinger equations in the Sobolev space of critical order. J. Funct. Anal. **150**, 364–380 (1998)
40. M. Nakamura, T. Ozawa, Global solutions in the critical Sobolev space for the wave equations with nonlinearity of exponential growth. Math. Z. **231**, 479–487 (1999)
41. M. Nakamura, T. Ozawa, The Cauchy problem for nonlinear Klein–Gordon equations in the Sobolev spaces. Publ. Res. Inst. Math. Sci. (Kyoto University) **37**, 255–293 (2001)
42. M. Nakao, K. Ono, Existence of global solutions to the Cauchy problem for the semilinear dissipative wave equations. Math. Z. **214**(2), 325–342 (1993)
43. T. Narazaki, L^p-L^q estimates for damped wave equations and their applications to semi-linear problem. J. Math. Soc. Jpn. **56**(2), 585–626 (2004)
44. K. Nishihara, L^p-L^q estimates of solutions to the damped wave equation in 3-dimensional space and their application. Math. Z. **244**(3), 631–649 (2003)
45. K. Nishihara, Sugaku **62**(2), 20–37 (2010) (in Japanese)
46. J. Shatah, M. Struwe, Well-posedness in the energy space for semilinear wave equations with critical growth. Internat. Math. Res. Notices **1994**(7), 303ff., approx. 7pp. (1994)
47. R.S. Strichartz, A note on Trudinger's extension of Sobolev's inequalities. Indiana Univ. Math. J. **21**, 841–842 (1971/1972)
48. G. Todorova, B. Yordanov, Critical exponent for a nonlinear wave equation with damping. J. Differ. Equ. **174**(2), 464–489 (2001)
49. H. Yang, J. Chen, A result on Hardy-Sobolev critical elliptic equations with boundary singularities. Commun. Pure Appl. Anal. **6**(1), 191–201 (2007)
50. Q.S. Zhang, A blow-up result for a nonlinear wave equation with damping: the critical case. C. R. Acad. Sci. Paris Ser. I Math. **333**(2), 109–114 (2001)

Global Existence Results for a Semilinear Wave Equation with Scale-Invariant Damping and Mass in Odd Space Dimension

Alessandro Palmieri

Abstract We consider a semilinear wave equation with scale-invariant damping and mass and power nonlinearity. For this model we prove some global (in time) existence results in odd spatial dimension n, under the assumption that the multiplicative constants μ and ν^2, which appear in the coefficients of the damping and of the mass terms, respectively, satisfy an interplay condition which makes the model somehow "wave-like". Combining these global existence results with a recently proved blow-up result, we will find as critical exponent for the considered model the largest between suitable shifts of the Strauss exponent and of Fujita exponent, respectively. Besides, the competition among these two kind of exponents shows how the interrelationship between μ and ν^2 determines the possible transition from a "hyperbolic-like" to a "parabolic-like" model. Nevertheless, in the case $n \geq 3$ we will restrict our considerations to the radial symmetric case.

1 Introduction

In the last years several papers have been devoted to the study of the semilinear wave equation with scale-invariant damping and power nonlinearity

$$\begin{cases} u_{tt} - \Delta u + \frac{\mu}{1+t} u_t = |u|^p, & x \in \mathbb{R}^n,\ t > 0, \\ u(0,x) = u_0(x), & x \in \mathbb{R}^n, \\ u_t(0,x) = u_1(x), & x \in \mathbb{R}^n, \end{cases} \tag{1}$$

where μ is a positive constant.

For suitably large μ it has been proved that (1) and the corresponding linear problem are "parabolic" under the point of view of the critical exponent for the power of the nonlinearity and under the point of decay estimates, respectively.

A. Palmieri (✉)
Institute of Applied Analysis, TU Bergakademie Freiberg, Freiberg, Germany

M. D'Abbicco et al. (eds.), *New Tools for Nonlinear PDEs and Application*,
Trends in Mathematics, https://doi.org/10.1007/978-3-030-10937-0_12

Indeed, in [3] a global (in time) existence result for $p > p_{\mathrm{Fuj}}(n) \doteq 1 + \frac{2}{n}$ in dimensions $n = 1, 2$ in the case $\mu \geq \frac{5}{4}$ and $\mu \geq 3$, respectively, is derived in the energy space. Furthermore, in higher dimensions $n \geq 3$ a global existence result is obtained in exponentially weighted energy spaces provided that $p > p_{\mathrm{Fuj}}(n)$ and $\mu \geq n + 2$. Hence, combining these existence results with the blow-up results of [4], it results that the critical exponent for (1) is the Fujita exponent $p_{\mathrm{Fuj}}(n)$, when μ is sufficiently large.

Independently, in [27], using different techniques, $p_{\mathrm{Fuj}}(n)$ is proved to be critical, assumed that μ is grater than a given constant $\mu_0 \approx (p - p_{\mathrm{Fuj}}(n))^{-2}$. In particular, the test function method is employed to prove the blow-up of the solution for $1 < p \leq p_{\mathrm{Fuj}}(n)$ when $\mu \geq 1$, of course, under suitable assumptions on data.

Nevertheless, for small values of μ the situation is completely different. In [6] the special value $\mu = 2$ is studied. In fact, thanks to this choice of μ, (1) can be transformed in a semilinear free wave equation with nonlinearity $(1+t)^{-(p-1)}|u|^p$. Then, using Kato's lemma, the authors prove a blow-up result for

$$1 < p \leq \max\{p_{\mathrm{Fuj}}(n), p_0(n+2)\} = \begin{cases} p_{\mathrm{Fuj}}(n) & \text{if } n = 1, \\ p_0(n+2) & \text{if } n \geq 2, \end{cases}$$

in all spatial dimensions, under some sign conditions for compactly supported initial data, where $p_0(n)$ denotes the so-called Strauss exponent, that is, the positive root of the quadratic equation

$$(n-1)p^2 - (n+1)p - 2 = 0.$$

Moreover, the above upper bound is shown to be really the critical exponent in the cases $n = 1, 2$ and $n = 3$ for radial symmetric solutions (really recently, the radial symmetry assumption for $n = 3$ has been removed in [9, 15]). Afterwords, in [5] the sharpness of that blow-up result is shown also in odd dimensions $n \geq 5$ in the radial symmetric case. Because of the fact that the critical exponent seemed reasonably to be $p_0(n+2)$ for any $n \geq 3$, that is, the shift of the Strauss exponent of order exactly 2, namely, the coefficient of the damping term, in [6] it was also conjectured that the critical exponent for (1) could have been $p_0(n+\mu)$ when $\mu \in (0, 2)$ and $n \geq 3$.

Recently, in several works [8, 16, 25, 26], it has been studied the blow-up of solutions to (1) in the case in which the constant μ is small. More precisely, in [16] the blow-up of solutions is proved for $p_{\mathrm{Fuj}}(n) \leq p < p_0(n + 2\mu)$ and $n \geq 2$, provided that $0 < \mu < \frac{n^2+n+2}{2(n+2)}$ (this condition on μ guarantees the existence of admissible values for p) and that initial data are nonnegative and compactly supported. Such a result, together with the upper bound for the lifespan, is derived by using a Kato type lemma from [24].

Then, in [8] a blow-up result for the larger range $p_{\mathrm{Fuj}}(n) \leq p \leq p_0(n+\mu)$ is proved, assuming that $n \geq 1$ and $0 < \mu < \frac{n^2+n+2}{n+2}$ (clearly, also in this case the condition on μ implies the nonemptiness of the range for p) and that initial data are nonnegative and compactly supported in a ball around the origin with radius smaller than 1. The approach therein used is based on suitable self-similar solution of the conjugate linear equation.

Finally, in [25, 26] the authors improved the range for p which implies the blow-up of solutions to $1 < p \leq p_0(n+\mu)$ (again under suitable assumptions on data), by using a similar approach to the one in [16], obtaining a better upper bound for the life span of the solution in the subcritical case.

In this work we will focus on the Cauchy problem for semilinear wave equation with scale-invariant damping and mass and power nonlinearity

$$\begin{cases} u_{tt} - \Delta u + \frac{\mu}{1+t} u_t + \frac{\nu^2}{(1+t)^2} u = |u|^p, & x \in \mathbb{R}^n,\ t > 0, \\ u(0,x) = u_0(x), & x \in \mathbb{R}^n, \\ u_t(0,x) = u_1(x), & x \in \mathbb{R}^n, \end{cases} \tag{2}$$

under the assumption

$$\delta \doteq (\mu - 1)^2 - 4\nu^2 = 1. \tag{3}$$

Recently (2) has been studied in the case in which is "parabolic-like" in [17–19], in the case in which is "hyperbolic-like" in [17, 20] and in the case in which is "Klein-Gordon-like" in [7]. For a deeper analysis of how the quantity δ describes the interplay between the damping term $\frac{\mu}{1+t} u_t$ and the mass term $\frac{\nu^2}{(1+t)^2} u$ we refer to [22].

We point out, that (3) implies the possibility to link the solution to (2) with the solution to the semilinear Cauchy problem

$$\begin{cases} v_{tt} - \Delta v = (1+t)^{-\frac{\mu}{2}(p-1)} |v|^p, & x \in \mathbb{R}^n,\ t > 0, \\ v(0,x) = u_0(x), & x \in \mathbb{R}^n, \\ v_t(0,x) = u_1(x) + \frac{\mu}{2} u_0(x), & x \in \mathbb{R}^n, \end{cases} \tag{4}$$

via the transformation $u(t,x) = (1+t)^{-\frac{\mu}{2}} v(t,x)$.

Assuming the validity of (3), in [17] has been proved a blow-up result for

$$1 < p \leq \begin{cases} p_{\mathrm{Fuj}}\left(n + \frac{\mu}{2} - 1\right) & \text{if } n = 1, 2, \\ p_0(n+\mu) & \text{if } n \geq 3, \end{cases}$$

provided that data are nonnegative and compactly supported (for further details see also [17, Theorem 2.6]).

Purpose of this work is study the sufficiency part for this problem. In other words, we want to prove that

$$p_{\mathrm{crit}}(n,\mu) \doteq \max\{p_{\mathrm{Fuj}}\big(n+\tfrac{\mu}{2}-1\big),\, p_0(n+\mu)\}$$

is actually the critical exponent for (2) in the case in which (3) is fulfilled.

More specifically, for $n = 1$ we will follow the approach of [3] if $p \geq 2$, while in the case $p < 2$ we will use $L^1 - L^p$ estimates. On the other hand, for $n \geq 3$, n odd we will restrict our considerations to the radial symmetric case, following the approach developed in [1, 10–14] for the semilinear free wave equation with power nonlinearity and in [5, 6] for (1).

We stress the importance of the presence of the mass term in (2), since it makes wider the range of ps, for which we can prove a global (in time) existence result under the assumption that (3) is fulfilled, than in (1) when $\mu = 2$. Moreover, the presence of the mass term involves a more explicit understanding of how the magnitude of μ and ν^2 and their interaction, described through the quantity δ, influence either a "parabolic-like" or a "hyperbolic-like" behavior of the semilinear scale-invariant model concerning the critical exponent.

Finally, we mention that in the radial symmetric case $n \geq 3$ an upper bound for the coefficient μ has to be required. Naively speaking, this restriction is due to the fact that we will consider as solutions to our problem functions from a parameter dependent weighted space. Hence, in order to guarantee the possibility to choose properly this parameter, whose range depends both on p and μ, we have to assume a further μ-dependent condition on p. The compatibility between this condition and the lower bound for p implies the following upper bound for μ:

$$\mu \leq M(n) \doteq \tfrac{n-1}{2}\Big(1+\sqrt{\tfrac{n+7}{n-1}}\Big). \tag{5}$$

Therefore, although it is clear that the upper bound for μ is due to technical reasons, it is interesting to see that asymptotically

$$M(n) \sim n-1$$

for large n. The interesting fact is that $M(n)$ has exactly the same asymptotic behavior of the constant $\mu_* = \frac{n^2+n+2}{n+2}$ which is the upper bound for the coefficient μ in (1), that is considered in [8] to prove the blow-up result we mentioned above. So, we found the same restriction for μ, from the asymptotic point of view, working with tools which are suitable in two different but both "hyperbolic-like" cases, in the sense we have explained before.

This work is organized as follows: in Sect. 2 the critical exponent for the Cauchy problem (2) is derived, assuming the validity of (3) in the one dimensional case; in Sect. 3 the three dimensional case is considered in the radial symmetric case and the critical exponent is shown to be a suitable shift of Strauss exponent; finally, in Sect. 4 the general odd n-dimensional case, $n \geq 5$ is studied in the radial symmetric case and also in this case the critical exponent turns out to be a suitable shift of Strauss exponent.

2 One-Dimensional Case

Let us begin with the global existence results when $n = 1$ for the case in which μ and ν^2 fulfill (3). As we mentioned in the introduction, we have two different results for p above and below 2, respectively. In the Sect. 2.1 we study the case $p \geq 2$, while in Sect. 2.2 the case $p < 2$ will be considered.

2.1 One-Dimensional Case: $p \geq 2$

In this subsection we derive a global (in time) existence result, by using $L^2 - L^2$ estimates and requiring additional L^1 regularity for Cauchy data. Nevertheless, before starting with the proof of the main Theorem, we need to recall known results for decay estimates of the corresponding linear problem in general space dimension $n \geq 1$.

Proposition 2.1 *Let μ and ν^2 be nonnegative constants such that $\delta > 0$. Let us consider $(u_0, u_1) \in \big(L^1(\mathbb{R}^n) \cap H^1(\mathbb{R}^n)\big) \times \big(L^1(\mathbb{R}^n) \cap L^2(\mathbb{R}^n)\big) \doteq \mathcal{A}_1$. Then, for all $\kappa \in [0, 1]$ the energy solution u to*

$$\begin{cases} u_{tt} - \Delta u + \frac{\mu}{1+t} u_t + \frac{\nu^2}{(1+t)^2} u = 0, & x \in \mathbb{R}^n,\ t > 0, \\ u(0, x) = u_0(x), & x \in \mathbb{R}^n, \\ u_t(0, x) = u_1(x), & x \in \mathbb{R}^n, \end{cases} \tag{6}$$

satisfies the decay estimates

$$\|u(t, \cdot)\|_{\dot{H}^{\kappa}(\mathbb{R}^n)} \lesssim \|(u_0, u_1)\|_{\mathcal{A}_1} \begin{cases} (1+t)^{-\kappa - \frac{n+\mu}{2} + \frac{1+\sqrt{\delta}}{2}} & \text{if } \kappa < \frac{1+\sqrt{\delta}-n}{2}, \\ (1+t)^{-\frac{\mu}{2}} \ell(t) & \text{if } \kappa = \frac{1+\sqrt{\delta}-n}{2}, \\ (1+t)^{-\frac{\mu}{2}} & \text{if } \kappa > \frac{1+\sqrt{\delta}-n}{2}, \end{cases}$$

where

$$\ell(t) \doteq \big(1 + (\log(1+t))^{\frac{1}{2}}\big).$$

Moreover, $\|u_t(t, \cdot)\|_{L^2(\mathbb{R}^n)}$ satisfies the same decay estimate as $\|u(t, \cdot)\|_{\dot{H}^1(\mathbb{R}^n)}$.

Proposition 2.2 *Let μ and ν^2 be nonnegative constants such that $\delta > 0$. Let us assume $u_0 = 0$ and $u_1 \in L^1(\mathbb{R}^n) \cap L^2(\mathbb{R}^n)$. Then, the energy solution u to*

$$\begin{cases} u_{tt} - \Delta u + \frac{\mu}{1+t} u_t + \frac{\nu^2}{(1+t)^2} u = 0, & x \in \mathbb{R}^n,\ t > \tau \geq 0, \\ u(\tau, x) = u_0(x), & x \in \mathbb{R}^n, \\ u_t(\tau, x) = u_1(x), & x \in \mathbb{R}^n, \end{cases} \tag{7}$$

satisfies for $t \geq \tau$ and $\kappa \in [0, 1]$ the following estimates

$$\begin{aligned} \|u(t,\cdot)\|_{\dot{H}^\kappa(\mathbb{R}^n)} &\lesssim \Big(\|u_1\|_{L^1(\mathbb{R}^n)} + (1+\tau)^{\frac{n}{2}} \|u_1\|_{L^2(\mathbb{R}^n)} \Big) (1+\tau)^{\frac{1+\mu}{2} - \frac{\sqrt{\delta}}{2}} \\ &\quad \times \begin{cases} (1+t)^{-\kappa - \frac{n+\mu}{2} + \frac{1+\sqrt{\delta}}{2}} & \text{if } \kappa < \frac{1+\sqrt{\delta}-n}{2}, \\ (1+t)^{-\frac{\mu}{2}} \ell(t,\tau) & \text{if } \kappa = \frac{1+\sqrt{\delta}-n}{2}, \\ (1+t)^{-\frac{\mu}{2}} (1+\tau)^{-\kappa - \frac{n}{2} + \frac{1+\sqrt{\delta}}{2}} & \text{if } \kappa > \frac{1+\sqrt{\delta}-n}{2}, \end{cases} \end{aligned}$$

where

$$\ell(t,\tau) \doteq \left(1 + \left(\log\left(\tfrac{1+t}{1+\tau} \right) \right)^{\frac{1}{2}} \right).$$

Moreover, $\|u_t(t,\cdot)\|_{L^2(\mathbb{R}^n)}$ satisfies the same decay estimate as $\|u(t,\cdot)\|_{\dot{H}^1(\mathbb{R}^n)}$.

For the proof of Propositions 2.1 and 2.2 one can see [19, Theorems 4.6 and 4.7].

Theorem 2.3 *Let $n = 1$ and let $\mu \geq 2$ and ν^2 be nonnegative constants such that $\delta = 1$ is satisfied. Let $p \geq 2$ be such that*

$$p > p_{\mathrm{Fuj}}(\tfrac{\mu}{2}).$$

Then, there exists $\varepsilon_0 > 0$ such that for any initial data $(u_0, u_1) \in \mathcal{A}_1$, where $\mathcal{A}_1$ is defined as in the statement of Proposition 2.1, satisfying $\|(u_0, u_1)\|_{\mathcal{A}_1} \leq \varepsilon_0$ there is a uniquely determined solution

$$u \in \mathcal{C}([0,\infty), H^1(\mathbb{R})) \cap \mathcal{C}^1([0,\infty), L^2(\mathbb{R}))$$

to (2). Moreover, the solution satisfies the decay estimates

$$\|u(t,\cdot)\|_{\dot{H}^\kappa(\mathbb{R})} \lesssim \begin{cases} (1+t)^{-\frac{\mu}{2}+\frac{1}{2}-\kappa} \|(u_0,u_1)\|_{\mathcal{A}_1} & \text{if } \kappa < \frac{1}{2}, \\ (1+t)^{-\frac{\mu}{2}} \log(e+t) \|(u_0,u_1)\|_{\mathcal{A}_1} & \text{if } \kappa = \frac{1}{2}, \\ (1+t)^{-\frac{\mu}{2}} \|(u_0,u_1)\|_{\mathcal{A}_1} & \text{if } \kappa > \frac{1}{2}, \end{cases} \tag{8}$$

for any $\kappa \in [0, 1]$ *and* $\|u_t(t,\cdot)\|_{L^2(\mathbb{R})}$ *satisfies the same decay estimates as* $\|u(t,\cdot)\|_{\dot{H}^1(\mathbb{R})}$.

Remark 2.4 Let us point out that (3) implies necessarily $\mu \geq 2$ for nonnegative and nontrivial μ and ν^2. For this reason we require 2 as lower bound for μ in the previous theorem.

Proof We will use the estimates from Propositions 2.1 and 2.2 in the case $n = 1$ and $\delta = 1$. For any $T > 0$ we introduce the space

$$X(T) \doteq \mathcal{C}([0,T], H^1(\mathbb{R})) \cap \mathcal{C}^1([0,T], L^2(\mathbb{R}))$$

equipped with the norm

$$\|u\|_{X(T)} \doteq \sup_{t\in[0,T]} \Big((1+t)^{\frac{\mu}{2}-\frac{1}{2}} \|u(t,\cdot)\|_{L^2(\mathbb{R})} + (1+t)^{\frac{\mu}{2}} (\ell(t))^{-1} \|u(t,\cdot)\|_{\dot{H}^{\frac{1}{2}}(\mathbb{R})} + (1+t)^{\frac{\mu}{2}} \|(u_x,u_t)(t,\cdot)\|_{L^2(\mathbb{R})} \Big),$$

where $\ell(t)$ denotes the same function as in the statement of Proposition 2.1.

We define the operator N as follows:

$$u \to Nu(t,x) \doteq E_0(t,0,x) *_{(x)} u_0(x) + E_1(t,0,x) *_{(x)} u_1(x) + Fu(t,x)$$

for any $u \in X(T)$, where $E_0(t,\tau,x)$ and $E_1(t,\tau,x)$ denote the fundamental solution of (7) with initial condition $(u_0,u_1) = (\delta_0, 0)$ and $(u_0,u_1) = (0,\delta_0)$, respectively, taken at the initial time $\tau \geq 0$ and

$$Fu(t,x) \doteq \int_0^t E_1(t,\tau,x) *_{(x)} |u(\tau,x)|^p \, d\tau.$$

According to Duhamel's principle we know that u is a solution to (2) if and only if u is a fixed point of N.

If N satisfies for all data $(u_0,u_1) \in \mathcal{A}_1$ the inequalities

$$\|Nu\|_{X(T)} \lesssim \|(u_0,u_1)\|_{\mathcal{A}_1} + \|u\|^p_{X(T)}, \tag{9}$$

$$\|Nu - N\tilde{u}\|_{X(T)} \lesssim \|u-\tilde{u}\|_{X(T)} \Big(\|u\|^{p-1}_{X(T)} + \|\tilde{u}\|^{p-1}_{X(T)} \Big) \tag{10}$$

for any $u, \tilde{u} \in X(T)$ and uniformly with respect to $T > 0$, then, by contraction principle it follows the global (in time) existence of small data solutions.

In order to prove (9) and (10) we will employ the fractional Sobolev embedding

$$\|f\|_{L^q(\mathbb{R}^n)} \lesssim \|f\|_{\dot{H}^\kappa(\mathbb{R}^n)}, \quad \kappa = n\left(\tfrac{1}{2} - \tfrac{1}{q}\right), \tag{11}$$

for any $q \geq 2$, in the case $n = 1$. Applying Proposition 2.1, we find immediately

$$\|E_0(t,0,x) *_{(x)} u_0(x) + E_1(t,0,x) *_{(x)} u_1(x)\| \lesssim \|(u_0,u_1)\|_{\mathcal{A}_1}.$$

Hence, we have just to estimate Fu in order to prove (9). For any $\kappa \in [0,1]$, by using Proposition 2.2, we get

$$\|Fu(t,\cdot)\|_{\dot{H}^\kappa(\mathbb{R})} \lesssim \begin{cases} (1+t)^{-\kappa-\frac{\mu}{2}+\frac{1}{2}} \int_0^t (1+\tau)^{\frac{\mu}{2}} G(\tau,u(\tau,\cdot))d\tau & \text{if } \kappa < \frac{1}{2}, \\ (1+t)^{-\frac{\mu}{2}} \int_0^t \tilde{\ell}(t,\tau)(1+\tau)^{\frac{\mu}{2}} G(\tau,u(\tau,\cdot))d\tau & \text{if } \kappa = \frac{1}{2}, \\ (1+t)^{-\frac{\mu}{2}} \int_0^t (1+\tau)^{-\kappa+\frac{\mu}{2}+\frac{1}{2}} G(\tau,u(\tau,\cdot))d\tau & \text{if } \kappa > \frac{1}{2}, \end{cases}$$

where $G(\tau,u(\tau,\cdot)) \doteq \|u(\tau,\cdot)\|^p_{L^p(\mathbb{R})} + (1+\tau)^{\frac{1}{2}}\|u(\tau,\cdot)\|^p_{L^{2p}(\mathbb{R})}$.

Using (11) we have for $j = 1,2$

$$\|u(\tau,\cdot)\|_{L^{jp}(\mathbb{R})} \lesssim \|u(\tau,\cdot)\|_{\dot{H}^{\kappa_j}(\mathbb{R})},$$

where $\kappa_j = \frac{1}{2} - \frac{1}{jp}$. We underline that from the condition $\kappa_1 \geq 0$ yields the restriction $p \geq 2$. On the other hand, the condition $\kappa_2 \leq 1$ is satisfied without further assumptions, being $n = 1$.

It is clear that, thanks to the definition of norm in $X(T)$, we have for any $u \in X(T)$ a suitable decay for the L^2, $\dot{H}^{\frac{1}{2}}$ and $\dot{H}^1$ norms of $u(t,\cdot)$. Furthermore, being homogeneous Sobolev spaces interpolation spaces, we get

$$\|u(t,\cdot)\|_{\dot{H}^\kappa(\mathbb{R})} \leq (1+t)^{-\frac{\mu}{2}+\frac{1}{2}-\kappa}(\ell(t))^{2\kappa}\|u\|_{X(T)} \qquad \text{for } \kappa \in \left(0,\tfrac{1}{2}\right),$$

$$\|u(t,\cdot)\|_{\dot{H}^\kappa(\mathbb{R})} \leq (1+t)^{-\frac{\mu}{2}}(\ell(t))^{2(1-\kappa)}\|u\|_{X(T)} \qquad \text{for } \kappa \in \left(\tfrac{1}{2},1\right).$$

This means that the decay rate is a nondecreasing function with respect to the exponent κ, for $\kappa \in (0,\frac{1}{2})$, and for $\kappa \geq \frac{1}{2}$ the decay rate is the same for any κ, modulo a logarithmic term, which of course is negligible.

Being $p > p_{\text{Fuj}}(\frac{\mu}{2})$, it results $\kappa_1 > \frac{4-\mu}{2(\mu+4)}$ and $\kappa_2 > \frac{2}{\mu+4}$. Therefore,

$$\|u(t,\cdot)\|_{\dot{H}^{\kappa_1}(\mathbb{R})} \leq (1+t)^{-\frac{\mu}{2}+\frac{\mu}{\mu+4}}\log(e+t)\|u\|_{X(T)},$$

$$\|u(t,\cdot)\|_{\dot{H}^{\kappa_2}(\mathbb{R})} \leq (1+t)^{-\frac{\mu}{2}+\frac{\mu}{2(\mu+4)}}\log(e+t)\|u\|_{X(T)}.$$

Summarizing,

$$\begin{aligned} G(\tau, u(\tau,\cdot)) &\lesssim \|u(\tau,\cdot)\|^p_{\dot H^{\kappa_1}(\mathbb{R})} + (1+\tau)^{\frac{1}{2}} \|u(\tau,\cdot)\|^p_{\dot H^{\kappa_2}(\mathbb{R})} \\ &\lesssim \Big((1+\tau)^{-\frac{\mu p}{2} + \frac{\mu p}{\mu+4}} + (1+\tau)^{\frac{1}{2} - \frac{\mu p}{2} + \frac{\mu p}{2(\mu+4)}} \Big) (\ell(\tau))^p \|u\|^p_{X(T)}. \end{aligned}$$

Because of $p > p_{\mathrm{Fuj}}(\frac{\mu}{2})$, we find

$$\int_0^t (1+\tau)^{\frac{\mu}{2}} G(\tau, u(\tau,\cdot)) d\tau \lesssim 1,$$

being the exponents of integrands smaller than -1. Consequently, we get

$$\|Fu(t,\cdot)\|_{\dot H^{\kappa}(\mathbb{R})} \lesssim \begin{cases} (1+t)^{-\kappa - \frac{\mu}{2} + \frac{1}{2}} \|u\|^p_{X(T)} & \text{if} \quad \kappa < \frac{1}{2}, \\ (1+t)^{-\frac{\mu}{2}} \ell(t) \|u\|^p_{X(T)} & \text{if} \quad \kappa = \frac{1}{2}, \\ (1+t)^{-\frac{\mu}{2}} \|u\|^p_{X(T)} & \text{if} \quad \kappa > \frac{1}{2}. \end{cases} \tag{12}$$

The estimate (12), together with the remark that the time derivative has the same decay behavior of the gradient with respect to the spatial variables in Theorem 2.2, guarantees the validity of (9).

Moreover, combining such relation with the results for the linear equation, we get the estimate (8) once we proved the existence of the solution, since

$$\|u\|_{X(T)} \lesssim \|(u_0, u_1)\|_{\mathcal{A}_1}.$$

In order to prove (10), it is sufficient to combine the inequality

$$\big| |u|^p - |\tilde u|^p \big| \lesssim |u - \tilde u| \big(|u|^{p-1} + |\tilde u|^{p-1} \big) \tag{13}$$

with Hölder's inequality and to repeat the same estimates as before in the case in which the source term is $|u|^p - |\tilde u|^p$.

Remark 2.5 The idea of considering the embedding (11) instead of using Gagliardo-Nirenberg inequality, in order to avoid loss of decay for the L^p and L^{2p} norms of the solution, is taken from [3, Section 4].

2.2 *One-Dimensional Case: p < 2*

In the previous subsection we have employed a fractional Sobolev embedding in order to derive a global existence result for small data solutions by using $L^1 \cap L^2 - L^2$ estimates. However, the employment of that inequality requires the restriction $p \geq 2$.

In this section we are going to prove a global (in time) existence result for exponents $p_{\mathrm{Fuj}}(\frac{\mu}{2}) < p < 2$, whether this range is not empty, that is, for $\mu > 4$. The main tool that allows us to overcome the necessity of condition $p \geq 2$ is the use of $L^1 - L^p$ estimates for the linear free wave equation in spatial dimension 1.

More precisely, we are going to use $L^{1+\eta} - L^p$ estimates with $\eta > 0$ sufficiently small.

Let us state the result concerning the linear part for the free wave equation.

Proposition 2.6 *Let v be the solution to the Cauchy problem*

$$\begin{cases} v_{tt} - v_{xx} = 0, & t > 0,\ x \in \mathbb{R}, \\ v(0,x) = v_0(x), & x \in \mathbb{R}, \\ v_t(0,x) = v_1(x), & x \in \mathbb{R}. \end{cases}$$

Then, for any $1 < p \leq q < \infty$ the solution v satisfies for any $t \geq 0$ the following a priori estimate:

$$\|v(t,\cdot)\|_{L^q(\mathbb{R})} \lesssim (1+t)^{1-\frac{1}{p}+\frac{1}{q}} \big(\|v_0\|_{H^1_p(\mathbb{R})} + \|v_1\|_{L^p(\mathbb{R})} \big).$$

Proof Let us define for a fixed $t > 0$ the operators

$$A(t)v_0(x) \doteq \mathcal{F}^{-1}(\cos(t\xi)\hat{v}_0(\xi))(x), \quad B(t)v_1(x) \doteq \mathcal{F}^{-1}\bigg(\frac{\sin(t\xi)}{\xi}\hat{v}_1(\xi)\bigg)(x).$$

Using basic properties of inverse Fourier transformation, it follows:

$$\|A(t)v_0\|_{H^1_p(\mathbb{R})} = \tfrac{1}{2}\|\mathcal{F}^{-1}((e^{it\xi} + e^{-it\xi})\langle\xi\rangle\hat{v}_0(\xi))\|_{L^p(\mathbb{R})} \leq \|v_0\|_{H^1_p(\mathbb{R})}, \tag{14}$$

and, then, $A(t) \in L(H^1_p(\mathbb{R}) \to H^1_p(\mathbb{R}))$ with $\|A(t)\|_{L(H^1_p(\mathbb{R})\to H^1_p(\mathbb{R}))} \leq 1$.

On the other hand, in [23] for any finite $p > 1$ and for a fixed $t > 0$ it is proved that

$$\|B(t)v_1\|_{H^1_p(\mathbb{R})} \lesssim \|v_1\|_{L^p(\mathbb{R})}. \tag{15}$$

Therefore, $B(t) \in L(L^p(\mathbb{R}) \to H^1_p(\mathbb{R}))$.

For Bessel potential spaces it holds the embedding $H^1_p(\mathbb{R}) \hookrightarrow L^q(\mathbb{R})$ for any $1 < p \leq q < \infty$. Let us underline here that we have to exclude the case $p = 1$ from the statement, otherwise the previous embedding is no longer true. Then, taking $t = 1$ in (14), (15) and using the previously recalled embedding, we obtain

$$\|A(1)v_0\|_{L^q(\mathbb{R})} \lesssim \|v_0\|_{H^1_p(\mathbb{R})}, \tag{16}$$

$$\|B(1)v_1\|_{L^q(\mathbb{R})} \lesssim \|v_1\|_{L^p(\mathbb{R})}. \tag{17}$$

Up to now we neglected somehow the decay rates in the estimates for the families of operators $\{A(t)\}_{t\geq 0}$ and $\{B(t)\}_{t\geq 0}$. However, using a homogeneity argument, we can now determine sharp decay rates.

The solution of the free wave equation with data (v_0, v_1) is given by

$$v(t,x) = A(t)v_0(x) + B(t)v_1(x) = \mathcal{F}^{-1}\Big(\cos(t\xi)\hat{v}_0(\xi) + \frac{\sin(t\xi)}{\xi}\hat{v}_1(\xi)\Big)(x).$$

Using the homogeneity properties

$$\mathcal{F}^{-1}\Big(\frac{\sin(t\xi)}{\xi}\hat{v}_1(\xi)\Big)(x) = t\mathcal{F}^{-1}\Big(\frac{\sin\xi}{\xi}\mathcal{F}(v_1(t\,\cdot))(\xi)\Big)\Big(\frac{x}{t}\Big),$$
$$\mathcal{F}^{-1}(\cos(t\xi)\hat{v}_0(\xi))(x) = \mathcal{F}^{-1}\big(\cos\xi\,\mathcal{F}(v_0(t\,\cdot))(\xi)\big)\Big(\frac{x}{t}\Big),$$

from (17), we get

$$\Big\|\mathcal{F}^{-1}\Big(\frac{\sin(t\xi)}{\xi}\hat{v}_1(\xi)\Big)\Big\|_{L^q(\mathbb{R})} \lesssim t^{1-\frac{1}{p}+\frac{1}{q}}\|v_1\|_{L^p(\mathbb{R})},$$

while from (16) we have

$$\|\mathcal{F}^{-1}(\cos(t\xi)\hat{v}_0(\xi))\|_{L^q(\mathbb{R})} \lesssim t^{1-\frac{1}{p}+\frac{1}{q}}\|v_0\|_{\dot{H}^1_p(\mathbb{R})} + t^{-\frac{1}{p}+\frac{1}{q}}\|v_0\|_{L^p(\mathbb{R})},$$

where in both cases we used scaling properties of L^r and $\dot{H}^\sigma_r$. Combining the last two estimates with the uniform boundedness of the family of operators $\{A(t)\}_{t\geq 0}$ in $L(H^1_p(\mathbb{R}) \to L^q(\mathbb{R}))$, it follows the desired estimate.

Using the inverse transformation $u(t,x) = (1+t)^{-\frac{\mu}{2}}v(t,x)$ and the invariance of the classical wave equation with respect to time translations, we obtain immediately the following results for the scale-invariant wave model with damping and mass.

Corollary 2.7 *Let $n = 1$ and μ, ν^2 be nonnegative constants satisfying $\delta = 1$. Let u be the solution to the Cauchy problem* (6)*. Then, for any $1 < p \leq q < \infty$ the solution u satisfies for any $t \geq 0$ the following a priori estimate:*

$$\|u(t,\cdot)\|_{L^q(\mathbb{R})} \lesssim (1+t)^{-\frac{\mu}{2}+1-\frac{1}{p}+\frac{1}{q}}\Big(\|u_0\|_{H^1_p(\mathbb{R})} + \|u_1\|_{L^p(\mathbb{R})}\Big).$$

Corollary 2.8 *Let $n = 1$ and μ, ν^2 be nonnegative constants satisfying $\delta = 1$. Let $\tau \geq 0$ and let u be the solution to the Cauchy problem* (7) *with $u_0 = 0$. Then, for any $1 < p \leq q < \infty$ the solution u satisfies for any $t \geq \tau$ the following a priori estimate:*

$$\|u(t,\cdot)\|_{L^q(\mathbb{R})} \lesssim (1+t)^{-\frac{\mu}{2}}(1+t-\tau)^{1-\frac{1}{p}+\frac{1}{q}}(1+\tau)^{\frac{\mu}{2}}\|u_1\|_{L^p(\mathbb{R})}.$$

Finally, we can state the main result of this subsection, that is, the following global (in time) existence result.

Theorem 2.9 *Let $n = 1$ and let $\mu > 4$ and ν^2 be nonnegative constants satisfying $\delta = 1$. Let us assume*

$$p \in \big(p_{\mathrm{Fuj}}(\tfrac{\mu}{2}), 2\big).$$

Then, there exist two constants $\alpha = \alpha(p) > 0$ and $\varepsilon_0 > 0$ such that for any data $(u_0, u_1) \in \big(L^1(\mathbb{R}) \cap H^1_{1+\frac{\alpha}{p}}(\mathbb{R}) \cap H^1(\mathbb{R})\big) \times \big(L^1(\mathbb{R}) \cap L^2(\mathbb{R})\big) \doteq \mathcal{A}_2$, with $\|(u_0, u_1)\|_{\mathcal{A}_2} \leq \varepsilon_0$ there is a uniquely determined solution

$$u \in \mathcal{C}([0,\infty), H^1(\mathbb{R}) \cap L^p(\mathbb{R})) \cap \mathcal{C}^1([0,\infty), L^2(\mathbb{R}))$$

to the Cauchy problem (2). Moreover, the solution u satisfies for any $t \geq 0$ the decay estimates

$$\|u(t,\cdot)\|_{L^p(\mathbb{R})} \lesssim (1+t)^{-\frac{\mu}{2}+\frac{1}{p}+\frac{\alpha}{p+\alpha}} \|(u_0,u_1)\|_{\mathcal{A}_2},$$

$$\|u(t,\cdot)\|_{\dot{H}^\kappa(\mathbb{R})} \lesssim \begin{cases} (1+t)^{-\frac{\mu}{2}+\frac{1}{2}-\kappa} \|(u_0,u_1)\|_{\mathcal{A}_2} & \text{if } \kappa < \frac{1}{2}, \\ (1+t)^{-\frac{\mu}{2}} \log(e+t) \|(u_0,u_1)\|_{\mathcal{A}_2} & \text{if } \kappa = \frac{1}{2}, \\ (1+t)^{-\frac{\mu}{2}} \|(u_0,u_1)\|_{\mathcal{A}_2} & \text{if } \kappa > \frac{1}{2}, \end{cases}$$

with $\kappa \in [0,1]$, and $\|u_t(t,\cdot)\|_{L^2(\mathbb{R})}$ satisfies the same decay estimates as $\|u(t,\cdot)\|_{\dot{H}^1(\mathbb{R})}$.

Proof For any $T > 0$ let us define the space

$$X(T) \doteq \mathcal{C}([0,T], H^1(\mathbb{R}) \cap L^p(\mathbb{R})) \cap \mathcal{C}^1([0,T], L^2(\mathbb{R})),$$

with the norm

$$\begin{aligned} \|u\|_{X(T)} \doteq \sup_{t \in [0,T]} \Big(& (1+t)^{\frac{\mu}{2}-\frac{1}{p}-\frac{\alpha}{p+\alpha}} \|u(t,\cdot)\|_{L^p(\mathbb{R})} + (1+t)^{\frac{\mu}{2}-\frac{1}{2}} \|u(t,\cdot)\|_{L^2(\mathbb{R})} \\ & + (1+t)^{\frac{\mu}{2}} \|u(t,\cdot)\|_{\dot{H}^1(\mathbb{R})} + (1+t)^{\frac{\mu}{2}} \|u_t(t,\cdot)\|_{L^2(\mathbb{R})} \\ & + (1+t)^{\frac{\mu}{2}} \big(\ell(t)\big)^{-1} \|u(t,\cdot)\|_{\dot{H}^{1/2}(\mathbb{R})} \Big), \end{aligned}$$

where $\alpha > 0$ is a sufficiently small constant which depends on p. Throughout the proof we will prescribe the conditions that α has to fulfill.

Let us underline that the main difference with respect to Theorem 2.3 is the presence of $L^p(\mathbb{R})$ in the space $X(T)$.

We define formally the operator N exactly as in the proof of Theorem 2.3. Also in this case we want to show that N is a contracting mapping from a closed ball of

$X(T)$ into itself. By Corollary 2.7 and Proposition 2.1 we get immediately

$$\|E_0(t,0,x)*_{(x)}u_0(x)+E_1(t,0,x)*_{(x)}u_1(x)\|_{X(T)}\lesssim\|(u_0,u_1)\|_{\mathcal{A}_2},$$

since, trivially, $\mathcal{A}_2\hookrightarrow\mathcal{A}_1$. In order to estimate the $X(T)$ norm of Nu, employing the same notations as in Theorem 2.3, it remains to estimate the integral term Fu. Therefore, using $L^{1+\frac{\alpha}{p}}-L^p$ estimates from Corollary 2.8, we find

$$\begin{aligned}\|Fu(t,\cdot)\|_{L^p(\mathbb{R})}&\lesssim(1+t)^{-\frac{\mu}{2}}\int_0^t(1+t-\tau)^{\frac{1}{p}+\frac{\alpha}{p+\alpha}}(1+\tau)^{\frac{\mu}{2}}\||u(\tau,\cdot)|^p\|_{L^{1+\frac{\alpha}{p}}(\mathbb{R})}d\tau\\&\lesssim(1+t)^{-\frac{\mu}{2}+\frac{1}{p}+\frac{\alpha}{p+\alpha}}\int_0^t(1+\tau)^{\frac{\mu}{2}}\|u(\tau,\cdot)\|^p_{L^{p+\alpha}(\mathbb{R})}d\tau.\end{aligned}$$

Since we are working with $p<2$, we may choose $\alpha>0$ such that $p+\alpha<2$. Thus, by using Hölder's inequality and the definition of $\|\cdot\|_{X(\tau)}$, we arrive at

$$\begin{aligned}\|u(\tau,\cdot)\|_{L^{p+\alpha}(\mathbb{R})}&\le\|u(\tau,\cdot)\|^{1-\theta}_{L^p(\mathbb{R})}\|u(\tau,\cdot)\|^{\theta}_{L^2(\mathbb{R})}\\&\le(1+\tau)^{-\frac{\mu}{2}+\frac{\alpha}{p+\alpha}+\frac{1}{p+\alpha}}\|u\|_{X(\tau)},\end{aligned}$$

where $\frac{1}{p+\alpha}=\frac{1-\theta}{p}+\frac{\theta}{2}$. Consequently,

$$\|Fu(t,\cdot)\|_{L^p(\mathbb{R})}\lesssim(1+t)^{-\frac{\mu}{2}+\frac{1}{p}+\frac{\alpha}{p+\alpha}}\int_0^t(1+\tau)^{-\frac{\mu}{2}(p-1)+\frac{p}{p+\alpha}+\frac{\alpha p}{p+\alpha}}d\tau\,\|u\|^p_{X(t)}.$$

Let us show now that is possible to choose $\alpha>0$ so that the last integral is uniformly bounded with respect to t. Since we are assuming $p>p_{\mathrm{Fuj}}(\frac{\mu}{2})$, then, $-\frac{\mu}{2}(p-1)+2<0$. Hence, we can choose α sufficiently small in order to get

$$-\tfrac{\mu}{2}(p-1)+2+\tfrac{\alpha(p-1)}{p+\alpha}<0,$$

and, being this condition equivalent to require that the exponent of the integrand in the last integral is strictly smaller than -1, we finally obtain

$$\|Fu(t,\cdot)\|_{L^p(\mathbb{R})}\lesssim(1+t)^{-\frac{\mu}{2}+\frac{1}{p}+\frac{\alpha}{p+\alpha}}\|u\|^p_{X(t)}.$$

Let us estimate now the L^2 norm of $Fu(t,\cdot)$. According to Proposition 2.2, we find

$$\begin{aligned}&(1+t)^{\frac{\mu}{2}-\frac{1}{2}}\|Fu(t,\cdot)\|_{L^2(\mathbb{R})}\\&\quad\lesssim\int_0^t(1+\tau)^{\frac{\mu}{2}}\Big(\|u(\tau,\cdot)\|^p_{L^p(\mathbb{R})}+(1+\tau)^{\frac{1}{2}}\|u(\tau,\cdot)\|^p_{L^{2p}(\mathbb{R})}\Big)d\tau\end{aligned}\tag{18}$$

By using the definition of norm in $X(\tau)$, we can immediately determine the decay rate of the L^p norm of $u(\tau,\cdot)$ as follows:

$$\|u(\tau,\cdot)\|_{L^p(\mathbb{R})} \leq (1+\tau)^{-\frac{\mu}{2}+\frac{1}{p}+\frac{\alpha}{p+\alpha}} \|u\|^p_{X(\tau)}.$$

In order to find the decay rate for the L^{2p} norm of u, we can use the fractional Gagliardo-Nirenberg inequality. Thus,

$$\begin{aligned}\|u(\tau,\cdot)\|_{L^{2p}(\mathbb{R})} &\lesssim \|u(\tau,\cdot)\|^{\theta_{1/2}(2p)}_{\dot{H}^{\frac{1}{2}}(\mathbb{R})} \|u(\tau,\cdot)\|^{1-\theta_{1/2}(2p)}_{L^2(\mathbb{R})} \\ &\lesssim (1+\tau)^{-\frac{\mu}{2}+\frac{1}{2p}} \ell(t)^{\theta_{1/2}(2p)} \|u\|_{X(\tau)},\end{aligned}$$

where $\theta_{1/2}(2p) \doteq 1-\frac{1}{p}$. Also,

$$(1+t)^{\frac{\mu}{2}-\frac{1}{2}} \|Fu(t,\cdot)\|_{L^2(\mathbb{R})} \lesssim \int_0^t (1+\tau)^{-\frac{\mu}{2}(p-1)+1+\frac{\alpha p}{p+\alpha}} d\tau \; \|u\|^p_{X(t)}.$$

As we did in the estimate of $\|Fu(t,\cdot)\|_{L^p(\mathbb{R})}$, since $p > p_{\mathrm{Fuj}}(\frac{\mu}{2})$ we may assume without loss of generality that

$$-\tfrac{\mu}{2}(p-1)+1+\tfrac{\alpha p}{p+\alpha} < -1,$$

for α sufficiently small. Then, we proved

$$\|Fu(t,\cdot)\|_{L^2(\mathbb{R})} \lesssim (1+t)^{-\frac{\mu}{2}+\frac{1}{2}} \|u\|^p_{X(t)}.$$

Using Proposition 2.2 in the logarithmic case, for the $\dot{H}^{\frac{1}{2}}$ norm of integral term, we have

$$\begin{aligned}&(1+t)^{\frac{\mu}{2}} \|Fu(t,\cdot)\|_{\dot{H}^{\frac{1}{2}}(\mathbb{R})} \\ &\quad\lesssim \int_0^t \ell(t,\tau)(1+\tau)^{\frac{\mu}{2}}\Big(\|u(\tau,\cdot)\|^p_{L^p(\mathbb{R})} + (1+\tau)^{\frac{1}{2}}\|u(\tau,\cdot)\|^p_{L^{2p}(\mathbb{R})}\Big)d\tau \\ &\quad\lesssim \ell(t)\int_0^t (1+\tau)^{\frac{\mu}{2}}\Big(\|u(\tau,\cdot)\|^p_{L^p(\mathbb{R})} + (1+\tau)^{\frac{1}{2}}\|u(\tau,\cdot)\|^p_{L^{2p}(\mathbb{R})}\Big)d\tau.\end{aligned}$$

Because the last integral is exactly the same integral on the right-hand side of (18), we may conclude

$$\|Fu(t,\cdot)\|_{\dot{H}^{\frac{1}{2}}(\mathbb{R})} \lesssim (1+t)^{-\frac{\mu}{2}} \ell(t) \|u\|^p_{X(t)},$$

provided that $p > p_{\mathrm{Fuj}}(\frac{\mu}{2})$ and $\alpha > 0$ is sufficiently small.

Finally, let us study the $\dot{H}^1$ norm of $Fu(t,\cdot)$ and the L^2 norm of the time derivative of $Fu(t,\cdot)$. From Proposition 2.2 we have

$$\begin{aligned}(1+t)^{\frac{\mu}{2}}\big(\|Fu(t,\cdot)\|_{\dot{H}^1(\mathbb{R})}+\|\partial_t Fu(t,\cdot)\|_{L^2(\mathbb{R})}\big)\\ \lesssim \int_0^t (1+\tau)^{\frac{\mu}{2}-\frac{1}{2}}\Big(\||u(\tau,\cdot)|^p\|_{L^1(\mathbb{R})}+(1+\tau)^{\frac{1}{2}}\||u(\tau,\cdot)|^p\|_{L^2(\mathbb{R})}\Big)d\tau.\end{aligned}$$

Since the integral in the last line is dominated by the integral on the right-hand side of (18), assuming again that $p > p_{\mathrm{Fuj}}(\frac{\mu}{2})$ and $\alpha > 0$ is sufficiently small, it results

$$\|Fu(t,\cdot)\|_{\dot{H}^1(\mathbb{R})}+\|\partial_t Fu(t,\cdot)\|_{L^2(\mathbb{R})} \lesssim (1+t)^{-\frac{\mu}{2}}\|u\|^p_{X(t)}.$$

Summarizing, we proved

$$\|Nu\|_{X(T)} \lesssim \|(u_0,u_1)\|_{\mathcal{A}_2}+\|u\|^p_{X(T)}. \tag{19}$$

In order to get the Lipschitz condition, one can proceed as in Theorem 2.3, obtaining

$$\|Nu-Nv\|_{X(T)} \lesssim \|u-v\|_{X(T)}\Big(\|u\|^{p-1}_{X(T)}+\|v\|^{p-1}_{X(T)}\Big), \tag{20}$$

provided that p and $\alpha > 0$ satisfy the same conditions as before. Combining (19) and (20), it follows the existence of a unique fixed point $u \in X(T)$ for N by Banach's fixed point theorem, provided that the norm of initial data is smaller than a suitable constant $\varepsilon_0 > 0$. Finally, since the estimate for Nu and $Nu - Nv$ are uniform with respect to T, we can extend the solution for any time $t > 0$. It remains to prove the decay estimates for the solution as in the statement. Since $\|u\|_{X(t)} \lesssim \|(u_0,u_1)\|_{\mathcal{A}_2}$ for any $t > 0$, from the definition of the space $X(t)$ the estimates for the L^p, L^2, $\dot{H}^1$ and $\dot{H}^{\frac{1}{2}}$ norms of $u(t,\cdot)$ and the L^2 norm of $u_t(t,\cdot)$ follow immediately. We derive now the estimate for the $\dot{H}^\kappa$ norm of $u(t,\cdot)$ when $\kappa \in (0,\frac{1}{2}) \cup (\frac{1}{2},1)$. For $\kappa < \frac{1}{2}$ we have

$$\begin{aligned}(1+t)^{\frac{\mu}{2}-\frac{1}{2}+\kappa}\|u(t,\cdot)\|_{\dot{H}^\kappa(\mathbb{R})} \lesssim \|(u_0,u_1)\|_{\mathcal{A}_1}\\ +\int_0^t (1+\tau)^{\frac{\mu}{2}}\Big(\|u(\tau,\cdot)\|^p_{L^p(\mathbb{R})}+(1+\tau)^{\frac{1}{2}}\|u(\tau,\cdot)\|^p_{L^{2p}(\mathbb{R})}\Big)d\tau.\end{aligned}$$

Employing the usual estimates for $\|u(\tau,\cdot)\|_{L^p(\mathbb{R})}$, $\|u(\tau,\cdot)\|_{L^{2p}(\mathbb{R})}$, for the integral term we find

$$\begin{aligned}\int_0^t (1+\tau)^{\frac{\mu}{2}}\Big(\|u(\tau,\cdot)\|^p_{L^p(\mathbb{R})}+(1+\tau)^{\frac{1}{2}}\|u(\tau,\cdot)\|^p_{L^{2p}(\mathbb{R})}\Big)d\tau\\ \lesssim \|u\|^p_{X(t)} \lesssim \|(u_0,u_1)\|^p_{\mathcal{A}_2} \lesssim \|(u_0,u_1)\|_{\mathcal{A}_2}.\end{aligned}$$

In a similar way, for $\kappa > \frac{1}{2}$ we get

$$(1+t)^{\frac{\mu}{2}} \|u(t,\cdot)\|_{\dot{H}^\kappa(\mathbb{R})} \lesssim \|(u_0,u_1)\|_{\mathcal{A}_2}.$$

Hence, the proof is completed.

In the next two sections we will deal with the radial odd case $n \geq 3$. In particular, we will consider first the three dimensional case and, then, the general odd case $n \geq 5$.

3 Radial Three Dimensional Case

Let us begin with the case $n = 3$. Hereafter, we denote $r = |x|$ for $x \in \mathbb{R}^n$ and $\langle y \rangle \doteq 1 + |y|$ for any $y \in \mathbb{R}$. In this section we follow [6, Section 5].

Since we are assuming that μ and ν^2 satisfy (3), performing the change of variables $v(t,x) = \langle t \rangle^{\frac{\mu}{2}} u(t,x)$, we arrive at the Cauchy problem

$$\begin{cases} v_{tt} - \Delta v = \langle t \rangle^{-\frac{\mu}{2}(p-1)} |v|^p, & t > 0, \ x \in \mathbb{R}^3, \\ v(0,x) = f(x), & x \in \mathbb{R}^3, \\ v_t(0,x) = g(x), & x \in \mathbb{R}^3, \end{cases} \tag{21}$$

where $f(x) = u_0(x)$ and $g(x) = u_1(x) + \frac{\mu}{2} u_0(x)$.

We are interested in radial solutions, hence, we may rewrite (21) as

$$\begin{cases} v_{tt} - v_{rr} - \frac{2}{r} v_r = \langle t \rangle^{-\frac{\mu}{2}(p-1)} |v|^p, & t > 0, \ r \in \mathbb{R}, \\ v(0,r) = f(r), & r \in \mathbb{R}, \\ v_t(0,r) = g(r), & r \in \mathbb{R}. \end{cases} \tag{22}$$

Throughout this section we will assume for f and g the conditions

$$|f^{(j)}(r)| \leq \varepsilon \langle r \rangle^{-(\kappa+j)} \quad \text{for } j = 0, 1, \tag{23}$$

$$|g(r)| \leq \varepsilon \langle r \rangle^{-(\kappa+1)}, \tag{24}$$

where ε and κ are positive parameters that will be fixed afterwards.

In the following it is convenient to extend f, g by even reflection, that is,

$$\begin{cases} f(-r) = f(r) \\ g(-r) = g(r) \end{cases} \quad \text{for } r < 0.$$

The remaining part of this section is organized in the following way: firstly, we recall some known result for the radial linear equation when $n = 3$; then, after some preparatory results, we will derive the global (in time) existence result.

3.1 Radial Linear Wave Equation: 3-d Case

In this subsection the corresponding linear problem is considered. Let us begin with a definition.

Definition 3.1 *Let us consider the Cauchy problem*

$$\begin{cases} v_{tt} - v_{rr} - \frac{2}{r}v_r = 0, & t > 0, \ r \in \mathbb{R}, \\ v(0,r) = f(r), & r \in \mathbb{R}, \\ v_t(0,r) = g(r), & r \in \mathbb{R}. \end{cases} \tag{25}$$

We call $v \in C([0,\infty) \times \mathbb{R})$ a solution to Eq. (25)*, if v fulfills*

$$\begin{cases} r^2 v_{tt} - (r^2 v_{rr} + 2r v_r) = 0, & t > 0, \ r \in \mathbb{R}, \\ v(0,r) = f(r), & r \in \mathbb{R}, \\ v_t(0,r) = g(r), & r \in \mathbb{R} \end{cases}$$

and if $rv \in C^1([0,\infty) \times \mathbb{R})$, $r^2 v \in C^2([0,\infty) \times \mathbb{R})$.

Proposition 3.2 *Let $f \in C^2(\mathbb{R})$ and $g \in C^1(\mathbb{R})$ be even functions. Let us define the function*

$$\begin{aligned} v(t,r) &= \frac{\partial}{\partial t}\left(\int_{-1}^{1} H_f(t+r\sigma)d\sigma\right) + \int_{-1}^{1} H_g(t+r\sigma)d\sigma \\ &= \frac{1}{2r}\left((t+r)f(t+r) - (t-r)f(t-r)\right) + \frac{1}{r}\int_{t-r}^{t+r} H_g(\rho)d\rho, \end{aligned}$$

with

$$H_h(\rho) = \frac{\rho h(\rho)}{2} \quad \text{for } h = f, g.$$

Then, v is a solution of (25) *in the sense of the Definition* 3.1.

Proof Let H be a arbitrary function from $C^1(\mathbb{R})$. If we denote

$$w[H](t,r) \doteq \int_{-1}^{1} H(t+r\sigma)d\sigma = \frac{1}{r}\int_{t+r}^{t-r} H(\rho)d\rho,$$

then, $w[H]$ is a radial solution of the wave equation in the sense of Definition 3.1. Although this is a well-known fact (see for example [2] or [6, Section 5]), we write the computation, since we will use some intermediate steps afterwards. Also, for any $r \neq 0$ it holds

$$\partial_t w[H] = \int_{-1}^{1} H'(t+r\sigma)d\sigma = \frac{1}{r}\left(H(t+r) - H(t-r)\right),$$

$$\partial_t^2 w[H] = \frac{1}{r}\left(H'(t+r) - H'(t-r)\right), \tag{26}$$

$$\begin{aligned}
\partial_r w[H] &= \int_{-1}^{1} \sigma H'(t+r\sigma)d\sigma = \frac{1}{r}\int_{-1}^{1} \sigma\, \partial_\sigma H(t+r\sigma)d\sigma \\
&= \frac{1}{r}\left(H(t+r) + H(t-r)\right) - \frac{1}{r}\int_{-1}^{1} H(t+r\sigma)d\sigma \\
&= \frac{1}{r}\left(H(t+r) + H(t-r)\right) - \frac{1}{r} w[H],
\end{aligned}$$

$$\begin{aligned}
\partial_r^2 w[H] &= -\frac{1}{r^2}\left(H(t+r) + H(t-r)\right) + \frac{1}{r}\left(H'(t+r) - H'(t-r)\right) \\
&\quad + \frac{1}{r^2} w[H] - \frac{1}{r}\partial_r w[H] \\
&= \frac{1}{r}\left(H'(t+r) - H'(t-r)\right) - \frac{2}{r}\partial_r w[H].
\end{aligned} \tag{27}$$

Thus, combining (26) and (27), we get

$$\partial_t^2 w[H] - \partial_r^2 w[H] = \frac{2}{r}\partial_r w[H].$$

Set $v = \partial_t w[H_f] + w[H_g]$. Due to the linearity of the wave equation it is clear that v is a radial solution of the wave equation (in the sense of Definition 3.1). Let us show that v satisfies initial conditions:

$$v(0,r) = \frac{1}{r}\left(H_f(r) - H_f(-r)\right) + \frac{1}{r}\int_r^{-r} H_g(\rho)d\rho = f(r),$$

$$v_t(0,r) = \frac{1}{r}\left(H_f'(r) - H_f'(-r)\right) + \frac{1}{r}\left(H_g(r) - H_g(-r)\right) = g(r),$$

where we used that H_f, H_g are odd functions, being f, g even by assumption.

Now we want to derive decay estimates in a weighted $L_t^\infty L_r^\infty$ space for radial solutions of the Cauchy problem (25).

For this reason, it is useful to take account of the following formula:

$$\partial_r(r w[H]) = H(t+r) + H(t-r). \tag{28}$$

For any fixed $\kappa > 1$, we introduce the Banach space

$$X_\kappa \doteq \{v \in \mathcal{C}([0,\infty)\times\mathbb{R}) : v \text{ even in } r\,,\, \partial_r(rv) \in \mathcal{C}([0,\infty)\times\mathbb{R}) \text{ and } \|v\|_{X_\kappa} < \infty\},$$

equipped with the norm

$$\|v\|_{X_\kappa} \doteq \sup_{t\geq 0\,,\, r\in\mathbb{R}} \Big(|v(t,r)| + \langle r\rangle^{-1}|\partial_r(rv(t,r))|\Big)\langle t+|r|\rangle\langle t-|r|\rangle^{\kappa-1}.$$

Proposition 3.3 *Let us assume* $(f,g) \in \mathcal{C}^2(\mathbb{R})\times\mathcal{C}^1(\mathbb{R})$ *satisfying* (23) *and* (24) *for some* $\kappa > 1$ *and* $\varepsilon > 0$*. Then,*

$$\|v\|_{X_\kappa} \lesssim \varepsilon.$$

Proof Let us consider first the case $v = w[H_g]$. Using the definition of H_g we get $|H_g(\rho)| \leq \varepsilon\langle\rho\rangle^{-\kappa}$. Thanks to (28), we obtain immediately

$$|\partial_r(rv(t,r))| \lesssim \varepsilon\langle t-|r|\rangle^{-\kappa}.$$

We estimate this last term in two different cases. If $t \geq 2|r|$, then, $\langle t\rangle \approx \langle t+|r|\rangle \approx \langle t-|r|\rangle$. Hence, using $\langle r\rangle \geq 1$, we find

$$|\partial_r(rv(t,r))| \lesssim \varepsilon\langle t-|r|\rangle^{-\kappa+1}\langle t+|r|\rangle^{-1} \leq \varepsilon\langle t-|r|\rangle^{-\kappa+1}\langle t+|r|\rangle^{-1}\langle r\rangle.$$

When $t \leq 2|r|$, then, $\langle t+|r|\rangle \lesssim \langle r\rangle$. Consequently, we have

$$|\partial_r(rv(t,r))| \lesssim \varepsilon\langle t-|r|\rangle^{-\kappa}\langle t+|r|\rangle^{-1}\langle r\rangle \leq \varepsilon\langle t-|r|\rangle^{-\kappa+1}\langle t+|r|\rangle^{-1}\langle r\rangle.$$

Therefore, we proved

$$\|\langle r\rangle^{-1}\langle t+|r|\rangle\langle t-|r|\rangle^{\kappa-1}\partial_r(rv(t,r))\|_{L^\infty([0,\infty)_t\times\mathbb{R}_r)} \lesssim \varepsilon.$$

Now we observe that

$$|v(t,r)| \leq \frac{1}{|r|}\int_{t-|r|}^{t+|r|}|H_g(\rho)|d\rho \lesssim \frac{\varepsilon}{|r|}\int_{t-|r|}^{t+|r|}\langle\rho\rangle^{-\kappa}d\rho,$$

where we used the growth condition for g to estimate the integrand. For $t \geq 2|r|$, since the integrand function takes its maximum for $\rho = t-|r|$ and $\langle t+|r|\rangle \approx \langle t-|r|\rangle$, we find

$$|v(t,r)| \leq \varepsilon\langle t-|r|\rangle^{-\kappa} \approx \varepsilon\langle t-|r|\rangle^{-\kappa+1}\langle t+|r|\rangle^{-1}.$$

When $t \leq 2|r|$ we consider two different subcases. If $|r| \leq 1$, since $\langle t - |r|\rangle \approx \langle t + |r|\rangle \approx 1$ and $\langle \rho\rangle^{-\kappa} \leq 1$, then, we get

$$|v(t,r)| \leq \frac{\varepsilon}{|r|}\int_{t-|r|}^{t+|r|} d\rho \lesssim \varepsilon \approx \varepsilon\langle t - |r|\rangle^{-\kappa+1}\langle t + |r|\rangle^{-1}.$$

On the other hand, if $|r| \geq 1$, then, $\langle t + |r|\rangle \leq 3\langle r\rangle$ and $|r| \approx \langle r\rangle$. Hence, for $t \geq |r|$

$$\begin{aligned}|v(t,r)| &\lesssim \frac{\varepsilon}{\langle r\rangle}\int_{t-|r|}^{t+|r|}\langle \rho\rangle^{-\kappa} d\rho \lesssim \varepsilon\langle t + |r|\rangle^{-1}\int_{t-|r|}^{t+|r|}\langle \rho\rangle^{-\kappa} d\rho \\ &\lesssim \varepsilon\langle t + |r|\rangle^{-1}\langle t - |r|\rangle^{-\kappa+1},\end{aligned}$$

thanks to $\kappa > 1$. Let us underline that in the case $t \leq |r|$, due to the oddness of H_g, we may represent v as follows:

$$v(t,r) = \frac{1}{r}\int_{r-t}^{t+r} H_g(\rho)d\rho.$$

Consequently, we obtain the same estimate as before, modifying properly the domain of integration. Note that we can assume without loss of generality that $r \geq 0$, being $v = w[H_g]$ even with respect to r. Hence, we proved

$$\|\langle t + |r|\rangle\langle t - |r|\rangle^{\kappa-1} v(t,r)\|_{L^\infty([0,\infty)_t\times\mathbb{R}_r)} \lesssim \varepsilon.$$

Now we study the case $v = \partial_t w[H_f]$. As in the first case, we have

$$|H_f(\rho)| \leq \varepsilon\langle \rho\rangle^{-(\kappa+j-1)} \qquad \text{for } j = 0, 1.$$

We know that v can be written also in the form

$$v(t,r) = \frac{1}{2r}\left(H_f(t+r) - H_f(t-r)\right).$$

Consequently, for any $t \geq 2|r|$, being $\langle t + |r|\rangle \approx \langle t - |r|\rangle$ and using the mean value theorem, it follows:

$$|v(t,r)| \lesssim \varepsilon\langle t - |r|\rangle^{-\kappa} \approx \varepsilon\langle t - |r|\rangle^{-\kappa+1}\langle t + |r|\rangle^{-1}.$$

When $t \leq 2|r|$ we distinguish two cases. For $|r| \leq 1$, since $\langle t + |r|\rangle \approx \langle t - |r|\rangle \approx 1$, using once again the mean value theorem, we have

$$|v(t,r)| \lesssim \varepsilon \lesssim \varepsilon\langle t - |r|\rangle^{-\kappa+1}\langle t + |r|\rangle^{-1}.$$

On the other hand, for $|r| \geq 1$ using the above representation formula for v, we find

$$|v(t,r)| \lesssim \frac{\varepsilon}{|r|}\left(\langle t+|r|\rangle^{-\kappa+1} + \langle t-|r|\rangle^{-\kappa+1}\right) \lesssim \frac{\varepsilon}{|r|}\langle t-|r|\rangle^{-\kappa+1}$$
$$\lesssim \varepsilon\langle t-|r|\rangle^{-\kappa+1}\langle t+|r|\rangle^{-1},$$

where we used the relation $|r| \approx \langle r\rangle \gtrsim \langle t+|r|\rangle$ in the last inequality. It is clear that

$$\partial_r(rv(t,r)) = \frac{1}{2}\left(H_f'(t+r) + H_f'(t-r)\right),$$

and, then,

$$|\partial_r(rv(t,r))| \leq \varepsilon\langle t-|r|\rangle^{-\kappa}.$$

Repeating the same estimates seen in the first case $v = w[H_g]$, we arrive at

$$|\partial_r(rv(t,r))| \lesssim \varepsilon\langle t-|r|\rangle^{-\kappa}\langle t+|r|\rangle^{-1}\langle r\rangle \leq \varepsilon\langle t-|r|\rangle^{-\kappa+1}\langle t+|r|\rangle^{-1}\langle r\rangle.$$

Summarizing, we have proved that $\|v\|_{X_\kappa} \lesssim \varepsilon$ also for $v = \partial_t w[H_f]$. Using the triangular inequality, we get the desired estimate for $\partial_t w[H_f] + w[H_g]$.

3.2 *Preliminary Results*

In the previous section we introduced the space X_κ in which we will consider the solutions to (22). In this subsection we provide some preliminary estimates that will play a fundamental role in next subsection.

By Duhamel's principle we know that the solution of the inhomogeneous Cauchy problem

$$\begin{cases} v_{tt} - v_{rr} - \frac{2}{r}v_r = F(t,r), & t>0,\ r\in\mathbb{R}, \\ v(0,r) = 0, & r\in\mathbb{R}, \\ v_t(0,r) = 0, & r\in\mathbb{R} \end{cases}$$

can be written as

$$v(t,r) = \int_0^t v_\tau(t,r)d\tau,$$

where v_τ is the solution of the homogeneous problem

$$\begin{cases} v_{tt} - v_{rr} - \frac{2}{r} v_r = 0, & t > 0,\ r \in \mathbb{R}, \\ v(\tau, r) = 0, & r \in \mathbb{R}, \\ v_t(\tau, r) = F(\tau, r), & r \in \mathbb{R}. \end{cases}$$

Being the Cauchy problem (25) invariant by time translation, we have

$$v_\tau(t, r) = \int_{-1}^{1} H_F[\tau](t - \tau + r\sigma) d\sigma = \frac{1}{r} \int_{t-\tau-r}^{t-\tau+r} H_F[\tau](\rho) d\rho$$

with $H_F[\tau](\rho) = \frac{\rho F(\tau,\rho)}{2}$. Thus, we obtain

$$v(t, r) = \int_0^t \int_{-1}^{1} H_F[\tau](t - \tau + r\sigma) d\sigma d\tau = \frac{1}{r} \int_0^t \int_{t-\tau-r}^{t-\tau+r} H_F[\tau](\rho) d\rho d\tau.$$

Let us define for any $v \in X_\kappa$ the operator

$$\begin{aligned} v \to Lv(t, r) &\doteq \int_0^t \langle \tau \rangle^{-\frac{\mu}{2}(p-1)} \int_{-1}^{1} \widetilde{H}_v[\tau](t - \tau + r\sigma) d\sigma d\tau \\ &= \frac{1}{r} \int_0^t \langle \tau \rangle^{-\frac{\mu}{2}(p-1)} \int_{t-\tau-r}^{t-\tau+r} \widetilde{H}_v[\tau](\rho) d\rho d\tau \end{aligned} \tag{29}$$

with

$$\widetilde{H}_v[\tau](\rho) \doteq \frac{\rho |v(\tau, \rho)|^p}{2}. \tag{30}$$

Hereafter, we denote by $\widetilde{H}_v[\tau]'(\rho)$ the derivative of $\widetilde{H}_v[\tau](\rho)$ with respect to ρ, considering τ as a parameter.

Proposition 3.4 *Let $v \in X_\kappa$. Then, $Lv \in X_\kappa$ and $r^2 Lv \in \mathcal{C}^2([0, \infty) \times \mathbb{R})$. Furthermore, Lv satisfies*

$$r^2(\partial_t^2 - \partial_r^2) Lv - 2r \partial_r Lv = \langle t \rangle^{-\frac{\mu}{2}(p-1)} r^2 |v(t, r)|^p, \qquad t > 0,\ r \in \mathbb{R} \tag{31}$$

with vanishing initial data.

Proof From the continuity of the function $\widetilde{H}_v[\tau](\rho)$, that follows from the assumption $v \in X_\kappa \subset \mathcal{C}([0, \infty) \times \mathbb{R})$, we obtain $Lv, \partial_r(rLv) \in \mathcal{C}([0, \infty) \times \mathbb{R})$. Furthermore, Lv is even in r. So, $Lv \in X_\kappa$.

Let us compute the time derivative of Lv. Being $\widetilde{H}_v[t](r\sigma)$ odd in σ, then,

$$\partial_t Lv = \frac{1}{r} \int_0^t \langle \tau \rangle^{-\frac{\mu}{2}(p-1)} \left(\widetilde{H}_v[\tau](t - \tau + r) - \widetilde{H}_v[\tau](t - \tau - r) \right) d\tau.$$

For the second time derivative, using the previous formula and (27), we get

$$\partial_t^2 Lv = \partial_r^2 Lv + \tfrac{2}{r}\partial_r Lv + \langle t\rangle^{-\frac{\mu}{2}(p-1)}|v(t,r)|^p.$$

Thus, Lv solves (31) and we get the continuity of the $r-$derivatives for $r^2 Lv$.

Having in mind the result obtained in Proposition 3.4, we introduce the following definition.

Definition 3.5 *We say that $v = v(t,r)$ is a radial solution to* (22) *in X_κ, if $v \in X_\kappa$ for some $\kappa > 1$ and satisfies the integral equation*

$$v = \partial_t w[H_f] + w[H_g] + Lv.$$

Since our goal is to prove the global (in time) existence through Banach's fixed point theorem, we prove the following preliminary result.

Proposition 3.6 *Let $p > p_0(3+\mu)$, $\mu \in [2, 1+\sqrt{5}]$ and let κ be such that:*

$$\max\left\{1, \tfrac{2}{p-1} - \tfrac{\mu}{2}\right\} < \kappa \le \left(\tfrac{\mu}{2}+1\right)(p-1) \quad \textit{if } p \in (p_0(3+\mu), p_{\mathrm{Fuj}}(\mu)), \tag{32}$$

$$\max\left\{1, \tfrac{\mu}{2}\right\} < \kappa \le \left(\tfrac{\mu}{2}+1\right)(p-1) \quad \textit{if } p = p_{\mathrm{Fuj}}(\mu), \tag{33}$$

$$\max\left\{1, \tfrac{1}{p-1}\right\} < \kappa \le \left(\tfrac{\mu}{2}+1\right)(p-1) \quad \textit{if } p > p_{\mathrm{Fuj}}(\mu). \tag{34}$$

Then, the following estimates hold for $v, w \in X_\kappa$:

$$\|Lv\|_{X_\kappa} \lesssim \|v\|_{X_\kappa}^p, \tag{35}$$

$$\|Lv - Lw\|_{X_\kappa} \lesssim \|v-w\|_{X_\kappa}\left(\|v\|_{X_\kappa}^{p-1} + \|w\|_{X_\kappa}^{p-1}\right). \tag{36}$$

Remark 3.7 We underline that we can actually find a $\kappa > 1$ satisfying the above conditions. This is possible since for $p > p_0(3+\mu)$ the relation $1 < \left(\frac{\mu}{2}+1\right)(p-1)$ is always true.

In fact the last inequality is equivalent to $p > p_{\mathrm{Fuj}}(\mu+2)$. However, it holds $p_0(3+\mu) > p_{\mathrm{Fuj}}(\mu+2)$, so, $p > p_{\mathrm{Fuj}}(\mu+2)$ is valid in this case.

The other conditions on κ are strongly related to the tools that we are using in order to prove our result and, therefore, they will be explained in detail after Lemma 3.10.

Remark 3.8 When $\mu = 1+\sqrt{5}$ we have $p_0(3+\mu) = p_{\mathrm{Fuj}}(\mu)$ (see also Remark 3.11), thus, we have just the condition (34) for κ in the above theorem.

Proof Due to the definition of the norm in X_κ, in order to prove (35) it is sufficient to show that for any $t \geq 0$, $r \in \mathbb{R}$ we have

$$|Lv(t,r)| \lesssim \langle t+|r|\rangle^{-1}\langle t-|r|\rangle^{-(\kappa-1)}\|v\|^p_{X_\kappa}, \tag{37}$$

$$|\partial_r(rLv)(t,r)| \lesssim \langle t+|r|\rangle^{-1}\langle t-|r|\rangle^{-(\kappa-1)}\langle r\rangle\|v\|^p_{X_\kappa}. \tag{38}$$

But Lv is even in r, therefore, we can restrict ourselves to consider nonnegative values of r.

Let us consider $v \in X_\kappa$. By using the relation $rv_r = \partial_r(rv) - v$, we get the following estimates:

$$|v(\tau,\rho)|^p \lesssim \|v\|^p_{X_\kappa}\langle \tau+|\rho|\rangle^{-p}\langle \tau-|\rho|\rangle^{-p(\kappa-1)},$$

$$\langle\rho\rangle^{-1}|\rho\partial_\rho|v(\tau,\rho)|^p| \lesssim \|v\|^p_{X_\kappa}\langle \tau+|\rho|\rangle^{-p}\langle \tau-|\rho|\rangle^{-p(\kappa-1)}.$$

Moreover, from (30) we obtain

$$|\widetilde{H}_v[\tau](\rho)| + |\widetilde{H}_v[\tau]'(\rho)| \lesssim \|v\|^p_{X_\kappa}\langle \tau+|\rho|\rangle^{-p}\langle \tau-|\rho|\rangle^{-p(\kappa-1)}\langle\rho\rangle. \tag{39}$$

For $r \geq 0$, applying the relation (39) to the definition of Lv, we get

$$|Lv(t,r)| \lesssim \frac{1}{r}\|v\|^p_{X_\kappa} I_0(t,r), \tag{40}$$

where

$$I_0(t,r) \doteq \int_0^t \langle\tau\rangle^{-\frac{\mu}{2}(p-1)}\int_{t-\tau-r}^{t-\tau+r}\langle \tau+|\rho|\rangle^{-p}\langle \tau-|\rho|\rangle^{-p(\kappa-1)}\langle\rho\rangle d\rho d\tau.$$

By using (28) and (39), we obtain

$$|\partial_r(rLv)(t,r)| \lesssim \|v\|^p_{X_\kappa}\sum_{\pm} I_{1,\pm}(t,r), \tag{41}$$

where

$$I_{1,\pm}(t,r) \doteq \int_0^t \langle\tau\rangle^{-\frac{\mu}{2}(p-1)}\langle \tau+|t-\tau\pm r|\rangle^{-p}\langle \tau-|t-\tau\pm r|\rangle^{-p(\kappa-1)}\langle t-\tau\pm r\rangle d\tau.$$

Analogously, in order to prove (36), we have to show

$$\langle t+|r|\rangle\langle t-|r|\rangle^{\kappa-1}|Lv(t,r)-Lw(t,r)| \lesssim \|v-w\|_{X_\kappa}\left(\|v\|^{p-1}_{X_\kappa}+\|w\|^{p-1}_{X_\kappa}\right), \tag{42}$$

$$\langle t+|r|\rangle\langle t-|r|\rangle^{\kappa-1}\langle r\rangle^{-1}|\partial_r(rLv)(t,r)-\partial_r(rLw)(t,r)| \lesssim \|v-w\|_{X_\kappa}\left(\|v\|_{X_\kappa}^{p-1}+\|w\|_{X_\kappa}^{p-1}\right). \tag{43}$$

Using (13) and (30), we arrive at

$$|\widetilde{H}_v[\tau](\rho)-\widetilde{H}_w[\tau](\rho)| \lesssim |\rho\|v(\tau,\rho)-w(\tau,\rho)|\left(|v(\tau,\rho)|^{p-1}+|w(\tau,\rho)|^{p-1}\right).$$

Therefore, from the definitions of L and of norm in X_κ it follows:

$$|Lv(t,r)-Lw(t,r)| \lesssim \frac{1}{r}\|v-w\|_{X_\kappa}\left(\|v\|_{X_\kappa}^{p-1}+\|w\|_{X_\kappa}^{p-1}\right)I_0(t,r).$$

In the same way, employing once again (13), we find

$$|\partial_r(rLv)(t,r)-\partial_r(rLw)(t,r)| \lesssim \|v-w\|_{X_\kappa}\left(\|v\|_{X_\kappa}^{p-1}+\|w\|_{X_\kappa}^{p-1}\right)\sum_{\pm} I_{1,\pm}(t,r).$$

Consequently, our next step will be to estimate the quantities $I_0(t,r)$ and $I_{1,\pm}(t,r)$. Hence, in order to conclude the proof we need to use Propositions 3.12 and 3.13, which are stated and proved in the last part of this subsection.

Remark 3.9 If $t\le r$, then, we can slightly modify the representation formula for Lv. Indeed, being $\widetilde{H}_v[\tau]$ an odd function, we have

$$\int_{(t-\tau)-r}^{r-(t-\tau)} \widetilde{H}_v[\tau](\rho)d\rho = 0.$$

Hence,

$$Lv(t,r) = \frac{1}{r}\int_0^t \langle\tau\rangle^{-\frac{\mu}{2}(p-1)}\int_{r-(t-\tau)}^{t-\tau+r} \widetilde{H}_v[\tau](\rho)d\rho d\tau.$$

For this reason, when $t\le r$ we may replace $I_0(t,r)$ by

$$\widetilde{I}_0(t,r) \doteq \int_0^t \langle\tau\rangle^{-\frac{\mu}{2}(p-1)}\int_{r-(t-\tau)}^{r+(t-v)} \langle\tau+\rho\rangle^{-p}\langle\tau-\rho\rangle^{-p(\kappa-1)}\langle\rho\rangle d\rho d\tau,$$

obtaining

$$|Lv(t,r)| \lesssim \frac{1}{r}\|v\|_{X_\kappa}^p \widetilde{I}_0(t,r). \tag{44}$$

The estimates for $I_0(t,r)$, $\widetilde{I}_0(t,r)$ and $I_{1,\pm}(t,r)$ are based on the next lemma.

Lemma 3.10 *Let $p > p_0(3+\mu)$, $\mu \in [2, 1+\sqrt{5}]$ and let κ be such that*

$$\tfrac{2}{p-1} - \tfrac{\mu}{2} \le \kappa \le \left(\tfrac{\mu}{2}+1\right)(p-1) \quad \text{if } p \in (p_0(3+\mu), p_{\mathrm{Fuj}}(\mu)), \tag{45}$$

$$\tfrac{\mu}{2} < \kappa \le \left(\tfrac{\mu}{2}+1\right)(p-1) \quad \text{if } p = p_{\mathrm{Fuj}}(\mu), \tag{46}$$

$$\tfrac{1}{p-1} \le \kappa \le \left(\tfrac{\mu}{2}+1\right)(p-1) \quad \text{if } p > p_{\mathrm{Fuj}}(\mu). \tag{47}$$

Then, for any $\xi \in \mathbb{R}$ it holds

$$I(\xi) \doteq \int_{-|\xi|}^{|\xi|} \langle \eta+\xi\rangle \langle \eta-\xi\rangle^{-\frac{\mu}{2}(p-1)} \langle \eta\rangle^{-p(\kappa-1)} d\eta \lesssim \langle \xi\rangle^{-(\kappa-p)}. \tag{48}$$

Proof We split the integral $I(\xi)$ into two parts

$$I_1(\xi) \doteq \int_{-\frac{|\xi|}{2}}^{\frac{|\xi|}{2}} \langle \eta+\xi\rangle \langle \eta-\xi\rangle^{-\frac{\mu}{2}(p-1)} \langle \eta\rangle^{-p(\kappa-1)} d\eta,$$

$$I_2(\xi) \doteq \left(\int_{-|\xi|}^{-\frac{|\xi|}{2}} + \int_{\frac{|\xi|}{2}}^{|\xi|}\right) \langle \eta+\xi\rangle \langle \eta-\xi\rangle^{-\frac{\mu}{2}(p-1)} \langle \eta\rangle^{-p(\kappa-1)} d\eta.$$

We begin with $I_1(\xi)$. Let $\eta \in [-|\xi|/2, |\xi|/2]$. Then, $\langle\xi\rangle \approx \langle \eta+\xi\rangle \approx \langle \eta-\xi\rangle$. Therefore, we have

$$\begin{aligned} I_1(\xi) &\lesssim \langle\xi\rangle^{1-\frac{\mu}{2}(p-1)} \int_{-\frac{|\xi|}{2}}^{\frac{|\xi|}{2}} \langle \eta\rangle^{-p(\kappa-1)} d\eta \\ &= 2\langle\xi\rangle^{1-\frac{\mu}{2}(p-1)} \int_0^{\frac{|\xi|}{2}} \langle \eta\rangle^{-p(\kappa-1)} d\eta. \end{aligned}$$

We study separately the cases $p(\kappa-1) \gtreqless 1$.

Case $\kappa > 1 + \frac{1}{p}$

Since the power of the integrand function is less than -1, we have

$$\int_0^{\frac{|\xi|}{2}} \langle \eta\rangle^{-p(\kappa-1)} d\eta \lesssim 1.$$

Consequently, $I_1(\xi) \lesssim \langle\xi\rangle^{1-\frac{\mu}{2}(p-1)} \le \langle\xi\rangle^{-(\kappa-p)}$ for $\kappa \le \left(\frac{\mu}{2}+1\right)(p-1)$.

Case $\kappa = 1 + \frac{1}{p}$

In this case a logarithmic term appears. Hence,

$$I_1(\xi) \lesssim \langle\xi\rangle^{1-\frac{\mu}{2}(p-1)} \log(\langle\xi\rangle/2) \le \langle\xi\rangle^{-(\kappa-p)}$$

for $\kappa < \left(\frac{\mu}{2}+1\right)(p-1)$. But $\kappa = 1+\frac{1}{p}$. So, the previous condition on κ is reduced to the inequality $1+\frac{1}{p} < \left(\frac{\mu}{2}+1\right)(p-1)$, which is equivalent to

$$(2+\mu)p^2 - (4+\mu)p - 2 > 0. \tag{49}$$

However, (49) is obviously satisfied, since $p > p_0(3+\mu)$. Let us remark that (49) is also useful to guarantee that we have a not empty range of admissible values of κ in the previous case.

Case $\kappa < 1+\frac{1}{p}$

Being the power of the integrand greater than -1, we get

$$\int_0^{\frac{|\xi|}{2}} \langle\eta\rangle^{-p(\kappa-1)} d\eta \lesssim \langle\xi/2\rangle^{-p(\kappa-1)+1}.$$

Therefore, $I_1(\xi) \lesssim \langle\xi\rangle^{2-\frac{\mu}{2}(p-1)-p(\kappa-1)} \leq \langle\xi\rangle^{-(\kappa-p)}$ for $\kappa \geq \frac{2}{p-1}-\frac{\mu}{2}$. We underline that also in this case the condition $p > p_0(3+\mu)$ implies that we have a not empty range of values for κ. Indeed, $\frac{2}{p-1}-\frac{\mu}{2} < 1+\frac{1}{p}$ is equivalent to (49).

Gluing together the previous conditions on κ, we obtain $I_1(\xi) \lesssim \langle\xi\rangle^{-(\kappa-p)}$ when

$$\kappa \in \left[\frac{2}{p-1}-\frac{\mu}{2}, \left(\frac{\mu}{2}+1\right)(p-1)\right].$$

Now, we want to estimate $I_2(\xi)$. We may write this term as sum of the following integrals:

$$I_{2,1}(\xi) \doteq \int_{\frac{|\xi|}{2}}^{|\xi|} \langle\eta+\xi\rangle\langle\eta-\xi\rangle^{-\frac{\mu}{2}(p-1)}\langle\eta\rangle^{-p(\kappa-1)} d\eta,$$

$$I_{2,2}(\xi) \doteq \int_{-|\xi|}^{-\frac{|\xi|}{2}} \langle\eta+\xi\rangle\langle\eta-\xi\rangle^{-\frac{\mu}{2}(p-1)}\langle\eta\rangle^{-p(\kappa-1)} d\eta.$$

We may reduce our considerations to the case $\xi \geq 0$, since for $\xi \leq 0$ the situation is in some sense symmetric between $I_{2,1}(\xi)$ and $I_{2,2}(\xi)$ with respect to the case $\xi \geq 0$. Let us begin with the estimate of $I_{2,1}(\xi)$. For $\eta \in [\xi/2, \xi]$ it holds the equivalences $\langle\eta+\xi\rangle \approx \langle\eta\rangle \approx \langle\xi\rangle$. Consequently,

$$I_{2,1}(\xi) \lesssim \begin{cases} \langle\xi\rangle^{1-p(\kappa-1)} & \text{if } \frac{\mu}{2}(p-1) > 1, \\ \langle\xi\rangle^{1-p(\kappa-1)}\log(\langle\xi/2\rangle) & \text{if } \frac{\mu}{2}(p-1) = 1, \\ \langle\xi\rangle^{2-p(\kappa-1)-\frac{\mu}{2}(p-1)} & \text{if } \frac{\mu}{2}(p-1) < 1. \end{cases}$$

Hence, we have $I_{2,1}(\xi) \lesssim \langle \xi \rangle^{-(\kappa-p)}$ if and only if

$$\begin{cases} \frac{1}{p-1} \leq \kappa & \text{if} \quad p > p_{\mathrm{Fuj}}(\mu), \\ \frac{\mu}{2} < \kappa & \text{if} \quad p = p_{\mathrm{Fuj}}(\mu), \\ \frac{2}{p-1} - \frac{\mu}{2} \leq \kappa & \text{if} \quad p < p_{\mathrm{Fuj}}(\mu). \end{cases}$$

We note that the condition $\frac{1}{p-1} \leq \kappa$ is stronger than $\frac{2}{p-1} - \frac{\mu}{2} \leq \kappa$, which is the lower bound for κ coming from the estimate of the term $I_1(\xi)$, exactly when $p > p_{\mathrm{Fuj}}(\mu)$.

We estimate now $I_{2,2}(\xi)$. When $\eta \in [-\xi, -\xi/2]$, we employ the equivalences $\langle \eta - \xi \rangle \approx \langle \eta \rangle \approx \langle \xi \rangle$. Thus, we get

$$\begin{aligned} I_{2,2}(\xi) &\lesssim \langle \xi \rangle^{-\frac{\mu}{2}(p-1)-p(\kappa-1)} \int_{-\xi}^{-\frac{\xi}{2}} \langle \xi + \eta \rangle d\eta \\ &\lesssim \langle \xi \rangle^{-\frac{\mu}{2}(p-1)-p(\kappa-1)+2} \leq \langle \xi \rangle^{-(\kappa-p)}, \end{aligned}$$

where again the last inequality is true for $\kappa \geq \frac{2}{p-1} - \frac{\mu}{2}$. This concludes the proof.

Remark 3.11 Let us explain why we may consider only the case $\mu \leq 1 + \sqrt{5}$ in Lemma 3.10. Firstly, for $\mu \leq 1 + \sqrt{5}$ it holds $p_0(3+\mu) \leq p_{\mathrm{Fuj}}(\mu)$. Thus, we have to consider three different cases for κ (of course, only one case for the limit value $\mu = 1 + \sqrt{5}$), as have seen in the proof.

Indeed, $p_0(3+\mu) < p_{\mathrm{Fuj}}(\mu)$ is equivalent to

$$\frac{\sqrt{(\mu+4)^2+8(\mu+2)}}{2(\mu+2)} < 1 + \frac{2}{\mu} - \frac{\mu+4}{2(\mu+2)}.$$

However, for nonnegative μ the right-hand side is always positive, thus, we can take the squared powers of each side, obtaining after some straightforward computations the inequality $\mu^3 - 8(\mu+1) < 0$. The polynomial

$$P(x) = x^3 - 8(x+1) = (x+2)\big(x - 1 + \sqrt{5}\big)\big(x - 1 - \sqrt{5}\big)$$

has $1 + \sqrt{5}$ as unique positive root and $P(x) < 0$ for $x \in (0, 1+\sqrt{5})$. But this is exactly the previous condition for μ.

Secondly, for $\mu > 1 + \sqrt{5}$ we have $p_0(3+\mu) > p_{\mathrm{Fuj}}(\mu)$. Such relation follows immediately from the property $P(x) > 0$ for $x > 1 + \sqrt{5}$.

Even though one could think that for $\mu > 1 + \sqrt{5}$ it might be sufficient to replace (45), (46) and (47) by

$$\frac{1}{p-1} \leq \kappa \leq \left(\frac{\mu}{2} + 1\right)(p-1) \quad \text{if} \quad p > p_0(3+\mu).$$

then, one should guarantee that the interval

$$\left[\tfrac{1}{p-1}, \left(\tfrac{\mu}{2}+1\right)(p-1)\right]$$

is not empty, that is, the inequality $\left(\frac{\mu}{2}+1\right)p^2-(\mu+2)p+\frac{\mu}{2}\geq 0$ should be satisfied. But this inequality is equivalent to

$$p \geq \widetilde{p}(\mu) \doteq 1+\sqrt{\tfrac{2}{\mu+2}}.$$

By straightforward computations, one can prove that $p_0(3+\mu) < \widetilde{p}(\mu)$ is equivalent to

$$(\mu+2)(\mu^2-2\mu-4) > 0.$$

However, the above relation is fulfilled just for $\mu > 1+\sqrt{5}$.

Summarizing, for $\mu \in [2, 1+\sqrt{5}]$ we have the chain of inequalities

$$\widetilde{p}(\mu) < p_0(3+\mu) < p_{\mathrm{Fuj}}(\mu),$$

which guarantees for $p > p_0(3+\mu)$ the nonemptiness of the range for κ in (45), (46) and (47). On the other hand, for $\mu > 1+\sqrt{5}$ we have the converse chain of inequalities $p_{\mathrm{Fuj}}(\mu) < p_0(3+\mu) < \widetilde{p}(\mu)$, which guarantees suitable assumptions on κ only when $p \geq \widetilde{p}(\mu)$, with the case $p \in (p_0(3+\mu), \widetilde{p}(\mu))$ left open.

Proposition 3.12 *Let $p > p_0(3+\mu)$. Let $\mu \in [2, 1+\sqrt{5}]$ and let κ be as in Proposition* 3.6. *Then, it holds*

$$I_0(t,r) \lesssim \begin{cases} r\langle t+r\rangle^{-\kappa} & \text{if } \; t \geq 2r \; \text{ or } \; 0 \leq r \leq 1, \\ \langle t-r\rangle^{-(\kappa-1)} & \text{if } \; r \leq t \leq 2r \; \text{ and } \; r \geq 1. \end{cases}$$

Moreover,

$$\widetilde{I}_0(t,r) \lesssim \langle t-r\rangle^{-(\kappa-1)} \qquad \text{if } \; t \leq r \; \text{ and } \; r \geq 1.$$

In particular, the estimates (37) *and* (42) *hold.*

Proof We consider first $I_0(t,r)$. Since $|t-\tau-r| \leq t-\tau+r$ (we are working only with $r \geq 0$), it follows:

$$I_0(t,r) \leq 2\int_0^t \langle\tau\rangle^{-\frac{\mu}{2}(p-1)} \int_{(t-\tau-r)_+}^{t-\tau+r} \langle\tau+\rho\rangle^{-p}\langle\tau-\rho\rangle^{-p(\kappa-1)}\langle\rho\rangle d\rho\, d\tau,$$

where we use the notation $(y)_+ \doteq \max\{y, 0\}$ for any $y \in \mathbb{R}$.

We perform the change of variables $\xi = \tau + \rho, \eta = \rho - \tau$. Being $\rho, \tau \geq 0$, we get $|\eta| \leq \xi$. On the other hand, the condition $\rho \in [(t - \tau - r)+, t - \tau + r]$ implies $(t - r)_+ \leq \xi \leq t + r$. Therefore,

$$
\begin{aligned}
I_0(t,r) &\lesssim \int_{(t-r)_+}^{t+r} \langle\xi\rangle^{-p} \int_{-\xi}^{\xi} \langle\eta+\xi\rangle\langle\eta-\xi\rangle^{-\frac{\mu}{2}(p-1)}\langle\eta\rangle^{-p(\kappa-1)} d\eta\, d\xi \\
&= \int_{(t-r)_+}^{t+r} \langle\xi\rangle^{-p} I(\xi)\, d\xi \lesssim \int_{(t-r)_+}^{t+r} \langle\xi\rangle^{-\kappa} d\xi, \qquad (50)
\end{aligned}
$$

where in the last inequality we used Lemma 3.10. Now we estimate the last integral in different zones of the (t, r)-plane.

Zone $t \geq 2r$

For $\xi \in [t - r, t + r]$ we have the equivalence $\langle\xi\rangle \approx \langle t + r\rangle$. Then, from (50) it follows $I_0(t, r) \lesssim r\langle t + r\rangle^{-\kappa}$.

Zone $0 \leq r \leq 1$ *and* $t \leq 2r$

In this zone $\langle t + r\rangle \approx 1$. So, it is sufficient to show that $I_0(t, r) \lesssim r$, but this is clear, being $\kappa > 0$.

Zone $r \geq 1$ *and* $r \leq t \leq 2r$

Being $\kappa > 1$ by (50), we get

$$
I_0(t,r) \lesssim \int_{(t-r)_+}^{t+r} \langle\xi\rangle^{-\kappa} d\xi \lesssim \langle t - r\rangle^{-(\kappa-1)}.
$$

Now we estimate $\widetilde{I}_0(t, r)$ for $r \geq 1$ and $t \leq r$. We consider as before the change of variables $\xi = \tau + \rho, \eta = \rho - \tau$ for the integral

$$
\widetilde{I}_0(t,r) = \int_0^t \langle\tau\rangle^{-\frac{\mu}{2}(p-1)} \int_{r-(t-\tau)}^{r+(t-\tau)} \langle\tau+\rho\rangle^{-p}\langle\tau-\rho\rangle^{-p(\kappa-1)}\langle\rho\rangle d\rho\, d\tau.
$$

From $\tau \geq 0$ and $r - (t - \tau) \leq \rho \leq r + (t - \tau)$ we have $r - t \leq \eta \leq \xi$ and $r - t \leq \xi \leq r + t$. Thus, we find

$$
\widetilde{I}_0(t,r) \lesssim \int_{r-t}^{r+t} \langle\xi\rangle^{-p} \int_{r-t}^{\xi} \langle\eta+\xi\rangle\langle\eta-\xi\rangle^{-\frac{\mu}{2}(p-1)}\langle\eta\rangle^{-p(\kappa-1)} d\eta\, d\xi.
$$

Moreover, $[r - t, \xi] \subset [-\xi, \xi]$ and, then, applying Lemma 3.10, we have

$$
\widetilde{I}_0(t,r) \lesssim \int_{r-t}^{r+t} \langle\xi\rangle^{-p} I(\xi)\, d\xi \lesssim \int_{r-t}^{r+t} \langle\xi\rangle^{-\kappa} d\xi \lesssim \langle t - r\rangle^{1-\kappa},
$$

where we used again the condition $\kappa > 1$.

Finally, we prove (37). We know that (40) and (44) hold for $t \geq r$ and $t \leq r$, respectively. For $t \geq 2r$ or $0 \leq r \leq 1$, using the relation $\langle t+r\rangle \geq \langle t-r\rangle$ and combining the estimates for $I_0(t,r)$ and $\widetilde{I}_0(t,r)$, we find

$$|Lv(t,r)| \lesssim \langle t+r\rangle^{-\kappa} \|v\|_{X_\kappa}^p \leq \langle t+r\rangle^{-1} \langle t-r\rangle^{-(\kappa-1)} \|v\|_{X_\kappa}^p.$$

If $t \leq 2r$ and $r \geq 1$, then, we have

$$|Lv(t,r)| \lesssim \frac{1}{r} \langle t-r\rangle^{-(\kappa-1)} \|v\|_{X_\kappa}^p \lesssim \langle t+r\rangle^{-1} \langle t-r\rangle^{-(\kappa-1)} \|v\|_{X_\kappa}^p,$$

where we used the equivalence $r \approx \langle r\rangle \approx \langle t+r\rangle$ in the last inequality.

Since

$$|Lv(t,r) - Lw(t,r)| \lesssim \frac{1}{r} \|v-w\|_{X_\kappa} \left(\|v\|_{X_\kappa}^{p-1} + \|w\|_{X_\kappa}^{p-1}\right) I_0(t,r) \qquad \text{for} \quad t \geq r,$$

$$|Lv(t,r) - Lw(t,r)| \lesssim \frac{1}{r} \|v-w\|_{X_\kappa} \left(\|v\|_{X_\kappa}^{p-1} + \|w\|_{X_\kappa}^{p-1}\right) \widetilde{I}_0(t,r) \qquad \text{for} \quad t \leq r,$$

in the same way one can prove (42). This concludes the proof.

Proposition 3.13 *Let $p > p_0(3+\mu)$. Let $\mu \in [2, 1+\sqrt{5}]$ and let κ be as in Proposition* 3.6. *Then, we have*

$$I_{1,-}(t,r) \lesssim \begin{cases} \langle t-r\rangle^{-\kappa} & \text{if} \quad t \geq 2r, \\ \langle t-r\rangle^{-(\kappa-1)} & \text{if} \quad t \leq 2r, \end{cases}$$

and

$$I_{1,+}(t,r) \lesssim \langle t+r\rangle^{-\kappa}.$$

In particular, the estimates (38) *and* (43) *hold.*

Proof We begin with the estimate of $I_{1,-}(t,r)$. Also in this case we divide the (t,r)-plane in different zones.

Zone $t \geq 2r$

If $\tau \in [t-r, t]$, then, $\tau + |t-\tau-r| \approx t-r$. On the other hand, for $\tau \in [0, t-r]$ we have $\tau + |t-\tau-r| = t-r$. Consequently,

$$\begin{aligned} I_{1,-}(t,r) &\lesssim \langle t-r\rangle^{-p} \int_0^t \langle \tau\rangle^{-\frac{\mu}{2}(p-1)} \langle \tau - |t-\tau-r|\rangle^{-p(\kappa-1)} \langle t-\tau-r\rangle d\tau \\ &= \langle t-r\rangle^{-p} (Q_- + Q_+), \end{aligned}$$

where

$$Q_- \doteq \int_0^{t-r} \langle\tau\rangle^{-\frac{\mu}{2}(p-1)}\langle 2\tau - t + r\rangle^{-p(\kappa-1)}\langle t-\tau-r\rangle d\tau,$$

$$Q_+ \doteq \langle t-r\rangle^{-p(\kappa-1)} \int_{t-r}^{t} \langle\tau\rangle^{-\frac{\mu}{2}(p-1)}\langle t-\tau-r\rangle d\tau.$$

For Q_+ we have

$$\begin{aligned} Q_+ &\leq \langle t-r\rangle^{-p(\kappa-1)-\frac{\mu}{2}(p-1)} \int_{t-r}^{t} \langle t-\tau-r\rangle d\tau \\ &\lesssim \langle t-r\rangle^{-p(\kappa-1)-\frac{\mu}{2}(p-1)}\langle r\rangle^2 \lesssim \langle t-r\rangle^{2-p(\kappa-1)-\frac{\mu}{2}(p-1)}. \end{aligned}$$

Since $\kappa \geq \frac{2}{p-1} - \frac{\mu}{2}$, we find the desired estimate $Q_+ \lesssim \langle t-r\rangle^{p-\kappa}$.

Using the change of variables $\eta = -\frac{t-r}{2} + \tau$ and Lemma 3.10, we have

$$\begin{aligned} Q_- &\lesssim \int_{-\frac{t-r}{2}}^{\frac{t-r}{2}} \left\langle \eta + \tfrac{t-r}{2}\right\rangle^{-\frac{\mu}{2}(p-1)} \langle\eta\rangle^{-p(\kappa-1)} \left\langle \eta - \tfrac{t-r}{2}\right\rangle d\eta \\ &= I\left(\tfrac{r-t}{2}\right) \lesssim \langle t-r\rangle^{p-\kappa}. \end{aligned} \tag{51}$$

Combining the estimate for Q_+ and Q_- we get $I_{1,-}(t,r) \lesssim \langle t-r\rangle^{-\kappa}$.

Zone $t \leq 2r$

We divide $I_{1,-}(t,r)$ into two integrals $I_{1,-}(t,r) = \tilde{Q}_- + \tilde{Q}_+$ with

$$\begin{aligned} \tilde{Q}_- &\doteq \int_0^{(t-r)_+} \langle\tau\rangle^{-\frac{\mu}{2}(p-1)}\langle t-r\rangle^{-p}\langle 2\tau - t + r\rangle^{-p(\kappa-1)}\langle t-\tau-r\rangle d\tau \\ &= \langle t-r\rangle^{-p} Q_-, \\ \tilde{Q}_+ &\doteq \int_{(t-r)_+}^{t} \langle\tau\rangle^{-\frac{\mu}{2}(p-1)}\langle 2\tau - t + r\rangle^{-p}\langle t-r\rangle^{-p(\kappa-1)}\langle t-\tau-r\rangle d\tau \\ &= \langle t-r\rangle^{-p(\kappa-1)} Q_+^{\#}, \end{aligned}$$

where

$$Q_+^{\#} \doteq \int_{(t-r)_+}^{t} \langle\tau\rangle^{-\frac{\mu}{2}(p-1)}\langle 2\tau - t + r\rangle^{-p}\langle t-\tau-r\rangle d\tau.$$

Let us observe that (51) holds for any $t \geq r$, so we obtain $\tilde{Q}_- \lesssim \langle t-r\rangle^{-\kappa}$.

On the other hand, being $p > 1$ and $\kappa > 1$, in order to show the estimate $\tilde{Q}_+ \lesssim \langle t-r\rangle^{-(\kappa-1)}$, it is sufficient to prove that $Q_+^{\#}$ is bounded.

Since $2\tau-(t-r)\geq\tau-(t-r)$, we have

$$Q_+^\# \lesssim \int_{(t-r)_+}^t \langle\tau\rangle^{-\frac{\mu}{2}(p-1)}\langle\tau-t+r\rangle^{-(p-1)}d\tau.$$

If $t\geq r$ and $\tau\in[0,(t-r)/2]$, then, $\langle\tau-(t-r)\rangle\geq\langle\tau\rangle$. Consequently, we get $\langle\tau-(t-r)\rangle^{-(p-1)}\leq\langle\tau\rangle^{-(p-1)}$.

When $t\geq r$ and $\tau\geq\frac{t-r}{2}$, it holds $\langle\tau-(t-r)\rangle\leq\langle\tau\rangle$. Hence, the inequality $\langle\tau-(t-r)\rangle^{-(p-1)}\geq\langle\tau\rangle^{-(p-1)}$ holds.

Summarizing, for $t\geq r$ we have

$$\begin{aligned}Q_+^\# &\lesssim \int_0^{(t-r)/2}\langle\tau\rangle^{-(\frac{\mu}{2}+1)(p-1)}d\tau+\int_{(t-r)/2}^t\langle t-\tau-r\rangle^{-(\frac{\mu}{2}+1)(p-1)}d\tau\\ &\lesssim \int_0^\infty\langle\tau\rangle^{-(\frac{\mu}{2}+1)(p-1)}d\tau.\end{aligned}$$

On the other hand, for $t\leq r$, from $\langle\tau-(t-r)\rangle\geq\langle\tau\rangle$ it follows:

$$Q_+^\# \lesssim \int_{(t-r)_+}^t \langle\tau\rangle^{-\frac{\mu}{2}(p-1)}\langle\tau-t+r\rangle^{-(p-1)}d\tau \leq \int_0^\infty\langle\tau\rangle^{-(\frac{\mu}{2}+1)(p-1)}d\tau.$$

However, for $p>p_0(3+\mu)$ we have already shown in Remark 3.7 the inequality $1<\left(\frac{\mu}{2}+1\right)(p-1)$. Thus, $\langle\tau\rangle^{-(\frac{\mu}{2}+1)(p-1)}\in L^1([0,\infty))$ and, consequently, we may estimate $Q_+^\#$ by a constant, as we expected.

Now, combining the estimates for $\tilde{Q}_-$, $\tilde{Q}_+$ and $Q_+^\#$, for $t\leq 2r$ we find

$$\begin{aligned}I_{1,-}(t,r)&=\tilde{Q}_-+\tilde{Q}_+=\langle t-r\rangle^{-p}Q_-+\langle t-r\rangle^{-p(\kappa-1)}Q_+^\#\\ &\lesssim\langle t-r\rangle^{-p-\kappa}+\langle t-r\rangle^{-p(\kappa-1)}\lesssim\langle t-r\rangle^{-(\kappa-1)}.\end{aligned}$$

The estimate for $I_{1,+}(t,r)$ is simpler to get. In fact, because of $t-\tau+r\geq 0$

$$I_{1,+}(t,r)=\langle t+r\rangle^{-p}\int_0^t\langle\tau\rangle^{-\frac{\mu}{2}(p-1)}\langle 2\tau-t-r\rangle^{-p(\kappa-1)}\langle t-\tau+r\rangle d\tau.$$

After the change of variables $\eta=\tau-\frac{t+r}{2}$, applying again Lemma 3.10, we find

$$\begin{aligned}I_{1,+}(t,r)&\lesssim\langle t+r\rangle^{-p}\int_{-\frac{t+r}{2}}^{\frac{t+r}{2}}\left\langle\eta+\tfrac{t+r}{2}\right\rangle^{-\frac{\mu}{2}(p-1)}\langle\eta\rangle^{-p(\kappa-1)}\left\langle\eta-\tfrac{t+r}{2}\right\rangle d\eta\\ &=\langle t+r\rangle^{-p}I\left(-\tfrac{t+r}{2}\right)\lesssim\langle t+r\rangle^{-\kappa}.\end{aligned}$$

Finally, we prove (38) and (43). For $t \geq 2r$, using $\langle t+r\rangle \approx \langle t-r\rangle$ and $\langle r\rangle \geq 1$, we have

$$\begin{aligned}|\partial_r(rLv)(t,r)| &\lesssim \|v\|^p_{X_\kappa}\left(I_{1,-}(t,r)+I_{1,+}(t,r)\right) \lesssim \|v\|^p_{X_\kappa}\left(\langle t-r\rangle^{-\kappa}+\langle t+r\rangle^{-\kappa}\right)\\ &\lesssim \langle r\rangle\langle t+r\rangle^{-1}\langle t-r\rangle^{-(\kappa-1)}\|v\|^p_{X_\kappa}.\end{aligned}$$

On the other hand, if $t \leq 2r$, then, $\langle t+r\rangle \approx \langle r\rangle$. Therefore,

$$\begin{aligned}|\partial_r(rLv)(t,r)| &\lesssim \|v\|^p_{X_\kappa}\left(I_{1,-}(t,r)+I_{1,+}(t,r)\right)\\ &\lesssim \|v\|^p_{X_\kappa}\langle t-r\rangle^{-(\kappa-1)}\\ &\lesssim \langle r\rangle\langle t+r\rangle^{-1}\langle t-r\rangle^{-(\kappa-1)}\|v\|^p_{X_\kappa}.\end{aligned}$$

In the same way one can prove (43). Hence, the proof is completed.

3.3 Semilinear Model: 3-d Case

Finally, in this section we can state the global (in time) existence result in the case $n=3$.

Theorem 3.14 *Let $n=3$. Let us assume $\mu \in [2, 1+\sqrt{5}]$ and $\nu^2 \geq 0$ satisfying the relation $\delta = 1$ and let*

$$p > p_0(3+\mu). \tag{52}$$

Then, there exist $\varepsilon_0 > 0$ and $\kappa_2 > \kappa_1 \geq 1$ such that for any $\varepsilon \in (0,\varepsilon_0)$ and any radial data $u_0 \in C^2(\mathbb{R}^3)$, $u_1 \in C^1(\mathbb{R}^3)$, satisfying

$$\begin{aligned}|d_r^j u_0(r)| &\leq \varepsilon\langle r\rangle^{-(\bar\kappa+j)} \quad \textit{for} \quad j=0,1,\\ |u_1(r)+\tfrac{\mu}{2}u_0(r)| &\leq \varepsilon\langle r\rangle^{-(\bar\kappa+1)},\end{aligned}$$

for some $\bar\kappa \in (\kappa_1,\kappa_2]$, the Cauchy problem (2) admits a uniquely determined radial solution $u \in C([0,\infty), C^2(\mathbb{R}^3\setminus\{0\}))$ in the sense that $v(t,r)=\langle t\rangle^{\frac{\mu}{2}}u(t,r)$ satisfies Definition 3.5 for any $\kappa \in (\kappa_1,\bar\kappa]$.

Furthermore, the following decay estimates hold for any $t \geq 0$, $r > 0$ and $\kappa \in (\kappa_1,\bar\kappa]$:

$$\begin{aligned}|u(t,r)| &\lesssim \varepsilon\,\langle t\rangle^{-\frac{\mu}{2}}\langle t-r\rangle^{-\kappa+1}\langle t+r\rangle^{-1},\\ |\partial_r u(t,r)| &\lesssim \varepsilon\,\langle t\rangle^{-\frac{\mu}{2}}\langle t-r\rangle^{-\kappa+1}\langle t+r\rangle^{-1}.\end{aligned}$$

Proof Let consider the transformation $v = \langle t\rangle^{\frac{\mu}{2}}u$. Then, in order to prove the existence of a radial global (in time) solution for (2), it is sufficient to find a global (in time) solution v to (22). For this purpose, we will use a standard contraction argument. Let us define the sequence $\{v_n\}_{n\in\mathbb{N}}$ of successive approximations

$$v_0 = \partial_t w[H_f] + w[H_g], \qquad v_{n+1} = v_0 + Lv_n.$$

From Propositions 3.3 and 3.6 we get

$$\|v_{n+1}\|_{X_\kappa} \leq \|v_0\|_{X_\kappa} + C_1\|v_n\|^p_{X_\kappa} \leq C_0\varepsilon + C_1\|v_n\|^p_{X_\kappa},$$

$$\|v_{n+1} - v_n\|_{X_\kappa} \leq C_2\|v_n - v_{n-1}\|_{X_\kappa}\left(\|v_n\|^p_{X_\kappa} + \|v_{n-1}\|^p_{X_\kappa}\right),$$

where C_0, C_1, C_2 suitable positive constants. Let us consider

$$\varepsilon_0 = \min\left\{\left(2^{\frac{p}{p-1}}C_0C_1^{\frac{1}{p-1}}\right)^{-1}, \left(2^{\frac{p+1}{p-1}}C_0C_2^{\frac{1}{p-1}}\right)^{-1}\right\},$$

then, for $\varepsilon_0 < \varepsilon$ we find via an induction argument

$$\|v_n\|_{X_\kappa} \leq 2C_0\varepsilon, \tag{53}$$

$$\|v_{n+1} - v_n\|_{X_\kappa} \leq 2^{-n}\|v_1 - v_0\|_{X_\kappa}. \tag{54}$$

By (54) we have that $\{v_n\}_{n\in\mathbb{N}}$ is a Cauchy sequence. Denote by v the limit of this sequence in X_κ. Being the operator L locally Lipschitz, we have that v is solution of the equation

$$v = v_0 + Lv.$$

Then, using Proposition 3.4, we have the searched solution of (22). The decay estimate follows passing to the limit in (53) and using the backward transformation $u = \langle t\rangle^{-\frac{\mu}{2}}v$.

4 Radial Odd Case in Higher Dimensions

In this section we consider the case $n \geq 5$, n odd. Although this section has some common points to Sect. 3, the approach is slightly different. We will follow [5, 11]. As in previous sections, because of (3), carrying out the transformation $v(t, x) =$

$\langle t \rangle^{\frac{\mu}{2}} u(t,x)$, we find the semilinear Cauchy problem

$$\begin{cases} v_{tt} - \Delta v = \langle t \rangle^{-\frac{\mu}{2}(p-1)} |v|^p, & t > 0, \ x \in \mathbb{R}^n, \\ v(0,x) = f(x), & x \in \mathbb{R}^n, \\ v_t(0,x) = g(x), & x \in \mathbb{R}^n, \end{cases} \tag{55}$$

where $f(x) = u_0(x)$ and $g(x) = u_1(x) + \frac{\mu}{2} u_0(x)$. Since we are interested in radial solutions, we will look for solution of (55) that solves

$$\begin{cases} v_{tt} - v_{rr} - \frac{n-1}{r} v_r = \langle t \rangle^{-\frac{\mu}{2}(p-1)} |v|^p, & t > 0, \ r > 0, \\ v(0,r) = f(r), & r > 0, \\ v_t(0,r) = g(r), & r > 0, \end{cases} \tag{56}$$

possibly allowing a singular behavior of solutions and their r-derivatives as $r \to 0^+$. As we have done in Sect. 3, we will require some decay properties for the radial initial data. One difficulty with respect to the case $n = 3$ consists in the more complicate integral representation formula for the solution of the corresponding linear problem, as we will see in the upcoming proofs.

The remaining part of this section is organized as follows: we begin recalling some known results for the corresponding linear equation; hence, after some preliminary lemmas, we will provide the global existence result. Nonetheless, in this case a different argument from the usual contraction principle is employed.

4.1 Radial Linear Wave Equation: Odd Case in Higher Dimensions

We recall now some known results for radial solutions of the linear free wave equation. According to [13, Lemma 3.1] the function v^0 defined by

$$v^0(t,r) \doteq \int_{|t-r|}^{t+r} g(\lambda) K(\lambda,t,r)\, d\lambda + \frac{\partial}{\partial t} \int_{|t-r|}^{t+r} f(\lambda) K(\lambda,t,r)\, d\lambda, \tag{57}$$

where the kernel function is defined by

$$K(\lambda,t,r) \doteq \frac{1}{2m!} \left(\frac{\lambda}{r}\right)^{2m+1} \left(\frac{\partial}{\partial \lambda} \frac{1}{2\lambda}\right)^m \phi^m(\lambda,t,r),$$

$$\phi(\lambda,t,r) \doteq (\lambda - (t-r))(\lambda - (t+r)), \qquad m \doteq \tfrac{n-3}{2},$$

is a solution to the to linear wave equation

$$\begin{cases} v_{tt} - v_{rr} - \frac{n-1}{r} v_r = 0, & t > 0, \ r > 0, \\ v(0, r) = f(r), & r > 0, \\ v_t(0, r) = g(r), & r > 0, \end{cases} \tag{58}$$

provided that $f \in \mathcal{C}^2((0, \infty))$ and $g \in \mathcal{C}^1((0, \infty))$.

In the following lemma we recall some known estimates for the kernel function $K(\lambda, t, r)$ that will be helpful afterwards. For the proof see [13, Lemma 2.3].

Lemma 4.1 *Let* $(t, r) \in [0, \infty) \times (0, \infty)$ *such that* $|t - r| \leq \lambda \leq t + r$. *Then, it holds*

$$|K(\lambda, t, r)| \lesssim r^{-m-1} \lambda^{m+1}. \tag{59}$$

Furthermore, if $t \leq 2r$, *then, we get*

$$|\partial_r K(\lambda, t, r)| \lesssim r^{-m-1} \lambda^{m}. \tag{60}$$

Since we are going to consider a suitable weighted supremum norm for the solution of the semilinear problem, it is reasonable to require some decay conditions for f, g.

Namely, we assume

$$|f^{(j)}(r)| \leq \varepsilon \langle r \rangle^{-(\kappa + j)}, \qquad \text{for} \quad j = 0, 1, 2, \tag{61}$$

$$|g^{(j)}(r)| \leq \varepsilon \langle r \rangle^{-(\kappa + 1 + j)}, \qquad \text{for} \quad j = 0, 1, \tag{62}$$

where ε, κ are positive parameters we will fix afterwards.

In order to define suitably the weight functions in the norm for the solution space, we have firstly to derive some a priori estimates for solutions to the corresponding linear homogeneous Cauchy problem. For the proof of the following two lemmas one can see [11, Lemma 2.4 and Lemma 2.5].

Lemma 4.2 *Let* $v^0 = v^0(t, r)$ *be the solution to the linear Cauchy problem* (58) *defined by* (57). *Let* $(t, r) \in [0, \infty) \times (0, \infty)$ *be such that* $t \geq 2r$. *Then, we have for* $\alpha = 0, 1$

$$|\partial_r^\alpha v^0(t, r)| \lesssim r^{-m-\alpha} \int_{t-r}^{t+r} \left(\sum_{j=0}^{m-1} \lambda^{m-j-1} |G_j(\lambda)| + |G_m(\lambda)| \right) d\lambda, \tag{63}$$

where

$$G_j(\lambda) \doteq \lambda^{j+1} g(\lambda) + (\lambda^{j+1} f(\lambda))', \qquad 0 \leq j \leq m - 1,$$

$$G_m(\lambda) \doteq -(\lambda^{m+1} g(\lambda))' + (\lambda^{m+1} f(\lambda))''.$$

Lemma 4.3 *Let $v^0 = v^0(t, r)$ be the solution to the linear Cauchy problem* (58) *defined by* (57)*. Let $(t, r) \in [0, \infty) \times (0, \infty)$ be such that $t \leq 2r$. Then, we have*

$$|v^0(t,r)| \lesssim r^{-m-1} \int_{|t-r|}^{t+r} \Big(\lambda^{m+1}|g(\lambda)| + \lambda^m |f(\lambda)|\Big) d\lambda + \sum_{\pm} r^{-m-1} |t \pm r|^{m+1} |f(|t \pm r|)|, \tag{64}$$

$$|\partial_r v^0(t,r)| \lesssim r^{-m-1} \int_{|t-r|}^{t+r} \Big(\lambda^{m}|g(\lambda)| + \lambda^{m-1} |f(\lambda)|\Big) d\lambda + \sum_{\pm} r^{-m-1} |t \pm r|^{m} |f(|t \pm r|)| + \sum_{\pm} r^{-m-1} |t \pm r|^{m+1} \Big(|g(|t \pm r|)| + |f'(|t \pm r|)|\Big). \tag{65}$$

Finally, we recall some decay estimates for the solution to the linear Cauchy problem (58).

Proposition 4.4 *Let $v^0 = v^0(t, r)$ be the solution to the linear Cauchy problem* (58) *defined by* (57)*, with $f \in \mathcal{C}^2((0, \infty))$, $g \in \mathcal{C}^1((0, \infty))$ satisfying* (61) *and* (62) *for some $\kappa > m + 1$ and $\varepsilon > 0$.*

Then, we have for any $(t, r) \in [0, \infty) \times (0, \infty)$

$$|v^0(t,r)| \leq C_0\, \varepsilon\, r^{1-m} \langle r \rangle^{-1} \psi_\kappa(t,r), \tag{66}$$

$$|\partial_r v^0(t,r)| \leq C_0\, \varepsilon\, r^{-m} \psi_\kappa(t,r), \tag{67}$$

where C_0 is a positive constant that is independent of (t, r) and the function $\psi_\kappa(t, r)$ is defined by

$$\psi_\kappa(t,r) \doteq \langle t + r \rangle^{-1} \langle t - r \rangle^{-\kappa+m+1}. \tag{68}$$

Proof Combining the results of Lemmas 4.2 and 4.3, we get the desired estimates (for further details see also [11, Proposition 2.3]).

4.2 Preliminary Lemmas

Let us deal with the semilinear Cauchy problem (56). Firstly, we clarify to what kind of solutions we are interested in. Let us introduce the following parameter dependent Banach space for solutions:

$$X_\kappa \doteq \big\{ v \in \mathcal{C}\big([0, \infty), \mathcal{C}^1(0, \infty)\big) : \|v\|_{X_\kappa < \infty} \big\},$$

where

$$\|v\|_{X_\kappa} \doteq \sup_{t\geq 0\,,\,r>0} \big(r^{m-1}\langle r\rangle |v(t,r)| + r^m |\partial_r v(t,r)|\big)\psi_\kappa(t,r)^{-1} \tag{69}$$

for a suitable constant $\kappa > m+1$, being ψ_κ defined by (68).

Thanks to Proposition 4.4, it follows immediately

$$\|v^0\|_{X_\kappa} \leq C_0\,\varepsilon, \tag{70}$$

where $v^0 = v^0(t,r)$ is the solution of the corresponding linear and radial Cauchy problem defined by (57).

Let us consider now the integral operator L defined for any $v \in X_\kappa$ by

$$\begin{aligned} v \to Lv(t,r) &\doteq \int_0^t \langle\tau\rangle^{-\frac{\mu}{2}(p-1)}\omega(t,\tau,r)d\tau \\ &= \int_0^t \langle\tau\rangle^{-\frac{\mu}{2}(p-1)} \int_{|\lambda_-|}^{\lambda_+} |v(\tau,\lambda)|^p K(\lambda, t-\tau, r)d\lambda d\tau \end{aligned} \tag{71}$$

with $\lambda_\pm \doteq t-\tau \pm r$. Being the model (58) invariant by time translations, according to Duhamel's principle Lv is the solution to the Cauchy problem

$$\begin{cases} w_{tt} - w_{rr} - \frac{n-1}{r}w_r = \langle t\rangle^{-\frac{\mu}{2}(p-1)}|v|^p, & t>0,\ r>0, \\ w(0,r) = 0, & r>0, \\ w_t(0,r) = 0, & r>0. \end{cases}$$

Therefore, it is natural to introduce the following definition:

Definition 4.5 *We say that $v = v(t,r)$ is a radial solution to* (55) *in X_κ, if $v \in X_\kappa$ for some $\kappa > m+1$ and*

$$v(t,r) = v^0(t,r) + Lv(t,r) \qquad \textit{for any} \quad t\geq 0\,,\ r>0.$$

Before stating the global (in time) existence result for the semilinear model (56), it is useful to prove some preliminary results. Let us start with the n dimensional variant of Lemma 3.10.

Lemma 4.6 *Let $n \geq 5$ be an odd integer. Let us denote*

$$I(\xi) \doteq \int_{-\xi}^{\xi} \langle\eta-\xi\rangle^{-\frac{\mu}{2}(p-1)}\langle\eta+\xi\rangle^{1-m(p-1)}\langle\eta\rangle^{-p(\kappa-(m+1))}d\eta,$$

where $m = \frac{n-3}{2}$. Let us assume $p > p_0(n+\mu)$ and $\mu \in [2, M(n)]$, where $M(n)$ is defined as in (5). *Let us consider κ in the following way:*

(i) *for* $\mu \in [2, n-1]$

$$\tfrac{2}{p-1} - \tfrac{\mu}{2} \leq \kappa \leq (\tfrac{n-1}{2} + \tfrac{\mu}{2})p - (\tfrac{\mu}{2} + 1) \quad \textit{if} \quad p \in \big(p_0(n+\mu), p_{\mathrm{Fuj}}(\tfrac{n-1}{2} + \tfrac{\mu}{2})\big),$$
$$\tfrac{n-1}{2} < \kappa \leq (\tfrac{n-1}{2} + \tfrac{\mu}{2})p - (\tfrac{\mu}{2} + 1) \quad \textit{if} \quad p \geq p_{\mathrm{Fuj}}(\tfrac{n-1}{2} + \tfrac{\mu}{2});$$

(ii) *for* $\mu \in [n-1, M(n)]$

$$\tfrac{2}{p-1} - \tfrac{\mu}{2} \leq \kappa \leq (\tfrac{n-1}{2} + \tfrac{\mu}{2})p - (\tfrac{\mu}{2} + 1) \quad \textit{if} \quad p \in \big(p_0(n+\mu), p_{\mathrm{Fuj}}(\mu)\big),$$
$$\tfrac{\mu}{2} < \kappa \leq (\tfrac{n-1}{2} + \tfrac{\mu}{2})p - (\tfrac{\mu}{2} + 1) \quad \textit{if} \quad p = p_{\mathrm{Fuj}}(\mu),$$
$$\tfrac{1}{p-1} \leq \kappa \leq (\tfrac{n-1}{2} + \tfrac{\mu}{2})p - (\tfrac{\mu}{2} + 1) \quad \textit{if} \quad p \in \big(p_{\mathrm{Fuj}}(\mu), p_{\mathrm{Fuj}}(n-1)\big),$$
$$\tfrac{n-1}{2} < \kappa \leq (\tfrac{n-1}{2} + \tfrac{\mu}{2})p - (\tfrac{\mu}{2} + 1) \quad \textit{if} \quad p \geq p_{\mathrm{Fuj}}(n-1).$$

Then, the integral $I(\xi)$ *can be estimated for any* $\xi \geq 0$ *as follows:*

$$I(\xi) \lesssim \langle \xi \rangle^{-(\kappa - m - p)}. \tag{72}$$

Remark 4.7 Let us underline that the choice of κ in Lemma 4.6 guarantees in all subcases that

$$\kappa > m + 1 = \tfrac{n-1}{2}. \tag{73}$$

Furthermore, the assumptions made on κ imply always the lower bound

$$\kappa \geq \tfrac{2}{p-1} - \tfrac{\mu}{2}. \tag{74}$$

We are going to use several times the lower bounds (73) and (74) for κ in the upcoming proofs.

Remark 4.8 Let us underline that $M(3) = 1+\sqrt{5}$. So, we found an upper bound for μ which is consistent with the one obtained in Sect. 3 for the case $n = 3$. Moreover, the conditions on κ when $\mu \in [n-1, M(n)]$ are exactly those we have seen in the statement of Lemma 3.10.

Proof Following the approach of [6], we split the integral into three subintegrals $I(\xi) = I_1(\xi) + I_2(\xi) + I_3(\xi)$, where

$$I_1(\xi) \doteq \int_{-\frac{\xi}{2}}^{\frac{\xi}{2}} \langle \eta + \xi \rangle^{1-m(p-1)} \langle \eta - \xi \rangle^{-\frac{\mu}{2}(p-1)} \langle \eta \rangle^{-p(\kappa-(m+1))} d\eta,$$

$$I_2(\xi) \doteq \int_{\frac{\xi}{2}}^{\xi} \langle \eta + \xi \rangle^{1-m(p-1)} \langle \eta - \xi \rangle^{-\frac{\mu}{2}(p-1)} \langle \eta \rangle^{-p(\kappa-(m+1))} d\eta,$$

$$I_3(\xi) \doteq \int_{-\xi}^{-\frac{\xi}{2}} \langle \eta + \xi \rangle^{1-m(p-1)} \langle \eta - \xi \rangle^{-\frac{\mu}{2}(p-1)} \langle \eta \rangle^{-p(\kappa-(m+1))} d\eta.$$

Let us start with $I_1(\xi)$. Since for $\eta \in [-\frac{\xi}{2}, \frac{\xi}{2}]$ we may use the equivalences $\langle \eta + \xi \rangle \approx \langle \xi \rangle \approx \langle \eta - \xi \rangle$, we get

$$I_1(\xi) \lesssim \begin{cases} \langle \xi \rangle^{1-(m+\frac{\mu}{2})(p-1)} & \text{if} \quad k > \frac{1}{p} + m + 1, \\ \langle \xi \rangle^{1-(m+\frac{\mu}{2})(p-1)} \log\langle \xi \rangle & \text{if} \quad k = \frac{1}{p} + m + 1, \\ \langle \xi \rangle^{2-(m+\frac{\mu}{2})(p-1)-p(\kappa-(m+1))} & \text{if} \quad k < \frac{1}{p} + m + 1. \end{cases}$$

In the first case, we have $\langle \xi \rangle^{1-(m+\frac{\mu}{2})(p-1)} \lesssim \langle \xi \rangle^{-(\kappa-m-p)}$ if and only if $\kappa \leq \left(m + \frac{\mu}{2} + 1\right) p - \left(\frac{\mu}{2} + 1\right)$. Therefore, we obtain

$$\kappa \in \left(\frac{1}{p} + m + 1, (m + \frac{\mu}{2} + 1)p - (\frac{\mu}{2} + 1)\right],$$

as range for κ in the first case. Using the condition $p > p_0(n + \mu)$, we have that is actually possible to choose κ in such a way. For $\kappa = \frac{1}{p} + m + 1$, we find $\langle \xi \rangle^{1-(m+\frac{\mu}{2})(p-1)} \log\langle \xi \rangle \lesssim \langle \xi \rangle^{-(\kappa-m-p)}$ thanks to $p > p_0(n + \mu)$. In the third case we may estimate $\langle \xi \rangle^{2-(m+\frac{\mu}{2})(p-1)-p(\kappa-(m+1))} \lesssim \langle \xi \rangle^{-(\kappa-m-p)}$, assuming $\kappa \geq \frac{2}{p-1} - \frac{\mu}{2}$, that is we consider

$$\kappa \in \left[\frac{2}{p-1} - \frac{\mu}{2}, \frac{1}{p} + m + 1\right).$$

Employing again $p > p_0(n + \mu)$, we find that this range for κ is not empty.

Summarizing, gluing together all three intervals for κ, we have the inequality $I_1(\xi) \lesssim \langle \xi \rangle^{-(\kappa-m-p)}$ provided that

$$\kappa \in \left[\frac{2}{p-1} - \frac{\mu}{2}, (m + \frac{\mu}{2} + 1)p - (\frac{\mu}{2} + 1)\right].$$

Now we estimate $I_2(\xi)$. For $\eta \in \left[\frac{\xi}{2}, \xi\right]$ we have $\langle \eta + \xi \rangle \approx \langle \xi \rangle \approx \langle \eta \rangle$, so,

$$I_2(\xi) \lesssim \begin{cases} \langle \xi \rangle^{1-m(p-1)-p(\kappa-(m+1))} & \text{if} \quad p > p_{\text{Fuj}}(\mu), \\ \langle \xi \rangle^{1-m(p-1)-p(\kappa-(m+1))} \log\langle \xi \rangle & \text{if} \quad p = p_{\text{Fuj}}(\mu), \\ \langle \xi \rangle^{2-(m+\frac{\mu}{2})(p-1)-p(\kappa-(m+1))} & \text{if} \quad p < p_{\text{Fuj}}(\mu). \end{cases}$$

Analogously as in the estimate of $I_1(\xi)$, we find that $I_2(\xi) \lesssim \langle \xi \rangle^{-(\kappa-m-p)}$, provided that

$$\begin{cases} \kappa \geq \frac{1}{p-1} & \text{if} \quad p > p_{\text{Fuj}}(\mu), \\ \kappa > \frac{\mu}{2} & \text{if} \quad p = p_{\text{Fuj}}(\mu), \\ \kappa \geq \frac{2}{p-1} - \frac{\mu}{2} & \text{if} \quad p < p_{\text{Fuj}}(\mu). \end{cases}$$

Finally, it remains to consider $I_3(\xi)$. In the case in which $\eta \in \left[-\xi, -\frac{\xi}{2}\right]$ it holds $\langle \eta - \xi \rangle \approx \langle \xi \rangle \approx \langle \eta \rangle$ and, then,

$$I_3(\xi) \lesssim \begin{cases} \langle \xi \rangle^{-\frac{\mu}{2}(p-1)-p(\kappa-(m+1))} & \text{if} \quad p > p_{\text{Fuj}}(m), \\ \langle \xi \rangle^{-\frac{\mu}{2}(p-1)-p(\kappa-(m+1))} \log\langle \xi \rangle & \text{if} \quad p = p_{\text{Fuj}}(m), \\ \langle \xi \rangle^{2-(m+\frac{\mu}{2})(p-1)-p(\kappa-(m+1))} & \text{if} \quad p < p_{\text{Fuj}}(m), \end{cases}$$
$$\lesssim \langle \xi \rangle^{-(\kappa-m-p)},$$

where, in order to allow the validity of the last inequality, we have to make the following assumptions on κ:

$$\begin{cases} \kappa \geq m - \frac{\mu}{2} & \text{if} \quad p > p_{\text{Fuj}}(m), \\ \kappa > m - \frac{\mu}{2} & \text{if} \quad p = p_{\text{Fuj}}(m), \\ \kappa \geq \frac{2}{p-1} - \frac{\mu}{2} & \text{if} \quad p < p_{\text{Fuj}}(m). \end{cases}$$

In particular, if we assume that $\kappa > m + 1$ and $\kappa \geq \frac{2}{p-1} - \frac{\mu}{2}$, then, the estimate $I_3(\xi) \lesssim \langle \xi \rangle^{-(\kappa-m-p)}$ is always fulfilled.

Since we need to require always the validity of the condition $\kappa > m + 1$ for κ, it is useful to find the values of the exponent p such that the lower bounds for κ, that is $\frac{2}{p-1} - \frac{\mu}{2}$ or $\frac{1}{p-1}$, are exactly equal to $m + 1$. These bounds are given by $p_{\text{Fuj}}(\frac{n-1+\mu}{2})$ and $p_{\text{Fuj}}(n-1)$, respectively.

Thanks to these remarks, we get

$$\max\left\{m+1, \frac{2}{p-1} - \frac{\mu}{2}\right\} = \begin{cases} m+1 & \text{if} \quad p \geq p_{\text{Fuj}}(\frac{n-1+\mu}{2}), \\ \frac{2}{p-1} - \frac{\mu}{2} & \text{if} \quad p \leq p_{\text{Fuj}}(\frac{n-1+\mu}{2}), \end{cases}$$

and

$$\max\left\{m+1, \frac{1}{p-1}\right\} = \begin{cases} m+1 & \text{if} \quad p \geq p_{\text{Fuj}}(n-1), \\ \frac{1}{p-1} & \text{if} \quad p \leq p_{\text{Fuj}}(n-1). \end{cases}$$

Moreover, the following chain of inequalities are satisfied:

$$p_{\text{Fuj}}(n-1) \leq p_0(n+\mu) < p_{\text{Fuj}}(\tfrac{n-1+\mu}{2}) \leq p_{\text{Fuj}}(\mu) \quad \text{for} \quad \mu \in [2, \tfrac{(n-1)^2}{n+1}]$$
$$p_0(n+\mu) \leq p_{\text{Fuj}}(n-1) \leq p_{\text{Fuj}}(\tfrac{n-1+\mu}{2}) \leq p_{\text{Fuj}}(\mu) \quad \text{for} \quad \mu \in [\tfrac{(n-1)^2}{n+1}, n-1],$$
$$p_0(n+\mu) \leq p_{\text{Fuj}}(\mu) \leq p_{\text{Fuj}}(\tfrac{n-1+\mu}{2}) \leq p_{\text{Fuj}}(n-1) \quad \text{for} \quad \mu \in [n-1, M(n)].$$

Among the previous inequalities the relations between the shift of the Strauss exponent and the shifts of Fujita exponent are less straightforward to prove than the others. Let us show how to prove that $p_0(n+\mu) < p_{\mathrm{Fuj}}(\mu)$ for $\mu \in [2, M(n)]$. By definition of the Strauss exponent, it follows immediately that $p_0(n+\mu) < p_{\mathrm{Fuj}}(\mu)$ is equivalent to require:

$$(n+\mu-1)p_{\mathrm{Fuj}}(\mu)^2 - (n+\mu+1)p_{\mathrm{Fuj}}(\mu) - 2 > 0,$$

that is,

$$\mu^2 - (n-1)\mu - 2(n-1) < 0.$$

The previous quadratic equation has positive discriminant for all $n \geq 5$ and its two roots have different sign, according to Descartes' rule. Thus, we get the condition $p_0(n+\mu) < p_{\mathrm{Fuj}}(\mu)$ provided that $\mu \in [2, M(n)]$. In particular, we see how the upper bound for μ comes into play. Analogously, one can prove that the condition

$$p_{\mathrm{Fuj}}(\tfrac{n-1+\mu}{2}) > p_0(n+\mu)$$

is always fulfilled for any $\mu \geq 2$ and that

$$p_{\mathrm{Fuj}}(n-1) \geq p_0(n+\mu) \quad \text{for} \quad \mu \geq \tfrac{(n-1)^2}{n+1}.$$

Combining the restrictions on κ coming from the estimates of I_1, I_2 and I_3 with all possible dispositions of the exponents $p_0(n+\mu)$, $p_{\mathrm{Fuj}}(\mu)$, $p_{\mathrm{Fuj}}(\frac{n-1+\mu}{2})$, $p_{\mathrm{Fuj}}(n-1)$ on the real line, we find (72).

Remark 4.9 In the statement of Lemma 4.6 in each subcase for the choice of κ in a suitable interval, that depends on μ and p, one can see that the range of admissible κs is not empty.

The nonemptiness of the interval $\left[\frac{2}{p-1} - \frac{\mu}{2}, (\frac{n-1}{2} + \frac{\mu}{2})p - (\frac{\mu}{2}+1)\right]$ is equivalent to the requirement $(n+\mu-1)p^2 - (n+\mu+1)p - 2 > 0$, which is true in all cases we are dealing with, since we are considering $p > p_0(n+\mu)$.

Also in the case of the interval $\left[\frac{n-1}{2}, (\frac{n-1}{2} + \frac{\mu}{2})p - (\frac{\mu}{2}+1)\right]$, we have a not empty set because of $p > p_0(n+\mu)$. Indeed, the nonemptiness of such range for κ is equivalent to $p > p_{\mathrm{Fuj}}(n-1+\mu)$. But $p_0(n+\mu) > p_{\mathrm{Fuj}}(n-1+\mu)$ for any $\mu \geq 0$, so, thanks to the assumption $p > p_0(n+\mu)$, we are always working with $p > p_{\mathrm{Fuj}}(n-1+\mu)$.

Finally, let us study the nonemptiness of the interval

$$\left[\tfrac{1}{p-1}, (\tfrac{n-1}{2} + \tfrac{\mu}{2})p - (\tfrac{\mu}{2}+1)\right],$$

which is equivalent to the solvability of the quadratic equation

$$(n+\mu-1)p^2-(n+2\mu+1)p+\mu\geq 0.$$

The positive solution to the quadratic equation related to the previous inequality is given by

$$\widetilde{p}(\mu,n)\doteq\frac{n+1+2\mu+\sqrt{(n+1)^2+8\mu}}{2(n+\mu-1)}.$$

One can prove that $p_0(n+\mu)\geq\widetilde{p}(\mu,n)$ when $\mu\leq M(n)$, with, of course, reversed inequality if we consider μ above $M(n)$. Therefore, in this last case we find a not empty range of values for κ. Let us point out that, in order to study the condition $p_0(n+\mu)\geq\widetilde{p}(\mu,n)$, one has to solve the cubic inequality

$$\mu^3-(n^2-1)\mu-2(n-1)^2>0.$$

Once we remark that a quadratic factor of the above cubic polynomial is $\mu^2-(n-1)\mu-2(n-1)$ (which we met in the study of the condition $p_0(n+\mu)<p_{\mathrm{Fuj}}(\mu)$ in the proof of Lemma 4.6), the desired condition follows immediately.

Remark 4.10 Let us explain now why we are not able to prove Lemma 4.6 for $p>p_0(n+\mu)$, if we deal with the case $\mu>M(n)$. In this case we have that $p_0(n+\mu)>p_{\mathrm{Fuj}}(\mu)$ and, therefore, one could think to change the assumption on κ in Lemma 4.6 in the following way:

$$\begin{aligned}&\tfrac{1}{p-1}\leq\kappa\leq(\tfrac{n-1}{2}+\tfrac{\mu}{2})p-(\tfrac{\mu}{2}+1) &&\text{if}\quad p_0(n+\mu)<p<p_{\mathrm{Fuj}}(n-1),\\ &\tfrac{n-1}{2}<\kappa\leq(\tfrac{n-1}{2}+\tfrac{\mu}{2})p-(\tfrac{\mu}{2}+1) &&\text{if}\quad p\geq p_{\mathrm{Fuj}}(n-1).\end{aligned}$$

Nevertheless, as we observed in Remark 4.9, now we are in the case in which $\widetilde{p}(\mu,n)>p_0(n+\mu)$ and there would be an empty range of admissible κs for exponents p close to $p_0(n+\mu)$. Namely, if we worked with the approach of this section even for $\mu>M(n)$, then, we would get a global existence result only for $p>\widetilde{p}(\mu,n)$.

In the next result we prove some estimates that we will use in order to deal with the nonlinearity in the weighted $L^\infty_t L^\infty_r$ space that we will consider in the next section as space for solutions.

Lemma 4.11 *Let $n\geq 5$ be an odd integer. Let us assume*

$$p\in\big(p_0(n+\mu),1+\tfrac{2}{m}\big)$$

and $\mu\in[2,M(n)]$. Let us consider κ as in Lemma 4.6.

Then, for any $(t,r)\in[0,\infty)\times(0,\infty)$ the following estimates hold:

$$I(t,r)\doteq\int_0^t\langle\tau\rangle^{-\frac{\mu}{2}(p-1)}\int_{|\lambda_-|}^{\lambda_+}\langle\lambda\rangle^{-m(p-1)+1}\psi_\kappa(\tau,\lambda)^p d\lambda\,d\tau\lesssim r\psi_\kappa(t,r),\qquad(75)$$

$$J(t,r)\doteq\int_0^t\langle\tau\rangle^{-\frac{\mu}{2}(p-1)}\int_{|\lambda_-|}^{\lambda_+}\lambda^{m-(m-1)p}\langle\lambda\rangle^{1-p}\psi_\kappa(\tau,\lambda)^p d\lambda\,d\tau\lesssim r\psi_\kappa(t,r).\qquad(76)$$

Moreover, if $t\geq 2r$, then, we get

$$J'(t,r)\doteq\int_0^{t-2r}\langle\tau\rangle^{-\frac{\mu}{2}(p-1)}\int_{\lambda_-}^{\lambda_+}\lambda^{-m(p-1)+1}\psi_\kappa(\tau,\lambda)^p d\lambda\,d\tau\lesssim r\psi_\kappa(t,r).\qquad(77)$$

Furthermore, if $r\geq 1$, then, we obtain

$$P_+(t,r)\doteq\int_0^t\langle\tau\rangle^{-\frac{\mu}{2}(p-1)}\langle\lambda_+\rangle^{-m(p-1)+1}\psi_\kappa(\tau,\lambda_+)^p d\tau\lesssim r\psi_\kappa(t,r),\qquad(78)$$

$$P_-(t,r)\doteq\int_0^t\langle\tau\rangle^{-\frac{\mu}{2}(p-1)}\langle\lambda_-\rangle^{-m(p-1)+1}\psi_\kappa(\tau,|\lambda_-|)^p d\tau\lesssim r\psi_\kappa(t,r).\qquad(79)$$

Proof In order to prove the estimates for $I(t,r)$, $J(t,r)$, $J'(t,r)$, $P_+(t,r)$ and $P_-(t,r)$, we need to split the (t,r)-plane in different zones.

Estimate for $I(t,r)$

We perform the change of variables $\xi=\lambda+\tau$, $\eta=\lambda-\tau$. Therefore,

$$\begin{aligned}I(t,r)&\lesssim\int_{|t-r|}^{t+r}\int_{-\xi}^{\xi}\langle\xi-\eta\rangle^{-\frac{\mu}{2}(p-1)}\langle\xi+\eta\rangle^{-m(p-1)+1}\langle\xi\rangle^{-p}\langle\eta\rangle^{-p(\kappa-(m+1))}\,d\eta\,d\xi\\&\lesssim\int_{|t-r|}^{t+r}\langle\xi\rangle^{-p}\int_{-\xi}^{\xi}\langle\xi-\eta\rangle^{-\frac{\mu}{2}(p-1)}\langle\xi+\eta\rangle^{-m(p-1)+1}\langle\eta\rangle^{-p(\kappa-(m+1))}\,d\eta\,d\xi\\&=\int_{|t-r|}^{t+r}\langle\xi\rangle^{-p}I(\xi)\,d\xi\lesssim\int_{|t-r|}^{t+r}\langle\xi\rangle^{-\kappa+m}\,d\xi,\end{aligned}$$

where in the last inequality we employed Lemma 4.6. Let us consider separately three different subcases.

Case $t\geq 2r$ or $r\leq 1$

$$I(t,r)\lesssim\int_{t-r}^{t+r}\langle\xi\rangle^{-\kappa+m}\,d\xi\lesssim\langle t-r\rangle^{-\kappa+m}r\lesssim r\psi_\kappa(t,r),$$

where in the second last inequality we used that $\langle\xi\rangle^{-\kappa+m}$ is decreasing on the domain of integration and in the last inequality the relation $\langle t-r\rangle \approx \langle t+r\rangle$ is employed.

Case $\frac{r}{2} \leq t \leq 2r$ *and* $r \geq 1$

Being $\kappa > m+1$, then,

$$I(t,r) \lesssim \int_{|t-r|}^{t+r} \langle\xi\rangle^{-\kappa+m}\, d\xi \lesssim \langle t-r\rangle^{-\kappa+m+1} \approx r\psi_\kappa(t,r),$$

where in the last estimate the equivalence $r \approx \langle t+r\rangle$ is used.

Case $t \leq \frac{r}{2}$ *and* $r \geq 1$

$$I(t,r) \lesssim \int_{r-t}^{t+r} \langle\xi\rangle^{-\kappa+m}\, d\xi \lesssim \langle r-t\rangle^{-\kappa+m} t \lesssim r\psi_\kappa(t,r),$$

where the relation $\langle r-t\rangle \approx \langle t+r\rangle$ is employed in the last inequality. Thus, we proved (75).

Estimate for $J(t,r)$

Since for $\lambda \geq 1$ it holds $\lambda \approx \langle\lambda\rangle$, then,

$$J(t,r) \lesssim \int_\Omega \langle\tau\rangle^{-\frac{\mu}{2}(p-1)} \int_{|\lambda_-|}^{\min(1,\lambda_+)} \lambda^{m-(m-1)p}\langle\lambda\rangle^{1-p}\psi_\kappa(\tau,\lambda)^p\, d\lambda\, d\tau + I(t,r),$$

where we denote $\Omega \doteq \{\tau \in [0,t] : |\lambda_-| \leq 1\}$. For $\lambda \in [0,1]$ and $\tau \geq 0$ it holds $\langle\tau+\lambda\rangle \approx \langle\tau\rangle \approx \langle\tau-\lambda\rangle$, therefore, it results $\psi_\kappa(\tau,\lambda) \lesssim \langle\tau\rangle^{-\kappa+m}$. Consequently,

$$J(t,r) \lesssim \widetilde{J} + I(t,r),$$

where

$$\widetilde{J} \doteq \int_\Omega \langle\tau\rangle^{-\frac{\mu}{2}(p-1)-(\kappa-m)p} \int_{|\lambda_-|}^{\min(1,\lambda_+)} \lambda^{m-(m-1)p}\, d\lambda\, d\tau.$$

In order to show (76), since we have already shown (75), it is sufficient to prove that $\widetilde{J} \lesssim r\psi_\kappa(t,r)$. We consider separately the case $t \geq 2r$ or $r \leq 1$ and the case $t \leq 2r$ and $r \geq 1$.

Case $t \geq 2r$ *or* $r \leq 1$

Since $\Omega = [t-r-1, t-r+1] \cap \{\tau \geq 0\}$, then, $\langle\tau\rangle \approx \langle t-r\rangle$ for $\tau \in \Omega$. Therefore,

$$\widetilde{J} \lesssim \langle t-r\rangle^{-\frac{\mu}{2}(p-1)-(\kappa-m)p} \int_\Omega \int_{|\lambda_-|}^{\min(1,\lambda_+)} \lambda^{m-(m-1)p}\, d\lambda\, d\tau.$$

If $m-(m-1)p>0$, being the function $\lambda^{m-(m-1)p}$ bounded on $[0,1]$ and $\mathrm{meas}(\Omega)\leq 2$, we obtain

$$\begin{aligned}\widetilde{J}&\lesssim \langle t-r\rangle^{-\frac{\mu}{2}(p-1)-(\kappa-m)p}\int_{\Omega}\int_{|\lambda_-|}^{\min(1,\lambda_+)} d\lambda\, d\tau\\&\lesssim r\langle t-r\rangle^{-\frac{\mu}{2}(p-1)-(\kappa-m)p}.\end{aligned}$$

On the other hand, when $m-(m-1)p<0$, we get

$$\begin{aligned}\widetilde{J}&\lesssim \langle t-r\rangle^{-\frac{\mu}{2}(p-1)-(\kappa-m)p}\int_{\Omega}|\lambda_-|^{m-(m-1)p}\int_{|\lambda_-|}^{\min(1,\lambda_+)} d\lambda\, d\tau\\&\lesssim \langle t-r\rangle^{-\frac{\mu}{2}(p-1)-(\kappa-m)p}(\lambda_+-|\lambda_-|)\int_{t-r-1}^{t-r+1}|t-\tau-r|^{m-(m-1)p}\,d\tau\\&\lesssim r\langle t-r\rangle^{-\frac{\mu}{2}(p-1)-(\kappa-m)p},\end{aligned}$$

where in the last inequality we used the condition

$$m-(m-1)p>-1, \tag{80}$$

in order to guarantee the finiteness of the last integral.

Let us underline that the condition (80) follows from the assumptions on p we are considering in this theorem. Indeed, for $n\geq 7$ the upper bound $p<1+\frac{2}{m}$ is a stronger condition on p than (80), while (80) is trivially fulfilled in the case $n=5$. Concluding, from (73), we have for $t\geq 2r$ or $r\leq 1$

$$\widetilde{J}\lesssim r\langle t-r\rangle^{-\frac{\mu}{2}(p-1)-(\kappa-m)p}\lesssim r\langle t-r\rangle^{-(\kappa-m)}\approx r\psi_\kappa(t,r),$$

where in the last step we used $\langle t+r\rangle\approx\langle t-r\rangle$ for $t\geq 2r$ or $r\leq 1$.

Case $t\leq 2r$ and $r\geq 1$

Firstly, we observe that $\Omega=[t-r-1,t-r+1]\cap\{\tau\geq 0\}$ is nonempty if and only if $t\geq r-1$. Using again (80), we find

$$\begin{aligned}\widetilde{J}&\lesssim \int_{\Omega}\langle\tau\rangle^{-\frac{\mu}{2}(p-1)-(\kappa-m)p}\int_0^1\lambda^{m-(m-1)p}d\lambda\, d\tau\\&\lesssim \int_{\Omega}\langle\tau\rangle^{-\frac{\mu}{2}(p-1)-(\kappa-m)p}\,d\tau.\end{aligned}$$

For $t\geq r+1$, since the exponent for $\langle\tau\rangle$ in the last integral is negative, we get

$$\widetilde{J}\lesssim\int_{t-r-1}^{t-r+1}\langle\tau\rangle^{-\frac{\mu}{2}(p-1)-(\kappa-m)p}\,d\tau\leq 2\langle t-r-1\rangle^{-\frac{\mu}{2}(p-1)-(\kappa-m)p}.$$

Moreover, $\langle t-r-1\rangle \gtrsim \langle t-r\rangle$. Hence, for $t \geq r+1$, it holds

$$\widetilde{J} \lesssim \langle t-r\rangle^{-\frac{\mu}{2}(p-1)-(\kappa-m)p} \lesssim \langle t-r\rangle^{-(\kappa-m)} \lesssim r\psi_\kappa(t,r),$$

where in the last step we used $\langle t+r\rangle \lesssim \langle r\rangle \approx r$ for $t \leq 2r$ and $r \geq 1$. Otherwise, if $t \in [r-1, r+1]$, then,

$$\widetilde{J} \lesssim \int_{t-r-1}^{t-r+1} \langle\tau\rangle^{-\frac{\mu}{2}(p-1)-(\kappa-m)p}\, d\tau \lesssim 1 \lesssim \langle t-r\rangle^{-(\kappa-m)},$$

being $|t-r| \in [0,1]$. Finally, using the inequality $r \gtrsim \langle t+r\rangle$ for $t \leq 2r$ and $r \geq 1$, we find again the estimate $\widetilde{J} \lesssim r\psi_\kappa(t,r)$ also in the subcase $t \in [r-1, r+1]$. Consequently, combining all subcases, we got $\widetilde{J} \lesssim r\psi_\kappa(t,r)$ and, hence, we proved (76).

Estimate for $J'(t,r)$

Now we want to prove the estimate for $J'(t,r)$ when $t \geq 2r$. Let us remark that we can reduce our considerations to the subcase $r \leq 1$.

Indeed, if $t \geq 2r$ and $r \geq 1$, then, $\lambda_- \geq r \geq 1$ for $\tau \in [0, t-2r]$, and, consequently, $J'(t,r) \lesssim I(t,r) \lesssim r\psi_\kappa(t,r)$. So, let us assume $t \geq 2r$ and $r \leq 1$. Analogously to what we did for $J(t,r)$, we may estimate

$$J'(t,r) \lesssim \int_{\Omega'} \langle\tau\rangle^{-\frac{\mu}{2}(p-1)} \int_{\lambda_-}^{\min(\lambda_+,1)} \lambda^{-m(p-1)+1} \psi_\kappa(\tau,\lambda)^p\, d\lambda\, d\tau + I(t,r),$$

where $\Omega' \doteq \{\tau \in [0, t-2r] : \lambda_- \leq 1\} = [0, t-2r] \cap \{\tau : \tau > t-r-1\}$. Thanks to $r \leq 1$, it follows that Ω' is not empty and $\Omega' = [(t-r-1)_+, t-2r]$. As we have already seen, for $\lambda \in [0,1]$ it holds $\psi_\kappa(\tau,\lambda) \lesssim \langle\tau\rangle^{-\kappa+m}$. Hence,

$$J'(t,r) \lesssim \widetilde{J}' + I(t,r),$$

where

$$\widetilde{J}' \doteq \int_{(t-r-1)_+}^{t-2r} \langle\tau\rangle^{-\frac{\mu}{2}(p-1)-(\kappa-m)p} \int_{\lambda_-}^{\min(\lambda_+,1)} \lambda^{-m(p-1)+1}\, d\lambda\, d\tau.$$

Also, if we prove that $\widetilde{J}' \lesssim r\psi_\kappa(t,r)$, then, it follows immediately (77). We distinguish two subcases.

Case $t \geq r+1$ and $r \leq 1$

Being the exponent for $\langle\tau\rangle$ negative in the previous integral and using the inequality $\langle t-r-1\rangle \gtrsim \langle t-r\rangle$, we get

$$\widetilde{J}' \lesssim \langle t-r\rangle^{-\frac{\mu}{2}(p-1)-(\kappa-m)p} \int_{t-r-1}^{t-2r} \int_{\lambda_-}^{\min(\lambda_+,1)} \lambda^{-m(p-1)+1}\, d\lambda\, d\tau. \tag{81}$$

When $-m(p-1)+1 \geq 0$, then, (81) implies

$$\begin{aligned}\widetilde{J}' &\lesssim r\langle t-r\rangle^{-\frac{\mu}{2}(p-1)-(\kappa-m)p}\int_{t-r-1}^{t-2r} \lambda_+^{-m(p-1)+1}\,d\tau \\ &\lesssim r\langle t-r\rangle^{-\frac{\mu}{2}(p-1)-(\kappa-m)p}(1+2r)^{-m(p-1)+1}(1-r) \\ &\lesssim r\langle t-r\rangle^{-\frac{\mu}{2}(p-1)-(\kappa-m)p}.\end{aligned}$$

Otherwise, in the case $-m(p-1)+1<0$, by (81) it follows:

$$\begin{aligned}\widetilde{J}' &\lesssim r\langle t-r\rangle^{-\frac{\mu}{2}(p-1)-(\kappa-m)p}\int_{t-r-1}^{t-2r} \lambda_-^{-m(p-1)+1}\,d\tau \\ &\lesssim r\langle t-r\rangle^{-\frac{\mu}{2}(p-1)-(\kappa-m)p},\end{aligned}$$

where in the last inequality we used the condition $-m(p-1)+1>-1$ (which is equivalent for the upper bound of p in the statement), in order to guarantee the boundedness of the integral.

Summarizing, we proved that $\widetilde{J}' \lesssim r\langle t-r\rangle^{-\frac{\mu}{2}(p-1)-(\kappa-m)p}$, so, using again $\langle t-r\rangle \approx \langle t+r\rangle$ for $t \geq 2r$ and (73), eventually, we find

$$\widetilde{J}' \lesssim r\langle t-r\rangle^{-\frac{\mu}{2}(p-1)-(\kappa-m)p} \lesssim r\langle t-r\rangle^{-(\kappa-m)} \approx r\psi_\kappa(t,r).$$

Case $2r \leq t \leq r+1$ *and* $r \leq 1$

Since in this case $\langle t-r\rangle \approx \langle t+r\rangle \approx 1$, it is sufficient to show that $\widetilde{J}' \lesssim r$ in order to obtain the same estimate as before. Being $-\frac{\mu}{2}(p-1)-(\kappa-m)p<0$, then,

$$\widetilde{J}' \lesssim \int_0^{t-2r}\int_{\lambda_-}^{\min(\lambda_+,1)} \lambda^{-m(p-1)+1}\,d\lambda\,d\tau.$$

If $-m(p-1)+1 \geq 0$, being $\lambda^{-m(p-1)+1}$ a bounded function on the domain of integration, we have

$$\widetilde{J}' \lesssim r(t-2r) \lesssim r(1-r) \lesssim r,$$

else when $-m(p-1)+1<0$, using again the upper bound for p, we obtain

$$\widetilde{J}' \lesssim r\int_0^{t-2r} \lambda_-^{-m(p-1)+1}\,d\tau \lesssim r.$$

So, we proved $\widetilde{J}' \lesssim r\psi_\kappa(t,r)$ for $t \geq 2r$ too and, in turn, (77).

Estimate for $P_-(t,r)$

Let us write down $\psi_\kappa(\tau, |\lambda_-|)$ more explicitly

$$\psi_\kappa(\tau, |\lambda_-|) = \begin{cases} \langle t-r\rangle^{-1}\langle t-r-2\tau\rangle^{-\kappa+m+1} & \text{if} \quad \tau \leq t-r, \\ \langle t-r-2\tau\rangle^{-1}\langle t-r\rangle^{-\kappa+m+1} & \text{if} \quad \tau \geq t-r. \end{cases} \tag{82}$$

We consider separately three different cases.

Case $t \geq 2r$ and $r \geq 1$

We use (82). Therefore,

$$P_-(t,r) \leq \langle t-r\rangle^{-p} R_- + \langle t-r\rangle^{-(\kappa-m)p} R_+,$$

being

$$R_- \doteq \int_0^{t-r} \langle\tau\rangle^{-\frac{\mu}{2}(p-1)}\langle t-\tau-r\rangle^{-m(p-1)+1}\langle t-r-2\tau\rangle^{(-\kappa+m+1)p} d\tau,$$

$$R_+ \doteq \int_{t-r}^{t} \langle\tau\rangle^{-\frac{\mu}{2}(p-1)}\langle t-\tau-r\rangle^{-m(p-1)+1} d\tau,$$

where we used for $t \geq 2r$ and $\tau \geq t-r$ the relation $\langle t-r-2\tau\rangle \geq \langle t-r\rangle$.

We estimate now R_+. Since $p < 1 + \frac{2}{m}$ and $t-r \geq r$, then,

$$\begin{aligned} R_+ &\lesssim \langle t-r\rangle^{-\frac{\mu}{2}(p-1)} \int_{t-r}^{t} \langle t-\tau-r\rangle^{-m(p-1)+1} d\tau \\ &\lesssim \langle t-r\rangle^{-\frac{\mu}{2}(p-1)}\langle r\rangle^{-m(p-1)+2} \lesssim \langle t-r\rangle^{-\frac{\mu}{2}(p-1)-m(p-1)+2}. \end{aligned}$$

Hence, being $r \geq 1$, it holds

$$\begin{aligned} \langle t-r\rangle^{-(\kappa-m)p} R_+ &\lesssim \langle t-r\rangle^{-(\kappa-m)p-\frac{\mu}{2}(p-1)-m(p-1)+2} \\ &\lesssim \langle t-r\rangle^{-(\kappa-m)} \lesssim r\langle t-r\rangle^{-(\kappa-m)}, \end{aligned}$$

where in the second inequality we used the condition (74) on κ.

Let us deal with the integral R_-. Carrying out the change of variables $t-r-2\tau = \eta$, we obtain

$$\begin{aligned} R_- &\lesssim \int_{-(t-r)}^{t-r} \langle t-r-\eta\rangle^{-\frac{\mu}{2}(p-1)}\langle t-r+\eta\rangle^{-m(p-1)+1}\langle\eta\rangle^{(-\kappa+m+1)p} d\eta \\ &\lesssim \langle t-r\rangle^{-(\kappa-m-p)}, \end{aligned}$$

where in the last estimate we employed Lemma 4.6. Also,

$$\langle t-r\rangle^{-p} R_- \lesssim \langle t-r\rangle^{-(\kappa-m)} \lesssim r\langle t-r\rangle^{-(\kappa-m)}.$$

Summarizing, for $t \geq 2r$ and $r \geq 1$, we have

$$P_-(t,r) \lesssim r\langle t-r\rangle^{-(\kappa-m)} \lesssim r\psi_\kappa(t,r),$$

where in the last inequality we used the relation $\langle t-r\rangle \approx \langle t+r\rangle$ for $t \geq 2r$.

Case $r \leq t \leq 2r$ and $r \geq 1$

By (82) it follows:

$$P_-(t,r) = \langle t-r\rangle^{-p} R_- + \langle t-r\rangle^{(-\kappa+m+1)p} R'_+,$$

where R_- is defined as before and

$$R'_+ \doteq \int_{t-r}^{t} \langle\tau\rangle^{-\frac{\mu}{2}(p-1)} \langle t-\tau-r\rangle^{-m(p-1)+1} \langle t-r-2\tau\rangle^{-p} d\tau.$$

We point out that the term R_- can be estimated exactly as in the previous case, so, $R_- \lesssim \langle t-r\rangle^{-(\kappa-m-p)}$.

Now we deal with R'_+. Being $\langle \tau-t+r\rangle \leq \langle 2\tau-t+r\rangle$ and $\langle \tau-t+r\rangle \leq \langle\tau\rangle$ for $\tau \in [t-r,t]$, it results

$$R'_+ \lesssim \int_{t-r}^{t} \langle \tau-t+r\rangle^{-\frac{\mu}{2}(p-1)-(m+1)(p-1)} d\tau \lesssim 1.$$

In the last step we employed the condition

$$-(m+1+\tfrac{\mu}{2})(p-1) < -1, \tag{83}$$

in order to guarantee the uniform boundedness of the integral. Indeed, (83) is equivalent to require the lower bound $p > p_{\mathrm{Fuj}}(n+\mu-1)$. However, we are assuming $p > p_0(n+\mu)$ and $p_0(n+\mu) > p_{\mathrm{Fuj}}(n+\mu-1)$ so, in particular, the condition $p > p_{\mathrm{Fuj}}(n+\mu-1)$ is fulfilled.

Summarizing, for $1 \leq r \leq t \leq 2r$ we obtained the estimate

$$\begin{aligned} P_-(t,r) &\lesssim \langle t-r\rangle^{-p} R_- + \langle t-r\rangle^{(-\kappa+m+1)p} R'_+ \\ &\lesssim \langle t-r\rangle^{-p} \langle t-r\rangle^{-(\kappa-m-p)} + \langle t-r\rangle^{(-\kappa+m+1)p} \\ &\lesssim \langle t-r\rangle^{-\kappa+m+1} \approx r\psi_\kappa(t,r), \end{aligned}$$

where in the last line we used the relation $r \approx \langle t+r\rangle$ for $t \leq 2r$ and $r \geq 1$.

Case $t \le r$ and $r \ge 1$

Using again (82) and the chain of inequalities $\langle \tau \rangle \le \langle \tau - t + r \rangle \le \langle 2\tau - t + r \rangle$ for $0 \le \tau \le t \le r$, we obtain

$$\begin{aligned} P_-(t,r) &\lesssim \langle t-r \rangle^{(-\kappa+m+1)p} \int_0^t \langle \tau \rangle^{-\frac{\mu}{2}(p-1)} \langle \tau - t + r \rangle^{-(m+1)(p-1)} d\tau \\ &\lesssim \langle t-r \rangle^{(-\kappa+m+1)p} \int_0^t \langle \tau \rangle^{-(m+1+\frac{\mu}{2})(p-1)} d\tau \\ &\lesssim \langle t-r \rangle^{-\kappa+m+1} \approx r\psi_\kappa(t,r), \end{aligned}$$

where we estimated uniformly the integral by a constant, because of the condition (83), and in the last step the condition $\langle t + r \rangle \approx r$ for $t \le r$ and $r \ge 1$ is used. Combining the estimates for $P_-(t,r)$ in the subcases $t \ge 2r$, $r \le t \le 2r$ and $t \le r$, we have (79).

Estimate for $P_+(t,r)$

We consider separately the cases $t \ge 2r$ and $t \le 2r$.

Case $t \ge 2r$ and $r \ge 1$

Since $\psi_\kappa(\tau, \lambda_+) = \langle t + r \rangle^{-1} \langle t + r - 2\tau \rangle^{-\kappa+m+1}$, then,

$$P_+(t,r) = \langle t+r \rangle^{-p} \int_0^t \langle \tau \rangle^{-\frac{\mu}{2}(p-1)} \langle \lambda_+ \rangle^{-m(p-1)+1} \langle t + r - 2\tau \rangle^{-p(\kappa-m-1)} d\tau.$$

Performing the change of variables $\eta = t + r - 2\tau$, $\tau \in [0,t]$ implies

$$\eta \in [r-t, t+r] \subset [-(r+t), r+t]$$

and, consequently, using the equivalence $\langle t+r \rangle \approx \langle t-r \rangle$ for $t \ge 2r$ and Lemma 4.6, we get for $r \ge 1$

$$P_+(t,r) \lesssim \langle t+r \rangle^{-p} I(t+r) \lesssim \langle t+r \rangle^{-(\kappa-m)} \lesssim r\psi_\kappa(t,r).$$

Case $t \le 2r$ and $r \ge 1$

We can repeat the same estimate seen in the subcase $t \ge 2r$, obtaining again $P_+(t,r) \lesssim \langle t+r \rangle^{-\kappa+m}$. Finally, since $\langle t-r \rangle \le \langle t+r \rangle$ and $r \ge 1$, we have

$$P_+(t,r) \lesssim \langle t+r \rangle^{-(\kappa-m)} \lesssim \langle t-r \rangle^{-\kappa+m+1} \langle t+r \rangle^{-1} \lesssim r\psi_\kappa(t,r).$$

Thus, combining the estimates for the cases $t \ge 2r$ and $t \le 2r$ we have (78). The proof is completed.

4.3 Semilinear Model: Odd Case in Higher Dimensions

In order to prove that the operator

$$N : v \in X_\kappa \longrightarrow Nv = v^0 + Lv$$

has a unique determined fixed point for some $\kappa > m + 1$, which is a radial solution to (55) in X_κ according to Definition 4.5, we are going to prove the following result.

Theorem 4.12 *Let $n \geq 5$ be an odd integer. Let $p \in \left(p_0(n+\mu), 1+\frac{2}{m}\right)$ and $\mu \in [2, M(n)]$. Let us consider κ as in Lemma 4.6. If $v \in X_\kappa$, then,*

$$\|Lv\|_{X_\kappa} \leq C_1 \|v\|_{X_\kappa}^p. \tag{84}$$

Furthermore, if we define on X_κ the norm

$$|||v|||_{X_k} \doteq \sup_{t \geq 0\,,\, r > 0} r^m |v(t,r)| \psi_\kappa(t,r)^{-1},$$

then, for any $p < 2$ and any $v, w \in X_\kappa$ we get

$$\|Lv - Lw\|_{X_\kappa} \leq C_2(M_1 + M_2), \tag{85}$$

$$|||Lv - Lw|||_{X_\kappa} \leq C_3 M_3, \tag{86}$$

while for any $p \geq 2$ and any $v, w \in X_\kappa$ we get

$$\|Lv - Lw\|_{X_\kappa} \leq C_4 M_1, \tag{87}$$

where

$$M_1 \doteq \|v - w\|_{X_\kappa} \left(\|v\|_{X_\kappa}^{p-1} + \|w\|_{X_\kappa}^{p-1}\right), \tag{88}$$

$$M_2 \doteq |||v - w|||_{X_\kappa}^{p-1} \left(\|v\|_{X_\kappa} + \|w\|_{X_\kappa}\right), \tag{89}$$

$$M_3 \doteq |||v - w|||_{X_\kappa} \left(\|v\|_{X_\kappa}^{p-1} + \|w\|_{X_\kappa}^{p-1}\right). \tag{90}$$

Here $C_1, \cdots, C_4$ denote positive constants which are independent of t and r.

Proof We want to prove first (84). Let $v \in X_\kappa$. Using the definition of the norm $\|\cdot\|_{X_\kappa}$, we get

$$|v(\tau, \lambda)|^p \lesssim \lambda^{-(m-1)p} \langle\lambda\rangle^{-p} \psi_\kappa(\tau, \lambda)^p \|v\|_{X_\kappa}^p, \tag{91}$$

$$|\partial_\lambda |v(\tau, \lambda)|^p| \lesssim \lambda^{-(m-1)p-1} \langle\lambda\rangle^{-(p-1)} \psi_\kappa(\tau, \lambda)^p \|v\|_{X_\kappa}^p. \tag{92}$$

Let us start with the case $r \leq 1$. Using the representation formula (71), we have for $\alpha = 0, 1$

$$\partial_r^\alpha Lv(t,x) = \int_0^t \langle\tau\rangle^{-\frac{\mu}{2}(p-1)} \partial_r^\alpha \omega(t,\tau,r)d\tau = A_\alpha + B_\alpha,$$

where A_α and B_α denote the integral over $[0,(t-2r)_+]$ and $[(t-2r)_+, t]$, respectively.

Let us begin with the estimate of the term A_α. Of course, it makes sense to consider this term only in the case $t \geq 2r$. Using (91), (92) and Lemma 4.2 when the initial time is shifted from 0 to τ and the first data is identically zero, we obtain

$$\begin{aligned}
|A_\alpha| &\lesssim \int_0^{t-2r} \langle\tau\rangle^{-\frac{\mu}{2}(p-1)} \left| \partial_r^\alpha \int_{\lambda_-}^{\lambda_+} |v(\tau,\lambda)|^p K(\lambda, t-\tau, r) d\lambda \right| d\tau \\
&\lesssim \int_0^{t-2r} \langle\tau\rangle^{-\frac{\mu}{2}(p-1)} r^{-m-\alpha} \int_{\lambda_-}^{\lambda_+} \lambda^m \big(|v(\tau,\lambda)|^p + \lambda |\partial_\lambda |v(\tau,\lambda)|^p| \big) d\lambda\, d\tau \\
&\lesssim r^{-m-\alpha} J(t,r) \, \|v\|_{X_\kappa}^p \lesssim r^{-m+1-\alpha} \psi_\kappa(t,r) \, \|v\|_{X_\kappa}^p,
\end{aligned}$$

where in the last line we have used (76) from Lemma 4.11.

For the estimate of the term B_α, we consider separately the cases $\alpha = 0$ and $\alpha = 1$.

We begin with the estimate of B_0 in the case $t \geq 2r$. Hence,

$$\begin{aligned}
|B_0| &\lesssim \int_{t-2r}^{t} \langle\tau\rangle^{-\frac{\mu}{2}(p-1)} \left| \int_{|\lambda_-|}^{\lambda_+} |v(\tau,\lambda)|^p K(\lambda, t-\tau, r) d\lambda \right| d\tau \\
&\lesssim \int_{t-2r}^{t} \langle\tau\rangle^{-\frac{\mu}{2}(p-1)} r^{-m-1} \int_{|\lambda_-|}^{\lambda_+} \lambda^{m+1} |v(\tau,\lambda)|^p d\lambda\, d\tau \\
&\lesssim r^{-m-1} \int_{t-2r}^{t} \langle\tau\rangle^{-\frac{\mu}{2}(p-1)} \int_{|\lambda_-|}^{\lambda_+} \lambda^{m+1-(m-1)p} \psi_\kappa(\tau,\lambda)^p d\lambda\, d\tau \, \|v\|_{X_\kappa}^p,
\end{aligned}$$

where in the second inequality we used (64) from Lemma 4.3, when the initial time is shifted from 0 to τ and the first data is identically zero, and in the third inequality we employed (91). Since for τ, λ belonging to the domain of integration in the last integral $|\lambda_-| \leq \lambda \leq \lambda_+ \leq 3r$ implies $4\langle\tau-\lambda\rangle \geq \langle\tau\rangle$, then, $\langle\tau+\lambda\rangle \geq \langle\tau-\lambda\rangle \gtrsim \langle\tau\rangle$. Also, the previous chain of inequalities implies

$$\psi_\kappa(\tau,\lambda) \lesssim \langle\tau\rangle^{-\kappa+m} \qquad \text{for } \tau \in [t-2r, t] \text{ and } \lambda \in [|\lambda_-|, \lambda_+]. \tag{93}$$

Consequently,

$$|B_0| \lesssim r^{-m-1} \int_{t-2r}^{t} \langle\tau\rangle^{-\frac{\mu}{2}(p-1)-(\kappa-m)p} \int_0^{3r} \lambda^{m+1-(m-1)p} d\lambda\, d\tau \, \|v\|_{X_\kappa}^p.$$

Let us remark that the exponent of λ in the internal integral is positive, thanks to (80). Therefore, we can estimate the λ-integral by $3r$, thus,

$$|B_0| \lesssim r^{-m} \int_{t-2r}^{t} \langle\tau\rangle^{-\frac{\mu}{2}(p-1)-(\kappa-m)p} d\tau \, \|v\|_{X_\kappa}^p. \tag{94}$$

The upper bound for p is equivalent to $m(p-1)-2<0$. Hence, thanks the lower bound for κ given in (74), we find

$$\begin{aligned} -\tfrac{\mu}{2}(p-1)-(\kappa-m)p &\leq -\tfrac{\mu}{2}(p-1)-(\kappa-m)+2-\kappa(p-1) \\ &\leq -(\kappa-m). \end{aligned} \tag{95}$$

Then,

$$|B_0| \lesssim r^{-m} \int_{t-2r}^{t} \langle\tau\rangle^{-(\kappa-m)} d\tau \, \|v\|_{X_\kappa}^p \lesssim r^{-m} \langle t-2r\rangle^{-(\kappa-m)} 2r \, \|v\|_{X_\kappa}^p.$$

Because of $\langle t-2r\rangle \gtrsim \langle t-r\rangle$ and $\langle t-2r\rangle \gtrsim \langle t+r\rangle$ for $r \in (0,1]$, we may conclude

$$|B_0| \lesssim r^{-m+1} \psi_\kappa(t,r) \, \|v\|_{X_\kappa}^p. \tag{96}$$

Let us estimate now the term B_0 in the case in which $t \leq 2r$. Even in this case we want to get the same estimate as in the case $t \geq 2r$. However, for $t \leq 2r$ and $r \leq 1$ it holds $\psi_\kappa(t,r) \approx 1$ and, then, it is sufficient to prove that $|B_0| \lesssim r^{-m+1} \|v\|_{X_\kappa}^p$ in order to get the same estimate as before. We can repeat the same estimates of the previous case provided that we substitute the first extreme of integration $t-2r$ with 0, since the chain of inequalities $|\lambda_-| \leq \lambda \leq \lambda_+ \leq t+r = 3r$ is still true for τ and λ in the domain of integration. Also,

$$\begin{aligned} |B_0| &\lesssim \int_0^t \langle\tau\rangle^{-\frac{\mu}{2}(p-1)} \left| \int_{|\lambda_-|}^{\lambda_+} |v(\tau,\lambda)|^p K(\lambda, t-\tau, r) d\lambda \right| d\tau \\ &\lesssim \int_0^t \langle\tau\rangle^{-\frac{\mu}{2}(p-1)} r^{-m-1} \int_{|\lambda_-|}^{\lambda_+} \lambda^{m+1} |v(\tau,\lambda)|^p d\lambda \, d\tau \\ &\lesssim r^{-m-1} \int_0^t \langle\tau\rangle^{-\frac{\mu}{2}(p-1)} \int_{|\lambda_-|}^{\lambda_+} \lambda^{m+1-(m-1)p} \psi_\kappa(\tau,\lambda)^p d\lambda \, d\tau \, \|v\|_{X_\kappa}^p \\ &\lesssim r^{-m-1} \int_0^t \langle\tau\rangle^{-\frac{\mu}{2}(p-1)-(\kappa-m)p} \int_0^{3r} \lambda^{m+1-(m-1)p} d\lambda \, d\tau \, \|v\|_{X_\kappa}^p \\ &\lesssim r^{-m} \int_0^t \langle\tau\rangle^{-\frac{\mu}{2}(p-1)-(\kappa-m)p} d\tau \, \|v\|_{X_\kappa}^p \lesssim r^{-m} t \, \|v\|_{X_\kappa}^p \lesssim r^{-m+1} \|v\|_{X_\kappa}^p, \end{aligned}$$

where in the fourth inequality we use that $\psi_\kappa(\tau,\lambda) \lesssim \langle\tau\rangle^{-(\kappa-m)}$, since the relations $\langle\tau+\lambda\rangle \geq \langle\tau\rangle$ and $4\langle\tau-r\rangle \geq \langle\tau\rangle$ are satisfied on the domain of integration in this case, thanks to $\lambda \leq 3r$.

Let us consider now the estimate of B_1. As for B_0, it is convenient to study the case $t \geq 2r$ first. Using (65) from Lemma 4.3 when the initial time is shifted from 0 to τ and the first data is identically zero, we get

$$
\begin{aligned}
|B_1| &\lesssim \int_{t-2r}^{t} \langle\tau\rangle^{-\frac{\mu}{2}(p-1)} \left| \partial_r \int_{|\lambda_-|}^{\lambda_+} |v(\tau,\lambda)|^p K(\lambda, t-\tau, r) d\lambda \right| d\tau \\
&\lesssim \int_{t-2r}^{t} \langle\tau\rangle^{-\frac{\mu}{2}(p-1)} r^{-m-1} \int_{|\lambda_-|}^{\lambda_+} \lambda^m |v(\tau,\lambda)|^p d\lambda\, d\tau \\
&\quad + \sum_{\pm} \int_{t-2r}^{t} \langle\tau\rangle^{-\frac{\mu}{2}(p-1)} r^{-m-1} |\lambda_\pm|^{m+1} |v(\tau, |\lambda_\pm|)|^p \, d\tau.
\end{aligned}
$$

The estimate of the double integral is analogous to that one for the term B_0, the only difference is the power of λ in the internal integral. Thus, we can proceed as follows:

$$
\begin{aligned}
&\int_{t-2r}^{t} \langle\tau\rangle^{-\frac{\mu}{2}(p-1)} r^{-m-1} \int_{|\lambda_-|}^{\lambda_+} \lambda^m |v(\tau,\lambda)|^p d\lambda\, d\tau \\
&\quad \lesssim r^{-m-1} \int_{t-2r}^{t} \langle\tau\rangle^{-\frac{\mu}{2}(p-1)-(\kappa-m)p} \int_0^{3r} \lambda^{m-(m-1)p} d\lambda\, d\tau \, \|v\|_{X_\kappa}^p \\
&\quad \lesssim r^{-m-1} \int_{t-2r}^{t} \langle\tau\rangle^{-\frac{\mu}{2}(p-1)-(\kappa-m)p} \, d\tau \, \|v\|_{X_\kappa}^p ,
\end{aligned}
$$

where in the last step we used the condition $m-(m-1)p > -1$ to control the λ-integral with a constant. We remark that we obtained exactly the integral that appears in the right-hand side of (94). So, repeating the same steps as before, we arrive at

$$
\int_{t-2r}^{t} \langle\tau\rangle^{-\frac{\mu}{2}(p-1)} r^{-m-1} \int_{|\lambda_-|}^{\lambda_+} \lambda^m |v(\tau,\lambda)|^p d\lambda\, d\tau \lesssim r^{-m} \psi_\kappa(t,r) \, \|v\|_{X_\kappa}^p ,
$$

here, as we have already explained, the different power for r with respect to the final estimate of B_0 is due to the fact that we estimated the λ-integral by a constant.

It remains to estimate the terms

$$
\int_{t-2r}^{t} \langle\tau\rangle^{-\frac{\mu}{2}(p-1)} r^{-m-1} |\lambda_\pm|^{m+1} |v(\tau, |\lambda_\pm|)|^p \, d\tau.
$$

By using (91) and (93), we get

$$|v(\tau, |\lambda_\pm|)|^p \lesssim |\lambda_\pm|^{-(m-1)p} \langle \lambda_\pm \rangle^{-p} \psi_\kappa(\tau, |\lambda_\pm|)^p \, \|v\|^p_{X_\kappa}$$
$$\lesssim |\lambda_\pm|^{-(m-1)p} \langle \tau \rangle^{-(\kappa-m)p} \, \|v\|^p_{X_\kappa}.$$

Hence,

$$\int_{t-2r}^{t} \langle \tau \rangle^{-\frac{\mu}{2}(p-1)} r^{-m-1} |\lambda_\pm|^{m+1} |v(\tau, |\lambda_\pm|)|^p \, d\tau$$
$$\lesssim \int_{t-2r}^{t} \langle \tau \rangle^{-\frac{\mu}{2}(p-1)-(\kappa-m)p} r^{-m-1} |\lambda_\pm|^{m+1-(m-1)p} \, d\tau \, \|v\|^p_{X_\kappa}$$
$$\lesssim r^{-m-1} \int_{t-2r}^{t} \langle \tau \rangle^{-\frac{\mu}{2}(p-1)-(\kappa-m)p} \, d\tau \, \|v\|^p_{X_\kappa},$$

where in the last inequality we used the fact that the exponent of $|\lambda_\pm|$ is positive and that $0 \leq |\lambda_-| \leq \lambda_+ \leq 3r \leq 3$ for $t - 2r \leq \tau \leq t$. Using again (95), we find

$$\int_{t-2r}^{t} \langle \tau \rangle^{-\frac{\mu}{2}(p-1)} r^{-m-1} |\lambda_\pm|^{m+1} |v(\tau, |\lambda_\pm|)|^p \, d\tau$$
$$\lesssim r^{-m-1} \int_{t-2r}^{t} \langle \tau \rangle^{-(\kappa-m)} \, d\tau \, \|v\|^p_{X_\kappa} \lesssim r^{-m} \psi_\kappa(t, r) \, \|v\|^p_{X_\kappa},$$

where in the last step we used the inequality $\langle t - 2r \rangle^{-(\kappa-m)} \leq \psi_\kappa(t, r)$ as in (96). Summarizing, if we combine the previous two estimates, we have for $t \geq 2r$

$$|B_1| \lesssim r^{-m} \psi_\kappa(t, r) \, \|v\|^p_{X_\kappa}.$$

The next step is to prove that $|B_1| \lesssim r^{-m} \, \|v\|^p_{X_\kappa}$ for $t \leq 2r \leq 2$. Indeed, thanks to $\psi_\kappa(t, r) \approx 1$ which holds in this zone, the previous inequality implies $|B_1| \lesssim r^{-m} \psi_\kappa(t, r) \, \|v\|^p_{X_\kappa}$ as in the case $t \geq 2r$.

As in the above case, we can estimate B_1 with the same two terms. More precisely, replacing the first extreme of integration with 0 in place of $t - 2r$, we obtain

$$|B_1| \lesssim \int_0^t \langle \tau \rangle^{-\frac{\mu}{2}(p-1)} r^{-m-1} \int_{|\lambda_-|}^{\lambda_+} \lambda^m |v(\tau, \lambda)|^p d\lambda \, d\tau$$
$$+ \sum_\pm \int_0^t \langle \tau \rangle^{-\frac{\mu}{2}(p-1)} r^{-m-1} |\lambda_\pm|^{m+1} |v(\tau, |\lambda_\pm|)|^p \, d\tau.$$

Let us estimate in first place the integral

$$\widetilde{B}_1 \doteq \int_0^t \langle \tau \rangle^{-\frac{\mu}{2}(p-1)} r^{-m-1} \int_{|\lambda_-|}^{\lambda_+} \lambda^m |v(\tau, \lambda)|^p d\lambda \, d\tau.$$

Following the same approach we used for the estimate of B_0 in the case $t \leq 2r \leq 2$, we arrive at

$$\begin{aligned}\widetilde{B}_1 &\lesssim r^{-m-1} \int_0^t \langle \tau \rangle^{-\frac{\mu}{2}(p-1)} \int_{|\lambda_-|}^{\lambda_+} \lambda^{m-(m-1)p} \psi_\kappa(\tau,\lambda)^p d\lambda \, d\tau \, \|v\|^p_{X_\kappa} \\ &\lesssim r^{-m-1} \int_0^t \langle \tau \rangle^{-\frac{\mu}{2}(p-1)-(\kappa-m)p} \int_0^{3r} \lambda^{m-(m-1)p} d\lambda \, d\tau \, \|v\|^p_{X_\kappa} \\ &\lesssim r^{-m-1} \int_0^t \langle \tau \rangle^{-\frac{\mu}{2}(p-1)-(\kappa-m)p} d\tau \, \|v\|^p_{X_\kappa} \lesssim r^{-m} \, \|v\|^p_{X_\kappa},\end{aligned}$$

where we estimated the λ-integral by a constant thanks to (80) and the other steps are analogous to the above cited situation.

Let us deal with the term

$$\widehat{B}_1 \doteq \sum_{\pm} \int_0^t \langle \tau \rangle^{-\frac{\mu}{2}(p-1)} r^{-m-1} |\lambda_\pm|^{m+1} |v(\tau, |\lambda_\pm|)|^p \, d\tau.$$

We can follow the steps we have employed in the case $t \geq 2r$, since the only difference consists of the domain of integration with respect to τ. Indeed, we have already seen that $\psi_\kappa(\tau,\lambda) \lesssim \langle \tau \rangle^{-(\kappa-m)}$ for $t \leq 2r \leq 2$, $\tau \in [0,t]$ and $\lambda \in [|\lambda_-|, \lambda_+]$. This yields

$$\begin{aligned}\widehat{B}_1 &\lesssim \sum_{\pm} \int_0^t \langle \tau \rangle^{-\frac{\mu}{2}(p-1)-(\kappa-m)p} r^{-m-1} |\lambda_\pm|^{m+1-(m-1)p} \, d\tau \, \|v\|^p_{X_\kappa} \\ &\lesssim r^{-m-1} \int_0^t \langle \tau \rangle^{-\frac{\mu}{2}(p-1)-(\kappa-m)p} \, d\tau \, \|v\|^p_{X_\kappa}.\end{aligned}$$

Also in this case we used that the functions $|\lambda_\pm|^{m+1-(m-1)p}$ are bounded on the domain of integration, being $m+1-(m-1)p > 0$ and $|\lambda_-| \leq \lambda_+ \leq 3$. By using again (95), we get

$$\widehat{B}_1 \lesssim r^{-m-1} \int_0^t \langle \tau \rangle^{-(\kappa-m)} \, d\tau \, \|v\|^p_{X_\kappa} \lesssim r^{-m-1} t \, \|v\|^p_{X_\kappa} \lesssim r^{-m} \, \|v\|^p_{X_\kappa}.$$

Summarizing, we obtained for $t \leq 2r \leq 2$

$$|B_1| \lesssim \widetilde{B}_1 + \widehat{B}_1 \lesssim r^{-m} \, \|v\|^p_{X_\kappa},$$

as we expected.

Up to now we considered only the case $r \in (0, 1]$. Now we have to study the case $r \geq 1$. Using (59), (91) and $\lambda \leq \langle\lambda\rangle$, we have

$$|Lv(t,r)| \lesssim r^{-m-1} \int_0^t \langle\tau\rangle^{-\frac{\mu}{2}(p-1)} \int_{|\lambda_-|}^{\lambda_+} |v(\tau,\lambda)|^p \lambda^{m+1} \, d\lambda \, d\tau$$

$$\lesssim r^{-m-1} J(t,r) \, \|v\|_{X_\kappa} \lesssim r^{-m} \psi_\kappa(t,r) \, \|v\|_{X_\kappa},$$

where in the last inequality we have used (76).

Now we deal with the term $\partial_r Lv(t,r)$ for $r \geq 1$. As in the case $r \leq 1$, we split $\partial_r Lv(t,r)$ into two integrals

$$\partial_r Lv(t,x) = \int_0^t \langle\tau\rangle^{-\frac{\mu}{2}(p-1)} \partial_r \omega(t,\tau,r) d\tau = A_1 + B_1.$$

The estimate of the term A_1 is exactly the same seen in the case $r \leq 1$. So, when $t \geq 2r$, we find $|A_1| \lesssim r^{-m} \psi_\kappa(t,r) \|v\|^p_{X_\kappa}$ even for $r \geq 1$.

Let us consider the term B_1. Differentiating the λ-integral, it results

$$\begin{aligned} B_1 = &\int_{(t-2r)_+}^t \langle\tau\rangle^{-\frac{\mu}{2}(p-1)} \int_{|\lambda_-|}^{\lambda_+} |v(\tau,\lambda)|^p \partial_r K(\lambda, t-\tau, r) \, d\lambda \, d\tau \\ &+ \int_{(t-2r)_+}^t \langle\tau\rangle^{-\frac{\mu}{2}(p-1)} |v(\tau,\lambda_+)|^p K(\lambda_+, t-\tau, r) \, d\tau \\ &- \int_{(t-2r)_+}^t \langle\tau\rangle^{-\frac{\mu}{2}(p-1)} \underbrace{\partial_r |\lambda_-|}_{\pm 1} |v(\tau, |\lambda_-|)|^p K(|\lambda_-|, t-\tau, r) \, d\tau. \end{aligned}$$

Therefore, using (60), (59) for $\lambda = |\lambda_\pm|$ and (91), it follows:

$$\begin{aligned} |B_1| \lesssim\; & r^{-m-1} \int_{(t-2r)_+}^t \langle\tau\rangle^{-\frac{\mu}{2}(p-1)} \int_{|\lambda_-|}^{\lambda_+} |v(\tau,\lambda)|^p \lambda^m \, d\lambda \, d\tau \, \|v\|^p_{X_\kappa} \\ &+ r^{-m-1} \sum_\pm \int_{(t-2r)_+}^t \langle\tau\rangle^{-\frac{\mu}{2}(p-1)} |v(\tau, |\lambda_\pm|)|^p |\lambda_\pm|^{m+1} \, d\tau \, \|v\|^p_{X_\kappa} \\ \lesssim\; & r^{-m-1} \big(J(t,r) + P_+(t,r) + P_-(t,r) \big) \, \|v\|^p_{X_\kappa} \lesssim r^{-m} \psi_\kappa(t,r) \, \|v\|^p_{X_\kappa}, \end{aligned}$$

where in the second step we estimate $|\lambda_\pm|^{m+1-(m-1)p} \leq \langle\lambda_\pm\rangle^{m+1-(m-1)p}$ and in the last estimate we used (76), (78) and (79).

Combining all estimates we have obtained up to now, we have shown

$$\begin{aligned} |Lv(t,r)| &\lesssim r^{-m+1} \psi_\kappa(t,r) \|v\|^p_{X_\kappa} \qquad \text{for } r \in (0,1], \\ |Lv(t,r)| &\lesssim r^{-m} \psi_\kappa(t,r) \|v\|^p_{X_\kappa} \qquad \text{for } r \geq 1, \\ |\partial_r Lv(t,r)| &\lesssim r^{-m} \psi_\kappa(t,r) \|v\|^p_{X_\kappa}. \end{aligned}$$

The previous inequalities imply (84), according to the definition of $\|\cdot\|_{X_\kappa}$.

Now we prove (85) and (87). If $v, w \in X_\kappa$, then,

$$\begin{aligned} Lv(t,r) - Lw(t,r) &= \int_0^t \langle\tau\rangle^{-\frac{\mu}{2}(p-1)} \widetilde{\omega}(t,\tau,r)\,d\tau \\ &= \int_0^t \langle\tau\rangle^{-\frac{\mu}{2}(p-1)} \int_{|\lambda_-|}^{\lambda_+} \big(|v(\tau,\lambda)|^p - |w(\tau,\lambda)|^p\big) K(\lambda, t-\tau, r)\,d\lambda\,d\tau. \end{aligned} \tag{97}$$

By using the definition of the norm $\|\cdot\|_{X_\kappa}$, (13) and (88), we get

$$\big||v(\tau,\lambda)|^p - |w(\tau,\lambda)|^p\big| \lesssim \lambda^{-(m-1)p} \langle\lambda\rangle^{-p} \psi_\kappa(\tau,\lambda)^p M_1. \tag{98}$$

For the λ-derivative, when $p < 2$, if we denote $F(u) = p|u|^{p-2}u$, then,

$$\begin{aligned} &\big|\partial_\lambda \big(|v(\tau,\lambda)|^p - |w(\tau,\lambda)|^p\big)\big| \\ &\quad \lesssim |F(v(\tau,\lambda))||\partial_\lambda v(\tau,\lambda) - \partial_\lambda w(\tau,\lambda)| + |F(v(\tau,\lambda)) - F(w(\tau,\lambda))||\partial_\lambda w(\tau,\lambda)| \\ &\quad \lesssim |v(\tau,\lambda)|^{p-1} |\partial_\lambda v(\tau,\lambda) - \partial_\lambda w(\tau,\lambda)| + |v(\tau,\lambda) - w(\tau,\lambda)|^{p-1} |\partial_\lambda w(\tau,\lambda)| \\ &\quad \lesssim \lambda^{-(m-1)p-1} \langle\lambda\rangle^{-(p-1)} \psi_\kappa(\tau,\lambda)^p M_1 + \lambda^{-mp} \psi_\kappa(\tau,\lambda)^p M_2, \end{aligned} \tag{99}$$

where in the second inequality we used that F is Hölder continuous and M_1, M_2 are defined by (88) and (89), respectively.

On the other hand, when $n = 5$ and $p \in [2,3)$, then, using the mean value theorem, we get

$$\begin{aligned} &\big|\partial_\lambda \big(|v(\tau,\lambda)|^p - |w(\tau,\lambda)|^p\big)\big| \\ &\quad \lesssim |F(v(\tau,\lambda))||\partial_\lambda v(\tau,\lambda) - \partial_\lambda w(\tau,\lambda)| + |F(v(\tau,\lambda)) - F(w(\tau,\lambda))||\partial_\lambda w(\tau,\lambda)| \\ &\quad \lesssim |v(\tau,\lambda)|^{p-1} |\partial_\lambda v(\tau,\lambda) - \partial_\lambda w(\tau,\lambda)| \\ &\qquad\quad + |v(\tau,\lambda) - w(\tau,\lambda)| (|v(\tau,\lambda)| + |w(\tau,\lambda)|)^{p-2} |\partial_\lambda w(\tau,\lambda)| \\ &\quad \lesssim \lambda^{-(m-1)p-1} \langle\lambda\rangle^{-(p-1)} \psi_\kappa(\tau,\lambda)^p M_1. \end{aligned} \tag{100}$$

Since the decay rate for $|v(\tau,\lambda)|^p$ in (91) is the same decay rate of $|v(\tau,\lambda)|^p - |w(\tau,\lambda)|^p$ in (98) and the decay rate for $\partial_\lambda |v(\tau,\lambda)|^p$ in (92) is the same decay rate of $\partial_\lambda \big(|v(\tau,\lambda)|^p - |w(\tau,\lambda)|^p\big)$ in (100), in the case $p \geq 2$ we can show (87) exactly as we showed (84), replacing $\|v\|_{X_\kappa}^p$ by M_1.

Proving (85), we should pay attention to the addend $\lambda^{-mp}\psi_\kappa(\tau,\lambda)^p M_2$, when we employ (99) to estimate $\partial_\lambda \big(|v(\tau,\lambda)|^p - |w(\tau,\lambda)|^p\big)$.

Checking the previous estimates we have done to prove (84), we see that we need to control the λ-derivative of the source only in the estimates of $|A_\alpha|$ for $t \geq 2r$. So, let us focus on the corresponding part in the case in which we work with $Lv - Lw$

in place of Lv. We use the notation

$$\widetilde{A}_\alpha \doteq \int_0^{(t-2r)_+} \langle\tau\rangle^{-\frac{\mu}{2}(p-1)} \partial_r^\alpha \widetilde{\omega}(t,\tau,r)d\tau \qquad \text{for} \quad \alpha = 0, 1.$$

If we proceed as we did in the estimate of A_α, employing (98) and (99), we obtain

$$\begin{aligned}
|\widetilde{A}_\alpha| &\lesssim \int_0^{t-2r} \langle\tau\rangle^{-\frac{\mu}{2}(p-1)} \left| \partial_r^\alpha \int_{\lambda_-}^{\lambda_+} \big(|v(\tau,\lambda)|^p - |w(\tau,\lambda)|^p\big) K(\lambda, t-\tau, r) d\lambda \right| d\tau \\
&\lesssim r^{-m-\alpha} \int_0^{t-2r} \langle\tau\rangle^{-\frac{\mu}{2}(p-1)} \int_{\lambda_-}^{\lambda_+} \lambda^{m-(m-1)p} \langle\lambda\rangle^{-(p-1)} \psi_\kappa(\tau,\lambda)^p d\lambda\, d\tau\, M_1 \\
&\quad + r^{-m-\alpha} \int_0^{t-2r} \langle\tau\rangle^{-\frac{\mu}{2}(p-1)} \int_{\lambda_-}^{\lambda_+} \lambda^{-(m-1)p+1} \psi_\kappa(\tau,\lambda)^p d\lambda\, d\tau\, M_2 \\
&\lesssim r^{-m-\alpha}\big(J(t,r)M_1 + J'(t,r)M_2\big) \lesssim r^{-m+1-\alpha} \psi_\kappa(t,r)\,(M_1+M_2),
\end{aligned}$$

where in the last estimate we used (76) and (77).

In all other possible cases, it is sufficient to use (98). Therefore, we may repeat exactly the same estimates seen for Lv also for $Lv - Lw$, replacing $\|v\|_{X_\kappa}^p$ by M_1, since we use (98) instead of (91). In this way, we find

$$\begin{aligned}
|Lv(t,r) - Lw(t,r)| &\lesssim r^{-m+1} \psi_\kappa(t,r)\,(M_1+M_2) && \text{for } r \in (0,1], \\
|Lv(t,r)_L w(t,r)| &\lesssim r^{-m} \psi_\kappa(t,r)\,(M_1+M_2) && \text{for } r \geq 1, \\
|\partial_r Lv(t,r) - \partial_r Lw(t,r)| &\lesssim r^{-m} \psi_\kappa(t,r)\,(M_1+M_2), &&
\end{aligned}$$

from which (85) follows immediately.

Finally, we prove (86). Similarly to what we have done in the previous estimate, we need first to determine a decay estimate by using the definition of the norms $\|\cdot\|_{X_\kappa}$, $|||\cdot|||_{X_\kappa}$. Using (13) and (90) we have

$$\big||v(\tau,\lambda)|^p - |w(\tau,\lambda)|^p\big| \lesssim \lambda^{-(m-1)p-1} \langle\lambda\rangle^{-(p-1)} \psi_\kappa(\tau,\lambda)^p\, M_3.$$

Employing the previous decay estimate in (97) together with (59), we find

$$\begin{aligned}
&|Lv(t,r) - Lw(t,r)| \\
&\quad \lesssim r^{-m-1} \int_0^t \langle\tau\rangle^{-\frac{\mu}{2}(p-1)} \int_{|\lambda_-|}^{\lambda_+} \lambda^{m-(m-1)p} \langle\lambda\rangle^{-(p-1)} \psi_\kappa(\tau,\lambda)^p\, d\lambda\, d\tau\, M_3 \\
&\quad \lesssim r^{-m-1} J(t,r) M_3 \lesssim r^{-m} \psi_\kappa(t,r) M_3,
\end{aligned}$$

where in the last step (76) is used. According to the definition of $|||\cdot|||_{X_\kappa}$, the above estimate implies immediately (86). This concludes the proof.

Theorem 4.13 *Let $n \geq 5$ be an odd integer. Let us assume $\mu \in [2, M(n)]$ and $\nu^2 \geq 0$ satisfying the relation $\delta = 1$, where $M(n)$ is defined as in (5), and let*

$$p \in \big(p_0(n+\mu), p_{\mathrm{Fuj}}(\tfrac{n-3}{2})\big). \tag{101}$$

Then, there exist $\varepsilon_0 > 0$ and $\kappa_2 > \kappa_1 \geq m+1$, with $m = \frac{n-3}{2}$, such that for any $\varepsilon \in (0, \varepsilon_0)$ and any radial data $u_0 \in \mathcal{C}^2(\mathbb{R}^n)$, $u_1 \in \mathcal{C}^1(\mathbb{R}^n)$, satisfying

$$|d_r^j u_0(r)| \leq \varepsilon \langle r\rangle^{-(\bar{\kappa}+j)} \quad \text{for} \quad j = 0, 1, 2,$$
$$|d_r^j (u_1(r) + \tfrac{\mu}{2} u_0(r))| \leq \varepsilon \langle r\rangle^{-(\bar{\kappa}+1+j)} \quad \text{for} \quad j = 0, 1,$$

for some $\bar{\kappa} \in (\kappa_1, \kappa_2]$, the Cauchy problem (2) admits a uniquely determined radial solution $u \in \mathcal{C}([0,\infty), \mathcal{C}^1(\mathbb{R}^n \setminus \{0\}))$, in the sense that $v(t,r) = \langle t\rangle^{\frac{\mu}{2}} u(t,r)$ satisfies Definition 4.5 for any $\kappa \in (\kappa_1, \bar{\kappa}]$.

Furthermore, the following decay estimates hold for any $t \geq 0$, $r > 0$ and $\kappa \in (\kappa_1, \bar{\kappa}]$:

$$|u(t,r)| \lesssim \varepsilon\, r^{-m+1} \langle r\rangle^{-1} \langle t\rangle^{-\frac{\mu}{2}} \langle t-r\rangle^{-\kappa+m+1} \langle t+r\rangle^{-1},$$
$$|\partial_r u(t,r)| \lesssim \varepsilon\, r^{-m} \langle t\rangle^{-\frac{\mu}{2}} \langle t-r\rangle^{-\kappa+m+1} \langle t+r\rangle^{-1}.$$

Proof Let us fix a κ in $(\kappa_1, \bar{\kappa}]$. Considering the transformed Cauchy problem (56), according to our setting it is enough to prove that the operator

$$Nv = v^0 + Lv \qquad \text{for any} \quad v \in X_\kappa$$

admits a uniquely determined fixed point, provided that ε_0 is sufficiently small. In the case in which $n = 5$ and $p \in [2, 3)$, thanks to (84) and (87), using a standard contraction argument, we may derive the existence and the uniqueness of a fixed point for N in a closed ball of X_κ around the origin with sufficiently small radius. When $p < 2$ we cannot use Banach's fixed point theorem, so we have to modify our argument. We will follow the method employed in [14, Section 5]. Let us consider the sequence of successive approximations, that is,

$$v_0 = v^0, \qquad v_{j+1} = Nv_j = v^0 + Lv_j \quad \text{for any} \quad j \geq 0.$$

Let ε_0 be defined by

$$\varepsilon_0 = \min\Big\{\big(2^{\frac{p}{p-1}} C_0 C_1^{\frac{1}{p-1}}\big)^{-1}, \big(2^{\frac{p+1}{p-1}} C_0 C_3^{\frac{1}{p-1}}\big)^{-1}, \big(2^{\frac{p+1}{p-1}} C_0 C_2^{\frac{1}{p-1}}\big)^{-1}\Big\}. \tag{102}$$

Combining (70), (84) and (102), it follows that N maps the closed ball in X_κ with radius $2C_0\varepsilon$ around 0 into itself.

Now we show that $\{v_j\}_{j\in\mathbb{N}}$ is a Cauchy sequence in X_κ. We know that $\|v_j\|_{X_\kappa} \leq 2C_0\varepsilon_0$ for any $j \geq 0$. Therefore, by using (85) and (102), we get

$$\begin{aligned}\|v_{j+1} - v_j\|_{X_\kappa} &= \|Lv_j - Lv_{j-1}\|_{X_\kappa} \\ &\leq 2^{-1}\|v_j - v_{j-1}\|_{X_\kappa} + 4C_0C_2\varepsilon_0 \left|\!\left|\!\left| v_j - v_{j-1} \right|\!\right|\!\right|_{X_\kappa}^{p-1}.\end{aligned}$$

Now we use the Lipschitz condition

$$|||Lv - Lw|||_{X_\kappa} \leq 2^{-1}|||v - w|||_{X_\kappa}, \tag{103}$$

for $v, w \in X_\kappa$ such that $\|v\|_{X_\kappa}, \|w\|_{X_\kappa} \leq 2C_0\varepsilon_0$, that is obtained combining (86) and the condition (102). Also,

$$\left|\!\left|\!\left| v_j - v_{j-1} \right|\!\right|\!\right|_{X_\kappa} \leq 2^{-j+1}|||v_1 - v_0|||_{X_\kappa} = 2^{-j+1}|||Lv_0|||_{X_\kappa}.$$

If we denote $A = 2^{p+1}C_0C_2\varepsilon_0|||Lv_0|||_{X_\kappa}^{p-1}$, then,

$$\|v_{j+1} - v_j\|_{X_\kappa} \leq 2^{-1}\|v_j - v_{j-1}\|_{X_\kappa} + A2^{-j(p-1)}.$$

Thus, applying iteratively the previous inequality, we get

$$\|v_{j+1} - v_j\|_{X_\kappa} \leq 2^{-j}\|Lv_0\|_{X_\kappa} + A\, j\, 2^{-j(p-1)} \longrightarrow 0 \qquad \text{as} \quad j \to \infty.$$

Being $(X_\kappa, \|\cdot\|_{X_k})$ a Banach space, there exists $v = \lim_{j\to\infty} v_j$ in X_κ, and, since L is locally Hölder-continuous, in particular $Nv = v$.

Finally, the uniqueness of the fixed point for N in $\{v \in X_\kappa : \|v\|_{X_\kappa} \leq 2C_0\varepsilon_0\}$ follows immediately from (103). This concludes the proof.

Remark 4.14 In Theorem 4.13, according to Lemma 4.6, the lower bound and the upper for the parameter κ are given by $\kappa_1 = \max\{m + 1, \frac{2}{p-1} - \frac{\mu}{2}, \frac{1}{p-1}\}$ and $\kappa_2 = (m + 1 + \frac{\mu}{2})p - (\frac{\mu}{2} + 1)$, respectively.

5 Conclusions

Combining the result of Sects. 2, 3 and 4 with the blow-up result we mentioned in the introduction, we may conclude that:

- $p_{\mathrm{Fuj}}(\frac{\mu}{2})$ is the critical exponent for (2) under the assumption (3) in the one dimensional case;
- $p_0(n + \mu)$ is the critical exponent for (2) in the radial symmetric case, provided that (3) and (5) are satisfied, in the odd dimensional case $n \geq 3$.

It would be resonable to get the same result of Sect. 4 in the even dimensional case $n \geq 4$. Nevertheless, because of the fact that Huygens' principle is no longer valid in even dimension, the representation formula for the radial linear problem (58) is more complicate and, hence, a more delicate analysis is necessary for the treatment of the semilinear case. Very recently in [21] this case is consider, assuming (3) and the same upper bound for μ as in this paper, and $p_0(n+\mu)$ is shown to be critical in this case too.

Acknowledgements The author thanks his supervisor Michael Reissig, who introduced him first to the study of the model considered in this work. Furthermore, the author thanks Marcello D'Abbicco, who provided a crucial hint to overcome a difficulty in the one dimensional case.

References

1. F. Asakura, Existence of a global solution to a semilinear wave equation with slowly decreasing initial data in three space dimensions. Commun. Partial Differ. Equ. **11**(13), 1459–1487 (1986)
2. R. Courant, D. Hilbert, *Methods of Mathematical Physics*, vol. 2. (Interscience, New York, 1962)
3. M. D'Abbicco, The threshold of effective damping for semilinear wave equation. Math. Methods Appl. Sci. **38**(6), 1032–1045 (2015)
4. M. D'Abbicco, S. Lucente, A modified test function method for damped wave equation. Adv. Nonlinear Stud. **13**(4), 867–892 (2013)
5. M. D'Abbicco, S. Lucente, NLWE with a special scale invariant damping in odd space dimension. Discret. Contin. Dyn. Syst. Suppl. (2015). 10th AIMS Conference on Dynamical Systems, Differential Equations and Applications, pp. 312–319
6. M. D'Abbicco, S. Lucente, M. Reissig, A shift in the Strauss exponent for semi-linear wave equations with a not effective damping. J. Differ. Equ. **259**(10), 5040–5073 (2015)
7. M. D'Abbicco, A. Palmieri, $L^p - L^q$ estimates on the conjugate line for semilinear critical dissipative Klein-Gordon equations. Preprint
8. M. Ikeda, M. Sobajima, Life-span of solutions to semilinear wave equation with time-dependent critical damping for specially localized initial data. Math. Ann. (2018). https://doi.org/10.1007/s00208-018-1664-1
9. M. Kato, M. Sakuraba, Global existence and blow-up for semilinear damped wave equations in three space dimensions (2018). arXiv:1807.04327
10. H. Kubo, Asymptotic behaviour of solutions to semilinear wave equations with initial data of slow decay. Math. Methods Appl. Sci. **17**(12), 953–970 (1994)
11. H. Kubo, On critical decay and power for semilinear wave equations in odd space dimensions. Deptartment of Mathematics, Hokkaido University. Preprint series #274, ID 1404 (1994), pp. 1–24. http://eprints3.math.sci.hokudai.ac.jp/1404
12. H. Kubo, Slowly decaying solutions for semilinear wave equations in odd space dimensions. Nonlinear Anal. **28**(2), 327–357 (1997)
13. H. Kubo, K. Kubota, Asymptotic behaviors of radially symmetric solutions of $\Box u = |u|^p$ for super critical values p in odd space dimensions. Hokkaido Math. J. **24**(2), 287–336 (1995)
14. H. Kubo, K. Kubota, Asymptotic behaviors of radial solutions to semilinear wave equations in odd space dimensions. Hokkaido Math. J. **24**(1), 9–51 (1995)
15. N.A. Lai, Weighted $L^2 - L^2$ estimate for wave equation and its applications (2018). arXiv:1807.05109
16. N.A. Lai, H. Takamura, K. Wakasa, Blow-up for semilinear wave equations with the scale invariant damping and super-Fujita exponent. J. Differ. Equ. **263**(9), 5377–5394 (2017)

17. W. Nunes do Nascimento, A. Palmieri, M. Reissig, Semi-linear wave models with power non-linearity and scale-invariant time-dependent mass and dissipation. Math. Nachr. **290**(11–12), 1779–1805 (2017)
18. A. Palmieri, Global existence of solutions for semi-linear wave equation with scale-invariant damping and mass in exponentially weighted spaces. J. Math. Anal. Appl. **461**(2), 1215–1240 (2018)
19. A. Palmieri, M. Reissig, Semi-linear wave models with power non-linearity and scale-invariant time-dependent mass and dissipation, II. Math. Nachr. **291**(11–12), 1859–1892 (2018)
20. A. Palmieri, M. Reissig, A competition between Fujita and Strauss type exponents for blow-up of semi-linear wave equations with scale-invariant damping and mass. J. Differ. Equ. (2018). https://doi.org/10.1016/j.jde.2018.07.061.
21. A. Palmieri, A global existence result for a semilinear wave equation with scale-invariant damping and mass in even space dimension (2018). arxiv:1804.03978
22. A. Palmieri, Global in time existence and blow-up results for a semilinear wave equation with scale-invariant damping and mass. PhD thesis, TU Bergakademie Freiberg, 2018, 279pp
23. J. Peral, L^p estimates for the wave equation. J. Funct. Anal. **36**(1), 114–145 (1980)
24. H. Takamura, Improved Kato's lemma on ordinary differential inequality and its application to semilinear wave equations. Nonlinear Anal. **125**, 227–240 (2015)
25. Z. Tu, J. Lin, A note on the blowup of scale invariant damping wave equation with sub-Strauss exponent (2017). arxiv:1709.00866v2
26. Z. Tu, J. Lin, Life-span of semilinear wave equations with scale-invariant damping: critical strauss exponent case (2017). arxiv:1711.00223
27. Y. Wakasugi, Critical exponent for the semilinear wave equation with scale invariant damping, in *Fourier Analysis*. Trends Mathematics (Birkhäuser/Springer, Cham, 2014), pp. 375–390

Wave Equations in Modulation Spaces–Decay Versus Loss of Regularity

Maximilian Reich and Michael Reissig

Abstract Recently the study of partial differential equations in modulation spaces gained some relevance. A well-known Cauchy problem is that for the wave equation, where several contributions exist concerning the local (in time) well-posedness. We refer to Bényi et al. (J Func Anal 246.2:366–384, 2007), Cordero and Nicola (J Math Anal Appl 353.2:583–591, 2009) and Reich (Modulation spaces and nonlinear partial differential equations. PhD thesis, TU Bergakademie Freiberg, 2017). By taking advantage of some tools and concepts from the theory of partial differential equations the authors provide some time-dependent estimates of the solution $u = u(t, x)$ to the Cauchy problem of the free wave equation. The main result yields the possibility to consider more delicate problems concerning the wave equation in modulation spaces such as global (in time) well-posedness results.

1 Introduction

Classical modulation spaces got originally introduced by Feichtinger [5] as a family of Banach spaces controlling globally local frequency information of a function or distribution, respectively. Thus, modulation spaces are an important tool when discussing problems in time-frequency analysis. But it also turned out that modulation spaces find fruitful applications in the theory of partial differential equations (e.g. see [1, 3, 10, 20, 30–32]). Numerous practical applications concern the propagation of different kind of waves. Within the scope of this work we consider the corresponding abstract model, which is the Cauchy problem for the free wave equation

$$u_{tt} - \Delta u = 0, \quad u(0, x) = \varphi(x), \quad u_t(0, x) = \psi(x), \tag{1}$$

M. Reich · M. Reissig (✉)
Faculty for Mathematics and Computer Science, Technical University Bergakademie, Freiberg, Germany
e-mail: reichma@web.de; reissig@math.tu-freiberg.de

M. D'Abbicco et al. (eds.), *New Tools for Nonlinear PDEs and Application*, Trends in Mathematics, https://doi.org/10.1007/978-3-030-10937-0_13

where $(t,x) \in \mathbb{R}^{n+1}$. Our goal is to establish time-dependent estimates of the solution $u = u(t,x)$ to Cauchy problem (1). Note that there are already several contributions to the Cauchy problem for the wave equation, see, e.g., [1, 3] or generalized Schrödinger equations, see, e.g., [11]. In [1] the authors prove the following estimate (without any loss of regularity)

$$\|u(t,\cdot)\|_{M^0_{p,q}} \leq C(t)\big(\|\varphi\|_{M^0_{p,q}} + \|\psi\|_{M^0_{p,q}}\big), \tag{2}$$

whereas in [3] the local (in time) well-posedness of the corresponding semi-linear Cauchy problem

$$u_{tt} - \Delta u = f(u), \quad u(0,x) = \varphi(x), \quad u_t(0,x) = \psi(x),$$

with analytic source terms $f = f(u)$ is proved for initial data

$$(\varphi,\psi) \in M^s_{p,q}(\mathbb{R}^n) \times M^{s-1}_{p,q}(\mathbb{R}^n).$$

However, our main interest is devoted to the function $C = C(t)$ in (2). Following the computations in [1] this constant can be specified by $C(t) = Ce^{c|t|}$ with some positive constants C, c. Subsequently we will apply some fundamental tools, which are already well-established in the theory of partial differential equations, in order to propose a family of estimates instead of (2) only. In particular the goal is to find assumptions under which the right-hand side in (2) decays in time, i.e., we want to avoid the behavior $C(t) \to \infty$ as $t \to \infty$, but we allow a loss of regularity instead. Section 4 treats this problem. It turns out that the concept of working on the conjugate line with respect to the L_p-norm together with the method of stationary phase gives us the possibility to obtain the desired estimates. Extensive literature concerning this topic is provided by Reissig and Ebert [4, Chapter 16]. As it can be seen there it is of great advantage to work in so-called homogeneous spaces, which allow a better localization around the origin. Generally this turned out to be a fruitful tool in the recent study of partial differential equations. We refer to, e.g., [2, 12, 16, 21]. So we will establish a decomposition method providing a better localization around the origin in the modulation space norm, which is done in Sect. 3. Let us begin with some preliminaries.

2 Preliminaries

First we introduce some basic notation. We define $\langle\xi\rangle := (1+|\xi|^2)^{\frac{1}{2}}$. The notation $a \lesssim b$ is equivalent to $a \leq Cb$ with a positive constant C. Let X and Y be two Banach spaces. Then the symbol $X \hookrightarrow Y$ indicates that the embedding is continuous. The Fourier transform of an admissible function f is defined by

$$\mathbb{F}f(\xi) = \hat{f}(\xi) = (2\pi)^{-\frac{n}{2}} \int_{\mathbb{R}^n} f(x)e^{-\mathrm{i}x\cdot\xi}\,dx \quad (x,\xi \in \mathbb{R}^n).$$

Analogously, the inverse Fourier transform is defined by

$$\mathbb{F}^{-1}\hat{f}(x) = (2\pi)^{-\frac{n}{2}} \int_{\mathbb{R}^n} \hat{f}(\xi) e^{\mathrm{i}x\cdot\xi} \, d\xi \quad (x, \xi \in \mathbb{R}^n).$$

Feichtinger [5] originally defined modulation spaces by taking Lebesgue norms of the so-called short-time Fourier transform of a function f with respect to x and ξ. The short-time Fourier transform is a particular joint time-frequency representation. For the definition and mapping properties we refer to [8]. By introducing the following decomposition principle we adopt the idea of obtaining local frequency properties of a function f, which was also established in [5]. The so-called frequency-uniform decomposition was independently introduced by Wang (e.g., see [31]). Let $\rho : \mathbb{R}^n \mapsto [0, 1]$ be a Schwartz function, which is compactly supported in the cube

$$Q_0 := \Big\{ \xi \in \mathbb{R}^n : -\frac{3}{4} \leq \xi_i \leq \frac{3}{4}, \; i = 1, \cdots, n \Big\}.$$

Moreover, assume that $\rho(\xi) = 1$ if $|\xi_i| \leq 1/2$ for $i = 1, 2, \cdots n$. Then we define $\rho_k(\xi) := \rho(\xi - k)$, $\xi \in \mathbb{R}^n$, $k \in \mathbb{Z}^n$. Now the collection $\{\sigma_k\}_{k\in\mathbb{Z}^n}$ is a decomposition of $\mathbb{R}^n$ into uniform cubes, where

$$\sigma_k(\xi) := \rho_k(\xi) \Big(\sum_{k\in\mathbb{Z}^n} \rho_k(\xi) \Big)^{-1}, \quad \xi \in \mathbb{R}^n, \quad k \in \mathbb{Z}^n.$$

The operator

$$\square_k := \mathbb{F}^{-1}\big(\sigma_k \mathbb{F}(\cdot)\big), \quad k \in \mathbb{Z}^n,$$

is called uniform decomposition operator. In [5] Feichtinger also showed that there is an equivalent definition of modulation spaces defined in terms of the short-time Fourier transform and modulation spaces defined by means of the uniform decomposition operator. Subsequently we will work with the latter one.

Definition 2.1 *Let $1 \leq p, q \leq \infty$ and assume $s \in \mathbb{R}$ to be the weight parameter. Suppose the window $\rho \in \mathscr{S}(\mathbb{R}^n)$ is compactly supported. Then the weighted modulation space $M^s_{p,q}(\mathbb{R}^n)$ consists of all tempered distributions $f \in \mathscr{S}'(\mathbb{R}^n)$ such that their norm*

$$\|f\|_{M^s_{p,q}} = \Big(\sum_{k\in\mathbb{Z}^n} \langle k \rangle^{sq} \|\square_k f\|^q_{L^p} \Big)^{\frac{1}{q}}$$

is finite with obvious modifications when $p = \infty$ and/or $q = \infty$.

Moreover, essential tools to obtain our main results in Sect. 4 concern the theory of Fourier multipliers. Let us introduce the sets

$$L_p^q := \{T \in \mathbb{S}'(\mathbb{R}^n) : \|T * f\|_{L_q} \leq C\|f\|_{L_p} \text{ for all } f \in \mathbb{S}(\mathbb{R}^n)\}$$

and

$$M_p^q := \{\mathbb{F}(T) : T \in L_p^q\},$$

where M_p^q is the set of multipliers of type (p, q). If $p = q$, then we write M_p instead of M_p^p. For the theory of Fourier multipliers in L_p-spaces we refer to Hörmander [9], Mikhlin [14] and Lizorkin [13]. A general reference regarding Fourier multipliers is Grafakos [7]. Moreover, general Fourier multipliers of classical modulation spaces are treated in [6]. In the following we list some important results given in the literature, which will be of great use later on.

First let us state a version of the so-called Bernstein multiplier theorem, see [33, Proposition 1.11].

Proposition 2.2 *Assume $L > \frac{n}{2}$ to be an integer and $\partial_{x_i}^k \sigma \in L_2(\mathbb{R}^n)$, $i = 1, \cdots, n$, for $0 \leq k \leq L$. Then for $1 \leq p \leq \infty$ it holds*

$$\big\|\mathbb{F}^{-1}\big(\sigma \mathbb{F} f\big)\big\|_{L_p} \lesssim \|\sigma\|_{L_2}^{1-\frac{n}{2L}} \Big(\sum_{i=1}^{n} \|\partial_{x_i}^L \sigma\|_{L_2}\Big)^{\frac{n}{2L}} \|f\|_{L_p}$$

for all $f \in \mathscr{S}(\mathbb{R}^n)$.

A concept of radial Fourier multipliers is given by means of modified Bessel functions. For a detailed insight we refer to Narazaki, Reissig [15], where some basic properties and the proof of Lemma 2.3 below can be found. We define a compactly supported radial test function $\chi \in C_0^\infty(\mathbb{R}^n)$ such that

$$\chi(\xi) = \rho(|\xi|) = \begin{cases} 0 & \text{if } |\xi| \geq r, \\ 1 & \text{if } |\xi| \leq \frac{r}{2}, \\ 0 \leq \rho(|\xi|) \leq 1 & \text{if } \frac{r}{2} \leq |\xi| \leq r, \end{cases} \tag{3}$$

where $\rho \in C_0^\infty(\mathbb{R})$ and $r \geq 2$ is a fixed number, which will be specified later. Moreover, subsequently we write for the partial Fourier transform with respect to the x-variable $\mathbb{F}$ instead of $\mathbb{F}_{x \to \xi}$ and $\mathbb{F}^{-1}$ instead of $\mathbb{F}^{-1}_{\xi \to x}$, respectively.

Lemma 2.3 *Let the dimension $n \geq 2$ and $\alpha \in (0, 1]$. Assume $\chi \in C_0^\infty(\mathbb{R}^n)$ to be defined by* (3). *Then it holds*

$$\big\|\mathbb{F}^{-1}\big(\chi(\xi) e^{it|\xi|^\alpha}\big)\big\|_{L_1} \lesssim (1 + |t|)^{\frac{n+1}{2}}. \tag{4}$$

Remark 2.4

(i) In order to prove Lemma 2.3 we can let χ be any smooth radial function with compact support. Definition of χ in (3) is given for technical reasons with regard to later computations and Lemma 2.5 below.
(ii) The proof of Lemma 2.3, see [15], provides the Fourier multiplier estimate (4) for all multipliers of the form $m_t(\xi) = \chi(\xi)e^{t\mu(|\xi|)}$ as long as $\mu \in C^{n+1}(\mathbb{R} \setminus \{0\})$ is a complex-valued homogeneous function of order $\alpha > 0$, i.e., $\mu(s|\xi|) = s^\alpha \mu(|\xi|)$ for all $s > 0$ and all $\xi \neq 0$, and $m_t = m_t(\xi)$ is bounded for all $t > 0$ and all $\xi \in \mathbb{R}^n$.
(iii) Inequality (4) immediately yields $\chi(\cdot)e^{it|\cdot|^\alpha} \in M_p$ for all $1 \leq p \leq \infty$.

Some fundamental work was done in [1]. Lemma 2.5 provides a special case of Theorem 4 in [1]. There the authors proved the same result for the multiplier $\chi e^{i\mu}$ by assuming that $\mu \in C^{n+1}(\mathbb{R}^n \setminus \{0\})$ is homogeneous of order $\alpha > 0$.

Lemma 2.5 *Let the dimension $n \geq 1$. Let $\chi \in C_0^\infty$ be defined by* (3) *and $\alpha \in (0, 1]$. Then it holds*

$$\left\|\mathbb{F}^{-1}\big(\chi(\xi)e^{it|\xi|^\alpha}\big)\right\|_{L_1} \lesssim e^{c|t|} \tag{5}$$

with some positive constant $c > 0$. Consequently, $\chi(\cdot)e^{it|\cdot|^\alpha} \in M_p$ for all $1 \leq p \leq \infty$.

Remark 2.6

(i) The constant c in (5) can be specified by $c = (2r)^\alpha$.
(ii) Lemma 2.3 excludes the one-dimensional case $n = 1$ in contrast to Lemma 2.5.

Let us state another Fourier multiplier result. We refer to [9, Theorem 1.11]. Here we use the notation meas$\{G\}$ for the Lebesgue measure of a given set $G \subset \mathbb{R}^n$.

Proposition 2.7 *Let f be a measurable function such that*

$$meas\big\{\xi \in \mathbb{R}^n : |f(\xi)| \geq l\big\} \leq Cl^{-b}$$

with some positive constants C, $b \in (1, \infty)$ and all $l > 0$. If additionally $1 < p \leq 2 \leq q < \infty$ and $\frac{1}{p} - \frac{1}{q} = \frac{1}{b}$, then $f \in M_p^q$.

Eventually, we state a Littman type lemma, which is a fundamental tool in order to get the desired time decay estimates in Sect. 4.

Proposition 2.8 *Let $f \in C_0^\infty(\mathbb{R}^n)$ be such that supp $f \subset \big\{\xi \in \mathbb{R}^n : \frac{1}{2} \leq |\xi| \leq 2\big\}$. Assume τ_0 to be a large positive number. Then for all $\tau \geq \tau_0$ it holds*

$$\left\|\mathbb{F}^{-1}\big(e^{-i\tau|\xi|}f(\xi)\big)\right\|_{L_\infty} \leq C(1+\tau)^{-\frac{n-1}{2}} \sum_{|\alpha|\leq s} \|D_\xi^\alpha f(\xi)\|_{L_\infty},$$

where $s > \frac{n+3}{2}$.

Proof A proof and some further references can be found in [4], see Theorem 16.3.1.

Furthermore, we provide another important tool, which is covered by Theorem 1.7 in [9].

Proposition 2.9 *Let* $1 \leq p \leq \infty$ *and* m_1, m_2 *be two multipliers belonging to* M_p. *Then* $m_1 \cdot m_2 \in M_p$.

3 A Better Localization Around the Origin

Homogeneous spaces use a better localization of functions around the origin. The motivation of introducing homogeneous function spaces is well-known, see, e.g., Triebel [29]. Here we adopt the concept of localizing the origin in the modulation space norm although we cannot have homogeneity for modulation spaces with respect to dilation, see [22, Theorem 1.1]. This contrasts with, e.g., homogeneous Besov spaces $\dot{B}^s_{p,q}(\mathbb{R}^n)$, which satisfy the estimate

$$\|f(\lambda\cdot)\|_{\dot{B}^s_{p,q}} \leq c\lambda^{s-\frac{n}{p}}\|f\|_{\dot{B}^s_{p,q}}$$

for all $\lambda > 0$ and all $f \in \dot{B}^s_{p,q}(\mathbb{R}^n)$, where c is some positive constant, $1 \leq p, q \leq \infty$ and $s \in \mathbb{R}$. The proof can be found in [29] and is based on basic Fourier analysis, where one takes advantage of the dyadic decomposition.

For modulation spaces $M^s_{p,q}(\mathbb{R}^n)$ the idea is to refine the decomposition around the origin, i.e. around $\operatorname{supp}\sigma_0$, based on dyadic cubes, which was introduced by Triebel [26, 27], see also Runst, Sickel [19]. The set

$$K_j := \{\xi \in \mathbb{R}^n : |\xi_i| \leq 2^j,\ i = 1, \cdots, n\} \setminus \{\xi \in \mathbb{R}^n : |\xi_i| < 2^{j-1}, i = 1, \cdots, n\},$$

$j = 0, -1, -2, \cdots$, is divided into the 3 n-hyperplanes

$$\{\xi \in \mathbb{R}^n : \xi_i = 0\}, \quad \{\xi \in \mathbb{R}^n : \xi_i = 2^{j-1}\}, \text{ and } \quad \{\xi \in \mathbb{R}^n : \xi_i = -2^{j-1}\}$$

for $i = 1, \cdots, n$. Thus, we obtain congruent cubes $P_{j,l}$ with $l = 1, 2, \cdots, 4^n - 2^n$ and $j = 0, -1, -2, \cdots$. Let $\xi_{j,l}$ be the center of $P_{j,l}$. Then there exist smooth and compactly supported functions η_j on $\mathbb{R}$ such that

$$\phi_{j,0}(\xi) := \eta_j(\xi_1)\eta_j(\xi_2)\cdots\eta_j(\xi_n),$$

$\xi = (\xi_1, \xi_2, \cdots, \xi_n)$, $j = 0, -1, -2, \cdots$, satisfies

$$\operatorname{supp}\phi_{j,0} \subset \{\xi \in \mathbb{R}^n : |\xi_i| \leq 2^{j-1},\ i = 1, \cdots, n\}.$$

Moreover, we put $\phi_{j,l} := \phi_{j,0}(\xi - \xi_{j,l})$. Consequently,

$$\operatorname{supp} \phi_{j,l} \subset \{\xi \in \mathbb{R}^n : |\xi_i - (\xi_{j,l})_i| \le 2^{j-1},\ i = 1, \cdots, n\}.$$

Hence, we defined a decomposition $\{\phi_{j,l}\}_{j,l}$ of $[-1, 1]^n$ into dyadic cubes and we call the operator

$$\Delta_{j,l} := \mathbb{F}^{-1}\big(\phi_{j,l}\mathbb{F}(\cdot)\big), \quad l = 1, \cdots, 4^n - 2^n, \quad j = 0, -1, -2, \cdots,$$

dyadic decomposition operator. However, before defining the modified modulation spaces we need to overcome some technical difficulties in order to preserve the basic functional analytic structure. Let

$$Z(\mathbb{R}^n) = \{\phi \in \mathscr{S}(\mathbb{R}^n) : D^\alpha \mathbb{F}\phi(0) = 0 \text{ for all } \alpha \in \mathbb{N}^n\}$$

be a closed subspace of the Schwartz space $\mathscr{S}(\mathbb{R}^n)$. Hence, $Z(\mathbb{R}^n)$ is a Fréchet space. Let us consider the restriction of a bounded linear functional u on $\mathscr{S}(\mathbb{R}^n)$, i.e. $u \in \mathscr{S}'(\mathbb{R}^n)$, on $Z(\mathbb{R}^n)$. It holds

$$(u + p)(\phi) = u(\phi)$$

for all $\phi \in Z(\mathbb{R}^n)$, where p is some polynomial. Let $Z'(\mathbb{R}^n)$ denote the topological dual of $Z(\mathbb{R}^n)$. Conversely, a distribution $u \in Z'(\mathbb{R}^n)$ can be linearly and continuously extended from $Z(\mathbb{R}^n)$ to $\mathscr{S}(\mathbb{R}^n)$. We see that two different extensions u_1 and u_2 only differ by a polynomial, i.e., $\operatorname{supp} \mathbb{F}(u_1 - u_2) = \{0\}$. Hence, the space $Z'(\mathbb{R}^n)$ may be identified with the quotient space $\mathscr{S}'(\mathbb{R}^n)/\mathscr{P}$, where $\mathscr{P}$ is the collection of all polynomials p.

Definition 3.1 *Let $1 \le p, q \le \infty$ and $s \in \mathbb{R}$. Assume that $\{\sigma_k\}_{k\in\mathbb{Z}^n}$ is a uniform decomposition of unity (defined above). Moreover, let $\{\phi_{j,l}\}_{j,l}$, $l = 1, \cdots, 4^n - 2^n$ and $j = 0, -1, -2, \cdots$, be the decomposition of $[-1, 1]^n$ into dyadic cubes. Then the localized modulation space $K^s_{p,q}(\mathbb{R}^n)$ is the collection of all $f \in Z'(\mathbb{R}^n)$ such that*

$$\|f\|_{K^s_{p,q}} := \Big(\sum_{k\in\mathbb{Z}^n\setminus\{0\}} |k|^{sq} \|\Box_k f\|^q_{L_p} + \sum_{j=-\infty}^{0} \sum_{l=1}^{4^n-2^n} 2^{sjq} \|\Delta_{j,l} f\|^q_{L_p}\Big)^{\frac{1}{q}} < \infty$$

with obvious modification for $q = \infty$.

We find the following connections between the spaces $K^s_{p,q}(\mathbb{R}^n)$ and the classical modulation spaces $M^s_{p,q}(\mathbb{R}^n)$.

Proposition 3.2 *Let $1 \le p, q \le \infty$ and $s \in \mathbb{R}$. Then it holds*

$$L_p(\mathbb{R}^n) \cap K^s_{p,q}(\mathbb{R}^n) \hookrightarrow M^s_{p,q}(\mathbb{R}^n) \text{ for } p \in [1, \infty).$$

If $s > 0$, then

$$M^s_{p,q}(\mathbb{R}^n) \hookrightarrow K^s_{p,q}(\mathbb{R}^n).$$

Here we identify $g \in M^s_{p,q}(\mathbb{R}^n)$ with its equivalence class $[g]$ in $K^s_{p,q}(\mathbb{R}^n)$.
If $p = \infty$, then we have the a-priori estimate

$$\|g\|_{M^s_{\infty,q}} \leq C\big(\|[f]\|_{K^s_{\infty,q}} + \|g\|_{L^\infty}\big) \text{ for all } g \in L^\infty \cap [f].$$

Proof A basic Fourier multiplier argument immediately yields

$$\|f\|_{M^s_{p,q}} \lesssim \Big(\sum_{k\in\mathbb{Z}^n\setminus\{0\}} |k|^{sq}\|\Box_k f\|^q_{L_p} + \|\Box_0 f\|^q_{L_p}\Big)^{\frac{1}{q}} \lesssim \|f\|_{K^s_{p,q}} + \|f\|_{L_p}.$$

Moreover, for $j < 0$ it holds $\|\Delta_{j,l} f\|_{L_p} \lesssim \|\Box_0 f\|_{L_p}$ and due to $s > 0$ we, consequently, obtain

$$\begin{aligned}\|f\|_{K^s_{p,q}} &\lesssim \Big(\sum_{k\in\mathbb{Z}^n\setminus\{0\}} |k|^{sq}\|\Box_k f\|^q_{L_p} + \|\Box_0 f\|^q_{L_p} \sum_{j=-\infty}^{0}\sum_{l=1}^{4^n-2^n} 2^{sjq}\Big)^{\frac{1}{q}}\\ &\lesssim \Big(\sum_{k\in\mathbb{Z}^n} \langle k\rangle^{sq}\|\Box_k f\|^q_{L_p}\Big)^{\frac{1}{q}},\end{aligned}$$

which proves the desired result.

Let us define the operator

$$I_\mu f(x) = \mathbb{F}^{-1}\big(|\xi|^\mu \mathbb{F} f(\xi)\big)(x), \quad f \in Z'(\mathbb{R}^n)$$

for any real number μ. Then I_μ maps $Z'(\mathbb{R}^n)$ onto itself. Applying the theory of maximal functions and following the concepts explained by Triebel [28, 29] gives the following result.

Proposition 3.3 *Let $1 \leq p, q \leq \infty$, $s \in \mathbb{R}$ and $\mu \in \mathbb{R}$. Then I_μ maps $K^s_{p,q}(\mathbb{R}^n)$ isomorphically onto $K^{s-\mu}_{p,q}(\mathbb{R}^n)$. Moreover, $\|I_\mu f\|_{K^{s-\mu}_{p,q}}$ is an equivalent norm on $K^s_{p,q}(\mathbb{R}^n)$.*

4 Estimates for the Free Wave Equation

Let us consider the Cauchy problem for the free wave equation

$$u_{tt} - \Delta u = 0, \quad u(0,x) = \varphi(x), \quad u_t(0,x) = \psi(x), \tag{6}$$

with initial data φ, ψ belonging to the test function space $C_0^\infty(\mathbb{R}^n)$. Without loss of generality we only consider positive times t in the following. The solution u to Cauchy problem (6) can be represented by

$$\begin{aligned} &u(t,x) \\ &= \frac{1}{2}\mathbb{F}^{-1}\Big(\big(e^{i|\xi|t}+e^{-i|\xi|t}\big)\hat{\varphi}(\xi)\Big)(t,x)+\mathbb{F}^{-1}\Big(\big(e^{i|\xi|t}-e^{-i|\xi|t}\big)\frac{1}{2i|\xi|}\hat{\psi}(\xi)\Big)(t,x), \end{aligned}$$

where $(t,x) \in \mathbb{R}_+ \times \mathbb{R}^n$. Taking partial derivatives with respect to the time t we obtain

$$\begin{aligned} \partial_t^\ell u(t,x) = &\frac{i^\ell}{2}\mathbb{F}^{-1}\Big(\big(|\xi|^\ell e^{i|\xi|t}+(-1)^\ell|\xi|^\ell e^{-i|\xi|t}\big)\hat{\varphi}(\xi)\Big)(t,x) \\ &+\frac{i^{\ell-1}}{2}\mathbb{F}^{-1}\Big(\big(|\xi|^{\ell-1} e^{i|\xi|t}-(-1)^\ell|\xi|^{\ell-1} e^{-i|\xi|t}\big)\hat{\psi}(\xi)\Big)(t,x) \end{aligned} \tag{7}$$

for $\ell \in \mathbb{N}$. Moreover, we know that the weight parameter $s \in \mathbb{R}$ corresponds to the regularity of an element in the modulation space $M^s_{p,q}(\mathbb{R}^n)$, or respectively $K^s_{p,q}(\mathbb{R}^n)$. Due to Toft [25, Theorem 3.9'] and Proposition 3.3 we find the relation

$$\|\partial_x^\beta f\|_{M^s_{p,q}} \lesssim \|f\|_{M^{s+|\beta|}_{p,q}} \quad \text{and} \quad \|\partial_x^\beta f\|_{K^s_{p,q}} \lesssim \|f\|_{K^{s+|\beta|}_{p,q}}$$

for all $\beta \in \mathbb{N}^n$ and $f \in M^{s+|\beta|}_{p,q}(\mathbb{R}^n)$ or $f \in K^{s+|\beta|}_{p,q}(\mathbb{R}^n)$. Thus, in the subsequent computations we do not consider derivatives with respect to physical space variable x. Instead we only assume for the weight parameter $s \geq 0$ since this includes arbitrary derivatives of the function $u = u(t,x)$ with respect to x.

The first step consists in providing well-known estimates of the term $\partial_t^\ell u(t,\cdot)$ in the classical modulation space $M^s_{p,q}(\mathbb{R}^n)$, where we use standard Fourier multiplier results.

Theorem 4.1 *Let $1 \leq p,q \leq \infty$ and $s \geq 0$. Assume $n \geq 2$ and $u = u(t,x)$ to be the solution to Cauchy problem* (6). *It holds*

$$\|u(t,\cdot)\|_{M^s_{p,q}} \lesssim (1+t)^{\frac{n+1}{2}}\big(\|\varphi\|_{M^s_{p,q}} + t\|\psi\|_{M^{s-1}_{p,q}}\big)$$

and

$$\big\|\partial_t^\ell u(t,\cdot)\big\|_{M^s_{p,q}} \lesssim (1+t)^{\frac{n+1}{2}}\big(\|\varphi\|_{M^{s+\ell}_{p,q}} + \|\psi\|_{M^{s+\ell-1}_{p,q}}\big),$$

where $\ell \geq 1$. Moreover, if $n = 1$, then there exists a constant $c > 0$ such that

$$\big\|\partial_t^\ell u(t,\cdot)\big\|_{M^s_{p,q}} \lesssim e^{ct}\big(\|\varphi\|_{M^{s+\ell}_{p,q}} + \|\psi\|_{M^{s+\ell-1}_{p,q}}\big)$$

for all $\ell \in \mathbb{N}$.

Proof Due to representation (7) we see that we need to distinguish the cases $\ell = 0$ and $\ell \geq 1$. We proceed considering the case $\ell = 0$. Observe that

$$\|\Box_k u(t,\cdot)\|_{L_p} \leq \Big\|\mathbb{F}^{-1}\Big(\sigma_k(\xi)\big(e^{i|\xi|t} + e^{-i|\xi|t}\big)\hat{\varphi}(\xi)\Big)\Big\|_{L_p} + \Big\|\mathbb{F}^{-1}\Big(\sigma_k(\xi)\frac{\sin(|\xi|t)}{|\xi|}\hat{\psi}(\xi)\Big)\Big\|_{L_p} =: P_1 + P_2.$$

Obviously, the multipliers $\frac{\sin(|\xi|t)}{|\xi|}$ and $e^{\pm i|\xi|t}$ are not differentiable at the origin for fixed times $t > 0$. Therefore we investigate the terms P_1 and P_2 separately for small and large frequencies. Hence, we distinguish the cases $\|k\|_\infty \leq 2$ and $\|k\|_\infty \geq 3$. Moreover, we define χ by (3), where we put $r = 8\sqrt{n}$. First assume $n \geq 2$ for the dimension n. By using Young's inequality and Lemma 2.3 it follows

$$\begin{aligned} P_2 &\leq \sum_{\|l\|_\infty \leq 1} \Big\|\mathbb{F}^{-1}\Big(\sigma_{k+l}(\xi)\frac{\sin(|\xi|t)}{|\xi|}\sigma_k(\xi)\hat{\psi}(\xi)\Big)\Big\|_{L_p} \\ &\leq t\Big\|\mathbb{F}^{-1}\Big(\chi(\xi)\frac{\sin(|\xi|t)}{|\xi|t}\Big)\Big\|_{L_1}\|\Box_k\psi\|_{L_p} \\ &\lesssim t(1+t)^{\frac{n+1}{2}}\|\Box_k\psi\|_{L_p} \end{aligned}$$

for all $k \in \mathbb{Z}^n$ such that $\|k\|_\infty \leq 2$. Here we applied the ideas of the proofs of the results from Section 2 in [15] to verify the estimate

$$\Big\|\mathbb{F}^{-1}\Big(\chi(\xi)\frac{\sin(|\xi|t)}{|\xi|}\Big)\Big\|_{L_1} \lesssim t(1+t)^{\frac{n+1}{2}}.$$

For the same range of k we analogously obtain

$$P_1 \lesssim (1+t)^{\frac{n+1}{2}}\|\Box_k\varphi\|_{L_p}.$$

If the dimension $n = 1$, then Lemma 2.5 yields

$$P_1 \lesssim te^{ct}\|\Box_k\psi\|_{L_p} \qquad \text{and} \qquad P_2 \lesssim e^{ct}\|\Box_k\varphi\|_{L_p}$$

for some constant $c > 0$. Now we turn to all $k \in \mathbb{Z}^n$ such that $\|k\|_\infty \geq 3$. By Young's inequality and Proposition 2.2 it holds

$$\begin{aligned} &P_1 \lesssim (1+t)^{\frac{n}{2}}\|\Box_k\varphi\|_{L_p} \quad \text{and} \\ &P_2 \lesssim \sum_{\|\ell\|_\infty \leq 1} \Big\|\mathbb{F}^{-1}\Big(\sigma_{k+\ell}(\xi)\frac{e^{i|\xi|t}}{|\xi|}\sigma_k(\xi)\hat{\psi}(\xi)\Big)\Big\|_{L_p} \lesssim (1+t)^{\frac{n}{2}}\langle k\rangle^{-1}\|\Box_k\psi\|_{L_p}, \end{aligned}$$

which completes the proof for $\ell = 0$. By the same arguments we obtain the desired results for $\ell \geq 1$. We shall only add the comment that the multiplier $\chi(\cdot)|\cdot|^{\ell} \in M_p$ for all $1 \leq p \leq \infty$, which follows immediately, e.g., due to the proof of Lemma 2.5, see [1]. Thus, Lemma 2.5 together with Proposition 2.9 yields $\chi(\cdot)|\cdot|^{\ell} e^{\pm i|\cdot|t} \in M_p$, $1 \leq p \leq \infty$. Thus, the proof is complete.

In the estimates of Theorem 4.1 we observe that the solution does not loose regularity with respect to the initial data. However, the right-hand sides are increasing in time. A well-known concept in order to propose modified estimates to the estimates of Theorem 4.1 regarding the time-dependence of the right-hand sides is working on the conjugate line with respect to the L_p-norm. This gives us the opportunity to obtain an estimate of the solution $u = u(t, \cdot)$ to (6), where the right-hand side is independent of the time t. However, we have to accept a loss of regularity.

Theorem 4.2 *Suppose $s \geq 0$, $n \geq 2$ and let $u = u(t, x)$ be the solution to Cauchy problem* (6). *Let $1 \leq q \leq \infty$ and $1 < p \leq 2$ be such that $n(\frac{2}{p} - 1) = 1$. Then it holds*

$$\big\|\partial_t^{\ell} u(t, \cdot)\big\|_{M^s_{p',q}} \lesssim \|\varphi\|_{M^{s+1+\ell}_{p,q}} + \|\psi\|_{M^{s+\ell}_{p,q}}$$

for all $\ell \in \mathbb{N}$. Here p' denotes the conjugate exponent to p.

Proof First note that $\big|\frac{e^{i|\xi|t}}{|\xi|}\big| \geq l$ implies $|\xi| \leq \frac{1}{l}$ and therefore Proposition 2.7 yields

$$\mathbb{F}^{-1}\Big(\frac{e^{i|\xi|t}}{|\xi|}\Big) \in L_p^{p'}$$

if $1 < p \leq 2$ and $n\big(\frac{2}{p} - 1\big) = 1$. Then we find the estimates

$$\Big\|\mathbb{F}^{-1}\Big(\frac{e^{i|\xi|t}}{|\xi|}\Big) * \mathbb{F}^{-1}\big(\sigma_k(\xi)|\xi|\hat{\varphi}(\xi)\big)\Big\|_{L_{p'}} \lesssim \big\|\mathbb{F}^{-1}\big(\sigma_k(\xi)|\xi|\hat{\varphi}(\xi)\big)\big\|_{L_p} \lesssim \langle k\rangle \|\Box_k \varphi\|_{L_p}$$

and

$$\Big\|\mathbb{F}^{-1}\Big(\frac{e^{i|\xi|t}}{|\xi|}\Big) * \mathbb{F}^{-1}\big(\sigma_k(\xi)\hat{\psi}(\xi)\big)\Big\|_{L_{p'}} \lesssim \|\Box_k \psi\|_{L_p}.$$

Analogously, we proceed in the case $\ell \geq 1$, which completes the proof.

Moreover, it is desirable not only to get an estimate, where the right-hand side is independent of the time t, see Theorem 4.2, but also to obtain an estimate of the solution u to the free wave equation, where the right-hand side is even decaying in time. The subsequent classical method gives us some interplay between the decay rate with respect to the time t and the loss of regularity of the solution u. Thus, we introduce the so-called method of stationary phase, where we take advantage of working on the conjugate line with respect to the L_p-norm and a better localization around the origin. For our purposes this can be done by applying the localized modulation space $K^s_{p,q}(\mathbb{R}^n)$ introduced in Definition 3.1. We basically adopt the theory explained in [4, Chapter 16]. We define two zones in the extended phase space $(0,\infty)\times\mathbb{R}^n$. The pseudo-differential zone is given by

$$Z_{pd} := \big\{(t,\xi)\in(0,\infty)\times\mathbb{R}^n : t|\xi|\le 1\big\}$$

and the hyperbolic zone is defined by

$$Z_{hyp} := \big\{(t,\xi)\in(0,\infty)\times\mathbb{R}^n : t|\xi|\ge 1\big\}.$$

Moreover, we define an auxiliary function $\chi\in C^\infty(\mathbb{R}^n)$ satisfying $\chi(\xi)\equiv 0$ for $|\xi|\le\frac{1}{2}$, $\chi(\xi)\equiv 1$ for $|\xi|\ge\frac{3}{4}$, and $\chi(\xi)\in[0,1]$. With the help of this function we can treat the Fourier transformed solution to the free wave equation (6) separately in the pseudo-differential zone and in the hyperbolic zone. Thus, we rewrite $\partial_t^\ell u$ explained by (7) as follows:

$$\begin{aligned}\partial_t^\ell u(t,x) = {}&\frac{i^\ell}{2}\mathbb{F}^{-1}\Big(\big(e^{i|\xi|t}+(-1)^\ell e^{-i|\xi|t}\big)\frac{1-\chi(t\xi)}{|\xi|^{2r}}\mathbb{F}\big(|D|^{2r+\ell}\varphi\big)(\xi)\Big)(t,x)\\&+\frac{i^\ell}{2}\mathbb{F}^{-1}\Big(\big(e^{i|\xi|t}+(-1)^\ell e^{-i|\xi|t}\big)\frac{\chi(t\xi)}{|\xi|^{2r}}\mathbb{F}\big(|D|^{2r+\ell}\varphi\big)(\xi)\Big)(t,x)\\&+\frac{i^{\ell-1}}{2}\mathbb{F}^{-1}\Big(\big(e^{i|\xi|t}-(-1)^\ell e^{-i|\xi|t}\big)\frac{1-\chi(t\xi)}{|\xi|^{2r}}\mathbb{F}\big(|D|^{2r+\ell-1}\psi\big)(\xi)\Big)(t,x)\\&+\frac{i^{\ell-1}}{2}\mathbb{F}^{-1}\Big(\big(e^{i|\xi|t}-(-1)^\ell e^{-i|\xi|t}\big)\frac{\chi(t\xi)}{|\xi|^{2r}}\mathbb{F}\big(|D|^{2r+\ell-1}\psi\big)(\xi)\Big)(t,x),\end{aligned}$$

where we also used basic properties of the Fourier transform. Let us start with the localization in the pseudo-differential zone, i.e., we estimate the quantity

$$\Big\|\mathbb{F}^{-1}\Big(e^{-i|\xi|t}\frac{1-\chi(t\xi)}{|\xi|^{2r}}\mathbb{F}\big(|D|^{2r+\ell}\varphi\big)(\xi)\Big)\Big\|_{K^s_{p,q}}.$$

First note that $\big|e^{-i|\eta|}\frac{1-\chi(|\eta|)}{|\eta|^{2r}}\big|\ge l$ implies $|\eta|\le l^{-\frac{1}{2r}}$ and, therefore, Proposition 2.7 yields

$$\mathbb{F}^{-1}\Big(e^{-i|\eta|}\frac{1-\chi(|\eta|)}{|\eta|^{2r}}\Big)\in L^m_p$$

if $1 < p \le 2 \le m < \infty$ and $2r \le n(\frac{1}{p} - \frac{1}{m})$. Hence, taking Definition 3.1 into account, it follows

$$\begin{aligned}
&\Big\| \mathbb{F}^{-1}\Big(\sigma_k(\xi)e^{-i|\xi|t}\frac{1-\chi(t\xi)}{|\xi|^{2r}}\mathbb{F}\big(|D|^{2r+\ell}\varphi\big)(\xi)\Big)\Big\|_{L_m} \\
&\quad = t^{2r-n+\frac{n}{m}}\Big\| \mathbb{F}^{-1}\Big(e^{-i|\eta|}\frac{1-\chi(|\eta|)}{|\eta|^{2r}}\Big) * \mathbb{F}^{-1}\Big(\sigma_k\Big(\frac{\eta}{t}\Big)\mathbb{F}\big(|D|^{2r+\ell}\varphi\big)\Big(\frac{\eta}{t}\Big)\Big)\Big\|_{L_m} \\
&\quad \lesssim t^{2r-n+\frac{n}{m}}\Big\| \mathbb{F}^{-1}\Big(\sigma_k\Big(\frac{\eta}{t}\Big)\mathbb{F}\big(|D|^{2r+\ell}\varphi\big)\Big(\frac{\eta}{t}\Big)\Big)\Big\|_{L_p} \\
&\quad = t^{2r-n+\frac{n}{m}+n-\frac{n}{p}}\big\|\Box_k\big(|D|^{2r+\ell}\varphi\big)\big\|_{L_p} \\
&\quad = t^{2r-n(\frac{1}{p}-\frac{1}{m})}\big\|\Box_k\big(|D|^{2r+\ell}\varphi\big)\big\|_{L_p}
\end{aligned} \tag{8}$$

for $k \in \mathbb{Z}^n \setminus \{0\}$, where we also used the change of variables $\eta = t\xi$ and $tz = x$ together with some basic properties of the Fourier transform. In the same way we obtain

$$\begin{aligned}
&\Big\| \mathbb{F}^{-1}\Big(\phi_{j,l}(\xi)e^{-i|\xi|t}\frac{1-\chi(t\xi)}{|\xi|^{2r}}\mathbb{F}\big(|D|^{2r+\ell}\varphi\big)(\xi)\Big)\Big\|_{L_m} \\
&\quad \lesssim t^{2r-n(\frac{1}{p}-\frac{1}{m})}\big\|\Delta_{j,l}\big(|D|^{2r+\ell}\varphi\big)\big\|_{L_p}
\end{aligned} \tag{9}$$

with $l = 1, 2, \cdots, 4^n - 2^n$ and $j = 0, -1, -2, \cdots$. We shall remark that the left-hand side of (8) vanishes for large times t due to the support properties of the window function σ_k. The previous estimates also hold if we replace the term $e^{-i|\xi|t}$ by $e^{i|\xi|t}$.

Now let us treat the localization in the hyperbolic zone. We define a test function $\rho \in C_0^\infty(\mathbb{R}^n)$ such that $\operatorname{supp}\rho \subset \{\xi \in \mathbb{R}^n : \frac{1}{2} \le |\xi| \le 2\}$. Moreover, we put $\rho_\kappa(\xi) := \rho(2^{-\kappa}\xi)$ for $\kappa \ge 1$ and $\rho_0(\xi) := 1 - \sum_{\kappa=1}^\infty \rho_\kappa(\xi)$. By Young's inequality we obtain

$$\begin{aligned}
&\Big\| \mathbb{F}^{-1}\Big(e^{-i|\xi|t}\frac{\chi(t|\xi|)\rho_\kappa(t|\xi|)}{|\xi|^{2r}}\sigma_k(\xi)\mathbb{F}\big(|D|^{2r+\ell}\varphi\big)(\xi)\Big)\Big\|_{L_\infty} \\
&\quad \le \Big\| \mathbb{F}^{-1}\Big(e^{-i|\xi|t}\frac{\chi(t|\xi|)\rho_\kappa(t|\xi|)}{|\xi|^{2r}}\Big)\Big\|_{L_\infty}\big\|\Box_k\big(|D|^{2r+\ell}\varphi\big)\big\|_{L_1}.
\end{aligned}$$

Let us fix some sufficiently large number $\kappa_0 \in \mathbb{N}$. Then there exists a constant $C = C(\kappa_0)$ such that for all $\kappa \leq \kappa_0$ it follows

$$\begin{aligned}
&\Big\|\mathbb{F}^{-1}\Big(e^{-i|\xi|t}\frac{\chi(t|\xi|)\rho_\kappa(t|\xi|)}{|\xi|^{2r}}\Big)\Big\|_{L_\infty} \\
&\quad = t^{2r}\Big\|\mathbb{F}^{-1}\Big(e^{-i|\xi|t}\frac{\chi(t|\xi|)\rho_\kappa(t|\xi|)}{(t|\xi|)^{2r}}\Big)\Big\|_{L_\infty} \leq Ct^{2r-n}
\end{aligned}$$

due to support properties. For the case $\kappa \geq \kappa_0$ we introduce the change of variables $t\xi = 2^\kappa \eta$ and obtain

$$\begin{aligned}
&\Big\|\mathbb{F}^{-1}\Big(e^{-i|\xi|t}\frac{\chi(t|\xi|)\rho_\kappa(t|\xi|)}{|\xi|^{2r}}\Big)\Big\|_{L_\infty} \\
&\quad = 2^{\kappa n-2\kappa r}t^{2r-n}\Big\|\mathbb{F}^{-1}\Big(e^{-i2^\kappa|\eta|}\frac{\chi(2^\kappa|\eta|)\rho_\kappa(2^\kappa|\eta|)}{|\eta|^{2r}}\Big)\Big\|_{L_\infty} \\
&\quad = 2^{\kappa(n-2r)}t^{2r-n}\Big\|\mathbb{F}^{-1}\Big(e^{-i2^\kappa|\eta|}\frac{\rho(|\eta|)}{|\eta|^{2r}}\Big)\Big\|_{L_\infty} \\
&\quad \lesssim (1+2^\kappa)^{-\frac{n-1}{2}}2^{\kappa(n-2r)}t^{2r-n} \\
&\quad \lesssim 2^{\kappa(\frac{n}{2}+\frac{1}{2}-2r)}t^{2r-n},
\end{aligned}$$

where we also used the Littman type lemma with $\tau := 2^\kappa$, see Proposition 2.8. Now we want to deduce an $L_2 - L_2$-estimate. By Plancherel's identity and Hölder's inequality it follows

$$\begin{aligned}
&\Big\|\mathbb{F}^{-1}\Big(e^{-i|\xi|t}\frac{\chi(t|\xi|)\rho_\kappa(t|\xi|)}{|\xi|^{2r}}\sigma_k(\xi)\mathbb{F}\big(|D|^{2r+\ell}\varphi\big)(\xi)\Big)\Big\|_{L_2} \\
&\quad \lesssim \Big\|e^{-i|\xi|t}\frac{\chi(t|\xi|)\rho_\kappa(t|\xi|)}{|\xi|^{2r}}\Big\|_{L_\infty}\big\|\Box_k\big(|D|^{2r+\ell}\varphi\big)\big\|_{L_2}.
\end{aligned}$$

Taking account of support properties we obtain

$$\Big\|e^{-i|\xi|t}\frac{\chi(t|\xi|)\rho_\kappa(t|\xi|)}{|\xi|^{2r}}\Big\|_{L_\infty} = t^{2r}\Big\|e^{-i|\xi|t}\frac{\chi(t|\xi|)\rho_\kappa(t|\xi|)}{(t|\xi|)^{2r}}\Big\|_{L_\infty} \lesssim 2^{-2\kappa r}t^{2r}.$$

Consequently, the well-known Riesz-Thorin interpolation theorem yields the estimate

$$\begin{aligned}
&\Big\|\mathbb{F}^{-1}\Big(e^{-i|\xi|t}\frac{\chi(t|\xi|)\rho_\kappa(t|\xi|)}{|\xi|^{2r}}\sigma_k(\xi)\mathbb{F}\big(|D|^{2r+\ell}\varphi\big)(\xi)\Big)\Big\|_{L_{p'}} \\
&\quad \lesssim 2^{\kappa((\frac{n}{2}+\frac{1}{2})(\frac{2}{p}-1)-2r)}t^{2r-n(\frac{2}{p}-1)}\big\|\Box_k\big(|D|^{2r+\ell}\varphi\big)\big\|_{L_p}
\end{aligned} \tag{10}$$

for $k \in \mathbb{Z}^n \setminus \{0\}$ and $1 < p \leq 2$. Here p' denotes the conjugate exponent to p. The same computations yield

$$\begin{aligned} &\Big\|\mathbb{F}^{-1}\Big(e^{-i|\xi|t}\frac{\chi(t|\xi|)\rho_\kappa(t|\xi|)}{|\xi|^{2r}}\phi_{j,l}(\xi)\mathbb{F}\big(|D|^{2r+\ell}\varphi\big)(\xi)\Big)\Big\|_{L_{p'}} \\ &\qquad\lesssim 2^{\kappa((\frac{n}{2}+\frac{1}{2})(\frac{2}{p}-1)-2r)}t^{2r-n(\frac{2}{p}-1)}\big\|\Delta_{j,l}\big(|D|^{2r+\ell}\varphi\big)\big\|_{L_p} \end{aligned} \tag{11}$$

with $l = 1, 2, \cdots, 4^n - 2^n$ and $j = 0, -1, -2, \cdots$. We shall note that the left-hand side of (11) vanishes for small times t due to the support properties of the window function $\phi_{j,l}$. Remark that the estimates (10) and (11) also hold if we replace the term $e^{-i|\xi|t}$ by $e^{i|\xi|t}$.

After proving the Fourier multiplier estimates on the conjugate line in the pseudo-differential and the hyperbolic zone we are able to estimate the term $\partial_t^\ell u$, where u is the solution to the free wave equation (6). First we consider small times $t \in (0, 1]$. Combining (8) to (11) and putting $2r := n(\frac{2}{p} - 1)$ yields

$$\big\|\mathbb{F}^{-1}\big(|\xi|^\ell e^{-i|\xi|t}\sigma_k(\xi)\mathbb{F}(\varphi)(\xi)\big)\big\|_{L_{p'}} \lesssim \big\|\square_k\big(|D|^{n(\frac{2}{p}-1)+\ell}\varphi\big)\big\|_{L_p}$$

and

$$\big\|\mathbb{F}^{-1}\big(|\xi|^\ell e^{-i|\xi|t}\phi_{j,l}(\xi)\mathbb{F}(\varphi)(\xi)\big)\big\|_{L_{p'}} \lesssim \big\|\Delta_{j,l}\big(|D|^{n(\frac{2}{p}-1)+\ell}\varphi\big)\big\|_{L_p}$$

for all $1 < p \leq 2$. For times $t \geq 1$ we choose $2r := \frac{n+1}{2}(\frac{2}{p} - 1)$ in (8) to (11). This gives

$$\begin{aligned} &\big\|\mathbb{F}^{-1}\big(|\xi|^\ell e^{-i|\xi|t}\sigma_k(\xi)\mathbb{F}(\varphi)(\xi)\big)\big\|_{L_{p'}} \\ &\qquad\lesssim (1+t)^{-\frac{n-1}{2}(\frac{2}{p}-1)}\big\|\square_k\big(|D|^{\frac{n+1}{2}(\frac{2}{p}-1)+\ell}\varphi\big)\big\|_{L_p} \end{aligned}$$

and

$$\begin{aligned} &\big\|\mathbb{F}^{-1}\big(|\xi|^\ell e^{-i|\xi|t}\phi_{j,l}(\xi)\mathbb{F}(\varphi)(\xi)\big)\big\|_{L_{p'}} \\ &\qquad\lesssim (1+t)^{-\frac{n-1}{2}(\frac{2}{p}-1)}\big\|\Delta_{j,l}\big(|D|^{\frac{n+1}{2}(\frac{2}{p}-1)+\ell}\varphi\big)\big\|_{L_p} \end{aligned}$$

for all $1 < p \leq 2$. In the same way we can show

$$\begin{aligned} &\big\|\mathbb{F}^{-1}\big(|\xi|^{\ell-1} e^{-i|\xi|t}\sigma_k(\xi)\mathbb{F}(\psi)(\xi)\big)\big\|_{L_{p'}} \\ &\qquad\lesssim (1+t)^{-\frac{n-1}{2}(\frac{2}{p}-1)}\big\|\square_k\big(|D|^{n(\frac{2}{p}-1)+\ell-1}\psi\big)\big\|_{L_p} \end{aligned}$$

as well as

$$\big\|\mathbb{F}^{-1}\big(|\xi|^{\ell-1}e^{-i|\xi|t}\phi_{j,l}(\xi)\mathbb{F}(\psi)(\xi)\big)\big\|_{L_{p'}}$$
$$\lesssim (1+t)^{-\frac{n-1}{2}(\frac{2}{p}-1)}\big\|\Delta_{j,l}\big(|D|^{n(\frac{2}{p}-1)+\ell-1}\psi\big)\big\|_{L_p}$$

for all times $t > 0$ and for all $1 < p \leq 2$. Here p' denotes the conjugate exponent to p. Moreover, the term $e^{-i|\xi|t}$ can be replaced by $e^{i|\xi|t}$. Summarizing the previous estimates and taking account of Proposition 3.3 we have proved the following theorem.

Theorem 4.3 *Let* $1 \leq q \leq \infty$ *and* $1 < p \leq 2$. *Assume* $\varphi, \psi \in C_0^\infty(\mathbb{R}^n)$ *and let* $u = u(t,x)$ *be the solution to Cauchy problem* (6). *Then it holds*

$$\big\|\partial_t^\ell u(t,\cdot)\big\|_{K^s_{p',q}} \lesssim (1+t)^{-\frac{n-1}{2}(\frac{2}{p}-1)}\big(\|\varphi\|_{K^{s+M_p+\ell}_{p,q}} + \|\psi\|_{K^{s+M_p+\ell-1}_{p,q}}\big),$$

where $M_p \geq n(\frac{2}{p}-1) \geq 1$, $s \geq 0$ *and* $\ell \in \mathbb{N}$. *Here* p' *denotes the conjugate exponent to* p.

Remark 4.4 Let us give some remarks to the statements of Theorems 4.1, 4.2, and 4.3. In Theorem 4.1 we give an a-priori estimate for solutions of the Cauchy problem for the free wave equation in scales of modulation spaces, where in these estimates we do not have any loss of regularity. Here the time-dependent coefficient is not essential because one would use such an estimate only for getting well-posedness results. The estimatcs from Theorems 4.2 and 4.3 are on the conjugate line. Here we have on the one hand a loss of regularity and on the other hand decaying time dependent coefficients. The application of stationary phase method provides an almost optimal estimate. Optimality should be verified by showing that exactly this loss of regularity together with the decay, and not a better decay, appears. But this is another story.

Theorem 4.3 points out that localized modulation spaces $K^s_{p,q}(\mathbb{R}^n)$ provide new estimates of the solution u to the free wave equation and its derivatives $\partial_t u$ concerning the decay rate of the time t. Furthermore, we immediately notice the connection between the decay rate and the loss of regularity, which is described by the parameter M_p. So the faster the decay in time the higher the loss of regularity, which is a rather natural observation. Nevertheless the spaces $K^s_{p,q}(\mathbb{R}^n)$ can only serve as auxiliary spaces since our main goal is to obtain estimates in classical modulation spaces $M^s_{p,q}(\mathbb{R}^n)$. Thus, we use the connection between these two spaces, see Proposition 3.2, together with some embedding results. But first let us recall an estimate for the solution to the Cauchy problem (6) in Lebesgue spaces. We refer to Theorem 16.6.4 and Theorem 16.6.6 in [4]. Here we use the Sobolev spaces $H^s_p = H^s_p(\mathbb{R}^n)$ of fractional order for $p \in (1,\infty)$ and $s \in \mathbb{R}$. They are defined as follows:

$$H^s_p(\mathbb{R}^n) := \big\{f \in S'(\mathbb{R}^n) : \|f\|_{H^s_p} = \big\|F^{-1}\big(\langle\xi\rangle^s F(f)\big)\big\|_{L^p} < \infty\big\}.$$

Proposition 4.5 *Let $1 < p \leq 2$. Assume $\varphi, \psi \in C_0^\infty(\mathbb{R}^n)$ and let $u = u(t,x)$ be the solution to Cauchy problem* (6). *It holds*

$$\big\|\partial_t^\ell u(t,\cdot)\big\|_{L_{p'}} \lesssim (1+t)^{-\frac{n-1}{2}(\frac{2}{p}-1)}\big(\|\varphi\|_{H_p^{M_p+\ell}} + \|\psi\|_{H_p^{M_p+\ell-1}}\big),$$

where $M_p \geq n(\frac{2}{p}-1) \geq 1$. Here p' denotes the conjugate exponent to p.

Remark 4.6 Note that in Theorem 4.3 and Proposition 4.5 we only need to assume $M_p \geq n(\frac{2}{p}-1) \geq 0$ if $\ell \geq 1$.

Moreover, we mention two embeddings between modulation spaces and fractional Sobolev spaces, see [23, Proposition 2.9] and [24], respectively. We denote the conjugated exponent of p and q by p' and q', respectively.

Proposition 4.7 *Let $1 \leq p, q \leq \infty$ and $s \in \mathbb{R}$. Then it holds*

$$M_{p,q}^s(\mathbb{R}^n) \hookrightarrow H_p^{s+\mu n\theta(p,q)}(\mathbb{R}^n)$$

for any $\mu > 1$, where

$$\theta(p,q) := \min\big\{0; q^{-1} - \max\big\{p^{-1}; p'^{-1}\big\}\big\}.$$

Moreover, if $1 \leq q \leq p \leq r \leq q'$, then

$$M_{p,q}^s(\mathbb{R}^n) \hookrightarrow H_r^s(\mathbb{R}^n).$$

The estimate of Theorem 4.3 in the auxiliary spaces $K_{p,q}^s(\mathbb{R}^n)$ combined with Proposition 4.5, taking account of the embeddings in Proposition 4.7, yield the following main result due to Proposition 3.2.

Theorem 4.8 *Let $1 \leq q \leq \infty$ and the dimension $n \geq 2$. Let $1 < p \leq \frac{2n}{n+1}$. Assume $\varphi, \psi \in C_0^\infty(\mathbb{R}^n)$ and let $u = u(t,x)$ be the solution to Cauchy problem* (6).

(i) *If $q \leq p$, then it holds*

$$\big\|\partial_t^\ell u(t,\cdot)\big\|_{M_{p',q}^s} \lesssim (1+t)^{-\frac{n-1}{2}(\frac{2}{p}-1)}\big(\|\varphi\|_{M_{p,q}^{s+M_p+\ell}} + \|\psi\|_{M_{p,q}^{s-1+M_p+\ell}}\big),$$

where $M_p \geq n(\frac{2}{p}-1)$, $s > 0$ and $\ell \in \mathbb{N}$.

(ii) *If $q > p$, then it holds*

$$\big\|\partial_t^\ell u(t,\cdot)\big\|_{M_{p',q}^s} \lesssim (1+t)^{-\frac{n-1}{2}(\frac{2}{p}-1)}\big(\|\varphi\|_{M_{p,q}^{s+M_p+\ell}} + \|\psi\|_{M_{p,q}^{s-1+M_p+\ell}}\big),$$

where $M_p \geq n(\frac{2}{p}-1)$, $s > n(\frac{1}{p}-\frac{1}{q})$ and $\ell \in \mathbb{N}$.

Remark 4.9

(i) The condition $p \le \frac{2n}{n+1}$ ensures that $M_p \ge n(\frac{2}{p} - 1) \ge 1$. This together with $p > 1$ gives that our result is only valid for dimensions $n \ge 2$.
(ii) The remark under Proposition 4.5 applies here as well. So if $\ell \ge 1$, then we can assume $1 < p \le 2$.

Summarizing, Theorem 4.1 and Theorem 4.8 provide estimates of the form

$$\big\|\partial_t^\ell u(t,\cdot)\big\|_{M^{s_0}_{p',q}} \le C(t)\big(\|\varphi\|_{M^{s_1+\ell}_{p,q}} + \|\psi\|_{M^{s_1-1+\ell}_{p,q}}\big) \tag{12}$$

for $\ell \in \mathbb{N}$ and some weights s_0, s_1. Remark that modulation spaces increase with their integrability parameters, see, e.g., [18, Corollary 2.7]. For this reason we can replace the $M^s_{p,q}$ norms at the left-hand sides of the estimates in Theorem 4.1 by the $M^s_{p',q}$ norms. Here we take account of $p' > p$. Consequently, a standard interpolation argument between the statements of Theorems 4.1 and 4.8 gives

$$\big\|\partial_t^\ell u(t,\cdot)\big\|_{M^{s}_{p',q}} \lesssim (1+t)^\rho\big(\|\varphi\|_{M^{s+\eta+\ell}_{p,q}} + \|\psi\|_{M^{s-1+\eta+\ell}_{p,q}}\big), \tag{13}$$

where for $\theta \in [0, 1]$ we have

$$\rho = \rho(\theta, n, p) = \theta\frac{n+3}{2} - (1-\theta)\frac{n-1}{2}\Big(\frac{2}{p} - 1\Big) \quad \text{if} \quad \ell = 0,$$

$$\rho = \rho(\theta, n, p) = \theta\frac{n+1}{2} - (1-\theta)\frac{n-1}{2}\Big(\frac{2}{p} - 1\Big) \quad \text{if} \quad \ell \ge 1, \quad \text{and}$$

$$\eta = \eta(\theta, n, p) = (1-\theta)M_p \ge (1-\theta)n\Big(\frac{2}{p} - 1\Big).$$

Example 4.10 As explained in the introduction it is desirable that the right-hand side in (13) is bounded with respect to the time t. In other words, it is desirable to have $\rho \le 0$ while the parameter η describes the loss of regularity of $u = u(t,\cdot)$. A first result for $\rho = 0$ is obtained by Theorem 4.2. Thus, (13) modifies this result if

$$p > \frac{2n^2+2}{n^2+n+2} \quad \text{or} \quad p > \frac{2n^2+4n+2}{n^2+3n+4},$$

respectively. Note that we can even choose θ such that $C(t) \to 0$ as $t \to \infty$ in (12).

Acknowledgements Both authors thank both referees for their valuable proposals. In particular, they pointed out some incorrectness in Proposition 3.2 in the submitted version.

References

1. Á. Bényi, K. Gröchenig, K.A. Okoudjou, L.G. Rogers, Unimodular Fourier multipliers for modulation spaces. J. Func. Anal. **246.2**, 366–384 (2007)
2. F.M. Christ, M.I. Weinstein, Dispersion of small amplitude solutions of the generalized Korteweg-de Vries equation. J. Func. Anal. **100.1**, 87–109 (1991)
3. E. Cordero, F. Nicola, Remarks on Fourier multipliers and applications to the wave equation. J. Math. Anal. Appl. **353.2**, 583–591 (2009)
4. M.R. Ebert, M. Reissig, *Methods for Partial Differential Equations* (Birkhäuser, Cham, 2018)
5. H.G. Feichtinger, Modulation spaces of locally compact Abelian groups, in *Proceedings of International Conference on Wavelets and Applications*, ed. by R. Radha, M. Krishna, S. Thangavelu (New Delhi Allied Publishers, Chennai, 2003), pp. 1–56
6. H.G. Feichtinger, G. Narimani, Fourier multipliers of classical modulation spaces. Appl. Comput. Harmon. Anal. **21.3**, 349–359 (2006)
7. L. Grafakos, *Classical and Modern Fourier Analysis* (Pearson Education, Inc., Upper Saddle River, 2004)
8. K. Gröchenig, *Foundations of Time-Frequency Analysis* (Birkhäuser, Boston, 2001)
9. L. Hörmander, Estimates for translation invariant operators in L^p spaces. Acta Math. **104.1**, 93–140 (1960)
10. T. Iwabuchi, Navier-Stokes equations and nonlinear heat equations in modulation spaces with negative derivative indices. J. Differ. Equ. **248**, 1972–2002 (2009)
11. K. Johansson, Generalized free time-dependent Schrödinger equation with initial data in Fourier Lebesgue spaces. J. Pseudo-Differ. Oper. Appl. **2**, 543–556 (2011)
12. C.E. Kenig, G. Ponce, L. Vega, Well-posedness and scattering results for the generalized Korteweg-de Vries equation via the contraction principle. Commun. Pure Appl. Math. **46.4**, 527–620 (1993)
13. P.I. Lizorkin, Multipliers of Fourier integrals and bounds of convolution in spaces with mixed norms. Applications. Math. USSR-Izvestiya **4.1**, 225 (1970)
14. S.G. Mikhlin, On the multipliers of Fourier integrals. Doklady Akademii nauk SSSR **109**, 701–703 (1956)
15. T. Narazaki, M. Reissig, L^1 estimates for oscillating integrals related to structural damped wave models, in *Studies in Phase Space Analysis with Applications to PDEs*, ed. by M. Cicognani, F. Colombini, D. Del Santo (Birkhäuser, Basel, 2013), pp. 215–258
16. A. Palmieri, M. Reissig, Semi-linear wave models with power non-linearity and scale-invariant time-dependent mass and dissipation, II. Math. Nachr. **291**, 1859–1892 (2018)
17. M. Reich, Modulation Spaces and Nonlinear Partial Differential Equations. PhD thesis, TU Bergakademie Freiberg (2017)
18. M. Reich, W. Sickel, Multiplication and composition in weighted modulation spaces, in *Mathematical Analysis, Probability and Applications – Plenary Lectures: ISAAC 2015*, Macau. Springer Proceedings in Mathematics and Statistics, ed. by L. Rodino, T. Qian (Springer, Cham, 2016), pp. 103–149
19. T. Runst, W. Sickel, Sobolev Spaces of Fractional Order, Nemytskij Operators, and Nonlinear Partial Differential Equations. Walter de Gruyter (1996)
20. M. Ruzhansky, M. Sugimoto, W. Wang, Modulation spaces and nonlinear evolution equations, in *Evolution Equations of Hyperbolic and Schrödinger Type*, ed. by M. Ruzhansky, M. Sugimoto, J. Wirth. Progress in Mathematics (Springer, Basel, 2012), pp. 267–283
21. G. Staffilani, The initial value problem for some dispersive differential equations. PhD thesis, University of Chicago (1995)
22. M. Sugimoto, N. Tomita, The dilation property of modulation spaces and their inclusion relation with Besov spaces. J. Funct. Anal. **248.1**, 79–106 (2007)
23. J. Toft, Continuity properties for modulation spaces, with applications to pseudo-differential calculus, II. Ann. Global Anal. Geom. **26**, 73–106 (2004)

24. J. Toft, Convolution and embeddings for weighted modulation spaces. Adv. Pseudo-Differ. Oper. **155**, 165–186 (2004)
25. J. Toft, Pseudo-differential operators with smooth symbols on modulation spaces. Cubo **11.4**, 87–107 (2009)
26. H. Triebel, Fourier analysis and function spaces. Teubner-Texte Math. **7**, Teubner, Leipzig (1977)
27. H. Triebel, Spaces of Besov-Sobolev-Hardy type. Teubner-Texte Math. **9**, Teubner, Leipzig (1978)
28. H. Triebel, Modulation spaces on the Euclidean n-space. Z. Anal. Anwend. **2.5**, 443–457 (1983)
29. H. Triebel, *Theory of Function Spaces* (Birkhäuser, Basel, 1983)
30. B. Wang, C. Huang, Frequency-uniform decomposition method for the generalized BO, KdV and NLS equations. J. Differ. Equ. **239.1**, 213–250 (2007)
31. B. Wang, H. Hudzik, The global Cauchy problem for the NLS and NLKG with small rough data. J. Differ. Equ. **232**, 36–73 (2007)
32. B. Wang, Z. Lifeng, G. Boling, Isometric decomposition operators, function spaces $E^{\lambda}_{p,q}$ and applications to nonlinear evolution equations. J. Funct. Anal. **233.1**, 1–39 (2006)
33. B. Wang, Z. Huo, C. Hao, Z. Guo, *Harmonic Analysis Method for Nonlinear Evolution Equations, I* (World Scientific, Singapore/Hackensack, 2011)

GPSR Compliance
The European Union's (EU) General Product Safety Regulation (GPSR) is a set of rules that requires consumer products to be safe and our obligations to ensure this.

If you have any concerns about our products, you can contact us on

ProductSafety@springernature.com

In case Publisher is established outside the EU, the EU authorized representative is:

Springer Nature Customer Service Center GmbH
Europaplatz 3
69115 Heidelberg, Germany

www.ingramcontent.com/pod-product-compliance
Ingram Content Group UK Ltd.
Pitfield, Milton Keynes, MK11 3LW, UK
UKHW021919270726
14059UKWH00002B/96

* 9 7 8 3 0 3 0 1 0 9 3 6 3 *